D0944439

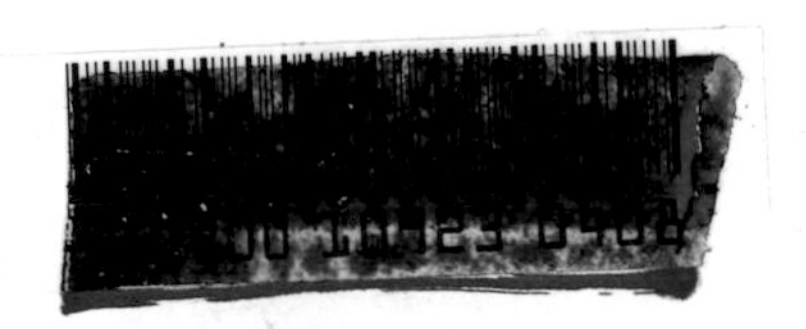

Environmental Disasters

Anthropogenic and Natural

Springer
London
Berlin
Heidelberg
New York
Barcelona
Hong Kong
Milan
Paris
Santa Clara
Singapore
Tokyo

Kirill Ya. Kondratyev, Alexei A. Grigoryev
and Costas A. Varotsos

Environmental Disasters

Anthropogenic and Natural

Professor K. Ya. Kondratyev
Counsellor of the Russian Academy of Sciences
Scientific Research Centre for Ecological Safety
Nansen Foundation for Environment and Remote Sensing
St Petersberg
Russia

Professor A. A. Grigoryev
St Petersberg State University
St Petersberg
Russia

Dr. C. A. Varotsos
Associate Professor
Department of Physics
Division of Applied Physics
University of Athens
Greece

SPRINGER–PRAXIS BOOKS IN ENVIRONMENTAL SCIENCES
SUBJECT *ADVISORY EDITOR*: John Mason B.Sc., Ph.D.

ISBN 3-54043-303-1 Springer-Verlag Berlin Heidelberg New York

Die Deutsche Bibliothek – CIP-Einheitsaufnahme
Kondrat'ev, Kirill Ja.:
Environmental disasters : anthropogenic and natural / Kirill Ya. Kondratyev, Alexei A. Grigoryev and Costas A. Varotsos. – London; Berlin; Heidelberg; New York; Barcelona; Hong Kong; Milan; Paris; Santa Clara; Singapore; Tokyo: Springer; Chichester, UK: Praxis Publ., 2002
(Springer Praxis books in environmental sciences)
ISBN 3-540-43303-1
0101 deutsche buecherei

Library of Congress Cataloging-in-Publication Data
A catalogue record for this book is available from the Library of Congress

Printed by MPG Books Ltd, Bodmin, Cornwall, UK

Cover design: Jim Wilkie
Project management: Originator, Gt Yarmouth, Norfolk, UK

Printed on acid-free paper supplied by Precision Publishing Papers Ltd, UK

Contents

Preface

The role of disasters in the evolution of nature is very well known. The development of civilization gave rise to new problems: on the one hand, the sensitivity of humans and their economic activity to various external influences increased; on the other, the phenomena of human-induced disasters appeared, whose consequences were sometimes catastrophic (as was, e.g., the case of the Chernobyl accident). Potential global environmental catastrophes acquire increasingly threatening scales; among these, the first position (from the viewpoint of a long-term perspective) should certainly belong to the human-induced violation of closedness (closure) of global biogeochemical cycles; however, the speculative idea of "global warming" has attracted much more attention so far. As the influences of environmental disasters on human life and economic activity become stronger, so the assessment of the risk of such catastrophic events becomes an urgent matter. Relevant studies have not been sufficient in number or reliable enough. The experience gained from such studies is discussed in this book.

Environmental catastrophes are defined as extreme, disastrous situations in the vital and economic activity of the population, caused by substantial, unfavourable changes in the environment. They can be provoked by human-made (technological) causes (accidents at factories, plants, transport, etc.) or by dangerous natural phenomena (earthquakes, volcanic eruptions, etc.). A third category is represented by environmental catastrophes due to complex, interrelated human- and nature-induced factors (e.g. the case of desertification).

There are diverse kinds of natural disasters that are of utmost importance such as forest and grassland fires, earthquakes and tsunamis, floods, tropical cyclones, drought, landslides, snowstorms, ozone and climate changes, desertification and deforestation, oil spills, etc. Over 10 years ago, the General Assembly of the United Nations declared the 1990s an International Decade for Natural Disaster Reduction (IDNDR). The IDNDR initiative was later further developed into a new institutional arrangement entitled an International Strategy for Disaster Reduction (ISDR), whose fundamental principle was broad-based cooperation and partnership.

Disasters are mainly assessed from the standpoint of danger and damage to population. We shall, however, also dwell on environmental calamities, whose consequences are very important not only for humans, but for other living organisms as well (the best known examples of the latter are oil spills resulting from supertanker accidents).

Chapter 1 of this monograph is devoted to the problem of risk caused by natural and human-induced environmental disasters and to the analysis of regular features of their space and time distribution. The most important factors of environmental risk are: (1) dangerous natural and technological phenomena; (2) population vulnerability; (3) the social and natural background of the development of dangerous phenomena; (4) reaction of the population to dangerous phenomena, its degree of awareness and readiness. However, other classifications are also quite possible, especially those that take into account the fact that the risk of a catastrophe is determined by two basic factors: the most dangerous phenomenon (its specific features, scale, etc.) and the vulnerability of the population (its reaction, the organization of warning measures, etc.). The latter depends on a whole number of circumstances (i.e. economic, social, ethno-cultural, psychological, etc.). All these factors are naturally variable in space and time. Therefore, they can be mapped. The use of risk maps makes it possible to plan measures to mitigate the consequences of dangerous phenomena. The most important element of risk mapping is the use of satellite monitoring data.

The consequences of environmental disasters, the peculiar features of their spatial distribution and genesis, and their temporal dynamics (chaotic and rythmic patterns) are treated in Chapter 1. Estimates are presented of population vulnerability to natural and human-induced disasters. The historic experience of surviving catastrophic events, the peculiar features of the perception of risk and of information about it, as well as risk management are considered.

Chapter 2 offers a survey of the most significant natural and human-induced disasters that took place in the historic past; the survey includes both a general analysis of the role of natural calamities in the development of humankind and a consideration of the specific features and consequences of environmental catastrophes. The chapter ends with a discussion of the experience accumulated over many centuries that is relevant to survival of catastrophic circumstances.

Three subsequent chapters (Chapters 3, 4, and 5) are devoted to consideration of the experience of studying various disastrous events; satellite monitoring of catastrophes and their consequences is also discussed there. Chapter 3 presents an analysis of results relevant to meteorological, hydrological and geological catastrophes. Chapter 4 contains a review of the results of studying human-induced disasters such as heavy pollution of the atmosphere, aquatic areas and soils, as well as forest fires, technological accidents and catastrophes.

Chapter 5 deals with problems relevant to slowly developing, regional- and global-scale disasters. Here, global problems of general acute interest are mainly treated such as changes in the stratospheric ozone layer and climate, and the problem of biodiversity. Among regional-scale problems considered, there are the central ones: those of deforestation, desertification, and variation in levels of inner water-bodies.

Chapter 6 presents a brief analysis of the present-day state and perspectives of satellite monitoring of environmental disasters.

In conclusion, it has been emphasized that the environmental risk of natural and human-induced catastrophes can only be reduced if socio-economic aspects are taken into account. The risk will clearly be close to zero if regions of population concentration and centres of dangerous phenomena are remote enough from each other. Otherwise (which is the case under real conditions), socio-economic and psychological factors become essential (or sometimes the most important) determinants of the scale of a given calamity. In those regions of the Earth where environmental risk is especially high, the fundamental principle of vital and economic activity should be "living with risk". Risk should be considered as an integral part of existence. It cannot be regarded as a rare or occasional phenomenon in a chain of events. Risk perception and the adoption of the above concept by people, followed by its practical realization, imply that the consideration of risk should be a component part in planning all kinds of vital and economic activity of the population, and their importance should be clearly understood by all economic, social (including educational and cultural), and political institutions in society.

An active role in reducing environmental risk should be played by information about it, which must include knowledge of the nature and specific features of dangerous phenomena, and of the actions needed in case of their development. Timely information about the threat and development of a dangerous phenomenon coupled with knowledge of how to behave in a critical situation can minimize the risk.

Of course, this book is far from comprehensive. Many aspects of the complex problem of disasters are beyond the scope of this book. For example, the range of problems relevant to so-called "cosmic waste" and its impact on circumterrestrial space and space vehicles have been omitted altogether; impacts on the environment from extraterrestrial factors that cause consequences of a catastrophic scale have hardly been touched upon; information concerning objects using nuclear or chemical technologies has been presented in a condensed form, etc. The authors will greatly appreciate any comments and criticism on the part of the readers.

Figures

Tables

Abbreviations

ATEP	Atlantic Tropical Experiment Programme
ASS	Assiros station
AE	Acoustic emission
BI	Basic index
BCS	Bardeen–Cooper–Schrieffer index
CEI	Extreme climate index
CAI	Bagrov climate anomaly index
CZCS	Coastal Zone Colour Scanner
CIS	Crystalline ionic solids
ECHAM	A specific general circulation model
ENSO	El Niño Southern Oscillation
EOF	Empirical orthogonal function
ETR	European Territory of Russia
EM	Electromagnetic
FAO	Food and agriculture organization
GARP	Global Atmospheric Research Programme
GWP	Gross world product
GNP	Gross national product
GHG	Greenhouse gasses
GEO	a Unep Report
GAC	General atmospheric circulation
GCRI	Greenhouse effect on climate index
GPS	Global positioning system
GIS	Geographical information system
GVEF	Gradual variation of the electric field
HIPC	Highly indebeted poor countries
IDNDR	International Decade for Natural Disaster Reduction
ISDR	International Strategy for Disaster Reduction
IMF	International Monetary Fund
IFRCRCS	International Forum of the Red Cross and Crescent

IAEA	International Atomic Energy Agency
IPCC	Intergovermental Panel on Climate Change
KI	Key indices
MJO	Madden–Julian oscillation
MJ	Jellium model
NDVI	Normalized vegetation index
OIA	Ioannina station
PV	Photovoltaic
PC	Principal component
REU	Relative equivalent units
RTFPF	Russian territory free from permafrost
SAT	Surface air temperature
SGP	Seasonal genesis parameter
SST	Sea surface temperature
TM	Thematic mapper
UNEP	United Nations Environment Programme
WB	World Bank
WHO	World Health Organization
WMO	World Meteorological Organization

1

Natural and anthropogenic environmental disasters. The problems of risk

The role of disasters in the evolution of nature is very well known. The development of civilization gave rise to new problems: on the one hand, the increased sensitivity of humans and their economic activity to various external forcings, and, on the other hand, the phenomenon of human-induced catastrophes, whose consequences are sometimes (as, e.g., in the case of Chernobyl) really disastrous (see, e.g., the entries in the Bibliography at the end of Chapter 1). Potential global environmental disasters have in recent decades been developing on a more threatening scale; among these, the first positions (from the standpoint of a long-term perspective) belongs to the human-induced violation of the closed nature of global biogeochemical cycles (Gorshkov, 1995; Kondratyev *et al.*, 1996; Kondratyev, 1998c, d; 1999). However, speculative considerations concerning human-induced global warming have been much more popular thus far (Boehmer-Christiansen, 1999; Demirchian and Kondratyev, 1999; Kondratyev, 1998c, 1999; Kondratyev and Demirchian, 2000; Berk and Fovell, 1999). As the impact of environmental disasters on human life and economic activity grows stronger, so assessment of the risk involved becomes a most urgent task. Relevant studies have not been numerous, and some of them are subject to debate.

Since anomalous phenomena represent strong deviations from background situations (stable states) (Kondratyev *et al.*, 1996; Kondratyev, 1998c; Kotlyakov *et al.*, 1997; Lavrov, 1999; Trofimov *et al.*, 1999a, b; Bossel, 1998; Brebbia and Uso, 1999, etc.), we shall start with a brief analysis of 'background' global ecodynamics during the second half of the 20th century, on the basis of recent overviews (Brown and Flavin, 1999a; Brown *et al.*, 1999b; *Global* . . . , 1999).

1.1 GLOBAL CHANGES ON THE BORDERLINE BETWEEN TWO MILLENNIA

1.1.1 Introduction

If we digress from the fantastic predictions of the end of the world that have proved untrue, we arrive at the pessimistic conclusion that humankind is moving toward an environmental catastrophe, even though serious relevant scientific literature does not provide an unambiguous assessment of the relationship between society and nature (Kondratyev and Losev, 2002). Regrettably, neither the Second UN Conference on Environment and Development (1992) nor the subsequent Special Session of the UN General Assembly 'Rio + 5' (1997) and World Summit on Sustainable Development 'Rio + 10' (2002) have brought about a consensus regarding priorities where the problem of global change (the interaction between ecodynamics and socio-economic development) is concerned. Doubtless, a matter for greatest concern is continuing population growth on a global scale in conditions of limited natural resources and the increasing controversy between production and consumption. In the context of the illusion of 'sustainable development', the necessity of a different paradigm of development that takes into account the realities of global ecodynamics and socio-economic development becomes increasingly evident (Gorshkov, 1995, 1998; Grigoryev, 1991; Grigoryev and Kondratyev, 1998, 2000; Gritsman, 1993; Demirchan and Kondratyev, 1999; Zavarzin, 1999; Zavarzin and Kotlyakov, 1999; Zadorozhnyuk and Zozolyuk, 1994; Zerbino, 1991; Izrael, 1996; Ivanov and Yasamanov, 1991; Kondratyev, 1992, 1998a; Kondratyev *et al.*, 1996; Adams, 1995; Hugget, 1997). In this connection, the concept of biotic regulation of the environment (Gorshkov, 1995) acquires key importance. The purpose of this survey is the analysis (based on the factual data of Brown and Flavin, 1999a; Brown *et al.*, 1999b; *Global*..., 1999; *Our*..., 1999) of global ecodynamics on the borderline between two millennia, as well as a perspective of the development paradigm.

Brown and Flavin (1999a) have been quite right to note that, on the basis of their own studies, they have come to the conclusion that the economic model, formed in the industrially developed West and spreading worldwide, constantly undermines itself. It will not be able to serve our basis for long in the 21st century. So, the question is how to find an alternative way of development... Governments and corporations are becoming increasingly aware of the necessity of a new economic model that would be able to ensure environmentally sustainable development. Regrettably, Brown and his co-authors have not been able to constructively answer their own questions. Instead, they did not go beyond standard statements about the necessity of inventing energy and material-saving wasteless technologies, as well as of expanding the system of environmental taxes to activate the 'the polluter pays' principle. A number of Russian publications (Gorshkov, 1995; Kondratyev, 1998c, d; Kondratyev *et al.*, 1996) demonstrate that such a formulation of the problem is insufficient and inadequate. We shall now dwell on the analysis of global tendencies of socio-economic development and ecodynamics.

1.1.2 Global dynamics of socio-economic development and the environment: updated statistics

The priority of the problem of population growth is no longer subject to debate. We shall therefore start our consideration of the problem.

1.1.2.1 Global population number

Table 1.1 illustrates the continuing growth in global population number; in spite of a minor drop in the annual absolute growth rate of the population number (Brown and Flavin, 1999a), its increase in 1998, for instance, was still equivalent to another Germany. According to the symbolic event, which took place in Sarajevo (Bosnia), the global population hit 6 billion on 12 October 1999. Over 80% of the global population (4.8 billion in 1998) inhabit developing countries, and it is this part of the world population that is rcsponsiblc for nearly the entire population growth rate.

The peak of the annual population growth rate was registered in 1963 (2.2%); it decreased to 1.4% by 1998 but still remained twice as large as that in the middle of the 20th century. It is a well-known fact that the dynamics of the population number

Table 1.1. World population, total and annual addition, 1950–98.

Year	Total (billion)	Annual addition (million)	Year	Total (billion)	Annual addition (million)
1950	2.556	38	1980	4.454	76
1955	2.780	53	1981	4.530	80
1960	3.039	41	1982	4.610	80
1965	3.345	70	1983	4.690	79
1966	3.416	70	1984	4.770	81
1967	3.485	72	1985	4.851	82
1968	3.557	75	1986	4.933	86
1969	3.631	75	1987	5.018	86
1970	3.707	77	1988	5.105	86
1971	3.784	77	1989	5.190	87
1972	3.861	76	1990	5.277	82
1973	3.937	76	1991	5.359	82
1974	4.013	73	1992	5.442	81
1975	4.086	72	1993	5.523	80
1976	4.158	72	1994	5.603	80
1977	4.231	72	1995	5.682	79
1978	4.303	75	1996	5.761	80
1979	4.378	76	1997	5.840	78
			1998*	5.919	78
			12 October 1999	6.000	–

* Preliminary.

is inhomogeneous. Whereas in Western Europe and Japan the population level was stable or even declined (which is certainly true as well for Russia, where life expectancy dropped below the level of 1950), in many African countries the population growth rate reached 3%, and the population doubled there during a little more than two decades. On the whole in the world, two persons out of five inhabit countries where a decline in population number can be expected because the birth rate there dropped below the level required for population growth (this is true, in particular, for superpowers, such as China and the United States).

The latest UN projections put world population in 2050 at 8.9 billion, a substantial reduction from the previous estimate of 9.4 billion. Birth rates began to decline almost everywhere in the world (as calculated per woman): for example, in South Korea from 7 (1970) to 1.7 (1998) children, in Bangladesh from 7 (1975) to 3.3 (the process has hardly affected Africa). This was partly due to family planning programmes.

Global life expectancy crept to a new high figure of 66 years in 1998, meaning that the average person born then will die in the year 2064. A person born today will live 20 years (43%) longer than a person born in 1950 (for 1997, this increase was only 0.2 year). Of course, such estimates are characteristic of great variability. For instance, whereas in industrially developed countries the lifespan averages 75 years, it is only 64 years in developing nations; in 1950 in Asia and Africa, it was only 40 years. Although, as a rule, life duration is essentially dependent on income level, this correlation is absent in some countries. For example, an average lifespan in one of the poorer countries, Cuba, is higher than in North America. On the average, women outlive men by 5 years. Of course, as lifespans extend, the human population is ageing. The share of the world's population over the age of 60 will more than double – from 9 to 21% – by 2050, with this share reaching 40% in several industrial nations (the situation is critical in Japan).

1.1.2.2 Global energy sources

The fundamental basis of socio-economic development is energy production, which is still very dependent on the use of fossil fuels: coal, oil, and natural gas (Table 1.2). Relevant trends at the end of the 20th century are characteristic of a minor growth (0.8%) in the use of oil and a drop (–2.5%) in that of coal. Since the consumption of coal in China (where it is the main energy carrier) was reduced by 7%, we may assume that the use of this most environmentally damaging fossil fuel is near its historical peak (however, in the USA the consumption of coal increased by 1.7% in 1999).

Growth in the use of oil, which accounts for 30% of world energy use, was held back by a 3.5% decline in Japan in 1998, and by slower growth throughout Asia. The stronger economies of Western Europe and North America, on the other hand, continued to expand their use of oil, particularly the USA. The substantial increases in oil prices that took place in 1999 was favourable for the development of the Russian economy.

Natural gas markets were steadier than oil markets in 1998, as gas use grew by

Table 1.2. World fossil fuel use, by type, 1950–98.

Year	Coal (million tons of oil equivalent)	Oil	Natural gas (million tons of oil equivalent)	Year	Coal (million tons of oil equivalent)	Oil	Natural gas (million tons of oil equivalent)
1950	1,043	436	187	1980	2,021	2,873	1,406
1955	1,234	753	290	1981	1,816	2,781	1,448
1960	1,500	1,020	444	1982	1,878	2,656	1,448
1965	1,533	1,485	661	1983	1,918	2,632	1,463
1966	1,559	1,591	721	1984	2,001	2,670	1,577
1967	1,480	1,696	774	1985	2,100	2,654	1,640
1968	1,554	1,849	847	1986	2,135	2,743	1,653
1969	1,601	2,025	928	1987	2,197	2,789	1,739
1970	1,635	2,189	1,022	1988	2,244	2,872	1,826
1971	1,632	2,313	1,097	1989	2,269	2,914	1,909
1972	1,629	2,487	1,150	1990	2,241	2,958	1,945
1973	1,668	2,690	1,184	1991	2,186	2,955	1,980
1974	1,691	2,650	1,212	1992	2,167	2,980	1,983
1975	1,709	2,616	1,199	1993	2,157	2,953	2,011
1976	1,787	2,781	1,261	1994	2,168	3,011	2,017
1977	1,835	2,870	1,283	1995	2,200	3,235	2,075
1978	1,870	2,962	1,334	1996	2,275	3,325	2,179
1979	1,991	2,998	1,381	1997	2,293	3,396	2,175
				1998*	2.236	3,423	2,210

* Preliminary.

1.6% (major gas discoveries in China have greatly contributed to the situation in oil markets).

Between 1996 and 1998, the total nuclear generating capacity declined (for only the second time since the 1950s), and nuclear energy production stabilized at the level of 343 megawatts (MW) (see Table 1.3). Compared with 1990, the increase was only 4.4% (because only a few governments support further development of nuclear power production). Only three new reactors – two in South Korea and one in Slovakia – were connected to the grid in 1998, and two reactors – one each in the USA and Japan – were permanently closed. The new German Government has a goal of phasing out the country's 19 nuclear power reactors. North America also appears close to abandoning its existing nuclear plants, though this is due to the high cost rather than the unpopularity of nuclear power. The total number of reactors that have been retired after an average service life of less than 18 years constituted 94; by the end of 1998, the total number of operating reactors was 429. Asia still remains the last region of growth for nuclear power (this is very relevant to South Korea), though the pace is slowing down there as well. China currently has three operating nuclear reactors and six under construction and has ambitious plans to build 50 additional reactors by 2020.

Table 1.3. World net installed electrical generating capacity of nuclear power plants, 1960–98.

Year	Energy production (GW)	Year	Energy production (GW)
1960	1	1979	121
1961	1	1980	135
1962	2	1981	155
1963	2	1982	170
1964	3	1983	189
1965	5	1984	219
1966	6	1985	250
1967	8	1986	276
1968	9	1987	297
1969	13	1988	310
1970	16	1989	320
1971	24	1990	328
1972	32	1991	325
1973	45	1992	327
1974	61	1993	336
1975	71	1994	338
1976	85	1995	340
1977	99	1996	343
1978	114	1997	343
		1998*	343

* Preliminary.

Although the advantage of development of nuclear power production are subject to heated debate (mainly due to the Chernobyl catastrophe), we can hardly hope that the world will be able to do without nuclear power: global fossil fuel resources are limited, and their use is becoming increasingly ecologically dangerous (the historically short period of 'carbon' energy production is surely approaching its end).

In this connection, data on the rate of development of wind and solar power production are of great interest. Their global contribution has not been large so far, but both have been developing at a record rate. The world wind energy production developed strongly in 1998 (see Table 1.4), having added 2,100 MW of new generating capacity. Wind energy production is developing very quickly (17 billion kilowatt-hours of electricity was produced in 1999). The 1999 boom in wind energy was led by Germany, which added 790 MW, pushing its wind energy capacity over 2,800 MW. Germany's wind industry is now producing more than 1% of the nation's electricity, and has reached 11% in the northernmost state of Schleswig-Holstein. Spain has also made great progress in the field by adding 380 MW of wind power, thus pushing its overall capacity up 84% to 830 MW. In the northern industrial province of Navarra, 20% of electricity already comes from wind

Table 1.4. World wind energy generating capacity, total and annual addition, 1980–98.

Year	Total (MW)	Annual addition (MW)
1980	10	5
1981	25	15
1982	90	65
1983	210	120
1984	600	390
1985	1,020	420
1986	1,270	250
1987	1,450	180
1988	1,580	130
1989	1,730	150
1990	1,930	200
1991	2,170	240
1992	2,510	340
1993	2,990	480
1994	3,680	720
1995	4,820	1,294
1996	6,115	1,290
1997	7,630	1,566
1998*	9,600	2,100

* Preliminary.

turbines. Government policy in support of wind energy production has been most favourable in these two countries.

Wind power installations grew rapidly in the USA in 1998, with some 226 MW of new capacity added in 10 different states. The largest projects are a 107-MW wind farm in Minnesota, one of 42 MW in Wyoming, and one of 25 MW in Oregon. Denmark, however, was a leader in the global wind power industry, adding 308 MW of capacity in 1998. Some 1,400 MW of wind power now generate more than 8% of the country's electricity. And Denmark's wind companies export heavily, accounting for more than half the new wind turbines installed worldwide in 1998. Altogether, the Danish wind industry had gross sales of just under $1 billion in 1998 – roughly equal to the combined sales of the nation's natural gas and fishing industries. The developing world could benefit most from further growth of the wind industry (India is the leader so far, with more than 900 MW of wind power).

Shipments of solar photovoltaic (PV) cells neared 152 MW in 1998, up 16% from 1990 and 21% from 1997. These silicon-based semiconductors turn sunlight directly into electricity. The USA and Japan are world leaders in the development of solar power production (54 and 49 MW, respectively). For comparison, the whole of Europe produced only 30 MW. Some progress has been marked in Australia, India,

Table 1.5. Trends in global energy use, by source, 1990–8.

Energy source	Annual rate of growth (%)
Wind power	22.2
Solar photovoltaics	15.9
Geothermal power	4.3
Hydroelectric power*	1.9
Oil	1.8
Natural gas	1.6
Nuclear power	0.6
Coal	0

* 1990–7 only.

China and Taiwan (a total of 19 MW). Some governments have strengthened support for PV use on rooftops of residential buildings. The EU and the USA have each announced Million Roofs programmes, which include solar heating as well as solar electric systems, to be completed by 2010.

Table 1.5 shows trends in global energy use for different sources of energy production.

1.1.2.3 Biosphere resources

The natural biosphere (forests, rivers, lakes, and oceans) is the source of food and renewable natural resources, whereas cultivated regions of the biosphere (agricultural land) are the main sources of food products. In the context of the concept of biotic regulation of the environment, the biosphere is the basis of global environmental equilibrium (Gorshkov, 1995; Kondratyev, 1999).

1.1.2.3.1 Agricultural production and fish catch

Paradoxical as it may seem, the global grain-harvested area shrank to 684 million hectares in 1998, a drop of more than 6 million hectares (see Table 1.6). Since the historic high in 1981, the grain-harvested area has declined by 48 million ha (7%). And the grain area harvested per person – at 0.12 ha – has plummeted to half the 1951 level. Nevertheless, grain production has nearly tripled since 1950, largely from yield increases. The grain-harvested area indicates the acreage of land reaped each year. Grains provide more than half the calories and protein eaten directly by humans, as well as the feedgrain for meat, milk, and egg production. Planted on roughly half the world's cropland, grains serve as a proxy for trends in all crops. Wheat area declined by nearly 2% in 1998, rice area barely changed, and corn area rose 2%, spurred by rising global feedgrain demand. But wheat dominates global grain acreage with 225 million ha, followed by rice area (150 million ha), and corn

Table 1.6. World grain-harvested area, 1950–98.

Year	Area harvested (million ha)	Area per person (ha)	Year	Area harvested (million ha)	Area per person (ha)
1950	587	0.23	1980	722	0.16
1955	639	0.23	1981	732	0.16
1960	639	0.21	1982	716	0.16
1965	653	0.20	1983	707	0.15
1966	655	0.19	1984	711	0.15
1967	665	0.19	1985	715	0.15
1968	670	0.19	1986	709	0.14
1969	672	0.18	1987	685	0.14
1970	663	0.18	1988	688	0.14
1971	672	0.18	1989	694	0.13
1972	661	0.17	1990	694	0.13
1973	688	0.18	1991	692	0.13
1974	691	0.17	1992	694	0.13
1975	708	0.17	1993	685	0.12
1976	717	0.17	1994	686	0.12
1977	714	0.17	1995	682	0.12
1978	713	0.17	1996	703	0.12
1979	711	0.16	1997	690	0.12
			1998*	684	0.12

* Preliminary.

area (140 million ha). And soybean area worldwide has tripled since 1950, steadily replacing grain on some of the best cropland.

According to UN assessment, on a global basis, 38% of cultivated land area has been damaged to some degree by agricultural mismanagement since 1950, with higher levels of degradation in Latin America and Africa. Present cultivated land losses range from 5 to 12 million ha per year, according to estimates available. In Pakistan, Ethiopia, Iran, and other nations, where area per person is already a fraction of the world average, projected population doublings or triplings coupled with the problem of food security become especially painful. Tables 1.7 and 1.8 show how acute the situation is with regard to world grain production.

In 1998, the world grain harvest totalled 1.85 billion tons, down 30 million tons from 1997 – a drop of nearly 2%. With the world population growing by some 78 million in 1998, the per capita grain supply dropped to 312 kilograms, down from 321 kg in 1997. This was a continuation of the trend that had begun in 1984 (see Table 1.7). The 30-million-ton drop in the world grain harvest in 1998 was due largely to severe drought and heat in Russia on top of an overall deterioration in the Russian economy. The Russian wheat harvest dropped from 44 million tons in 1997 to 27 million tons in 1998, a decrease of 39%.

Among the big three grains – wheat, rice, and corn – the drop in the wheat harvest was the largest. The 1998 harvest of 586 million tons was 23 million tons

Table 1.7. World grain production, 1950–98.

Year	Total (million tons)	Per person (kg)	Year	Total (million tons)	Per person (kg)
1950	631	247	1981	1,482	327
1955	759	273	1982	1,533	333
1960	824	271	1983	1,469	313
1965	905	270	1984	1,632	342
1966	989	289	1985	1,647	339
1967	1,014	291	1986	1,665	337
1968	1,053	296	1987	1,598	318
1969	1,063	293	1988	1,549	304
1970	1,079	291	1989	1,671	322
1971	1,177	311	1990	1,769	335
1972	1,141	295	1991	1,708	319
1973	1,253	318	1992	1,790	329
1974	1,204	300	1993	1,714	310
1975	1,237	303	1994	1,761	314
1976	1,342	323	1995	1,712	301
1977	1,319	312	1996	1,870	324
1978	1,445	336	1997	1,875	321
1979	1,410	322	1998*	1,845	312
1980	1,430	321			

* Preliminary.

below the 1997 record. The world rice harvest also dropped in 1998, falling to 378 million tons from 84 million tons the year before. This nearly 2% drop was largely the result of severe flooding in China's Yangtze River basin. At the same time, the production of corn climbed from 574 million tons in 1997 to a record 597 million tons in 1998 (a gain of 4%). Despite the 30-million-ton drop in the world grain harvest in 1998, grain prices are at their lowest level for more than a decade. The reason is that the demand for wheat has recently weakened substantially (see Table 1.8).

The world soybean harvest in 1998 (155 million tons) was approximately the same as the year before (156 million tons), just as the soybean per capita consumption: 27 kg in 1997 and 26 kg in 1998. During the 1990s, the demand for soybeans grew by 5% a year (mainly due to increased demand for soybean oil). Since 1950, the world soybean harvest has expanded from 17 million tons to 155 million tons, largely owing to increased land productivity, but also due to the expanded area of soybean planting (replacing grain cultures). The principal exporter of soybeans and soybean products is the USA, accounting for 23 million tons of the 39 million tons traded in 1998. The main importers are China, Europe, North Africa, and the Middle East.

An important source for further progress of agricultural production is irrigation (see Table 1.9). Some 40% of the world's food comes from the 17% of global

Table 1.8. World grain carryover stocks, 1961–99*.

Year	Stocks (million tons)	Stocks (days' use)	Year	Stocks (million tons)	Stocks (days' use)
1961	203	90	1980	315	81
1962	182	81	1981	288	72
1963	190	82	1982	307	77
1964	193	83	1983	356	88
1965	194	78	1984	305	73
1966	159	62	1985	366	85
1967	189	72	1986	434	100
1968	213	79	1987	466	104
1969	244	87	1988	405	89
1970	228	77	1989	314	70
1971	193	63	1990	297	64
1972	217	69	1991	339	72
1973	180	56	1992	326	69
1974	191	56	1993	363	76
1975	199	61	1994	318	66
1976	219	66	1995	306	63
1977	279	79	1996	255	53
1978	277	77	1997	291	
1979	326	85	1998	324	
			1999†	313	

* Data are for year when new harvest begins.
† Preliminary

cropland that is watered artificially. The world irrigated area rose by 4 million ha in 1996 (the last year for which global data are available). The 1.7% increase over 1995 is the largest increase during the decade, but it still runs well behind the peak growth rates of the 1970s. The per capita irrigated area was relatively stable. Despite its growing importance, irrigation faces a host of challenges, the most critical being water availability (already felt in some countries such as India, China, etc.). Another threat to irrigation is salination of land. According to recent estimates, by 2025 more than a billion people will be living in countries facing absolute water scarcity.

World meat production increased markedly in 1998 and totalled 216 million tons, up 2.4% from the 211 million tons of 1997 (see Table 1.10). Meat production per person continued to rise, but much more slowly, going from 36.1 kg in 1997 to 36.4 in 1998 (a gain of 1%). While the production of pork (91 million tons) expanded by 3.5%, and the production of poultry also increased substantially, that of beef remained essentially the same as in 1997 (54 million tons).

The world fish catch from marine and inland waters fell slightly to 93.7 million tons in 1997, the latest year for which global data are available. Before that, the world fish catch was increasing steadily (starting with 19 million tons in 1950).

Table 1.9. World irrigated area, 1961–96.

Year	Total (million ha)	Per person (ha per 1,000)	Year	Total (million ha)	Per person (ha per 1,000)
1961	139	45.0	1979	207	47.4
1962	141	44.9	1980	209	47.1
1963	144	44.9	1981	213	47.1
1964	147	44.8	1982	215	46.6
1965	150	44.9	1983	216	46.2
1966	153	44.9	1984	221	46.4
1967	156	44.8	1985	224	46.2
1968	159	44.9	1986	225	45.8
1969	164	45.1	1987	227	45.3
1970	167	45.2	1988	230	45.1
1971	171	45.3	1989	236	45.5
1972	174	45.2	1990	241	45.8
1973	180	45.8	1991	245	45.9
1974	183	45.8	1992	248	45.7
1975	189	46.3	1993	252	45.7
1976	194	46.7	1994	255	45.7
1977	198	47.0	1995	259	45.6
1978	204	47.4	1996	263	45.8

Although world supplies dropped by 1% from the all-time high in 1996, per capita supplies declined by 2.4% to 16 kg per person. It is important to know that the ecologically permissible limit for global fish catch is about 95 million tons.

1.1.2.3.2 Roundwood and paper production.

According to the UN Food and Agriculture Organization (FAO), global production of roundwood increased between 1950 and 1997 from 1.421 to 3.359 billion cubic metres, having exceeded the level of 3 billion m^3 in 1984 and by 1990 having reached the level of 'saturation'. The fall in roundwood production that followed after 1990 was mainly due to a drastic drop in Russia (from 386 million cubic metres in 1990 to just 111 million m^3 by 1997). About 55% of the roundwood cut today is used directly as fuelwood (up to 90% in developing countries), with the other 45% becoming 'industrial roundwood' intended for the production of lumber and panels for construction purposes, and paper. The main producers of industrial roundwood are the USA, Canada, and Russia. Temperate and boreal forests account for 83% of the total volume of industrial roundwood produced.

A disproportionate share of the world's industrial roundwood is consumed by industrial nations. In fact, 77% of the world's timber harvested for industrial purposes is used by 22% of the world's population, who live in these nations. Consumption per person in industrial nations is 12 times higher than in developing ones. Fuelwood is actually the only wood product used in the developing world,

Table 1.10. World meat production, 1950–98.

Year	Total (million tons)	Per person (kg)	Year	Total (million tons)	Per person (kg)
1950	44	17.2	1980	130	29.2
1955	58	20.7	1981	132	29.2
1960	64	21.0	1982	134	29.0
1965	81	24.2	1983	138	29.4
1966	84	24.4	1984	142	29.7
1967	86	24.5	1985	146	30.1
1968	88	24.8	1986	152	30.8
1969	92	25.4	1987	157	31.3
1970	97	26.2	1988	164	32.2
1971	101	26.7	1989	166	32.0
1972	106	27.4	1990	171	32.5
1973	105	26.8	1991	173	32.2
1974	107	26.6	1992	175	32.1
1975	109	26.6	1993	177	32.1
1976	112	26.9	1994	187	33.3
1977	117	27.6	1995	197	34.7
1978	121	28.2	1996	206	35.7
1979	126	28.8	1997	211	36.1
			1998*	216	36.4

* Preliminary.

and even then their consumption per person is less than twice that in industrial countries.

Paper now accounts for the largest single share of industrial wood use. Paper production has grown by 189% since 1965 (with sawnwood accounting for only 29% of production). Production of paper increased from 78 million tons in 1950 to 298 million tons in 1997, with per capita rise from 18 to 51 kg. The basic producers of paper and paper products are the USA, Japan, and China (31, 11 and 9% of global production, respectively). They are also the world's leading consumers of paper. Industrial nations use the lion's share of the world's paper: close to 75% in 1996, when per capita consumption was 160 kg in comparison with 16 kg in the developing world. But consumption is growing at a faster rate in developing nations, and by 2010 these countries are expected to use almost 33%.

Recycling has seen a major upsurge in the last two decades, rising from 23% of fibre supply in 1970 to 36% today. FAO predicts that, by 2010, recycled paper will account for over 45% of the fibre supply for paper. However, the paper industry has many ecological impacts (above all, emissions of toxic components into the environment).

Table 1.11. Gross world product, 1950–98.

Year	Total (US$ trillion 1997)	Per person (US$, 1997)	Year	Total (US$ trillion 1997)	Per person (US$, 1997)
1950	6.4	2,503	1980	23.5	5,276
1955	8.1	2,912	1981	24.0	5,292
1960	10.0	3,275	1982	24.2	5,261
1965	12.7	3,795	1983	25.0	5,320
1966	13.4	3,921	1984	26.1	5,465
1967	13.9	3,986	1985	27.0	5,557
1968	14.6	4,119	1986	27.9	5,650
1969	15.5	4,258	1987	28.9	5,754
1970	16.2	4,381	1988	30.1	5,900
1971	16.9	4,465	1989	31.0	5,974
1972	17.7	4,585	1990	31.6	5,993
1973	18.9	4,794	1991	31.8	5,927
1974	19.3	4,811	1992	32.1	5,902
1975	19.6	4,785	1993	33.0	5,971
1976	20.5	4,936	1994	34.3	6,115
1977	21.4	5,062	1995	35.5	6,245
1978	22.3	5,181	1996	37.0	6,412
1979	23.0	5,261	1997	38.5	6,583
			1998*	39.3	6,638

* Preliminary.

1.1.2.4 Global economic development

The gross world product (GWP), which is the sum of gross national products (GNPs), is the best index of global economic dynamics. As can be seen from Table 1.11, the global economy continued to grow in 1998, expanding by 2.2% despite the economic turmoil in East Asia, Russia, and Brazil. This growth, down by nearly one-half from the 4.2% global expansion in 1997, is the slowest since the 1.8% registered in 1991. The slower growth in 1998 stands in contrast to the last few years, when the global economy was easily growing twice as fast as population. The change in income per person among countries varied widely (more widely than ever before), ranging from a drop of 17% in Indonesia and 10% in Thailand to an increase of 10% in Georgia and 6.6% in Ireland.

Among the major industrial countries, the USA remained the pacesetter, expanding by 3.6%. At the other end of the spectrum, the Japanese economy contracted by 2.8%. Growth in the major countries in Europe was somewhat slower than in the USA: 3.0% in France, 2.7% in Germany, 2.6% in the UK, and 1.3% in Italy.

Some of the world's most dynamic economies in 1998 were in Central and Eastern Europe. Poland, Latvia, Lithuania, and Estonia each expanded by

Table 1.12. External debt and debt service of all developing countries, 1971–97.

Year	Debt (US$ billion 1997)	Debt service (US$ billion 1997)	Year	Debt (US$ billion 1997)	Debt service (US$ billion 1997)
1971	277	32	1984	1,319	180
1972	312	37	1985	1,428	186
1973	354	46	1986	1,533	190
1974	405	52	1987	1,701	187
1975	493	56	1988	1,646	206
1976	566	62	1989	1,668	193
1977	707	76	1990	1,720	191
1978	827	103	1991	1,746	183
1979	973	133	1992	1,776	182
1980	1,115	169	1993	1,899	187
1981	1,171	178	1994	2,015	203
1982	1,260	190	1995	2,119	238
1983	1,313	172	1996	2,134	267
			1997	2,171	269

roughly 6%. Bulgaria and Hungary grew at 5%. In contrast, Romania declined by 6%. The Central Asian and Transcaucasian regions also achieved impressive economic gains in 1998. Demonstrating a strong recovery, the region was led by rates of roughly 6–8% in Armenia, Azerbaijan, and the Kyrgyz Republic.

Asia recorded a 2.6% growth in 1998, well below Africa's 3.6% growth. Several African nations were in the 5–7% category, including Cameroon, Ghana, Morocco, Sudan, and some other countries. Economic conditions vary widely within Asia. Several economies in the region are shrinking, including Indonesia (−15%), Thailand (−8%), Malaysia (−8%), and South Korea (−7%). China's economy grew at a somewhat slower rate (7.2%). Other Asian countries are still expanding at a steady rate such as India, Pakistan, and Bangladesh (4.7%). In the Middle East, a region whose economy suffered from low oil prices, economic growth averaged only 3.3% in 1998, the slowest of any region. Egypt at 5% and Turkey at just over 4% were among stronger economies there.

Perhaps, the most distinguishing feature of 1998 is the difference in performance of regions compared with a few years ago. For example, Asia, the regional pacesetter in economic growth for many years, dropped from the average of close to 8% to 2.6%. Meanwhile, some of the world's highest national growth rates were recorded in Africa and Eastern Europe.

A problem of crucial importance for many developing nations and countries with economies in transition is that of their external debts and debt service (see Table 1.12). The indebtedness of developing countries rose to US$2.2 trillion in 1997 – up from US$2.1 trillion in 1996. Some 52% of this was owed to commercial creditors, 31% to other governments, and 17% to multilateral creditors, including the International Monetary Fund (IMF) and the World Bank (WB). Developing countries spend a large share of their annual export revenues servicing their growing debts – diverting scarce resources from other vital investments. In Kenya, for instance, some 25% of government revenue is spent on debt service payments, compared with 6.8% on education and 2.7% on health. In some countries of sub-Saharan Africa and Latin America, the external debt constituted 60–80% of their national income. The debt burden can also have serious implications for the environment.

Table 1.12 reveals gigantic and ever increasing amounts of money necessary for servicing the accumulated enormous debts. There is hardly any doubt that huge amounts of money will likely never be repaid. With this in mind, bilateral creditors have rescheduled debt payments and some even cancelled debts. For example, between 1990 and 1997, the US government forgave US$2.3 billion – about 37% – of loans to the world's most indebted countries. This is, of course, 'a drop in the ocean'. That is why in 1996 the World Bank and the IMF launched the Highly Indebted Poor Countries (HIPC) Initiative, the first effort to include all creditors in addressing the repayment problems of the 40 most indebted nations. As of late 1998, only 10 countries of the 40 had been considered under the HIPC Initiative, and of these only 7 were actually eligible for debt relief (i.e. they were prepared to implement the necessary rigid economic reforms).

Of course, a radical and simple solution could be an outright cancellation of debt payments. In Africa alone, this would save the lives of 7 million children per year and provide 90 million females with access to basic education. And forgiving this debt would cost industrial nations relatively little: for the USA, roughly the equivalent of two B-2 bombers or the accounting errors in 1 year of the Pentagon's budget. It should be mentioned in this context that, in 1998, advertising expenditure exceeded US$400 billion.

An essential index of global economy is the level of development of world trade, which declined 3.8% in 1998, falling to US$5.4 trillion. This was a sharp reversal from the 8.4% growth rate recorded the previous year. The decline can be traced largely to lower demand in Japan and other parts of East Asia as a result of the region's economic slump. Besides, between October 1997 and October 1998, fuel prices fell by 26%, agricultural prices by 18%, and metals and minerals by 16%. Here, it should be noted that growth in trade had generally outpaced overall economic growth for the past 250 years (except the period between 1913 and 1950, when two world wars and the Great Depression in the USA brought international commerce nearly to a standstill). Between 1950 and 1997, exports increased 15-fold, while the world economy expanded sixfold. While exports of goods accounted for only 6% of GWP in 1950, by 1997 this figure had climbed to 15%. A significant consequence of this situation was fast development of all kinds of transportation.

1.1.2.5 The environment

The rapid development of civilization (the growing population number and increasing scale of industrial and agricultural production) naturally has a negative

Table 1.13. Global average temperature, 1950–98, and atmospheric concentrations of carbon dioxide, 1960–98.

Year	Temperature (°C)	Carbon dioxide (ppm)	Year	Temperature (°C)	Carbon dioxide (ppm)
1950	13.84	–	1980	14.18	338.5
1955	13.91	–	1981	14.30	339.8
1960	13.96	316.8	1982	14.09	341.0
1965	13.88	319.9	1983	14.28	342.6
1966	13,96	321.2	1984	14.13	344.2
1967	14.00	322.0	1985	14.10	345.7
1968	13.94	322.9	1986	14.16	347.0
1969	14.03	324.5	1987	14.28	348.8
1970	14.02	325.5	1988	14.32	351.3
1971	13.93	326.2	1989	14.24	352.8
1972	14.01	327.3	1990	14.40	354.0
1973	14.11	329.5	1991	14.36	355.5
1974	13.92	330.1	1992	14.11	356.3
1975	13.94	331.0	1993	14.12	357.0
1976	13.81	332.0	1994	14.21	358.9
1977	14.11	333.7	1995	14.38	360.9
1978	14.04	335.3	1996	14.32	362.7
1979	14.08	336.7	1997	14.40	363.8
			1998*	14.57	366.7

* Preliminary.

impact on the environment. Regrettably, the analysis of this impact is traditionally largely reduced to consideration of the problem of 'global warming'; that is, a rise in the air temperature resulting from the enhanced greenhouse effect of the atmosphere due to increased emissions to the atmosphere of the so-called 'greenhouse' gases (GHGs), mainly carbon dioxide (Brown and Flavin, 1999; Brown *et al.*, 1999). In the past century, global warming (a rise in the global-averaged annual mean surface air temperature [SAT]) actually took place within 0.3–0.6°C (this broad range of uncertainty can be accounted for by the lack of sufficiently reliable observation data for the past century), and an increase in carbon dioxide was observed. This situation is reflected by Table 1.13, relevant to the second half of the 20th century. It can be seen that SAT reached a maximum (for the entire period under consideration) of 14.57°C in 1998, having jumped dramatically within a single year. The increase of 0.17°C was unusually large, particularly given the fact that it immediately followed a new record set the previous year. In 1998, the global average SAT was 0.66°C higher than its long-term mean value of 13.8°C. An unusually strong El Niño and an unprecedented surface temperature rise in the Indian Ocean doubtless contributed to the increased SAT. Data on the global distribution of SAT anomalies and precipitation for July of 1999 revealed the presence of strong spatial inhomogeneities.

During the last 25 years, climate was the warmest when compared with the rest of the over 100-year period of instrumental observations. In this connection, it is usually assumed (and the assumption is confirmed by numerical modelling data) that climate warming was human-induced ('greenhouse') and was due to the increased concentration of GHGs in the atmosphere. At the same time, Brown *et al.* (1999) state that the considerable increase in CO_2 concentration observed in 1998 was a *consequence* of climate warming. The uncertainty of predictions of global climate change can be illustrated, in particular, by W. S. Broecker's assumption that, in conditions of global warming, a violation of the conveyor circulation may happen in the Atlantic Ocean (followed by the Gulf Stream stopping to function), which will result in the coming of another ice age.

Not going into details (see the entries in the Bibliography at the end of Chapter 1), we shall only mention that the nature of global climate system change is very complicated; therefore, climate change cannot be attributed to the effect of any single factor. This is convincingly confirmed by numerical climate models that, however, are still far from being adequate. The oversimplified and therefore unacceptable notions about the reasons for global climate change in the past century, which form the basis of the 'greenhouse' stereotype, paved the way for not only speculative ideas about climate (e.g., the statement that meteorological catastrophes have been more frequent in conditions of global warming, which has not been proved [Kondratyev, 1999]; at the same time, we have to acknowledge that material damage from catastrophes had increased considerably), but also a number of misleading documents, one of which was the Kyoto Protocol (see Boehmer-Christiansen, 1999; Kondratyev, 1998b–d, 1999). The latter document offers recommendations for reducing carbon dioxide emissions to the atmosphere that are hardly feasible, especially in the developing world; besides, the realization of these recommendations may only lead to insignificant global climate changes. Boehmer-Christiansen (1999) convincingly reveals the circumstances that, being far from science, the 'greenhouse stereotype' had brought about considerations that resulted in biased assessments of global climate change causes. The latter served as the basis for working out the Kyoto recommendations, some of which are simply unacceptable ('emission trading', etc.).

Exaggeration of the importance of the problem of human-induced global climate change has reached a level where some scientists predict 'battles' around the problem will take place in the 21st century of the same strategic value as the 'hot' and 'cold' wars of the previous century. In the opinion of such experts, global climate change can soon become the environmental equivalent of a cold war.

1.1.3 Society and nature

In this context, the concept of biotic regulation of the environment formulated and substantiated by Gorshkov (1995) is of fundamental value. According to the concept, the state of the environment (and, to a great extent, of climate) is governed by the dynamics of the biosphere, which is characterized by the degree of closedness of global biogeochemical cycles (this, above all, relates to the carbon, nitrogen, and sulphur cycles). A real global environmental catastrophe would be not so much climate warming (a secondary phenomenon) as a violation of the closed cycles (i.e. the already happening destruction of the biosphere caused by the impossibly high level of use of biospheric resources). It is this incredibly high level that reveals the fact of Earth's overpopulation and the necessity of seeking a new socio-economic paradigm (some considerations on the matter can be found in a monograph by Kondratyev, 1999).

In this connection, analysis of the role of market mechanisms and state regulations is particularly significant. It should be remembered, for example, that subsidies to farmers in the industrially developed world reached US$284 billion in 1996 (according to Brown *et al.*, 1999b). Many governments (including the Russian government) still continue to fund the coal mining industry to a high level, although during the last few years, the funding has lessened considerably (especially in Russia – by 76%: from US$62.5 billion in 1990–1 to US$14.8 billion in 1995–6). Subsidizing, of all kinds, on a global scale used to exceed US$650 billion.

The following examples illustrate the deadlocked nature of the present-day development paradigm. If mankind follows the present US standards (one car per two persons, etc.), then, by the year 2050, when the world population is predicted to approach 10 billion people, 5 billion private cars will be required (against the present 501 million), 9 billion tons of grain (four times the present amount), 360 million barrels of petrol per day (instead of the present 67 million barrels). The unreality of these figures is quite obvious.

The recent UNEP data (see *Global*..., 1999) indicate that the development of solar and hydrogen energy sources will fundamentally change the current global situation in the field of energy production, making it similar to that in the food industry: today, most nations can fully provide food for themselves. Rejection of the 'carbon' economy eventually becomes a geopolitical imperative. In this regard, Iceland is setting an example by planning a transition to the 'hydrogen' economy in the course of the next 10–15 years.

However, solution of the problem of optimization of global socio-economic development will be extremely difficult. The part of the problem concerning the study of nature is made especially difficult by the fact that information on the changing environment remains insufficient: for example, conclusions in the field of climate diagnostics are very contradictory (see Adamenko and Kondratyev, 1999; Boehmer-Christiansen, 1999; Kondratyev, 1992). Hence follows the urgent necessity for further development of a global observing system (Kondratyev, 1999, 2000; Kondratyev and Cracknell, 1998) (paradoxical as it may seem, its weakest part, as a rule, are conventional ground-based observations).

The means of numerical modelling (as the main instrument for understanding processes taking place in the environment) are still very far from being adequate (due to numerous interaction processes that need to be taken account of) that sometimes we give up hope of achieving sufficiently reliable modelling (even more so with the so-called 'integral' modelling that takes into account socio-economic dynamics). All these problems can only be solved through unprecedented broad cooperation of experts in the area of natural and social sciences. Politicians should also join in these activities, all the more so because we live in a world that has accumulated a gigantic arsenal of nuclear weapons, which by 1997 amounted to 36,110 warheads (Brown and Flavin, 1999), taking account of the spread of nuclear weapons. The nuclear threat remains the most dangerous for mankind.

It has been justly noted in the UNEP Report *GEO-2000* (see *Global...*, 1999) that we live in a world of accelerating global change, in which worldwide environmental security measures lag behind the needs of socio-economic development. Ours is a world where achievements in the area of environmental security due to technological progress and environmental policy measures are practically reduced to zero because of the rate and scale of growth of population number and economic development. It can only be added that feasible solutions of these most complicated problems of interaction between society and environment must be found in the new century; otherwise, global civilization will face deadly danger. *GEO-2000* (*Global...*, 1999) gives a warning that should be taken into consideration: key aspects of environmental problems may arise in the 21st century as a result of sudden events or scientific discoveries, sudden, utterly unexpected transformations of problems known before, and also well-known problems which have not been properly dealt with from the standpoint of environmental policy measures. The latter circumstance is undoubtedly the most essential. According to *GEO-2000* (*Global...*, 1999), the following six aspects of life support are first-priority problems: freshwater shortage, environmental (mainly, chemical) pollution, intervention of dangerous species, weakening of immunity and resistive capacity in humans, critical state of fishing, shortage of food products. However, the problem of priorities is still far from being resolved.

1.2 SPACE AND TIME CHARACTERISTIC FEATURES OF DISASTERS

1.2.1 Definitions and classification of disasters

According to the theory of disasters, these are defined as spasmodic changes in a system that appear in the form of the system's sudden response to gradual changes in external conditions (Arnold, 1990). Environmental disasters are particularly important. These are defined as extraordinary, calamitous situations in the vital activity of population caused by essential unfavourable changes in the environment (Grigoryev, 1987; Ramade, 1987). Environmental disasters may be provoked by human-induced (technological) causes (accidents at plants and factories, in transport, etc.) or by dangerous natural phenomena (earthquakes, volcanic eruptions, etc.). A third category is formed by environmental disasters caused by complicated, interdependent human- or nature-induced factors (e.g., the case of desertification).

Anthropogenic (technological) changes of catastrophic nature are by no means necessarily accompanied by environmental disasters. For example, a fatal accident on an aircraft or submarine, an explosion in a mine, etc. will not cause large-scale changes in the environment, even though such accidents result in human deaths and great material losses. However, such situations are also classified as disasters (as adopted by the World Health Organization [WHO]). According to the WHO, a disaster is a situation (not necessarily an environmental one) presenting a sudden, serious and unforeseen threat to the health of society (to the extent that people are forced to ask for help from the outside) (Zerbino, 1991).

The generally accepted viewpoint is that a disaster is, above all, a calamity for the population. But, in the case of an environmental disaster, the notion should be much broader. For instance, disasters can be a threat to other living organisms, not humans.

There are different criteria of the dangerousness of natural phenomena. For example, it has been suggested to introduce the notion of a 'strike factor', which would characterize the excess in comparison with the stability threshold of a natural or social system for various parameters (Ivanov and Yasamanov, 1991). Out of all the existing criteria to determine the scale of a disaster, most often criteria are used that deal with estimates of the number killed, wounded or missing, and the amount of material damage. Such criteria are sufficiently universal and applicable to all kinds of calamities.

One insurance company in Zurich assumes that one of the following two conditions is the criterion of a calamity: (1) a minimum of 20 persons killed; (2) a minimum of US$16.2 million of material damage. According to a different approach (Smith, 1997), the following calamity criteria have been accepted: (1) a minimum of 100 persons killed; (2) a minimum of 100 persons wounded, or (3) a minimum of US$1 million of material damage. According to the UNEP report on the state of the environment (United . . . , 1991), natural calamities are events resulting in: (1) the death of at least 10 persons or (2) material damage above US$2 million.

In Russia, dangerous phenomena of natural or technological origin are called extraordinary situations if they satisfy one of the following conditions: (1) the number of victims is not less than 4 persons; (2) the number of wounded people is not less than 10–15 persons; (3) material damage exceeds 0.5 million roubles at the rouble rate for 1992 (Instruction . . . , 1992).

The classifications of environmental disasters are diverse (Kondratyev, 1998d; Ramade, 1987; Smith, 1997). The following kinds of disasters are generally singled out:

- Geophysical or geological–geomorphological disasters. Among these, the following types are marked out: earthquakes, volcanic eruptions, tsunamis, mudflows, landslides, and landslips.
- Climatic disasters such as droughts, tropical cyclones (storms, tornadoes), dust storms, severe frosts or heat periods. Here, anthropogenic impacts on the global climate as well as on the ozone layer are very important (Kondratyev, 1992; Kondratyev, 1998d; Kondratyev *et al.*, 1996).
- Hydrological disasters, including, in particular, river floods, coastal inundations,

slow but large-scale variations of levels of lakes, inland seas, riverbed displacements.

- Biological disasters such as plagues of various pests (e.g., locusts), epidemic outbreaks among humans or other living organisms. We should specifically mention the loss of biodiversity.
- Anthropogenic disasters of regional scale, among which the pollution of various natural media (most often, technological) is predominant. We should also mention here phenomena such as deforestation, desertification, soil erosion and salination (as a result of land reclamation), fires, the formation of most unfavourable ecological situations due to various technological constructions (dams, dikes, canals, water reservoirs, etc.).

It is hardly possible to separate, distinctly, one kind of environmental disaster from another because some of them are of mixed origin. A tsunami is simultaneously a geological (by origin) and hydrological (regarding its consequences) phenomenon. Desertification may be caused by climatic, man-made, or mixed activities. Deforestation, which may lead to local biological disaster, is human-induced. Finally, fires are increasingly often caused not by natural phenomena but by human activity, or are of mixed origin.

Anthropogenic influence is responsible for a considerable amount of specific disasters whose manifestations are 'delayed'. This concerns so-called 'slow-development' disasters such as loss of biodiversity, deforestation, and landscape degradation due to various construction activities. Of course, slow disasters are known to take place in nature as well: for example, droughts, periods of excessive heat or cold. But they are not numerous. At first sight, such phenomena cannot be classified as catastrophes (which are generally associated with events of an explosive nature). Eventually, 'slow-development' disasters have very serious and sometimes devastating consequences (droughts, desertification).

The above-presented classification only takes into account the most dangerous phenomena that may be conducive to environmental disasters resulting from natural, man-made, or mixed activities. Demographic 'explosions' are sometimes also classified as environmental disasters (Ramade, 1987). We consider it more reasonable to concentrate on the socio-economic nature of demographic dynamics.

1.2.2 Global features of disaster consequences

According to statistical data, three types of natural calamities are predominant on a global scale with regard to the number of victims: river floods, tropical cyclones (accompanied by coastal inundations), and earthquakes. This is in accordance with estimates for the 20th century presented in Table 1.14 (Hewitt, 1997). Between 1947 and 1967, the percentage of human victims from these three categories of natural calamities with respect to their total number was 39.2, 35.8, and 12.7%, respectively. The relationships change from one decade to another because the frequency of large-scale catastrophic events differs from year to year and from decade to decade.

Table 1.14. The number of natural calamities and human victims caused by them on a global scale between 1947 and 1967.

Natural calamities	Number of calamities	Number of victims	Percentage of victims in relation to their total number
Floods	209	173,110	139.2
Typhoons, hurricanes, tropical cyclones, tidal waves	153	158245	35.8
Earthquakes	86	56,100	12.7
Tornadoes	66	3,395	0.8
Storms, thunderstorms	32	20,940	4–7
Heat waves	16	4,675	1.1
Cold waves	13	3,370	0.8
Volcanic eruptions	13	7,220	1.6
Landslides	13	2,880	0.7
Heavy showers	10	1,100	0.2
Snow avalanches	9	3,680	0.8
Fogs	3	3,550	0.8
Frosts	2	–	–
Sand and dust storms	2	10	–
Total	*654*	*441,855*	*100*

Different authors present different statistical data on mortality from disasters. There are various reasons for this such as data confidentiality (e.g. in the former USSR the data on the number of victims of the Ashkhabad earthquake [110,000 killed] was kept secret and only became common knowledge very recently); the difficulty of estimating the number of victims (e.g. during the 1970 tropical cyclone in Bangladesh, the number of victims was evaluated as 265,000–500,000 people; during the 1908 earthquake in Italy, 42,000–150,000 people); or deliberate underestimation of the number of victims (e.g. during an earthquake in China, 242,000–400,000 people; during the 1988 earthquake in Armenia, 25,000–50,000 people). Therefore, the fantastic figures presented by certain experts are not very surprising (e.g. the figure of 30 million victims of the 1955–61 drought and starvation in China) (Davy, 1990). Erroneous data in the evaluation of the number of victims are especially characteristic of droughts. As a result, droughts are sometimes placed in third position (i.e. ahead of earthquakes). For instance, during the 1970s the annual number of victims of droughts was estimated as 23,110 persons (see Table 1.15; Hewitt, 1997).

The consequences of droughts and the starvation they caused became even more catastrophic during the subsequent decade, when the annual mean number of victims constituted 50% of their total number (see Table 1.14).

Among the largest-scale natural disasters with victims of 50,000 and more, three types of calamities can be singled out: earthquakes, tropical cyclones (with coastal

Table 1.15. Annual mean number of killed and wounded from natural calamities during the 1960s and 1970s.

	Number of events		Number of casualties (thousands)		Number of victims (thousands)	
Type of disaster	1960s	1970s	1960s	1970s	1960s	1970s
Droughts	5.2	9.7	18,500	24,400	1,010	23,110
Floods	15.1	22.2	5,200	15,400	2,370	4,680
Cyclones	12.1	14.5	2,500	2,800	10,750	34,360
Earthquakes	6.9	8.2	200	1,200	5,250	38,970
Other causes	10.8	19.5	200	500	2,890	12,960
Total	*50.1*	*73.2*	*26,600*	*44,300*	*22,270*	*114,080*

inundations), and river floods. During the 20th century, out of an annual average of 16 such events, there were 9 earthquakes, 4 cyclones, and 3 river floods with annual mean numbers of their victims of 1 billion people, about 600,000, and about 250,000, respectively (see Table 1.15). It is these disasters that caused the worst calamities both in the 20th century and the entire history of mankind, namely: the flood and cyclone of 1970 in Bangladesh (300,000 victims), the 1976 earthquake in China (242,000 victims), and the 1931 flood in China (140,000 victims).

The number of casualties (including displaced people, i.e. people deprived of their homes, property, etc.) is an important index of the seriousness of a disaster. According to data of the UN Department of Humanitarian Affairs for the 1963–92 period, droughts and river floods contributed most to natural calamities, namely: 33 and 32% of the total number of victims, respectively. Tropical cyclones are responsible for a smaller number of victims (20%), and earthquakes for a much smaller number (4%).

According to estimates of the International Forum of the Red Cross and Crescent (IFRCRCS) for a different time period (1968–92), the same four types of natural calamities were predominant, in the same order, namely: the number of victims of droughts and starvation was 58,973,495 persons, of floods 42,584,343 persons, of wind storms 9,431,063 persons, of earthquakes 1,764,831 persons (see Table 1.16).

So, during different decades of the 20th century, the above-mentioned four types of disasters take priority over all other natural calamities from the viewpoint of the number of victims they have caused. They are responsible for the majority of killed, wounded and missing.

According to IFRCRCS data, the role of anthropogenic (technological) catastrophes in global total estimates of disaster victims is insignificantly minor. For instance, during the same time period (1968–92), the annual mean number of killed was 580 (0.4%) and wounded 52,704 persons (0.5%). But, these values are underestimated and this is probably due to the incorrect, narrow understanding of

Table 1.16. The largest-scale natural disasters of the 20th century (with 50,000 or more victims), according to different sources. Some other data are given in parens in the right-hand column.

Year	Country, locality	Type of disaster	Number of victims (thousands)
1908	Italy (Messina)	Earthquake	120–150 (42–83)
1911	China (Yangtze River)	Flood	100
1920	China (Gansu)	Earthquake	180 (200)
1923	Japan (Tokyo, Yokogama)	Earthquake	140 (143)
1930	Turkey	Earthquake	50
1931	China (Yangtze River)	Flood	140
1939	Chile (Chillán)	Earthquake	50
1948	Turkmenistan (Ashkhabad)	Earthquake	110
1959–61	China	Drought	30 (from data of Western experts, uncertain)
1962	Bangladesh	Flood, cyclone	50
1963	Bangladesh	Flood, cyclone	50
1964	India (Calcutta)	Flood	50–70
1970	Bangladesh	Flood, cyclonc	300 (265–500)
1970	Peru, Uruguay	Earthquake	67 (66–700)
1976	China (Tianshan)	Earthquake	242 (up to 700)
1990	Iran	Earthquake	50 (40)
1991	Bangladesh	Flood, cyclone	139

the term 'technological' or 'anthropogenic' disaster. For example, the usual, inadequate understanding of the term becomes clear in the case of desertification. This is a phenomenon that is often accompanied by climatic desertification; in many of the Earth's regions, it develops independently. So, we have sound reasons to suppose that at least some of those who perished from droughts and starvation (one of the classification rubrics of the scale of global disasters offered by the IFRCRCS; namely, thousands or dozens of thousands of people), as well as some of those who were harmed (millions or dozens of millions people) were the victims of anthropogenic (technological) desertification.

Comparison of disasters by extent of material damage they caused reveals important features (see Table 1.17) (Smith, 1997; Thouret, 1990). The statistics of the largest-scale natural catastrophes between 1976 and 1995, which have caused material damage of US$1 billion and more, indicates the following. This kind of catastrophe includes precisely four types of disasters: earthquakes, river floods, tropical cyclones, and droughts. Within the above time period, earthquakes have been predominant, most devastating, and ruinous in comparison with other disaster types: 12 out of the total number of 30 catastrophic events, and the material damage they brought about was about US$105 billion. The damage caused by the 16 cyclones and floods that accompanied them was half that value, about US$55

Table 1.17. Annual mean number of victims of natural and anthropogenic disasters between 1968 and 1992 (according to IFRCRCS data).

Disaster type	Dead	Wounded	Harmed	Homeless
Droughts and starvation	73,606	–	58,973,495	22,720
Floods	12,067	10,725	42,584,343	2,870,831
Wind storms	28,534	7,468	9,431,063	989,544
Landslides	1,563	235	130,986	106,824
Volcanic eruptions	1,009	278	92,306	10,664
Earthquakes	22,956	36,003	1,764,831	224,006
Technological disasters	580	5,402	52,704	8,372
Total	*140,315*	*54,111*	*113,029,728*	*4,232,901*

billion; the 4 river floods were in the third place and caused damage of about US$29 billion.

It should be emphasized that the technological (anthropogenic) disasters of the 20th century are sometimes as catastrophic as the natural ones from the viewpoint of the material damage they cause. This is especially true of droughts. For instance, the contribution of the anthropogenic factor was doubtless substantial with regard to the material damage caused by droughts in 1984 in Canada (US$2.5 billion) or in 1988 in the USA (US$33 billion).

Material losses resulting from the environmental consequences of military operations are particularly large. As a result of the long-term war in Afghanistan, large-scale desertification of this country took place (at the moment, there are no relevant estimates available). The amount of material damage that resulted from the military operations in Iran and Kuwait in 1991 was huge – US$500 billion (see Table 1.18); this includes the 'environmental' component as well. The material losses from the aggression in Yugoslavia in 1999 were comparable in scale.

1.2.3 Regional (continental) features of disaster consequences

Comparison of large regions from the viewpoint of the size of the territory corresponding to a single large-scale (100 victims and more) disaster between 1963 and 1992 is quite revealing. North America dominates over the rest of the world here (1 disaster per 470,000 km^2). Africa, Europe, South America are somewhere in the middle (1 disaster per 270,000, 240,000 and 230000 km^2). Such regions of the Earth as Central America and the Caribbean Islands, as well as Asia, Australia, and Oceania have much smaller relevant territories (1 disaster per 150,000, 120,000 and 80,000 km^2, respectively). So, during the period under consideration, North America was the least affected by dangerous natural phenomena, while Australia and Oceania, Asia, Central America, and the Caribbean Islands were all subject to such impacts.

However, the risk of being affected by dangerous phenomena leading to

Table 1.18. The largest-scale disasters (from the viewpoint of the material damage they caused) – of US$1 billion and more – between 1976 and 1995.

Year	Country, location	Type of disaster	Number of victims	Damage (US$ billion)
1976	Guatemala	Earthquake	22,778	1.1
1976	Italy	Earthquake	978	3.6
1976	China	Earthquake	242,000	5.6
1979	Yugoslavia	Earthquake	131	2.7
1979	Caribbean islands, USA	Hurricane	1,400	2.0
1980	Algeria	Earthquake	2,590	3.0
1980	Italy	Earthquake	3,114	10.0
1984	Canada (western)	Drought	–	2.5 (1.1)
1985	Chile	Earthquake	200	1.2
1985	Mexico (Mexico City)	Earthquake	7,000 (10,000)	4.0
1986	Former USSR, Ukraine, Chernobyl	Explosion at nuclear power station	30,000 (several tens of thousands)	30.0
1986	Salvador	Earthquake	1,000	1.5
1986	Iran	Flood	500 (424)	1.5 (1.56)
1987	Bangladesh	Flood	1,600	1.6 (1.3)
1988	USA	Drought	–	39.0
1988	USA (Jamaica)	Hurricane Gilbert	1,000	11.0
1988	Armenia (Spitak)	Earthquake	25,000	11.0
1988	Bangladesh	Monsoon, flood	2,000	1.0
1989	Italy	Drought	–	1.5
1989	USA (Alaska)	Tanker accident	–	7.0
1989	USA (California)	Earthquake	61	7.0
1989	Hungary	Drought	–	2.0
1989	USA (Caribbean Islands)	Hurricane Hugo	100	8.2 (9.0)
1989	Italy	Oil spill	–	1.3
1990	Western Europe (UK, France)	Storms	140	1.5
1991	Bangladesh	Cyclone, flood	139,000	1.4
1991	Iraq, Kuwait	War, oil spills, fires	No data	500.0
1991	USA (California)	Forest and bush fires	–	1.2
1992	USA (Florida)	Hurricane Andrew	34	25.0 (15.0–16.0)
1992	Pakistan	Flood	3,000	1.0
1993	USA, Midwest	Flood	50	12.0 (15.0–20.0)
1993	India, Nepal, Bangladesh	Flood	3,000	12.6
1994	USA (North Ridge)	Earthquake	56	17.0
1994	China	Cyclone, flood	1,260	7.0
1995	Japan (Kobe)	Earthquake	5,000	40.0 (100.0)

Table 1.19. Average number of victims per single natural calamity for individual continents (and regions) between 1947 and 1967.

Region	Number of victims	Number of calamities	Average number of victims per single calamity
North America	7,965	210	37
Central America and the Caribbean Islands	14,820	49	302
South America	15,670	45	348
Africa	18,105	17	1,065
Europe (excluding former USSR)	19,575	85	230
Asia (excluding former USSR)	361,410	297	1,216
Australia	4,310	13	332
Total	*441,855*	*716*	*618*

catastrophic events does not always correlate with the scale of disasters taking place on the territories. The nature and scale of disasters, and particularly the number of their victims, are determined by other factors – mainly by measures intended to minimize the consequences of the disasters that are undertaken in relevant countries (see Table 1.19). Therefore, Australia and Oceania, for instance, which are most frequently affected by dangerous phenomena, are not among those regions with the greatest numbers of victims. As might be anticipated, Asia and Africa are far ahead there.

Among the continents, the greatest numbers of victims of large-scale natural calamities (100 persons and more) occur in Asia: between 1963 and 1992, there were 378 events, which is many times higher than relevant events in North America (41) and Europe (44). It is in Asia that 15 of the 17 largest-scale natural calamities (with 50,000 or more victims) of the 20th century took place (Table 1.20).

It is interesting to note that, from the point of view of the number of natural calamities that have caused huge amounts of casualties (1% of a nation's population or more), between 1963 and 1992, Asia (138 events) is preceded by Africa (181 events) (see Table 1.20). Quite the opposite was the situation in North America (the USA and Canada) and in Europe, where the number of such calamities was small (0 and 8, respectively).

We have no summarized data available on the consequences of technological (especially, environmental) catastrophes concerning the number of victims for individual continents. However, this fact does not reflect the seriousness of such consequences. For example, millions of Africans have suffered from the drought and human-induced desertification in the Sahel (1968–72) that affected more than a dozen different countries.

The environmental consequences of the Chernobyl accident in the Ukraine (1986) were continental in their scale (Izrael, 1996). According to the data of an independent association entitled 'The World's Hope', a high level of irradiation by

Table 1.20. Number of large-scale calamities for each continent between 1963 and 1992.

Region	Number killed (100 persons and more)	Damage (over 1% of GNP)	Number of victims (not less than 1% of population)	Total	Percentage
Asia	378	51	138	567	37.1
Europe	44	8	8	60	3.9
Africa	113	60	181	354	23.1
Central America and the Carribean Islands	32	59	65	156	10.2
North America	41	2	0	43	2.8
South America	77	31	51	159	10.4
Australia and Oceania	101	30	60	191	12.5
				1,530	100%

iodine-131 was found in hundreds of thousands of people. The threshold level of irradiation (30 R equivalent units – REU [Roentgen units]) was exceeded in several hundred thousand children. Hundreds of thousands of so-called 'accident liquidators' received high irradiation doses (over 100 REU) (Gritsman, 1993). However, according to International Atomic Energy Agency (IAEA) estimates (1991), information concerning the Chernobyl consequences for humans had not been confirmed. Such controversies indicate the necessity of another, non-biased analysis.

Data on the number of natural calamities between 1963 and 1992, taking account of disasters that caused material damage of 1% of GNP or more, reveal the extent of material damage on the continental (regional) scale. Here, the first three positions are occupied by Africa, Central America (plus the Caribbean Islands), and Asia – 60, 52, and 51 events, respectively. This is not surprising because many of the poor, developing states are situated in these regions and are unable to overcome the situation. The antipodes of such areas are the much more economically developed North America and Europe: 2 and 8 events, respectively.

We can get an idea of the material damage from largest-scale natural calamities (US$1 billion and more) for continents or big regions between 1976 and 1995 from Table 1.18. Of 31 such disasters, 9 took place in North America, 8 in Asia, and 5 in Europe. The material damage was approximately US$100, 80 and 20 billion, respectively. If we compare the number of victims of these catastrophes, we can see that the figure for Asia is huge – 400,000 people, whereas it is 4,000 for Europe and 1,000 for North America. This example spectacularly shows how much attention is paid in Europe and North America (on the whole) to the problem of protecting the population against natural calamities. Note that in Asia half of the entire material damage (US$40 billion out of 80) was caused by the 1995 earthquake in Japan – a country where, unlike most other Asian states, great efforts are taken to prevent calamitous consequences of natural disasters.

Material losses from technological catastrophes on a continental (regional) scale are much smaller than those from natural disasters; they are mostly felt in regions where large industrial countries are located (i.e. in Europe and North America). The extent of material damage from largest-scale technological disasters is shown in Table 1.18 such as the tanker accident near the North American shore (Alaska, 1969) at US$7 billion and the nuclear power plant accident in Europe (Ukraine, 1986) at US$14 billion.

1.2.4 Geographical distribution of disaster consequences

From the viewpoint of socio-economic consequences, the statistics on the calamities for individual states are most important. The list of countries having larger quantities of victims (2,000 persons and more) from a single disaster between 1968 and 1992 (from the IFRCRCS for 1994) is most revealing:

Ethiopia	48,465	Nicaragua	4,619
Bangladesh	40,536	Burundi	4,237
Cameroon	40,011	Peru	4,163
Somalia	21,697	Iran	3,315
Sudan	16,197	Colombia	2,970
China	12,682	Iraq	2,918
Mozambique	11,778	Lebanon	2,579
India	4,888	Philippines	2,131

The list comprises those states that are located in regions that are most often subjected to disasters. However, a number of countries whose situation is dangerous from the standpoint of potential natural calamities have not been included in the list. Among these, there is Japan affected by earthquakes, tsunamis, tropical cyclones, volcanic eruptions, and the USA often subject to earthquakes, river floods, hurricanes in the western, central, and eastern parts of the country, respectively. These are two rich countries, that invest a lot of money into monitoring disasters and minimizing of their consequences.

The order of sequence of states in accordance with mortality rates as a result of natural calamities is quite revealing. Here is a list of countries where, during the years 1947–81, there were 1,000 or more victims per 1 million inhabitants:

Bangladesh	3,958	Peru	1,309
Guatemala	3,174	New Guinea	1,283
Nicaragua	2,590	Haiti	1,189
Honduras	1,995	South Korea	1,021
Iran	1,539		

The above-mentioned states are situated in the most dangerous regions of the Earth, which are most often subject to various natural calamities. Apart from South Korea, all these countries belong to the developing world. Moreover, unlike the previous list of countries affected by the greatest numbers of disasters, these states are characterized by the highest density of population.

The combinations of disasters affecting various countries are very different. For example, in India it is a combination of droughts and floods, as well as desertification and deforestation; in Guatemala and Nicaragua, a combination of earthquakes and volcanic eruptions; in Bangladesh, tropical cyclones accompanied by inundations and river floods.

It is interesting to see how different relevant figures are for economically developed countries; below is a list of such countries and the maximum numbers of victims of natural disasters there:

Japan	276	Australia	11
UK	89	Germany	10
USA	51	Switzerland	9
France	19	South Africa	1
Canada	12		

We see that economically developed countries are the ones in the list of states with the minimum numbers of disaster victims. Even the USA and Japan, both located in very dangerous regions, are in the list. The level of mortality due to natural disasters is many times lower in these countries (especially in the USA) in comparison to developing states.

Coming back to the list of countries situated in the most dangerous areas of the Earth but the least protected from natural calamities due to poverty, we can see that Bangladesh holds the first position. In this connection, data on the flooding of the common Ganges–Brahmaputra delta caused by tropical cyclones between 1960 and 1985 are very interesting (Table 1.21) (Ramade, 1987).

During these 26 years, 20 floods of the densely populated river delta occurred. Some 489,200 people perished. This is a country located in one of the most dangerous regions of the Earth.

Table 1.21. Number of victims of floods in the Bay of Bengal (Bangladesh) caused by tropical cyclones (1960–85).

Date	Number of victims	Date	Number of victims
7–10 October 1960	5,000	1 October 1966	850
30–31 October 1960	15,000	17 April 1969	75
6–9 May 1961	1,000	22–23 October 1970	300
27–30 May 1961	10,466	12 November 1970	300,000
30 October 1962	50,000	6–9 December 1973	183
25–29 May 1963	50,000	24 November 1974	20
17–24 December 1964	1,000	10 December 1981	15
10–12 May 1965	12,000	15 October 1983	43
31 May–1 June 1965	19,279	9 November 1983	900
11–15 December 1965	1,000	14 May 1985	11,069

Well-developed, rich nations, such as the USA and Japan, suffer from the greatest material losses from disasters. For example, the damage to the economy of the USA caused by the drought of 1988 is estimated at US$39 billion, that to the economy of Japan as a result of the earthquakes in 1995 at US$40 billion. However, there are exceptions to the rule. For instance, the material losses in Armenia caused by the 1988 earthquake constituted US$11 billion (see Table 1.18). This happens when cities and towns are centres of catastrophic events. Usually, in poor nations, the number of victims of natural disasters is larger and the material damage smaller. For instance, such was the case in Bangladesh in 1981: the cyclone and subsequent flooding of the country killed 139,000 people, whereas material losses only amounted to US$1.4 billion (a large sum for a poor country).

1.2.5 Temporal dynamics of environmental disasters

This problem has been much less studied than the problem of spatial dynamics of disasters, probably because of the psychology of their perception. Many large-scale environmental disasters take place once every few decades at the same location. This is a great interval during the lifetime of a single generation; therefore, even the most dangerous phenomena are often perceived by people very wrongly as rare happenings.

One of the keys to understanding of development and even prediction of natural calamities is detection of their rhythmic constituent. There is no doubt that prediction of disasters mainly implies accounting for short-term rhythms (a few years, a few decades) that determine the development of 'slow' ('delayed') catastrophic events such as level variations of the lakes of inland seas. A very spectacular and important example of successful forecasting of this kind may be the prediction (made independently by several experts) of the change from regression of the Caspian Sea to transgression that occurred in 1977. Quite the opposite was the case when, ignoring the regular features of the rhythm system in the predictions of the development of the Great Lakes in the USA, experts failed to predict the change in sign in the variation of the level of the Great Salt Lake, which suddenly broke its banks and caused a calamity.

Sometimes, failure to account for the rhythm system in forecasts of level variations of lakes and inland seas leads to serious technological environmental disasters. The catastrophe in the Kara–Bogaz–Gol Bay, which happened after the bay was separated from the sea by a dam, serves as a good example. The detection of cyclic recurrence of disasters such as tropical cyclones over India is of great theoretical and practical interest. On the basis of statistical processing of data of observations of the Hindustan peninsula, cycles of 37 and 15 years have been found (Jayanthi and Krishnakumar, 1996). Earlier, the recurrence of cyclones over India was found to be 40 years, over the Bay of Bengal 30–45 years.

Long-term probability predictions of large-scale natural calamities are only theoretically significant, as are estimates of the development of disastrous floods on the Mississippi River, which occur once every 100 or even 500 years. Such estimates have been obtained in the USA, directly after catastrophic floods.

The determination of the rhythmic constituent in the dynamics of accidents, technological and natural disasters is of great interest. Analysis of 3,000 situations of this kind, which took place in Russia between 1990 and 1994, revealed the existence of a 16-month cycle in their development, as well as of 4-month cycles (Epov, 1994). In different classes of accidents and extreme situations, the cyclic pattern is different (stronger or weaker); it is stronger in the development of railway accidents and fires, and weaker in the cases of aircraft crashes.

As can be seen from statistical data, social tension has a strong influence on technological disasters. In particular, obvious peaks have been observed in recent years that correlate with the enhancement of social tension in Russia (i.e. in August 1991, October 1993, and January 1994). It is interesting to note that 8 out of 11 emergency situations at nuclear power plants in Russia occurred during dangerous periods, as singled out by analysing the entire disaster statistics. According to insurance companies' data, within the European part of Russia several zones can be singled out where the majority of technological and natural–technological disasters took place: for example, the St Petersburg area, the towns of Ufa and Sterlitamak, the Perm and Chelyabinsk regions (Epov, 1994). The cyclic pattern of 16- and 4-month intervals has been observed in the development of critical situations in these areas.

The rhythmic constituent in the development of natural disasters, which becomes increasingly noticeable in the development of technological catastrophes as well, makes doubtful the thesis about the chaotic nature and absolute suddenness of such phenomena put forward by many scientists (e.g. by Zebrowski, 1997). In the light of the statistical data presented above, we cannot negate the presence of a rhythmic pattern in the development of some types of disasters or deny the possibility (although limited) of their prediction. It looks like a combination of chaotic and rhythmic patterns is a characteristic feature of a disaster.

1.3 PROBLEMS OF RISK

Adams (1995) has been quite right to note that a comprehensive theory of risk is just as improbable as the theory of complete happiness. However, every human is a real 'expert' in the problem of risk in the initial sense of the word. Practical life is our teacher of risk management. The notion of risk is so difficult to define because it is a complex notion (see Bokov and Lushchik, 1998; Trofimov *et al.*, 1999a; *Ecological Encyclopedia*..., 1999; *Ecological Risk*..., 1998; Boroush, 1998; Dryzek, 1997; Ernst, 1999; *Global*..., 1999; Hewitt, 1997; Hugget, 1997; Kirchsteiger *et al.*, 1998; Knight, 1991; Koehn *et al.*, 1999; Kondratyev, 1998d; Kondratyev and Cracknell, 1998; Leslie, 1998; Lunine, 1998; Mannion, 1997; Platt *et al.*, 1999; Smith, 1999; Turner *et al.*, 1999). Although, in 1983, the British Royal Society in its report on the assessment of risk accepted very rigid relevant formulations, in 1992 a subsequent report on risk, its analysis, perception, and management did not offer any generally adopted definitions because there was no consensus between the experts involved who represented natural and social sciences (Adams, 1995).

The main difficulty is the necessity to distinguish the real, objective, and measurable risk, by following the formal laws of mathematical statistics, from the qualitatively perceived subjective risk. Doubtless, the problem of risk also requires analysis in the social and cultural context. This determines the limited possibility of using quantitative assessments.

The situation becomes even more complicated due to the subjectivity of assessments. Keely (see Adams, 1995) is quite right to mention that scientists are humans, and when they spot an opportunity of getting funding, they try to demonstrate that their research leads to 'correct' solutions. Today, it is politically correct to believe that smoking provokes lung cancer, and that man-made emissions of carbon dioxide to the atmosphere are responsible for the 'gathering speed' of global warming. Such a situation becomes dangerous when the scientific community is united to pursue their own objectives.

A separate category of risk, which, as a rule, can be assessed quantitatively, is environmental risk, determined as the possibility of unfavourable environmental consequences, caused by both man-made and dangerous natural factors. It is general knowledge that the United Nations Organization (UNO) declared the last decade of the 20th century a 'decade of mitigation of the consequences of natural calamities'. It is clear that this problem can be solved by decreasing the risk of natural calamities (including technological disasters). Here are the most important contributory factors to environmental risk:

- dangerous natural and technological phenomena;
- vulnerability of population;
- the social and natural background of the development of dangerous phenomena;
- the response of population to dangerous phenomena, the degree of preparedness to such phenomena.

However, other classifications are possible that take into account the fact that the risk from a disaster is determined by the following two factors: the most dangerous phenomenon (its specific features, scale, etc.) and population vulnerability (its response, the organization of warning measures, etc.). The latter factor depends on a whole number of circumstances such as economic, social, ethnic–cultural, psychological, etc. (Kondratyev, 1998b, c, 1999; *Natural*..., 1979; Hewitt, 1997; Smith, 1997). Naturally, all the above-mentioned factors vary in space and time and therefore can be mapped. The use of risk maps makes it possible to plan measures for mitigation of the consequences of dangerous phenomena.

1.3.1 Risk of environmental disasters

According to data of the UNEP between 1963 and 1992, the global risk from dangerous natural phenomena can be characterized as follows (*Global*..., 1999; United..., 1991): the greatest number of disasters with 100 or more victims related to floods (202 cases), tropical storms (153 cases), epidemics (133 cases), and earthquakes (102 cases).

The maximum number of calamities, each accompanied by losses equal to 1% or more of annual GNP, also related to floods and tropical storms (76 and 73 cases, respectively). Droughts occupy the third position (53 cases) and earthquakes the fourth (24 cases).The global distribution of disasters that have caused the death of 1% or more of the population of a certain country is somewhat different. From this point of view, droughts occupy the first position (167 cases), followed by floods (162), and tropical storms (100). The risk from such a threatening phenomenon as an earthquake appears to be much less. Earthquakes occupy the fourth position (20 cases) regarding their consequences, although the scale of some earthquakes (e.g. the August 1999 one in Turkey) is enormous.

In accordance with the two main factors of risk – the nature of the dangerous phenomenon and the vulnerability of the population – there are two basic concepts of decreasing the risk. According to the first concept (sometimes called 'behavioural' and predominant nowadays), the risk can be decreased by fighting the dangerous phenomena themselves and employing all possible technical aids for the purpose (Hewitt, 1997). The followers of this concept presume that it is these means (and these alone) that are able to 'improve' and 'correct' a dangerous phenomenon, and minimize the risk. The second concept, called 'structural', proceeds from the conviction that the solution of the problem of natural calamities should be based on optimization of socio-economic conditions and, consequently, decreasing the vulnerability of the population.

We shall now consider the first constituent of the risk of disasters – the peculiar features of a dangerous phenomenon. Risk is variable as a function of the genesis of an event (e.g. a large landslide or an explosion at a plant), the location of its development, its temporal dynamics, and, finally, the natural peculiar features and conditions in which an event in particular takes place.

1.3.2 Location of disasters

On a global scale, the general regular features of the spatial distribution of dangerous natural phenomena have been studied well enough. However, the most significant information – that about the precise location of a given event – is often lacking. For example, it is still impossible (in spite of some progress in this respect; see Zebrowski, 1997) to predict reliably enough the precise location of the epicentre of a future earthquake, the area of the zone of propagation (impact) of a volcanic eruption, the region of a river flood or seawater inundation. The lack of such precise information aggravates the after-effects of disasters of the scale of the catastrophic floods on the Mississippi and Missouri (USA) in 1972 and 1983. The water propagation zone was unexpectedly very broad, which caused a national-scale calamity. Later, it was established that floods of this scale happen once every 100 or 500 years.

Statistical data do not make it possible to predict reliably the moment when a disaster is beginning. For example, for floods on the Nime River in France, the statistically found interval between catastrophic inundations was determined as 150–180 years (Myagkov, 1995). However, between 1557 and 1858 (i.e. for 300 years) not a single flood occurred. At the same time, during the subsequent 130 years, three large-scale floods took place: in 1859, 1868, and 1988, two of them happening within 9 years.

The centres and sources of possible destabilization of the environment of technological origin are, as a rule, narrowly localized and therefore relatively well known in a number of cases. The majority of technological disasters occur in industrially developed countries, in places where factories and plants, especially chemical enterprises, are concentrated (i.e. mostly in cities). The distribution throughout the planet of the 339 most serious incidents of air pollution (between 1900 and 1990) caused by accidents at chemical enterprises, pipelines, and transport is quite revealing. Just a little over one-third of such events took place in Asia and Latin America; most accidents occurred in Western Europe, the USA, and Canada (Hewitt, 1997). However, the situation changed radically during recent years as a result of mass 'migration' of dangerous industries to developing countries (Grigoryev and Kondratyev, 1998).

Information on the localization of potential technological environmental disasters, zones of their spread, and their consequences are in most cases very scarce. For instance, some consequences of the Chernobyl catastrophe in 1986 such as radioactive pollution of soil, etc. appeared unexpected and sudden. Of course, the locations of leaks in oil pipelines cannot be predicted; hence, catastrophic oil spills occur from time to time (e.g. the oil spill in 1997 in the Komi Republic in the north of Russia.

1.3.3 Genesis of disasters

The consequences of natural and technological disasters are to a great extent determined by the genesis of the phenomenon. Peculiar features of the risk involved are very much dependent on the genesis of a dangerous phenomenon: volcanic eruption, explosion at a chemical plant, etc. The impact on the environment can be very different: a territory can be flooded with water or submerged under volcanic lava, etc. Secondary destruction can also be of various kinds.

Some dangerous natural phenomena cause calamities not so much by direct impact, as via the processes that accompany them. For instance, the town of Armero in Colombia was destroyed in 1985 during a volcanic eruption. However, this was not due to lava flows or ashfall but to a huge mudflow provoked by the eruption. In a similar way, it is not so much earthquakes themselves as huge ground displacements (landslides) that are ruinous for the population.

The risk of dangerous phenomena can sometimes be reduced as a result of simple preventive measures, such as construction of pipe-bends to divert the flows of mud or lava.

1.3.4 Temporal variability and development of disasters

Information about the time of origination of a disaster that has serious environmental consequences is an important factor in reducing the risk. Unfortunately, such information can only be obtained about certain types of natural calamities. Information of this kind is based on both data (direct signs of the appearance of an event) and statistics concerning the repetition rate of the events. It is, of course, impossible to predict a disaster on the basis of statistical data. However, the degree of danger of the risk of threatening phenomena may be assessed with accuracy up to a particular season of a certain year, or even better.

This is relevant, for instance, to potential fires on the Mediterranean coast of France. Between 40 and 70% of fires occur there in summer, in the period of dry weather and strong winds. Fires do not often happen in winter, in frosty and windy weather. As revealed by statistical data (Myagkov, 1995), most fires develop in the daytime, between 11 and 17 hours.

The risk of tropical cyclones reaching the Atlantic shore of north-western Europe varies within a year (Hewitt, 1997). According to observation data for 1959–69, it is greatest in winter months (50.6% of cases); it is less by a factor of 1.5 in autumn (32.9%), and still less in spring (14.1%). The risk is minimal in summer, but, if the cyclones happen in that season, they are sudden and very dangerous.

The risk of dangerous phenomena is greatly influenced by the suddenness, intensity, speed, duration, and frequency of their development. No temporal regularities, distinctly expressed or significant, in the development of earthquakes and volcanic eruptions have been revealed. Nor have such regularities been discovered in the development of technological disasters (although, of course, there are certain links between them and natural phenomena).

The development and scale of an environmentally dangerous natural or technological phenomenon are often dependent on its natural background, which can or cannot be favourable for its spread, and consequently can either enhance or mitigate its harmful effect.

1.3.5 Vulnerability of population

The vulnerability of the population, which determines the scale of a calamity to a great extent, depends on economic and social factors, as well as on psychological aspects of the perception of a dangerous phenomenon, information about it, timely protection measures, operational speed of actions required to mitigate the consequences of the unfavourable situation.

One of the characteristic features of catastrophic events in the 20th century on the whole (and its last quarter in particular) was their increased number and, thus, greater risk for the population of our planet. One of the main causes of such an enhanced risk from both natural and anthropogenic disasters is the growth in global population and the development of various infrastructures.

Intensification of risk is also caused by the concentration of population, especially in mega-cities. This factor, for example, played the decisive role in the consequences of the earthquake in Mexico in 1985. Although the epicentre of the earthquake was in the vicinity of the Pacific coast, the greatest number of victims (about 10,000 people) and havoc were registered in and around Mexico City.

The marked expansion of territories cultivated by humans and their settlement in regions dangerous for economic activity also enhanced the risk. At present, about half the global population live in coastal regions that are often subjected to environmental calamities.

Among the social reasons for enhanced risk of disasters, the two main ones are the poverty of people and the economic backwardness of states. The role of population poverty as a factor of risk may be convincingly illustrated by the consequences of the 1978 earthquake in Guatemala. Almost all the 59,000 houses ruined in the city of Guatemala were slums located in ravines near steep slopes, on unstable rocks (Hewitt, 1997).

A whole group of social factors determined the seriousness of the 1988 earthquake in Armenia. As is well known, it was almost as strong (6.7 on the Richter scale) as the seismic shock in California in 1989. In both cases, the disastrous events took place in densely populated regions. However, while in California about 400 people were killed, in Armenia 25,000 met their deaths. In the towns of Spitak and Leninabad, most deaths were caused by houses, built without due accounting for seismic activity, collapsing. The population in the calamity area were not informed about potential danger and necessary protection measures. Finally, the local administration was not prepared for rescue measures (the clearing of obstructions, searching for people buried under the ruins, etc.).

The political structure of society also influences the development of catastrophic events, their number, and scale (Pequy, 1990). A fragmented state is usually more affected by such phenomena than a centralized one. The long history of China shows that floods became real calamities in periods when state and society were weak. In contemporary Russia – unstable, with weaker management and economic power – the number of human-induced catastrophes (mine explosions, accidents in oil and gas pipes, etc.) increased markedly, and the consequences of natural calamities became much more serious and difficult to cope with, although the number of such disastrous events was the same. A spectacular example of the latter is the recent earthquake on the island of Sakhalin.

In this context, the egocentric behaviour of industrially developed countries should be revealed because they underestimate the importance of investment in prevention measures.

1.3.6 Historic experience of surviving disasters

This experience is mainly relevant to natural calamities and its value is usually underestimated. Meanwhile, in every region of the Earth various natural dangerous phenomena recur, having certain specific features. Consideration of historical experience is very helpful in our attempts to reduce the risk of calamities.

We mentioned above the sad consequences of the earthquake in Spitak, Armenia, where historic experience in assessing dangerous situations was ignored. There are historical data showing that this region of Armenia has already been subject to ruinous underground shocks; for example, this is how the ancient capital of Armenia, the town of Ani, was destroyed.

Geologists suppose that to estimate the danger from underground shocks it is also necessary to take into account the neotectonic situation in the area of interest, and not just tectonic activity in historic times but also that in prehistoric times (Bousquet and Philip, 1990). Geological and geomorphological studies reveal that powerful seismic shocks similar to those in Spitak had repeatedly taken place in a not too remote geological past (they can be detected from large displacements of rocks). Such data are in close correlation with seismic information about the Spitak earthquake. In the course of the last 17,000 years, two earthquakes similar to the Spitak one occurred in the area. These data proved to be more representative than information about the Earth's displacements obtained for the shorter historic period of time (during which much less powerful underground shocks were registered at 5.0–5.7 on the Richter scale).

The experience of the catastrophic flood on the Nime River in France on 3 October 1988 made Davy (1990) look at the historic past (Davy, 1990). The flood, caused by unusually heavy rains (420 mm in 6.5 hrs) and six times stronger than usual, created water courses in usually dry ravines. Water flows, whose depth varied from 1 m to several metres, flooded a town located at the foot of a low-mountain plateau. The short-term meteorological forecast (including satellite-derived information) proved to be inaccurate: it did not foresee either the scale or localization of the event. The protective dams proved unreliable and insufficient. And the worst thing was that the possibility of large-scale flooding in the region had been ignored, in spite of the fact that such floods had occurred there previously in 1557, 1859, and 1868. The basic conclusion is that, in order to reduce the risk of similar calamities in the Nime area, a whole complex of measures is required, taking account of historic experience in the locality. This is relevant to rebuilding the town and to due consideration of the peculiar features of the local land pattern and particularly its ravines. Due attention should also be paid to the construction of protective dams and tail races. Then, the risk of catastrophic floods in many places of the Mediterranean region with similar natural conditions can be minimized.

1.3.7 Risk perception and information about risk

Different population groups perceive the risk of dangerous phenomena in different ways. Much is dependent on the social position, education level, and the amount of relevant information available to the authorities. The worst thing that can happen is when they have an inadequate perception of the degree of risk for a town, city, or region.

A characteristic example of such a situation took place in Pakistan in November 1970. At that time, a powerful tropical cyclone drove a huge sea wave to the coastal zone of the country. The wave caused a catastrophic coastal flood. The cyclone itself and the possibility of a disastrous flood were predicted beforehand by the Pakistan Meteorological Service (on the basis of satellite data). However, the administration of the coastal zone did not adequately assess the danger and was not prepared for the disastrous situation.

Informing the population, its perception of potential risk, and the efficiency of local administrations are factors that determine, to a great extent, the scale of after-effects of natural calamities. It is the lack of adequate perception of the degree of risk on the part of both the population and the local administration that caused the deaths of 23,000 people in Colombia during the eruption of the Nevado del Ruiz volcano on 13 November 1985. The majority of deaths occurred in the town of Armero situated at a great distance (72 km) from the volcano. Although the town is outside the zone of the most dangerous lava and pyroclastic flows, it is within the region of possible spread of lahars – mudflows originated by volcanic eruptions. The lahars killed many people and caused much material damage. However, this potential was no secret in the region: before the catastrophic volcanic eruption, relevant studies had been made and a map of risk zones compiled (Thouret, 1990), areas of possible fallout of volcanic material (bombs, lapilli), propagation of pyroclastic flows and, most essentially, lahars, had been marked on the maps. The town of Armero was singled out as located in the region of possible spread of lahars. The map indicated that the mudflows (lahars) would not only reach the town of Armero but even expand over the entire area. The boundary of their propagation was given at the valley of the Magdalena River, situated at a distance of 100 km from the volcano. Comparison of the forecast map of risk zones and the map of environmental changes caused by the volcanic eruption confirmed the correctness of the forecast.

However, neither the administration nor the population of Armero were prepared to perceive the risk of a lahar as credible, in spite of the presence of the map of risk zones. Moreover, nobody responded to the alarm signals transmitted 2 hours before the volcanic eruption and 2 hours 25 minutes before the time the town was flooded by the mudflows (a message had already been sent about the ioncoming lahar). Geologists (Thouret, 1990) warn that the risk of further damage to the town of Armero by a new similar lahar remains until another, maybe much weaker volcanic eruption occurs, because only about half the water from melted ice and snow was taken up by the previous lahar. The remaining water is preserved in the snow and ice cap, in relief depressions.

Perception of the threat of a dangerous phenomenon is more adequate and less risky when the relevant administration and population are informed about it. It was the lack (and even concealment) of information about possible underground shocks in Armenia that was the major factor of the inadequate response to the catastrophic event.

The basic information about dangerous phenomena must be derived from environmental monitoring. Satellite observations should be its most important component. Underestimation of their role by some specialists, whose erroneous notions concern insufficient resolution of images and the lack of due periodicity of observations can only be attributed to their inadequate knowledge of relevant scientific efforts (Smith, 1997).

The perception of dangerous phenomena should be active (not passive). It should be remembered that dangerous (including catastrophic) phenomena are habitual in many regions of the Earth. They are a component part of environmental dynamics. People should and, as can be seen from experience, are able to adjust to some extent to dangerous phenomena, especially because they bring about not only negative but some positive consequences. Sometimes, these frightening and destructive natural phenomena have a constructive aspect: they may be favourable for the environmental situation and economic activity of people. Fertile soils are sometimes formed on volcanic rocks. Rivers as a result of floods put down fertile silt. Tropical cyclones can also be useful, at least in some regions of the globe. The soil erosion resulting from heavy rains causes the transfer of silt particles from mountain slopes to coastal plains. On the plains, inhabited by farmers, the land absorbs more water and the water table rises. A year without strong cyclones and heavy rains is a year of drought; therefore, the local population welcomes cyclones.

Of course, the population of those regions of the Earth that are most frequently beset by disasters is accustomed to coexisting with them. Strange as it may seem, the least adjusted to natural calamities are the most civilized societies and the most developed states. Such nations do not always find the right way of fighting the elements. It is essential to accept the concept that risk is inevitable. Such a concept is actively developed by French experts (Catastrophes..., 1990; Trzpit, 1990). The problem of overestimation and exaggeration of some anthropogenic impacts on the environment deserves a separate consideration (see Kondratyev, 1998b).

1.3.8 Management of risk due to natural and anthropogenic disasters

At first sight, the formulation of the problem may seem meaningless. The very name ‘disaster’ implies a sudden event. However, recent studies of their origination and development revealed some important factors determining catastrophic consequences. It has been found that these consequences are due not so much to anthropogenic or natural calamities as to specific features of the economic activity of people in regions that are subject to such events. Taking account of these circumstances forms a basis for working out a concept of risk management. Let us illustrate the conclusion by a few examples.

According to Jayanthi and Krishnakumar (1996), forest protection from fires can be achieved if forest management (their state and development) is implemented. The system of forest management includes a complex of measures for fire prevention (clearing of forests, construction of water intakes, etc.), the organization of forest monitoring (including satellite and aircraft observations) for timely detection of fire sites, combating the fires and undertaking special measures to restore the forest ecosystem disturbed by the fires (indirect unfavourable environmental consequences such as soil erosion, etc.).

At present, man is unable to prevent droughts or heavy winds causing forest fires. However, the risk of these can and should be reduced by implementing a system of forest management similar to the French one. Of course, it involves financial investment.

In a similar way, the risk of unfavourable after-effects of tropical cyclones may be reduced. It is not the question of actually combating the cyclones, but rather of risk management. We should proceed from the fact that tropical cyclones are inevitable. To reduce the risk of cyclone after-effects, the following is required: (1) information (from the data of observations and forecasts) about the motion and main characteristics of a particular cyclone; (2) timely notification about the approach of a cyclone. Both the administration and the population should know what to do. The results of appropriate actions will be better if the territory is divided into several regions, the zones of least risk are determined and shelter areas provided.

The experience of risk management in case of avalanches, so dangerous in the French Alps and the Pyrenees, deserves special attention. From 1960 to 1990, the number of victims of avalanches did not exceed 30 people (Valla, 1990). During the 1950s, the largest number of victims and the greatest damage from the avalanches were registered in settlements; nowadays, the situation is different. Most human settlements are reliably protected, and avalanches are now dangerous mostly for skiers and tourists.

The reduction of the risk of avalanches for towns and villages is the result of risk management. The management is based on the mapping of avalanche-dangerous regions of the Pyrenees and the Alps (mainly due to remote sounding data; e.g., in 1975 about 800,000 hectares of the territory were mapped), compilation of risk maps for human settlements, modelling of the development of avalanches, and, finally, organization of various protection measures.

The above examples are convincing in that it is possible to manage the risk of various catastrophic phenomena. Among the components of such risk management, for both natural and anthropogenic disasters, there must be: (1) territory subdivision into zones according to the degree of danger; (2) organization of economic development of territories taking account of risk (e.g. chemical enterprises must be located far from the population); (3) regular monitoring of dangerous phenomena; (4) adequate education, instruction, and informing of the population; (5) construction of protective means; (6) taking measures to counteract a dangerous event (using all possible means) on the part of the local administration, both in advance and during the development of the disaster.

1.3.9 Conclusion

The environmental risk of natural and man-made disasters can only be reduced if their socio-economic aspects are taken into consideration. It is obvious that the risk will be close to zero if regions of population concentration and the centres of dangerous phenomena are spatially remote from one another. Otherwise (and this is what happens in reality), socio-economic and psychological factors become extremely important and sometimes play the main role as factors determining the scale of a calamity. In regions of higher environmental risk, the fundamental principle of economic activity must be the concept of 'coexistence with risk'. Risk should be considered an integral part of life. It cannot be considered a rare or casual event.

The inculcation of this concept in the human consciousness and its practical realization mean that taking account of risk should be a constituent part of all human activities; its role must be clearly understood by the economic, social (including educational and cultural), and political institutions of the society. Information about the risk plays a significant role in reduction of environmental risk; this information must include knowledge of the nature and peculiar features of a dangerous phenomenon, and of actions to be taken in the course of its development. Timely information about the threat and development of a dangerous phenomenon, as well as awareness about necessary actions in a critical situation, can minimize the risk.

1.4 BIBLIOGRAPHY

Adamenko, V. N. and Kondratyev, K. Ya. (1999) Global climate change and its empirical diagnostics. *Anthropogenic Impact on the Nature of the North and Its Consequences* (Collected papers, pp. 17–35). Kola Scientific Centre Publ., Russian Acad. Sci., Apatity (in Russian).

Adams, J. (1995) *Risk*, 228 pp. UCL Press, London.

Algin, A. P. (1991) *Facets of Economic Risk* (63 pp.) Znaniye Publ., Moscow (in Russian).

Arnold, V. I. (1990) *The Theory of Disasters*. Nauka, Moscow (in Russian).

Ayres, R. U., Button, K., and Nijkamp, P. (1999) *Global Aspects of the Environment* (1280 pp., 2 volumes). Edward Elgar, Abingdon, UK.

Barthlett, W. and Winiger, M. (eds.) (1998) *Biodiversity. A Challenge for Development Research and Policy* (xxi + 429 pp.). Springer Verlag, Berlin,

Bate, R. (ed.) (1997) *What Risk?* (328 pp.). Butterworth-Heinemann, Oxford.

Bek, U. (1994) From the industrial society to the risky society. Thesis, No. 5, pp. 161–9 (in Russian).

Bell, R. G. and Wilson, J. (2000) How much is too much? Thoughts about the use of risk assessment for countries in transition and the developing world. *Resources* **140**, 10–108.

Belov, P. (2000) Security: Bureaucratic or systemic? *Zelyony Mir (The Green World* **19–20**, 18–19 (in Russian).

Berk, R. A. and Fovell, R. G. (1999) Public perceptions of climate change: A 'willingness to pay' assessment. *Clim. Change* **41**(3–4), 413–46.

Bertrand, M. (1990) Incendies de forests en zone Mediterranéenne – catastrophes peu naturels. *Catastrophes et Risques Naturels, Bull. de la Soc. Languedocienne de Geogr.* **24**(1–2), 217–35.

Birnie, P. and Boyle, A. (1999) *International Law and the Environment* (2nd edn, 575 pp.). Oxford University Press,

Blumenstein, O., Schachtzabel, H., Barsch, H., Bork, H.-R., and Kueppers, U. (1999) *Grundlagen der Geooekologie. Erscheinungen und Prozesse in unserer Umwelt* (250 pp.). Springer Verlag, Berlin.

Boehmer-Christiansen, S. (1999) Who and in what way determines the climate change policy? *Izvestiya of Russian Geograph. Soc.* **131**(3), 6–22 (in Russian).

Bokov, V. A. and Lushchik, A. V. (1998) *The Basis of Environmental Security* (222 pp.). SONAT Publ., Simpheropol (in Russian).

Boroush, M. (1998) *Understanding Risk Analysis. A Short Guide for Health, Safety, and Environmental Policy Making* (39 pp.). American Chemical Society, Washington, DC.

Bossel, H. (1998) *Earth at a Crossroads. Paths to a Sustainable Future* (354 pp.). Cambridge University Press,

Bousquet, J.-C. and Philip, H. (1990) Géologie et risques seismiques. Les enseignements du seisme de Spitac (Arménie, 7 dec. 1988). *Catastrophes et Risques Naturels, Bull. de la Soc. Languedocienne de Geogr.* **24**(1–2), 61–77.

Brebbia, C. A. and Uso, J. L. (eds) (1999) *Ecosystems and Sustainable Development* (Vol. II, 416 pp.). WIT Press, Southampton, UK.

Brebbia, C. A. (ed.) (2000) *Risk Analysis* (Vol. II, 400 pp.). WIT Press, Southampton, UK.

Brown, L. R., Flavin, C. *et al.* (1999a) *State of the World* (259 pp.). Earthscan, London,

Brown, L. R., Penner, M., and Halweil, B. (1999b) *Vital Signs 1999. The Environmental Trends that Are Shaping Our Future* (704 pp.). World Watch Inst./W. W. Norton, New York.

Campbell, G. S. and Norman, J. M. (1998) *An Introduction to Environmental Biophysics* (286 pp.). Springer Verlag, Berlin.

Catastrophes et risques naturels (1990) *Bull. de. la Soc. Languedocienne de Geogr.* **24**(1–2), 240.

Chepurnykh, N. V. and Novosiolov, A. L. (1996) *Economics and Ecology: Development, Catastrophes* (272 pp.). Nauka, Moscow.

Chuvieco, E. (1999) *Remote Sensing of Large Wildfires in the European Mediterranean Basin* (212 pp.). Springer Verlag, Berlin.

Courtillot, V. (1999) *Evolutionary Catastrophes: The Science of Mass Extinctions* (173 pp.). Cambridge University Press.

Danilov-Danilyan, V. I. (ed.) (1999) *Ecological Encyclopedia Dictionary* (932 pp.). Noosphera, Moscow (in Russian).

Davis, L. (1994) *Encyclopedia of Man-made Catastrophes* (434 pp.). Headline, London.

Davy, L. (1990) La catastrophe Nimoise du 3 octobre 1998, était-elle prévisible? Catastrophes et Risques Naturels. *Bull. de la Soc. Languedocienne de Geogr.* **24**(1–2), 133–62.

Demirchan, K. S. and Kondratyev, K. Ya. (1999) Energy production development and the environment. *Izvestiya Russian Acad. Sci. Energetics* **6**, 3–46 (in Russian).

Dobson, A. (1998) *Justice and the Environment. Conceptions of Environmental Sustainability and Dimensions of Social Justice* (292 pp.). Clarendon Press, Oxford.

Doos, B. R. and Shaw, R. (1999) Can we predict the future food production? A sensitivity analysis. *Global Environ. Change* **9**, 261–83.

Drummond, I. and Marsden, T. (1998) *The Condition of Sustainability* (242 pp.). Routledge, London.

Dryzek, J. S. (1997) *The Politics of the Earth* (Environmental discourses, 234 pp.). Oxford University Press.

Ecological Risk: Analysis, Assessment, Forecast (1998) National Conference Proceeding, Irkutsk, Russia (in Russian).

Edmonds, J. A. (1999) Beyond Kyoto: Toward a technology greenhouse strategy. *Consequences* **5**(1), 17–28.

Environmental disasters: Wildlife deaths are a grim wake-up call in Eastern Europe (2000) *Science* **287**(5459), 1737.

Epov, A. B. (1994) *Accidents, Disasters and Natural Calamities in Russia.* Finizdat, Moscow (in Russian).

Ernst, W. G. (ed.) (1999) *Earth System. Processes and Issues* (704 pp.). Cambridge University Press.

Evripidou, P. and Kallos, G. (1999) Monitoring and predicting Saharan desert dust events in the eastern Mediterranean. *Weather* **54**(11), 359–65.

First Meeting of the Inter-Agency Task Force for Disaster Reduction. (2000) *ISDR Information* **1**, 5–7.

Fischer, F. and Hajer, M. (eds) (1999) *Living with Nature. Environmental Politics and Cultural Discourse* (292 pp.). Oxford University Press,

Freeze, R. A. (2000) *The Environmental Pendulum. A Quest for the Truth about Toxic Chemical, Human Health, and Environmental Protection* (337 pp.). University of California Press, Berkeley.

Fyodorov, L. A. and Yablokov, A. V. (1999) *Pesticides – A Toxic Blow for the Biosphere and Man* (462 pp.). Nauka, Moscow (in Russian).

Gerrard, S., Turner, R. K., and Bateman, I. J. (eds) (2001) *Environmental Risk Planning and Management* (640 pp.). Edward Elgar, London.

Gidaspov, B. V., Kuzmin, I. I., Laskin, B. M., and Aziyev, R. G. (1990) Progress in science and technology, security and sustainable development of civilization. *J. Russian Chem. Soc.* **35**, 409–15 (in Russian).

Global Environment Outlook 2000 (1999) (398 pp.). UNEP/Earthscan, London.

Gorshkov, S. P. (1998) *The Conceptual Basis of Geoecology* (445 pp.). Smolensk University Publ. (in Russian).

Gorshkov, V. G. (1995) *Physical and Biological Bases of Life Stability* (470 pp.). VINITI, Moscow (in Russian).

Grigoryev, Al. A. (1991) *Review of the Book by F. Ramade (1987): Les catastrophes ecologiques* (318 pp.). McGraw-Hill, Paris. *Izv. Russian Geogr. Soc.* **123**, 292–295 (in Russian).

Grigoryev, Al. A. and Kondratyev K. Ya. (1998) Global natural resources. *Izvestiya Russian Geograph. Soc.* **130**(1), 5–16 (in Russian).

Grigoryev, Al. A. and Kondratyev K. Ya. (2000) Natural and anthropogenic environmental disasters: Problems of risk. *Studying the Earth from Space* **2**, 5–12 (in Russian).

Gritsman, Yu. (1993) Glaring contradictions should be cleared up and made public. *The Green World Newspaper*, No. 18, p. 5 (in Russian).

Hewitt, K. (1997) *Regions of Risk. A Geographical Introduction to Disasters* (389 pp.). Longman, London.

Hoffmann, H. J. (2000) The rise of life on Earth. *National Geographic* **198**(3), 100–13.

Hugget, R. J. (1997) *Environmental Change. The Evolving Ecosphere* (378 pp.). Routledge, London.

Huntington, S. P. (1998) *The Clash of Civilizations and the Remaking of World Order* (368 pp.). Touchstone Books, London.

Hurrell, A. and Woods, N. (eds) (1999) *Inequality, Globalization, and World Politics* (240 pp.). Oxford University Press.

Instruction on the Established Procedure of Exchanging Information about Extreme Situations over the Territory of the Russian Federation (1992) Moscow (in Russian).

Ivanov, O. P. and Yasamanov, N. A. (1991) The indomitable Earth. *The Earth and the Universe* **5**, 26–32 (in Russian).

Izrael, Yu. A. (1996) *Nuclear Fallout after Nuclear Explosions and Accidents* (356 pp.). Progress-Pogoda, St Petersburg (in Russian).

Jayanthi, N. and Krishnakumar, G. (1996) Trends and periodicities in tropical cyclones over the Bay of Bengal. *Proc. Int. Conf. on Disaster Mitigation, Madras, India* (pp. B1–27–31), Anna University, Madras.

Jorgensen, S. E. and Muller, F. (1999) *Handbook of Ecosystems* (660 pp.). Springer Verlag, Berlin.

Kapitsa, S. P. (1999) *A General Theory of Human Race Growth: How Many People Lived, Are Living and Will Live on Earth* (190 pp.). Nauka, Moscow (in Russian).

Karl, T. R., Nicholls, N., and Ghazi, A. (1999) CLIVAR/GCOS/WMO Workshop on indices and indicators for climate extremes. Workshop Summary. *Clim. Change* **42**(1), 3–7.

Kasischke, E. S. and Stocks, B. (eds) (1999) *Fire, Climate Change and Carbon Cycling in the Boreal Forest* (490 pp.). Springer Verlag, Berlin.

Kirchsteiger, C., Christou, M. D., and Papadakis, G. A. (eds) (1998) *Risk Assessment and Management in the Context of the Seveso II Directive* (560 pp.). Elsevier Science, Oxford.

Knight, F. H. (ed.) (1991) The meaning of risk and uncertainty. *Risk, Uncertainty, and Profit* (pp. 210–35). Houghton Mifflin, Boston.

Kochurov, B. I. and Mironyuk, S. G. (1993) Approaches to the definition and classification of ecological risk. *Geography & Natural Resources* **4**, 22–8 (in Russian).

Koehn, J., Gowdy, J., Hinterberger, F., and van der Straaten, S. (eds) (1999) *Sustainability in Question. The Search for a Conceptual Framework* (368 pp.). Edward Elgar, Abingdon, UK.

Kondratyev, K. Ya. (1992) *Global Climate* (359 pp.). Nauka, St Petersburg (in Russian).

Kondratyev, K. Ya. (1998a) The outcome of the UN General Assembly Special Session. *Vestnik of Russian Acad. Sci.* **1**, 30–40 (in Russian).

Kondratyev, K. Ya. (1998b) Ecological risk: Real and hypothetical. *Izvestiya of Russian Acad. Sci.* **130**(3), 13–24 (in Russian).

Kondratyev, K. Ya. (1998c) Ecodynamics and geopolicy: From regional to global scale. *Izvestiya of Russian Geograph. Soc.* **130** (in Russian).

Kondratyev, K. Ya. (1998d) *Multidimensional Global Change* (761 pp.). Wiley/Praxis, Chichester, UK.

Kondratyev, K. Ya. (1999) *Ecodynamics and Geopolicy. Volume 1. Global Problems* (780 pp.). Victoria, St Petersburg (in Russian).

Kondratyev, K. Ya. and Cracknell, A. P. (1998) *Observing Global Climate Change* (562 pp.). Taylor & Francis, London.

Kondratyev, K. Ya. and Demirchian, K. S. (2000) Global climate change and the carbon cycle. *Izvestiya of Russian Geograph. Soc.* **132**(4), 1–20 (in Russian).

Kondratyev, K. Ya., and Krapivin V. F. (2002) Biocomplexity and global geoinformational monitoring. *Studying the Earth from Space* **1**, 1–8 (in Russian).

Kondratyev, K. Ya. and Losev, K. S. (2002) Contemporary stage of the civilization development and its perspectives. *Studying the Earth from Space* **2**, 3–23 (in Russian).

Kondratyev, K. Ya., Donchenko V. K., Losev K. S., and Frolov A. K. (1996) *Ecology–Economics–Policy* (828 pp.). Russian Academy Science, St Petersburg (in Russian).

Kotlyakov, V. M. (1997) *Science. Society. Environment* (410 pp.). Nauka, Moscow (in Russian).

Kotlyakov, V. M. (ed.) (1999) *The Anatomy of Crises* (239 pp.). Nauka, Moscow (in Russian).

Kotlyakov, V. M., Trofimov, A. M., Khuzeyev, R. G., Borunov, A. K., Gnedenkov, L. N., and Seliverstov, Yu. P. (1993) A geographical approach to the theory of disasters. *Izvestiya of Russian Acad. Sci., Ser. Geography* **5**, 5–18 (in Russian).

Kotlyakov, V. M., Glazovsky, N. F., and Rudenko, L. G. (1997) Geographical approaches to the problem of sustainable development. *Izvestiya of Russian Acad. Sci., Ser. Geography* **6**, 8–15 (in Russian).

Kryshev, I. I. and Sazykina, T. T. (1998) Methodology of the analysis of ecological risk in nuclear power engineering. *Problems of Atmospheric Physics* (Collected papers, pp. 304–319). Gidrometeoizdat, St Petersburg (in Russian).

Kudryavtsev, A. (2000) *100 Great Catastrophes of the 20th Century* (463 pp.). Martin, Moscow,

Kurbatova, A. S., Myagkov, S. M., and Shnyparkov, A. L. (1997) *Natural Risk for Russia's Cities* (240 pp.). Research Institute on City Ecology Press, Moscow (in Russian).

Larichev, O. I. (1987) Problems of decision-making taking account of the factors of risk and security. *Vestnik of Russian Acad. Sci.* **11**, 38 (in Russian).

Larin, V. (1996) *The 'Mayak' Industrial Complex – Half a Century of Problems* (175 pp.). Moscow (in Russian).

Lavrov, S. B. (1978) Foreword to *Natural Calamities: Studies and Fighting Techniques* (translated from English, pp. 5–21). Progress, Moscow (in Russian).

Lavrov, S. B. (1998) In search of a new paradigm. *Geographical Problems of the End of the 20th Century* (Collected papers, pp. 5–11). Russian Geographical Society, St Petersburg (in Russian).

Lavrov, S. B. (1999) Realities of globalization and mirages of sustainable development. *Izvestiya of Russian Geograph. Soc.* **131**(3), 1–8 (in Russian).

Leslie, J. (1998) *The End of the World. The Science and Ethics of Human Extinction* (310 pp.). Routledge, London.

Lobanov, F. I. and Shapiro, M. M. (1990) Ecological risk in industry. Assessment and management. *J. Russian Chem. Soc.* **35**, 405–9 (in Russian).

Luman, N. (1994) The notion of risk. Thesis, No. 5, pp. 135–61 (in Russian).

Lunine, J. I. (1998) *Earth: Evolution of a Habitable World* (352 pp.). Cambridge University Press.

Mannion, A. M. (1997) *Global Environmental Change* (2nd edn, 387 pp.). Longman, London.

Meadows, D. H., Meadows, D. L., and Randers, J. (1995) *Beyond the Limits. Global Collapse or a Sustainable Future* (300 pp.). Earthscan, London.

Mikisha, A. M. and Smirnov, M. A. (1998) Disasters on the Earth caused by the fall of heavenly bodies. *Vestnik of Russian Acad. Sci.* **69**(4), 327–336 (in Russian).

Moeller, D. (ed.) (1999) *Atmospheric Environmental Research. Critical Decisions between Technological Progress and Preservation of Nature* (xiv + 200 pp.). Springer Verlag, Berlin.

Moiseyev, N. N. (1998) Interaction between nature and society – global problems. *Vestnik of Russian Acad. Sci.* **68**(2), 167–70 (in Russian).

Mooney, L. and Bate, R. (eds) (1999) *Environmental Health. Third World Problems – First World Preoccupations* (236 pp.). Butterwoth-Heinemann, Oxford.

Myagkov, S. M. (1991) Problems of the geography of risk. *Vestnik of Moscow State University, Ser. 5 – Geography* **4**, 3–8 (in Russian).

Myagkov, S. M. (1994) Natural risk: Peculiar features of its perception. *Vestnik of Moscow State University, Ser. 5 – Geography* **4**, 30–6 (in Russian).

Myagkov, S. M. (1995) *Geography of Natural Risk* (224 pp.). Moscow State University (in Russian).

Myagkov, S. M. and Kozlov, K. A. (1993) The spread of technological and natural extreme situations in Russia. *Vestnik of Moscow State University, Ser. 5 – Geography* **5**, 3–12 (in Russian).

Natural Calamities: Studies and Fighting Techniques (translated from English, 440 pp.) (1979) Progress, Moscow (in Russian).

Night, F. (1994) The notions of risk and uncertainty. Thesis, No. 5, pp. 12–29 (in Russian).

O'Riordan, T. and Voisey, H. (1998) *The Transition of Sustainability: The Politics of Agenda 21 in Europe* (320 pp.). Earthscan, London,

Osipov, V. I. (1995) Natural disasters are a centre of attraction for scientists. *Vestnik of Russian Acad. Sci.* **65**(6), 483–95 (in Russian).

Our Changing Planet. The FY 2000 (1999) (Global Change Research Program, A supplement to the President's Fiscal Year 2000 Budget, 100 pp.). Washington, DC.

Park, S. H. and Labys, W. C. (1998) *Industrial Development and Environmental Degradation* (208 pp.). Edward Elgar, Abingdon, UK.

Pearce, D. (1999) *Economics and Environment. Essays on Ecological Economics and Sustainable Development* (352 pp.). Edward Elgar, Abingdon, UK.

Pequy, C.-P. (1990) Bavures ou derapage? Catastrophes et risques naturels. *Bull. de la Soc. Languedocienne de Geogr.* **24**(1–2), 9–20.

Platt, R. *et al.* (1999) *Disasters and Democracy. The Politics of Extreme Natural Events* (343 pp.). Island Press, Washington, DC.

Puzachenko, Yu. G. (1993) Preventive measures against ecological disasters. *New Concepts in Geography and Forecasting* (pp. 21–35). Russian Academy Science, Nauka, Moscow (in Russian).

Ramade, F. (1987) *Les Catastrophes ecologiques* (318 pp.). McGraw-Hill, Paris,

Rao, U. R. (1996) *Space Technology for Sustainable Development* (564 pp.). McGraw-Hill, New Delhi.

Rezanov, I. A. (1999) Was the Earth subject to intensive meteorite bombardment in its early history? *Bull. Moscow Soc. of Nature Investigators, Geology Section* **74**(6), 14–19 (in Russian).

Scheidegger, A. E. (1981) *Physical Aspects of Natural Disasters* (232 pp.). Nedra, Moscow (in Russian).

Security of Russia. Legal, Socio-Economic and Scientific-Technological Aspects. Protection of Population and Territories in Extreme Situations of Natural and Technogenic Origin (592 pp.) (1999) Znaniye, Moscow (in Russian).

Seliverstov, Yu. P. (1994) The problem of global ecological risk. *Izvestiya of Russian Geograph. Soc.* **126**(2), 2–16 (in Russian).

Shah, A. (1999) *Ecology and Crisis of Overpopulation* (176 pp.). Edward Elgar, Abingdon, UK.

Shershakov, V. M., Vozzhennikov, O. I., Golubenkov, A. V., Kolomeyev, M. P., Korenev, A. I., Kosykh, and V. S., Svirkunov, P. N. (1998) Development of a system of operational analysis and prediction of danger for population and the environment from pollutant emissions resulting from nuclear and other technological accidents. *Problems of Atmospheric Physics* (Collected papers, pp. 273–303). Gidrometeoizdat, St Petersburg (in Russian).

Shoigu, S. K., Vorobyov, Yu. L., and Vladimirov, V. A. (1997) *Catastrophes and States.* Energoatomizdat, Moscow (in Russian).

Shoigu, S. K., Vladimirov, V. A., Vorobyov, Yu. L., Dolgin, N. N., Makeyev, V. A., and Shakhramanyan, M. A. (1999) *Protection of the Population and Territories from Natural and Human-Induced Extreme Situations* (592 pp.). Znaniye, Moscow (in Russian).

Smith, K. (1997) *Environmental Hazards. Assessing Risk and Reducing Disasters* (389 pp.). Longman, London.

Smith, V. K. (1999) *The Economics of Environmental Risk* (350 pp.). Edward Elgar, Abingdon, UK.

Stothers, R. B. (1999) Volcanic dry fogs, climate cooling, and plague pandemics in Europe and the Middle East. *Clim. Change* **42**(4), 713–23.

Svirezhev, Yu. M. (1987) *Nonlinear Waves, Dissipative Structures and Ecological Disasters* (368 pp.). Nauka, Moscow (in Russian).

Swanson, T. and Johnston, S. (1999) *Global Problems and International Environmental Agreements* (288 pp.). Edward Elgar, Abingdon, UK.

Thiele, L. P. (1999) *Environmentalism for a New Millennium* (288 pp.). Oxford University Press.

Thouret, J.-C. (1990) Activité volcanique explosive et calotte glaciaire: le cas des lahars du Nevado del Ruiz, Colombie (13 novembre 1985) et l'évalution des risques volcano-glaciaires. *Catastrophes et Risques Naturels. Bull. de la Soc. Languedocienne de Geogr.* **24**(1–2), 29–60.

Trofimov, A. M., Kotlyakov, V. M., Seliverstov, Yu. P., Panasyuk, M. V., Rubtsov, V. A., and Pudovik, E. M. (1999a) Balanced development – the stable state of geosystems. *Izvestiya of Russian Geograph. Soc.* **131**(3), 9–16 (in Russian).

Trofimov, A. M., Kotlyakov, V. M., Seliverstov, Yu. P., and Zaynullina, A. R. (1999b) Basic approaches to the solution of risk problems. *Izvestiya of Russian Geograph. Soc.* **131**(4), 1–8 (in Russian).

Trzpit, R. J. P. (1990) Vivre avec tempêtes dans l'Europe du Nordouest: De l'envènement météorologique à la realité vécue. *Catastrophes et Risques Naturels. Bull. de la Soc. Languedocienne de Geogr.* **24**(1–2), 163–202.

Turner, B. A. and Pidgern, N. F. (1997) *Man-Made Disasters* (2nd edn, 250 pp.). Butterworth-Heinemann, Oxford.

Turner, K., Button, K., and Nijkamp, P. (eds) (1999) *Ecosystems and Nature. Economics, Science and Policy* (560 pp.). Edward Elgar, Cheltenham, UK.

United Nations Environment Programme (1991) *Environmental Data Report* (3rd edn, 408 pp.). Basil Blackwell, London.

Valla, F. (1990) Les avalanches. *Catastrophes et Risques Naturels. Bull. de la Soc. Languedocienne de Geogr.* **24**(1–2), 88–94.

van den Bergh, J. C. J. M. (ed.) (1999) *Handbook of Environmental and Resource Economics* (1216 pp.). Edward Elgar, Abingdon, UK.

Varotsos, P., Sarlis, N. Lazaridou, M., and Kapiris, P. (1998) A possible explanation for detecting seismic electric signals at certain sites of the earth's surface. *Studying the Earth from Space* **1**, 123–9.

Vladimirov, V. A. and Izmalkov, V. I. (2000) *Catastrophes and Ecology* (380 pp.). Centre for Strategic Civil Protection Studies/Contact-Culture, Moscow.

Vlavianos-Arvanis, A. (ed.) (1996) *Biopolitics. The Bioenvironment* (Vol. V. B.I.O. 671 pp., Athens.

Volk, T. (1998) *Gaia's Body. Toward a Physiology of Earth* (xvii + 269 pp.). Springer Verlag, Berlin.

Vorobyov, Yu. L., Loktionov, N. I., Faleyev, M. I., Shakhramanyan, M. A., Shoigu, S. K., and Sholokh, V. P. (1997) *Disasters and Man. Volume 1. The Russian Experience of Facing Extreme Situations* (256 pp.). AST-LGD, Moscow (in Russian).

Zadorozhnyuk, I. E. and Zozolyuk, A. V. (1994) The phenomenon of risk and its present-day economico-psychological interpretations. *Psychological J.* **15**(2), 26–37 (in Russian).

Zavarzin, G. A. (ed.) (1999) *The Carbon Cycle over Russia's Territory* (329 pp.). Ministry of Science, Moscow (in Russian).

Zavarzin, G. A. and Kotlyakov, V. M. (1998) The strategy of studying the Earth in the light of global change. *Vestnik of Russian Acad. Sci.* **68**(1), 23–9 (in Russian).

Zebrowski, E. Jr (1997) *Perils of a Restless Planet. Scientific Perspectives on Natural Disasters* (306 pp.). Cambridge University Press.

Zerbino, D. D. (1991) *Anthropogenic Ecological Disasters* (136 pp.). Naukova Dumka, Kiev.

2

Natural and anthropogenic disasters in the history of civilization

2.1 NATURAL DISASTERS AND MANKIND

A retrospective look at the historical past convinces us that natural disasters were the cause of many human calamities. Earthquakes and floods, droughts and fires, and other unfavourable natural phenomena affected man in the past more than at present because mankind was less able to foresee and had lower technical possibilities to protect against disasters (Grigoryev, 1986, 1991).

Natural disasters resulted in human death, the destruction, complete or partial, of settlements, cities, and made agricultural activities difficult. Along with other causes, they contributed to the destruction of whole cities and even states. Forecasting and establishing the role of natural phenomena in bringing about an unfavourable natural situation for people, which would prevent or hamper their daily lives, is of interest for the present time as well.

This chapter examines only a few disasters – just those that caused the greatest damage to man. The two main groups that can be distinguished among these phenomena are the short-term (occurring suddenly) and long-term (extended in time) disasters.

2.1.1 Volcanic eruptions

Volcanic eruptions are related to one of most destructive types of natural disasters (Kondratyev, 1985; Kondratyev and Galindo, 1997). The areas of volcanic activity have long since attracted people, due to the rich soils that form on the volcanic rocks thrown out during eruptions. This is why there are lots of settlements on areas surrounding volcanoes, and even cities appeared around many of them. The slopes of volcanic mountains have been used for a long time as farming lands.

Many settlements and cities situated at the foots or in the environs of volcanoes were subject to the destructive effect of volcanic eruptions. Some of them were swept away by lava, others were buried under a layer of ash, pumice, and lava. Some cities disappeared from the face of the earth, and new volcanic landscapes appeared instead. People forgot, with the passage of time, about the location of buried settlements. Nature did its own work: the sun, wind, and water changed the landscape's appearance.

There are many legends told by different peoples of the world about the destructive force of the ancient volcanic eruptions that were experienced by their ancestors.

The scientific, geological, and archaeological evidence on the history of how volcanic eruptions affect people is still scant. For example, the discovery made by the staff of the Department of Ancient History at Cologne University is of interest. While making archaeological investigations in the Rhine River basin in Germany, they discovered, early for this area, human settlements buried under lava (Street, 1984) Their age is approximately 11,000 years. This was the time of the end of the last glaciation in Europe. The glacier had already receded to the north of Europe when a volcano erupted in the area of Lake Laakherzee in the Neuwieder Hollow, where the first tribes appeared. The lava layer spread over the Hollow and reached a depth of 15 m. The ash traces of this clearly major eruption have also been found in Poland.

The information on the volcanic eruption at the Rhine River and on its effect on the European population is very scant, which is not surprising since it took place about 11,000 years ago. Natural disasters of similar size that occurred much later, in the civilized world, were completely forgotten. Thus, even in densely populated Italy, the location of two earlier well-known cities – Pompeii and Herculaneum – was forgotten for a long time. In AD 79 a major eruption of Vesuvius (to be more exact, its predecessor, Somma Volcano) took place, in the vicinity of which the two cities were located (Waltham, 1982). It lasted for quite a short period of time – no more than 48 hours. Both Pompeii and Herculaneum disappeared. They were buried: Pompeii under a layer of ash and pumice and Herculaneum under a layer of dirt. After 1,600 years, in the late 1730s Herculaneum was found as a result of excavations. It was laying under an almost 20-m thick layer of sediments. Pompeii was found later on. They have not yet been excavated completely, but the work done has revealed the terrible tragedy experienced by these cities. The inhabitants of the Vesuvius environs seemed not to realize that this mountain was a volcano. It did not erupt until AD 79 according to the Romans and Etruscans. Though, approximately 15 years before the beginning of the eruption, earthquakes started in the volcano environs. Now, they would have been a signal of the impending disaster, but, at that time, this fact escaped the attention of people.

The devastation around Vesuvius took place in the district with a radius up to 18 km covering an area over 310 km^2. In other regions of the planet, volcanic eruptions covered more extended areas, and clearly they were more destructive for the population of the surrounding area. This particularly concerns some regions in Central America in which there are located individual volcanoes and the accumulations of active volcanoes. It is well known that volcano eruptions disturbed the everyday life of ancient inhabitants of Central America. These tragic events are reflected in the legends of American Indians and in fine art. The causes of the fall of Central American ancient civilizations have not yet been completely found out, though volcanic eruptions could be one of them.

One of the best-known accumulations are the Chalguapa archaeological monuments situated in Salvador. Here, in the Rio Pas the remains of one of the major cities-states of the ancient Maya were found. The excavations discovered lots of temples, original stepped pyramids, and other constructions hidden under a thin soil layer covered partially by vegetation. It is noteworthy that ash and pumice were present everywhere in the sediments. They were thrown out during the Ilopango volcanic eruption, located 75 km from the city of Chalguapa. The pumice layer of 2 m in depth is found around the volcano in a radius up to 50 km. Ash is scattered over a longer distance: almost 80 km from the volcano. Near the volcano, the depth of ash layers reaches tens of metres. Two ash layers, 20 m and 50 m in depth, have been found at a distance from it. The emissions of glowing noxious gases, hot ash, and tuff destroyed everything around the volcano including forest and bush, along with all villages and the city itself. The total area of land that was devastated by fire is estimated to be 3,000 km^2 (Gulyaev, 1981; Sheets, 1981). This is approximately 10 times larger than the area around Vesuvius that was affected after its eruption in AD 79.

Unlike the environs of Vesuvius, the layer of volcanic sediments around Ilopango volcano was not so deep, since there were no powerful flows of dirt and lava here, though the disaster was of larger scale and more destructive on the whole. It is not surprising that a vast territory became deserted. Judging by the layers, the volcanic eruption took place at three stages. So, those who did not die had time to escape. According to archaeological data, a considerable influx of population was observed at that time to districts located northward of Chalguapa. As approximately estimated, the population could be 30,000 in the eruption area. The district became deserted afterwards for several hundred years – until the early fifth century AD.

The most destructive eruption for the ancient Maya, who inhabited in former times the territory of Salvador, occurred in AD 260. As a result, the city of Tseren perished, which was situated at a distance of 25 km to the north-west of Ilopango volcano. Two volcanic eruptions can be distinguished among other large-scale volcanic eruptions, which took place subsequently. One of them is related to AD 600, and the other to about AD 900–1,000. The American archaeologist Sheets (1981) assumes that the above volcanic eruptions had an effect on the development of a Maya state in this region.

A powerful volcanic eruption probably involved the downfall of the so-called Minoan people on Crete in the Mediterranean Sea, though, at first sight, this statement might seem to be paradoxical, since neither active nor extinct volcanoes are present on the island. In the 1930s, the Greek geologist and archaeologist Marinatos (1973) expressed a hypothesis that the state on Crete was destroyed in the 16th century BC as a result of the eruption of the Santorini volcano on Thera Island. This island is much smaller in size than Crete and is located at a distance of 120 km to the north-east of Crete. The volcanic rocks found during the excavations on Crete helped to prove this hypothesis. The proof that such an event occurred allows us to speculate about the reality of other events, which were the basis of the myth on Deukalion Flood (Ilyinskaya, 1988), which is well known in the Mediterranean Sea area, in Ancient Greece.

According to this myth, approximately in 1530 BC, the Flood took place. The greater part of the coast was flooded on the Balkan Peninsula and many islands in the Mediterranean Sea. Later excavations on this island and geological investigations confirmed the correctness of this hypothesis (Manning, 1989). The fact that the Crete civilization ended unexpectedly in the mid-1600 BC was known earlier due to investigations made by the archaeologist A. Evans. He managed to prove that, approximately in 1520 BC, the eruption of Santorini volcano occurred on Thera Island. Archaeological investigations show that a buried city (Akrotiri) is situated under volcanic rocks on this island. The depth of the ash and pumice layer above the remains of the city exceeds 10 m (Evans, 1921–1935).

Geological investigations enabled us to obtain information on the latest geological history of the island, which fact allowed us to estimate the natural situation in a new fashion (Manning, 1989; Marinatos, 1973). There is now a huge volcanic crater with a diameter of about 11 km. The volcano itself, as a mountain, does not exist since it was destroyed during the eruptions and explosions. One of the eruptions took place in about 1520 BC, and the second in about 1460 BC. Recent findings (tree-ring growth and acidity peaks in ice cores) suggest that these eruptions took place in the mid-17th century BC (Fytikas *et al.*, 1990). The destruction of Akrotiri was related to these eruptions. Some time after the first volcanic eruption, the city revived but, after the 1460-BC eruption, it was completely ruined. Its destruction coincides in time with two major destructions on Crete. Herodot (1972) mentions two large-scale disasters on Crete. Marinatos (1973) explained them by the volcanic eruption on Thera Island.

Present-day studies have shown that, during the explosive eruption on Thera Island, every living thing around was destroyed in a radius up to 170 km (Manning, 1989). Glowing noxious gases and hot ash causing fires extended to Crete as well. Underwater investigations helped to discover volcanic ash from this eruption at the sea bottom near the Crete coast and at other parts of the bottom. Pumice was found in different places on the coast of Minor Asia and in Macedonia. The main transfer of volcanic ejections took place eastwards. So, these ejections were found at the sea bottom even 600 km far to the east of Thera Island This explains the very large amount of remains of volcanic ejections on Crete. The two eruptions on Thera Island caused, most probably, the destruction and death of all living things on Crete as a result of the spread of fires, noxious gases, as well as earthquakes, which accompanied them. Huge sea waves – tsunamis – completed the destruction of the Crete cities.

Great calamities are explained by volcanic eruptions in the historical past. However, the total number of victims and the scale of devastation as a consequence of them are much smaller than those from earthquakes, floods, or hurricanes. Nevertheless, volcanic eruptions, rather frequent in the historical past, involved some centres of large-scale disasters, which were followed by a great number of victims

Table 2.1. Largest volcanic eruptions in the history of mankind (the number of lives lost as a result of eruptions and their consequences is 5,000 or more) (from different sources).

Year	Country	Volcano that cost lives	Number of victims
1586	Indonesia	Kelut	10,000
1631	Italy	Vesuvius	18,000
1669	Italy	Etna	10,000
1783	Indonesia	Paradajan	9,340
1792	Japan	Unzen	15,190
1815	Indonesia	Tambor	92,000
1883	Indonesia	Krakatoa	36,420
1902	Martinique (France)	Montpellier	29,000 (30,000)
1902	Guatemala	Santa Maria	6,000
1919	Indonesia	Kelut	5,110 (5,000)
1985	Colombia	Nevado del Ruiz	22,000 (23,000)

and the devastation of vast territories (Table 2.1). The effect of volcanic eruptions on climate has been studied thoroughly (Kondratyev, 1985; Kondratyev and Galindo, 1997).

2.1.2 Earthquakes

Considerable devastation of densely populated regions of the Earth, the death and ruin of many thousands of people are caused by earthquakes. Documentary evidence about the most ancient of them that caused human death can be found in the Chinese catalogue of earthquakes, which was compiled over 3,000 years. Other countries have no similar chronicles about earthquakes that cover a long time period. However, the echoes of large-scale disasters were imprinted on the memory of people, were reflected in legends and myths, which fact requires special scientific interpretation. Maybe, the Bible stories of terrible disasters of the past caused by underground shocks are based on real facts.

The Bible tells us about the destruction, during an earthquake, of the ancient Jewish city Jericho that occurred in AD 1100, as well as the cities of Sodom and Gomorrah. Scientific interpretation of these events allowed Frezer (1986) to claim that at least some of them took place. However, we should be very careful when judging this, since the Bible information on the earthquake could be distorted. Nevertheless, Jericho, Sodom, and Gomorrah are settlements that existed in reality and they were located in a tectonically active area within the rift zone. Earthquakes in this region, though weak, are observed now as well. According to historical data, 2–3 major earthquakes a century took place in the past.

Thus, it is quite possible that the ruin of all the cities was a result of the earthquakes. Unlike the Jericho ruins, the remains of Sodom and Gomorrah have not yet been found. Bulgarian specialists (Trifonov and El Hair, 1988) assume that these are on the south coast of the Dead Sea. The earthquake that led to the destruction of these cities was followed by ignition of the gases that escaped (this is mentioned in the Bible). The traces of the cities were not preserved since this area along with part of the coast sank into the water. There is one more interesting version on the location and fate of the two cities. According to Trifonov and El Hair (1988), these cities were situated to the south-east of Damascus on the outskirts of an extended field of volcanic rocks, and not on the Dead Sea coast. Here, at different times two large accumulations of animal bones were found and studied, as well as the remains of two ancient deserted settlements. Both the animals and settlements perished from a volcanic eruption, which occurred approximately 4,000 years ago. There is convincing comparative evidence on the bedding of volcanic formations and the remains of settlement and animals. There is no convincing proof, though, that these were Sodom and Gomorrah.

Apart from the information on earthquakes and their consequences taken from legends, there are data as a result of archaeological and geological investigations. Thus, in the environs of the ancient major cities Timgad and Lambesis in North Africa, which were semi-buried under the ruins, aerial pictures showed the outlines of the remains of former irrigation systems. With the help of them, water was supplied to the fields around these cities. From which sources did the water come for the irrigation systems? According to Aprodov (1976), the complete disappearance of the sources that supplied water to the cities and their environs which resulted in their destruction, can be explained by active geological processes. It has been established through geological investigations that the region is subject to rock movements, which can affect the underground water regime by causing their redistribution. As a result of this, water stopped coming to the surface from the sources, which fact, together with drought, could possibly have made the inhabitants of Timgad and Lambesis leave the cities.

In some regions of the Earth, man was subject to the simultaneous influence of several natural forces. Different natural interacting forces are, of course, much more difficult to withstand. On the coasts of some seas (in tectonically active areas), man is often met with two forces – earthquakes and coastal floods. Tectonic motions lead to uplift or sinking of some coastal parts; in the latter case, parts of the land were flooded by seawater. Such a phenomenon, in particular, can be observed on the coasts of the Black and Mediterranean Seas. Some parts of their coasts had been flooded, sometimes, as a result of tectonic sinking, especially in periods of marine transgressions. Cities and settlements sank to the sea bottom together with them. One such part is near Naples. Here, in the Poccuoli district, at a depth of about 7 m, the remains of a Roman imperial palace were found at the site of the former ancient Roman city Baya (Karbonyin, 1985).

In the early 1980s, the ruins of flooded cities were discovered, including ports, on the coast of India. How did they find themselves at the sea bottom? Specialists at the National Oceanographic Institute in Goa and those at the Archaeological Department, State Tamilnad University, in particular, tried to answer this question. The drowned cities are at different depths – from 10 to 130 m. Some of them are mentioned in ancient Indian books. Thus, in particular, Bet-Dvorak is mentioned in the *Makhabharat*. In 1983, its remains were found off the Gujarat coast (Delgado,

1998). The largest flooded city was discovered off the Madras coast, 1 km from the coast. It is supposed to be the city Caveripatinam, a major, old, Indian commercial centre and port in the past. It is not clear as to why the cities earlier situated on the Indian coast are now at the sea bottom. Ancient sources, for example, inform us that Caverpatinam was unexpectedly flooded by a gigantic tidal wave. It should be emphasized that the coast of India experiences periodical flooding by oceanic waves, particularly by tsunamis, due to underwater earthquakes. However, after water recess from the coast, cities are restored and remain on the land. The above cities ended up being flooded, most probably, as a result of the coast sinking. Deep geological processes could cause different motions of parts of the coast, and hence may explain why the cities are found at different depths and distances from the coast. It should be possible to establish the true cause with the help of future underwater investigations.

Earthquakes are responsible for the flooding of many ancient settlements and cities that were located on the coast of Mediterranean Sea. The town of Cavtat is situated on the Adriatic coast of Croatia. The local inhabitants and specialists–archaeologists noticed that some structures in the town (e.g. roads and aqueduct, built by the Romans) suddenly break at the coast. The town appears to continue at the bay bottom. According to fishermen, the walls of underwater structures can be seen through the water depth. The outstanding English archaeologist Arthur John Evans, in particular, who discovered the Minoan civilization on Crete, knew of this fact. More than 100 years ago, Evans visited Cavtat (Evans, 1921–1935). He strongly believed that a drowned city is located at the bottom of Cavtat Bay. It was most probably, according to Evans, the ancient Roman city Epidaurus. The city was believed to have disappeared from the face of the Earth, but information about it remained: its name and location was mentioned in historical Greek manuscripts. The name of the city can also be read on vases. Epidaurus was founded as a Greek colony by emigrants from the city of the same name that was situated not far from Corinth in Greece. The city is also known to have existed for a long time. It changed from a Greek colony to a Roman city. With the fall of the Roman Empire, Epidaurus perished. It is believed to have been sacked by barbarians in the middle of the sixth century AD. A new settlement appeared later on its site, which has become the present Cavtat.

According to other sources, Epidaurus was under water. A small group of archaeologists, headed by Falkon-Barker in the 1950s carried out the first serious underwater investigation and discovered the city for science (Falkon-Barker, 1967). They found walls, columns, and streets that continued some roads of the modern Cavtat. Some historical sources, as well as chronicles, indicate that Epidaurus was ruined during an earthquake. According to one of the chronicles, the earthquake occurred in AD 365, covering an extended area in Italy, Germany, and Illyria (an acient region along the east coast of the Adriatic Sea, including Dalmatia, Montenegro and North Albania). During this disaster a large part of the coast, together with the city, sank and was buried under water. The city, to be more exact, its ruins turned out to be at a depth of about 20 m. Thus, the city, the centre of marine communications with a population of 40,000, was destroyed.

The ruins of one more ancient city, Apollonia, aroused great interest in archaeologists and geographers. This city perished long ago, and there is a small settlement of the same name on its site, or, more exactly, in the vicinity of it, lying among the ruins, on the coast of Libya. The Cambridge University expedition found numerous flooded city structures (Cleator, 1973; Fantechi and Margaris, 1986). Apollonia was founded in 7 BC. It was built as a city-port to link Cyrene, the capital of Cyrenaica, located at a distance of 20 miles, with Greece. City structures under water are found at a distance of up to 400 m from the coast. The city became a Roman colony in the 90s BC. Judging by coastal structures, it proved to be submerged in AD 6. The flooded port structures provide evidence that, in the Roman period, the sea level not only rose but also, temporarily, fell. This is, in particular, indicated by bottom-deepening operations to clear the canal (passage) between the islands for ships coming to the city. Geological investigations have shown that Apollonia turned out to be flooded due to tectonic descent of part of the coast and subsequent inundation of the district. The areas of land that submerged together with the city are at a depth of no more than 2 m.

Many settlements have suffered from tsunamis in the past. One of the largest calamities took place in 1833 during the Krakatoa eruption. At the same time, an earthquake produced a huge wave that reached the nearby densely populated coasts on Java and Sumatra and led to the death of about 30,000 persons, devastating the coastal areas. In Europe, the greatest disaster occurred in 1755 on the coast of Portugal. A tsunami with its centre near the coast reached Lisbon, the capital of the country. Figure 2.1 illustrates the joint destructive effect of the earthquake

Figure 2.1. The destruction of Lisbon during the earthquake on 1 November 1755 (the joint effect of earthquake, tsunami, and heavy fires) (a print from the 18th century: from *UNESCO Courier*, 1976, No. 6).

Table 2.2. Largest scale earthquakes in the history of mankind (number of dead equal 50,000 or more) from different sources (in parens).

Year	Location	Number of dead (1,000)
365	East Mediterranean Sea (Syria)	50
844	Syria (Damascus)	50
893	Armenia	100
893	India	180
1138	Syria	100
1268	Turkey (Semjia)	60
1290	China (Zhili)	100
1456	Italy (Naples)	60
1556	China (Shansi)	830
1626	Italy (Naples)	70
1667	Azerbaijan (Shemakha)	80
1668	China (Shandong)	50
1693	Italy (Sicily)	60
1727	Iran (Tabriz)	77
1730	Japan (Hokkaido)	137
1737	India (Bengaluru)	300
1739	China (Ninaya)	50
1755	Portugal (Lisbon)	60
1783	Italy (Calabria)	50
1868	Ecuador (Ibarra)	70
1908	Italy (Messina)	120 (47–82; 50)
1920	China (Gansu)	180
1923	Japan (Canto)	140
1932	China (Gansu)	70
1935	Pakistan (Quetta)	60
1948	Turkmenia (Ashkhabad)	110
1970	Peru	70
1976	China (Tianshan)	245 (242–700)
1990	Iran (West)	50

that brought a tsunami and fire to Lisbon. According to rough estimations, it resulted in the death of 50,000–70,000 of its citizens, ruining the city (Bolt *et al.*, 1978).

Among the most destructive earthquakes of the historical past that resulted in the greatest devastation (Table 2.2) and led to enormous numbers of victims, we should mention two earthquakes. One took place in 1737 in India in the district of Calcutta. Its victims numbered 300,000. An even greater earthquake occurred in 1556 in the area of Shensi in China, in which 830,000 persons were killed.

2.1.3 Climate change

The ecological situation in the past changed not only as a result of rapidly occurring natural events such as earthquakes and volcanic eruptions. The change in landscapes and areas where humans dwelled and played out their lives took place due to slow gradual natural processes, at first sight of a non-disastrous nature and having no appreciable everyday effect on people. However, the accumulation of the results of such impacts and their total effect could lead, with time, to a drastic change in the natural environment and produce unfavourable and even disastrous conditions for the population.

Climate change resulted in changing the appearance of landscapes of extensive areas and could lead to large-scale migrations of the population. According to L. N. Gumilev, many displacements in Middle and Central Asia and Kazakhstan are explained by climate variations, the alternation of humid and dry periods and their consequences: the shift of natural zone boundaries, in particular, steppes and deserts (Gumilev, 1972). In dry periods, steppes gradually became deserts; pastures were destroyed and settlements, cities were covered by sand. In wet climatic periods, life settled into its normal routine. The rock paintings of different animals and people, which were found in different parts of the Sahara, have aroused interest for more than a hundred years. A great number of rock paintings, discovered by French archaeologists just before the Second World War, were met with much interest (Lot, 1973). They were found in canyons dividing the Tassilin-Adjer plateau on the outskirts of Akhaggar. Here, in the centre of the Sahara, they saw drawings of wagons on wheels with harnessed bulls and, surprisingly, animals that do not inhabit these areas at present: antelopes, buffaloes, elephants, and hippopotami. These drawings made by artists many years ago prove that the Sahara was not always a desert. Some specialists doubted that the rock paintings actually existed, and others considered them to be false. Every new find of such drawings convinces us of their originality, and, at the present time, the sceptics are very few. Figure 2.2 presents one of numerous petroglyphic drawings found in the Sahara, which proves that this area was not always a desert.

The petroglyphic drawings were found in the area of Fessan, to the east of the Tassilin-Adjer plateau (Jaquet, 1988). The most ancient of them were drawn in the 70th century BC Tens of thousands of rock paintings found in different places in the Sahara indicate ancient artists depicting living things that actually existed at that time (not fictitious ones). The landscapes do not look like the present desert areas of the Sahara and resemble savannas or sparse forests. The natural environment, with conditions of wet tropical climate, was favourable for human life.

The changes in climatic conditions in the Sahara that occurred in the last several thousands of years is evidenced not only by archaeological finds. There are other proofs that pluvial, more humid periods existed in the Saharan past – geological, climatic, botanical, and paleontological – though they are not numerous. Among these are drawings of small crocodiles, which astonished the eyewitnesses, in the centre of Sahara, painted on rocks by ancient artists. They were found in valley-like relief depressions, which are dry for most of the year, but are filled with water for

Figure 2.2. The Sahara before its change, as a consequence of a climatically caused disaster, to the present desert. The prehistoric drawing in the area of Tassilin-Adjer plateau (Algeria, Central Sahara) is evidence that there used to be a savanna in place of the desert (from *UNESCO Courier*, 1979, No. 3).

a short period are proof of rare rains. These drawings, which survived by some miracle in the desert, are proof of the former wetness of the Sahara (Kapo-Rey, 1958).

New information on the ecological situation in the Sahara in the historical past has been obtained with the help of space survey data. Ordinary pictures taken in the visible spectrum region show the Sahara as a desert devoid of water courses, as it is. Unlike pictures taken in the visible spectrum region, radar pictures enable us also to see, albeit roughly, the buried terrain hidden under a thin layer of loose surface deposits. Such a discovery was made from radar space pictures of the Sahara (Eberhard, 1982; Shuttle..., 1983). An extensive area of the present desert in the south-west of Egypt and to the north of Sudan turned out to be crossed by a dense net of river valleys. Old river valleys, invisible on the surface, could be seen through the sand cover. Judging by the features of the radar signal reflected from buried valleys, they are located at a depth of 1 to 5 m. In these valleys, filled with river deposits, water run-off seems to occur even now, that is why they were reflected on the pictures. Some old valleys are as wide as the present River Nile, which crosses the Sahara and is one of the longest rivers on the continent and on the Earth.

The buried valleys, former river beds, and even deltas discovered on space pictures in different areas of the Sahara are important evidence of the existence of long rivers, more humid climate, and other ecological conditions in this area. In this connection, we must mention other information, which is considered problematic so far, on the existence of a large river in the centre of the Sahara in the historical past. Another Nile has been discovered on medieval maps, to the surprise of modern geographers (in particular, on the map prepared by the great Arab geographer Idrisi) and even on the maps of antique geographers (e.g. Ptolemy, prepared in about AD 150: Thompson, 1953).

Archaeological investigations performed in different areas of the Sahara, including south-west Egypt (Tassilin-Adjer plateau, the area of Fessan), prove that a prehistorical civilization existed in the Sahara, which was a savanna at that time, in the period from 9000 to 3000 BC. It was crossed by rivers and had large lakes. The population was engaged not only in cattle breeding and fishing, but also in farming, which fact is evidenced by data of archaeological investigation and petroglyphic drawings (Jaquet, 1988). Approximately 5,000 years ago, the Sahara started changing to desert due to climate drying. Its population decreased abruptly, the desert became almost uninhabited. It is possible that the population migrated from these areas to Mesopotamia, Syria, Egypt. Similar events of climate drying and, hence, unfavourable changes in the ecological situation, are observed in the historical past in other regions of the Earth as well.

According to paleoclimatic studies, two dry periods have been established in the East Mediterranean Sea region in the area of Greece and Turkey (Lamb, 1981). They cover the time period from 1200 to 750 BC and from AD 600 to AD 800. It is likely that the first drought not only caused a deterioration of the population's vital activity in the regions, but also affected their socio-economic development. The drought was only observed in north and south Greece and Turkey. At the same time, in central Greece and north Egypt the climate was damp. The pattern of pressure, rainfall, and temperature distribution, developed in 1955, has confirmed that just this very paleoclimatic situation could form in the given region. This climatic situation was unfavourable for farming. Population migration was also observed in the past from the dry parts of Greece, south Italy, and Turkey to the coast. It should be mentioned that, between 1200 and 750 BC, the so-called 'empty period' is fixed in the history of Greece. According to historical data, 1100 BC saw the fall of Mycenean power, which was located in the present territory of Greece, and the rise of the Hittites. At the present time, there is no evidence that great internal social events or foreign invasions took place at that time. Some specialists assume that a severe drought could undermine farming and the economy of these states, which resulted in their fall (Lamb, 1981).

One of the most ancient cities-citadels (fortresses) Jawa was discovered in the Black Desert in Syria. This desert (part of larger Syrian Desert) is a rocky plain consisting of basalts. French investigator and archaeologist Poidebare saw the city ruins when he flew over the Black Desert in the 1930s. Subsequent ground investigations have shown that the city was founded about 4,000 years ago (Helms, 1981). Its area reached 10,000 m^2. The population was about 5,000. The inhabitants were able to adapt to the natural environment and knew the hydrological phenomena well, and this was the basis of the existence of the city, one of largest populated centres of the early Bronze Age. Structures were found, now without water, in Jawa where the ancient citizens collected water. They drained the springs and river beds with the help of canals built in the surrounding area of flood periods. The traces of farming were found around the city such as the remains of numerous enclosures for animals. The city fell into ruin very suddenly. One assumption is that there was an abrupt change in the natural situation, with climate drying the cause of its destruction. In dry years, water was nowhere to be found and water basins remained empty. A conflict could have arisen at that time between the citizens and countrymen, who lived in the desert, due to the lack of water. Jawa may have been destroyed during that conflict.

Some investigators suggest that the destruction of the Toltec Empire was related to climate change (Metcalfe, 1987). This state, which was situated on the territory of present Mexico, prospered between AD 600 and AD 900. At that time, a beneficial, warm and humid climate prevailed there. It ensured the development of agriculture. However, a drought started in AD 12 to AD 13. The time of onset coincided with population decrease in this region. It might be that drought hampered farming activities and resulted in the Toltec Empire fall. If in some regions of the Earth long-term slow climatic changes led to an increase in dryness and the onset of desert, then in other areas they caused cooling and glacier onset. A very large island with the strange name 'Greenland' is situated between Europe and North America, closer to the latter. The second largest glacier on Earth, after Antarctica, is located here. Greenland belongs to Denmark. A few emigrants from Denmark live even now along the narrow coastal belt. They came to this, now icy, desert mainly early in the 20th century. However, there are data according to which Greenland was inhabited by people born in Scandinavia, long before the present colonists. This is mentioned first of all in Scandinavian sagas, which tell us that Vikings left Scandinavia for Iceland, Greenland, and even North America many years ago. One of the sagas describes that in 962, when Erik Torvaldson, or Erik the Red, reached the coast of a huge unknown country. He thoroughly investigated this region and found it to be quite suitable for founding settlements. Some years later, Erik the Red went to this land once more, but this time he was not alone. About 900 persons joined him in his voyage. Those who managed to reach the island became its first colonists.

It has now been proved that the legends of the voyage of Erik the Red are based on fact (Plumet, 1987). Some documents found in the archives of the Vatican (Plumet, 1987) confirm them. The traces of ancient settlements have also been found in aerial photographs. Their remains have been investigated by archaeologists. However, it was not understood, for a long time, why the ancient settlers disappeared suddenly from the chosen coast. It has been established that colonists lived on the island coast for about 500 years. Their population reached 5,000–6,000. They founded two cities: Esterbugden and Westerbugden, the latter was 400 km to the west of the former. The mild climate and fine meadows allowed colonists to breed cattle and to be engaged in farming and fishing. The settlers had contacts with Iceland and Norway, from where they had come. This was a period of climatic optimum.

In the 15th century, the ancient settlements disappeared and the icy silence reigned again on the island for hundreds of years. The cause of disappearance of the colonies and destruction of settlements was not clear for a long time. The study of old documents and archaeological investigations enabled us to establish the time of colonization decline in south Greenland. Westerbugden stopped existing in the late 1330s. Communication of the city with Iceland was broken at that time. The relatives of settlers who arrived at the city in 1341 found it empty. Wild cattle walked along the city streets. The other city, Esterbugden, existed for about 100 years more. In 1440, a sailing vessel sought shelter in the Greenland coast during a storm, in the area of this city. One sailor, nicknamed 'Greenlander', had visited in this place before and he recognized the locality and the city, which was deserted. He found the only 'citizen' in one of the streets, who lay on the ground and was dead (Vozgrin, 1987; Plumet, 1987).

What happened to the settlers in south Greenland? Some investigators assume that they died as a result of conflict with Eskimos. Actually, Eskimos – the inhabitants of north Greenland – visited the south part of the island from time to time where they came into contact with colonists. Other specialists consider that an epidemic could have been the cause of the colonists' disappearance. Their viability might have been affected by temporary break of contacts with Iceland and Europe, owing to political situations. From these areas they were brought wood, goods, metal, and clothes, all of which were much needed. However, the facts convince us that there were neither wars, nor social shocks in this region. Raids by pirates could not be the cause, because the houses were not robbed. Nature proved to be the cause of island devastation, as many specialists believe at present. Most likely, the decline and destruction of the colony in Greenland took place as a result of climate deterioration. It has been found that, in the period between the 14th and 15th centuries AD, cooling occurred here, which is explained by the displacement of geomagnetic pole to the area of north Siberia and weakening of cyclones (the so-called 'Little Ice Age') (Lamb, 1981).

It is not mere chance that Erik the Red called the land, where he disembarked, the green land. The belt of land in the south and south-west part of Greenland in summer surprises the eyewitnesses by its rich and green grass even now. The Vikings also saw it to be friendly. They could breed the cattle they brought with them on rich meadows over a long time. When cooling started, the meadows grew meagre. Farming also declined. The colonists moved to the North American coast at that time. The geomagnetic pole returned to the area of Greenland between 1650 and 1850, and climate warming came with it.

The deterioration of climatic conditions due to cooling was observed not only in Greenland, but also in many other areas of North Europe (Lamb, 1981). The start of this cooling, the effect of which was unfavourable for the subsistence of population, is dated as the late 13th century. It continued until approximately the early to mid-19th century. A considerable increase in rainfall and area of swamps were observed at this time. The growth and advance of glaciers was observed in mountain areas. Thus, in particular in Iceland, the maximum glacier advance was observed before 1700, and only since 1850 have the glaciers started receding. The historical data have shown that the population of many European countries (England, Iceland, Denmark, Germany) met with great difficulties due to climate deterioration (Steensberg, 1951). Growing of cereals was reduced. As a result, peasants left their villages. Hundreds and thousands of villages in North European countries remained deserted for a long time – 200–300 years. The abundance of rainfall and rise of the water table caused flooding of mines in Germany and England. Constant attempts to fight against water – until the early 15th century – failed and mines were closed.

Numerous examples show the role of climatic change as an important factor in the decline and prosperity of settlements and even cities. Some investigators explain critical moments for the subsistence of peoples and civilizations by an abrupt change of climate. Huntington (1922) suggested that more favourable climatic conditions than at a later time promoted the appearance of ancient states in Guatemala and Yucatán Peninsula. Climate deterioration, development of a number of diseases caused by climate like malaria, were, in his opinion, some of the causes of their decline (Huntington, 1907, 1922). There are also theories about similar (climatic) causes for the decline of ancient Indian civilizations and that in the Tar Desert (Lambrick, 1963), as well as for the centres of ancient Chinese civilizations.

2.1.4 River floods

Rivers have long attracted the attention of people, who settled along river banks, built villages, cities, and cultivated the land. Rivers were a cheap and convenient means of communication, the source of water, and their banks contained rich soil favourable for farming. However, rivers can cause many disasters. As a result of the displacement of river beds, which migrated over long distances, rivers moved away, sometimes over tens and hundreds of kilometers from the beds they left. The main damage to population was caused by floods during which vast tracts of agricultural land were washed away, famine and diseases started among people, settlements, and cities were destroyed.

Many historians and geographers associate the origination and development of great civilizations of the world with rivers. This mainly refers to such big rivers as the Nile, Tigris, Euphrates, Indus, Ganges, Hwang Ho, and Yangtze. The states of Ancient Egypt, Sumer, Babylon, Ancient India, and China sprang up and prospered on their banks. In some cases, rivers caused great disasters in the history of development of the states.

The repercussions of such disasters as floods were reflected in legends and myths on the Flood. The information presented in the Bible is best known, according to which this disaster occurred at God's will for the edification of people. During the Flood, floods overflowed all lands and ruined all people, except one family, which had been forewarned. Life was soon restored on the deserted and ruined territory.

The investigations of myths, legends, and chronicles, coupled with the information on the Flood carried out by archaeologists, historians, linguists, geographers, and geologists, have shown that they are based on real events. It has also been found that different versions of the Flood, sometimes very similar, have remained in legends of many peoples of the world from different continents. The Bible information on the Flood was found to be not the oldest and to be taken from more ancient sources.

The oldest version of the legend of the Flood was discovered during excavations in Nippur by specialists of Pennsylvania University (Frezer, 1986). They found a plate of unburnt clay. There was writing on the plate about the coming flood that God would grant to the Earth as a punishment for people. This Sumerian version on the Flood is dated not later than 2100 BC. According to Frezer (1986), the legend is actually older than the plate on which it was written. Sumerians must have talked about this legend much earlier because, by 2100 BC, they stopped existing as a nation, 'absorbed' among their neighbours.

At present, different specialists have arrived at the conclusion that there is no real information on the true Flood. At the same time, analysis of legends and the geographical situation of the regions of the populations who mention these myths shows that there were large-scale disasters caused by water. In some regions of the world, great disasters took place as a result of river floods. Such disasters remain for long in the memory of people, who mistakenly believe they affected the whole world, becoming a part of the ethos of many peoples of the world.

Lots of calamities were caused in the historical past by one of the two great rivers of China – the Hwang Ho. This river overflowed its banks, changed the direction of its flow, and caused extensive flooding of the surrounding territory. Figure 2.3 shows how the Hwang Ho river bed has moved over more than 2,500 years. The violent 'temperament' of the Hwang Ho, (the Yellow River), and disastrous floods in its basin are due, mainly, to two causes. The river, flowing across a loess plateau, catches a great number of loess particles that are deposited at the bed and silt it up, which causes flooding. The other cause of disastrous overflowing is the monsoon-like nature of rainfall in the river basin, which produces heavy showers. About 200,000,000 people live in the Hwang Ho basin, lots of cities are situated here, as well as vast tracts of farming land. Great floods of the Hwang Ho have caused large-scale calamities to the local population. About 1,600 major floods have been observed in its basin in the last 4,000 years (Muranov, 1957).

The people who lived on the banks of Hwang Ho and its tributaries built bank dams and walls to protect themselves from floods. The length of these structures reached many hundreds of kilometers even in the distant past. However, as a rule, during major floods the water of Hwang Ho broke through the protective structures and flooded densely populated areas. Large-scale devastation was caused by the river in its lower reaches where it has radically changed its bed several times. The Hwang Ho discharges into the Yellow Sea to the north of Shandong Peninsula. However, it changed direction sharply several times (in 1194–1289, 1289–1324, 1324–1852, 1887–1929) to the south-east and discharged into the sea, far to the south of Shandong Peninsula.

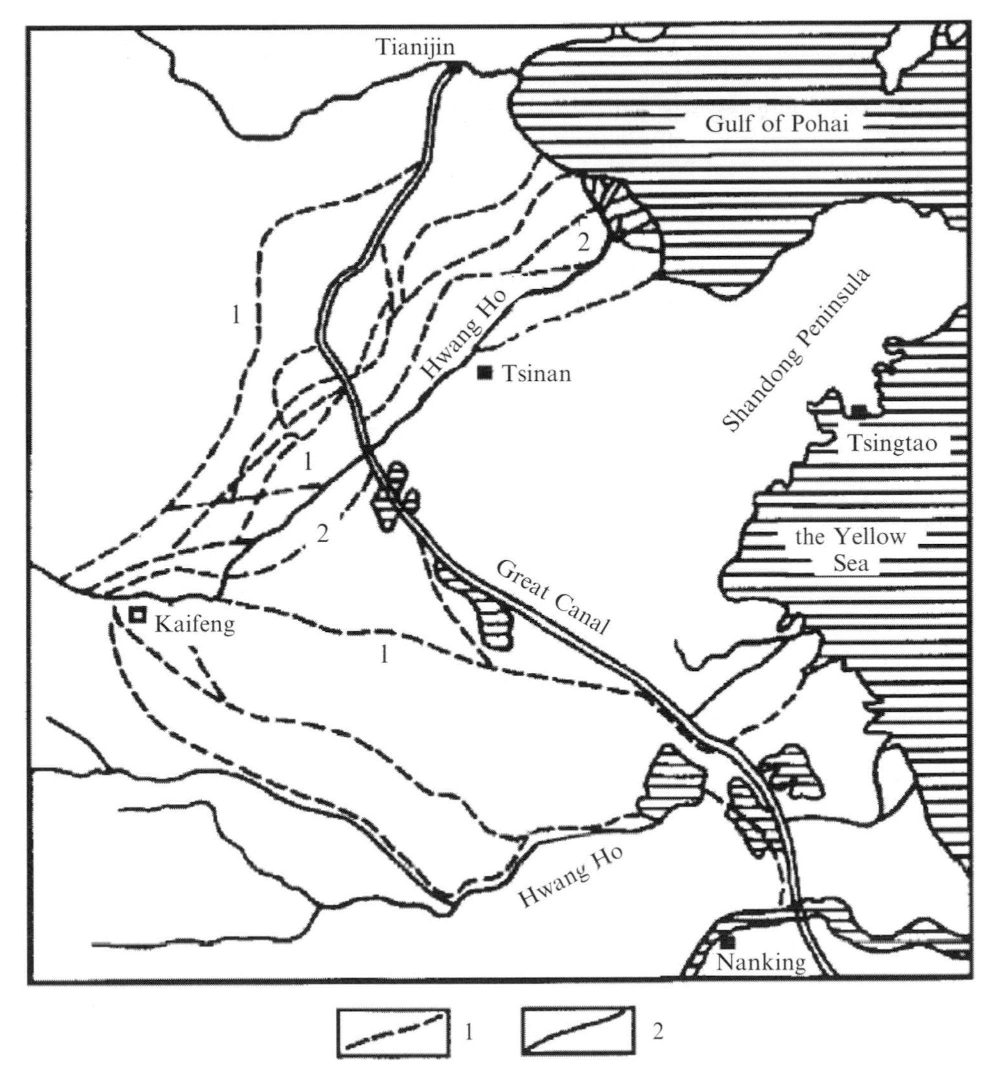

Figure 2.3. Historical displacements of the Hwang Ho river bed over 2,500 years, since the 7th century BC (1) and its present bed (2) (Muranov, 1959a).

Its new mouth was about 500–750 km to the south of the old one. The river moved several times to the north. The maximum distance of the new mouth to the old one was 800 km. As is clear from the data presented, the Hwang Ho flowed in its new beds for a long time.

It is well known that during the last 4,000 years the river broke protective dams from time to time and brought calamities. The worst flood took place in 2297 BC. The overflowing water of the Hwang Ho joined with the flooding water of the Yangtze – the other great river of China. As a result, the whole Great Plain of

China – the area in which most of the population lived – was flooded. The flood lasted over 9 years (Gumilev, 1972). It caused great calamities and required multi-year efforts by millions of people to fight against this natural event. Though there is no accurate information on the number of persons killed by this inundation, the number of victims must have reached hundreds of thousands. The Chinese population were used to calamities, though of smaller size, due to Yangtze overflows. Over 50 major river floods have occurred over the last 2,500 years, on average one every 50–55 years. The outskirts of such big centres of the country as Wuhan, Nanchang, Nanking have all been flooded (Muranov, 1959).

Large-scale, disastrous floods took place in the past in other regions of the Earth as well, including those areas where the most ancient civilizations originated and developed – in the basin of the rivers Indus and Ganges in the Hindustan Peninsula – and in the Mesopotamian lowland they were due to Tigris and Euphrates overflows. Information on the oldest disastrous floods of the past are not always reflected in historical documents. However, in some cases archaeological finds and paleogeographical data are evidence for them. The traces of such a flood were found around Ur – one of the oldest cities of Sumer (Cleator, 1973). Clay deposits 2.7–3.3 m in depth were found here among the ruins. Judging from the analysis results, they appeared to be due to Euphrates flooding; their age is from 4,000–3,000 BC to 2,800 BC. The river flooded the city to a depth of about 7.5 m. Such an overflow would have covered 100,000 km^2 of the territory between the city Elalit and the Syrian Desert hills. Maybe, this is the flood that remained in the memory of populations as the Flood.

The devastation the Tigris caused to Mesopotamian lowland inhabitants was even greater than that produced by the Euphrates. Baghdad, the largest city in Mesopotamia and since AD 762 the capital of the state, was subject to flooding and partial or complete destruction. In the 19th century, one of the recent large-scale floods in Mesopotamia was in 1892. The inundation was caused by Tigris and Euphrates overflows (Muranov, 1959). As a result of the inundation, huge tracts of agricultural land were flooded in the lower reaches of the two rivers. Baghdad, in which water remained for several weeks, was partly ruined once again.

The reason has not yet been finally established as to the destruction of one of the great ancient civilizations – the Harappa (or Indus). It occupied the vast territory of the Hindustan Peninsula from the Himalayas in the north to the present Bombay (Mumbai) in the south and from Karachi in the west almost to Delhi in the east. Over one thousand settlements have been excavated now (Possehl, 1988). The ancient state arose suddenly, existed for hundreds of years, and declined rapidly after its largest centre Mohenjo-daro ceased to exist. Long ago, its population was 40,000–50,000. Now, there are only ruins on the site of this city (at the edge of the Tar desert) among barkhans and takyrs.

No information has so far been found to prove that there were wars or other social reasons on the eve of Mohenjo-daro fall. Therefore, some specialists assume that the leading role in this state's ruin was played by natural factors (i.e. the River Indus). It is the major river in India and one of large rivers of the world. According to one hypothesis, Mohenjo-daro was flooded by Indus water (Possehl, 1988). The overflow was of multi-year duration, and not seasonal. The finds of lake deposits discovered both in the city and in its environs indicate that there was a vast lake basin in this area. These clay deposits were found on the slopes of hills at a height of about 9 m above the plain level. This was the depth of the old lake at its margins. Lake clays are found at three heights. This indicates that the reservoir first appeared, then water receded, and finally dried up. This occurred three times. Each time, the city was covered by water for approximately one hundred years. Some specialists, including archaeologist G. Dales (Lambrick, 1963), even believe that this region was flooded not less than five times. And the city was restored to life each time, except the last one.

According to archaeologist R. Rikes (Lambrick, 1963), the huge lake, which also flooded many other settlements of the Harappian state, could form as a consequence of tectonic rise in the lower reaches of the Indus. According to geological data, we can observe here a structure subject to active, though slow ascents. As a result, the Indus met an obstacle on its way and its overflowing water formed the lake. Time passed and the river cut through the obstacle once again and the lake descended.

There is a different point of view that denies the formation of a lake basin (Lambrick, 1963). The thin silts and clays found near the ruins of Mohenjo-daro could equally be deposited by the river during its highest floods. In such periods, the city was flooded for a long time and life stopped in it. However, it could have stopped not only due to disastrous overflows of the Indus, but also because of the shift of its bed away from the city. The Indus bed is very changeable. Traces of its wandering are well seen in space pictures. It is possible that, in the period of city prosperity, the river that ran in its vicinity moved farther from the city with time, which caused the drying of the district and undermined the development of agriculture. As a result of famine, the Harappian civilization fell.

Rivers cause devastation not only at times of overflows. Change in their flow direction and formation of a new river bed were unfavourable for population and, in particular, for cities that turned out to be on the bank of dry rivers. In some cases, this ruined cities. Some authors assume that it was in this way, due to a change in the Tarim river bed in AD 330, that Lulan and other cities along the Silk Road near Lobnor Lake were destroyed. It might be that river-bed wandering was the reason why the inhabitants left the oldest African city Djenne-Djeno – the capital of the ancient state of Timbuktu in West Africa. It was discovered quite recently close to the present city Djenne. Excavations have shown that the city appeared in the Iron Age in 200 BC and existed until AD 1400 (McIntosch and McIntosch, 1986). The area of the ruins of the ancient city, located 3 km from the present one, covers 3,000 m^2.

As a result of investigating this district with the aid of aerial photographs, one of possible causes of Djenne-Djeno destruction was suggested. The city ruins are in the delta of the Niger river. The aerial photographs show that the city was built on the bank of Vat – one of the tributaries of the Niger. The Vat has long since changed its flow direction and is now far from the city. We can see solonchaks (salt crusts) in the aerial photograph at the place where it once flowed. Some authors explain the devastation and destruction of ancient settlements and cities in Hindustan by river-bed migration (Tosi, 1986).

The most dangerous areas for population are river deltas: they are subject to large-scale floods and active river-bed wanderings on flat lowlands. Filling bays and lagoons with sediments, delta accretion, migrations of river channel all these can produce the situations that are is unfavourable for people and their settlements. As a result, settlements and cities could not exist and they would be buried under river sedimentations with time.

One interesting recent discovery was the Greek–Etruscan city Spina, which was not marked in old maps. The city remains were found on a heavily bogged-up plain near the Po river mouth, approximately 30 km from the city of Ravenna. Traces of the former existence of the city (ruins of an ancient acropolis), buried under the ground, were found as long ago as 1919 (Eudoix, 1967), when drainage work was done in the area. The city itself was discovered much later, in 1956. Its outlines were seen in aerial photographs that also show canals. As a result of airborne observations and field investigations made by archaeologist N. Alfieri and engineer V. Valvasori, different city buildings were found, as well as numerous graves of Etruscans. Spina was an important port in the past.

2.1.5 Sea level changes

People have always been attracted to the coasts of seas and oceans when settling. Many settlements, which later became big cities, appeared on their coasts, particularly in the convenient bays and gulfs of the Mediterranean, Baltic, Black, and North Seas. In the states that had an outlet to the sea, there were favourable conditions for developing trade, fishing, navigation. The sea has played an important role in the development of many powers such as Phoenicia, Ancient Greece, Turkey, Spain, Great Britain, Russia. However, the sea is not only positive but also has a negative side for states having access to the sea. Sea coasts, including densely populated ones, were always subject to the impact of sea events. Disasters often occurred on the coasts of low-lying land, which killed people and sometimes damaged the subsistence and economy of a country on the whole (Table 2.3).

Sea coasts have periodically been flooded for a short time during storms or as a result of tsunami development. Moreover, coastal areas have dried up or flooded due to slow fluctuations of sea level, its transgressions or regressions. The coasts of low lying land in many densely populated regions of the Earth have long since been subject to disastrous sea floods of various origins. Some of them appeared owing to the rise of seawater during monsoons. South-west monsoons, the strongest air currents moving towards the coast of India and Pakistan, and a sharp rise in the level of coastal water result in coastal flooding.

Cyclonic eddies accompanied by storms cause water level rise in the North Sea. In this area, the present coasts of Holland, North Germany, and Denmark have been flooded many times in the historical past. The following information on the largest floods are indicative of the scale of calamities due to sea floods and destruction of cities and villages in the north-west of Europe (Lamb, 1981): in 1099, 100,000 persons perished in south-east England and Holland; in 1164, in north-west

Table 2.3. Biggest floods of sea coasts in the history of mankind with 35,000 or more victims (from different sources).

Year	Country (location)	Flood cause	Number of deaths (1,000)
365	East Mediterranean	Tsunami	50
1099	England, Holland	Cyclone	100
1164	Germany (north-west)	Cyclone	100
1200	Holland (Frisland)	Cyclone	100
1212	Holland (north)	Cyclone	306
1421	Holland (Frisland)	Cyclone	100
1737	India (Ganges Delta)	Cyclone	300
1755	Portugal (Lisbon)	Tsunami and earthquake	60
1864	India (Calcutta)	Cyclone	50
1876	India (Bay of Bengal)	Cyclone	100
1883	Indonesia (Java, Sumatra)	Tsunami	36
1963	India (Bangladesh)	Cyclone	50
1964	India (Calcutta)	Cyclone	50
1970	Bangladesh (Ganges Delta)	Cyclone	300

Germany 100,000; in 1200, in Frisland 100,000; in 1212, in north Holland 306,000; in 1412, in Frisland and Germany 100,000, and 72 villages were destroyed.

Astonishing are the roads that seemingly lead nowhere, stopping abruptly at the sea coast. Such roads have been found, for instance, in the Crimea. Some settlements and even cities have been discovered at the bottom of seas or oceans near coasts. The remains of the ancient Greek city Dioskuria, covered by sediment, lies at the bottom of the Black Sea bay opposite the present Sukhumi. In the 6th century AD, this city, which had never been mentioned before, quite simply disappeared (Razumov and Khasin, 1978). A more detailed story on this city will be given below.

The arrangement of ancient defence works in the area of the present Sukhumi is similarly astonishing: the remains of their walls approach the sea and are then hidden under its sediments. It is of no doubt that here, at the Black Sea coast, a tragedy took place. How was Dioskuria ruined – suddenly or over a long time? What caused its demise? What are the true causes? These questions have no answers for the time being.

Lots of ancient settlements and cities have been found under water, not only in the Black Sea but also in all seas of other continents and near the coasts of oceans in shallow water. Dwelling places of people in caves in the early Stone Age and neolithic settlements have been found at a depth of several tens of meters (Flemming, 1959). Stone Age settlements lie at the bottom of seas near the coasts of Gibraltar, Italy, Florida, and California. About 100 cities are submerged in the Mediterranean Sea area. They are at a depth of 1–5 m. The ruins of many settlements have also been found by archaeologists near the Black Sea coast. In the early 1980s the remains of the site of an ancient town were found by St Petersburg archaeologists at the bottom of Kerch Strait in the Black Sea. First, its traces, in particular ceramic articles (fragments of amphoras), were found on the coast in a sand spit. Aerial photography has enabled us to look into the strait bottom in the water area adjacent to the spit. It proved to be divided into big squares which had once been blocks in ancient settlements. The size of the site of the ancient settlement is 300–400 m^2. Ceramic fragments were found here with the help of underwater investigations, as well as other traces of human activities, which were found even at a distance of about 180 km from the coast. Further, at the strait bottom, ancient anchors were buried. The collected finds allowed us to establish the time the settlement existed – the end of the first thousand BC. It was supposed to be Akkra, mentioned by the ancient authors Strabo and Ptolemy. The Black Sea level was at that time much lower (by 8–10 m) than the present one. Later, it started rising slowly. The coastal settlements, cities, and among them Akkra, were flooded and vanished from sight for a long time. Their number is several tens on the Black Sea coast (Argunov, 1987).

About 22,000 years ago, the Black Sea level decreased considerably and the Black Sea was isolated from the Mediterranean Sea and ocean. The relationship with the Mediterranean Sea was restored much later, after transgression, which started about 7,000–8,000 years ago (according to other sources, 10,000 years ago: (Nikiforov, 1975). This transgression continued throughout subsequent years. At the same time, the natural situation on the coast of the Black Sea, during the last several thousands of years, has depended on simultaneous small fluctuations, rather than on the general tendency of its level rise. The biggest of these fluctuations was the Fanagorian regression, which occurred 2,000–3,000 years ago. Its peak was in the 5th to 3rd centuries BC. During this regression, the sea level was on average 5 (from 3 to 7) m lower than at present. The regression of smaller size, Korsunian, was observed 1,500–500 years ago with its peak in the 14th and 15th centuries AD. The sea level decreased 2 m at that time.

Between the Fanagorian and Korsunian regressions 1,000–1,500 years ago in the Black Sea, the Nimpheyan transgression took place (Nikiforov, 1975). As a result of it, by the middle of the first thousand, the sea level was 1–2 m higher than at present. These fluctuations in Black Sea level, caused not only by fluctuations in Mediterranean Sea but also the whole World Ocean, could not but affect the nature of coastal territories. Some islands were washed away, others appeared, sand spits grew or disappeared, shoals and underwater sandbanks migrated. Bays, gulfs, estuaries disappeared and appeared. As a result, rivers left their old beds and bank lines migrated sharply, changing the outlines, particularly on shallow banks. Vast parts of the coast alternately sank to the sea bottom and then became again land.

The sea did not simply recede from or advance on the coast, it interacted constantly with the coast. Particularly complex processes occurred at places where rivers flowed into the sea, since the process of development of lowlands in deltas is related not only to sea activity and fluctuations in its level, but also to the activity of river and the amount of suspended matter brought by it. People settled in delta plains, their buildings were under the constant effect of these joint forces – sea and river. It is to such effects of natural events that Dioskuria was subject in the past.

People who lived on the east Caucasian coast of the Black Sea knew about the existence of the ancient Greek city Dioskuria from a lot of historical sources (Razumov and Khasin, 1978). It was founded 2,500 years ago by Greek colonists at the site of an older local settlement. Later on, the Romans gave this city a new name Sebastopolis. The city was abandoned both by the Greeks and the Romans. The causes of this were unclear for a long time. Moreover, the existence the city was long questionable since, in spite of numerous records of this city made by ancient authors, its location was unknown.

Using different sources, the location of Dioskuria (Sebastopolis) was determined more accurately in the area of Sukhumi – at the bottom of Sukhumi Bay. Information about the submerged ruined city Sebastopolis was passed down to us by an Italian traveller in the 16th century A. Lamberti. Actually, stories by local inhabitants about city buildings that could be seen sometimes at the bottom from boats began to appear (Shervashidze, 1967). Numerous finds of articles of ancient age in the shallow water evidenced in favour of the existence of a submerged city. Systematic investigations made by Abkhasian and Georgian archaeologists in the 1950s finally confirmed the existence of the submerged city. The ruins of building were found under water and the Sebastopolis tower was investigated. As a result of studying the relationship between the submerged buildings and the ruins that remain on the coast, it was established more accurately that these buildings existed from the 1st to the 4th century AD. Traces of the more ancient city Dioskuria (predecessor of Sebastopolis) were not found.

Geological and geomorphological investigations have shown that the destruction of Dioskuria was caused by natural forces. About 2,500 years ago, the delta of two Caucasian rivers – Gumista and Kelasuri – was at Sukhumi Bay (Tosi, 1986; Shervashidze, 1967). However, due to tectonic rise of the coast in the delta area, the Gumista bed shifted several kilometers aside. The delta stopped forming, and soon the coastal banks were destroyed, which protected it against the disastrous effect of sea waves. Then, the internal parts of the delta were washed away and, together with it, Dioskuria, the major part of which was at the external higher seaward part of the delta, ceased to be. Seawater broke through the gaps in the coastal banks and flooded the low, partly bogged up, central delta area. The gardens and farming land in the city environs were ruined in this way.

Sebastopolis was later built at the site of Dioskuria, and the events which destroyed Dioskuria happened again. The sea started washing out the delta, undermining the walls, which had to be strengthened with supports and then moved farther from the sea. In spite of all these efforts, Sebastopolis was submerged. By the 6th century AD, the ruins of Sebastopolis (on the site of the present Sukhumi) remained only on land outside the delta. The remaining part of the Roman city has been under water for many centuries. This explains the changing natural situation and destruction of Dioskuria and Sebastopolis as suggested by Shervashidze (1967).

According to Voronov's (1980) estimations, the width of land that sank to the sea bottom in this area reached 100 m over 2000 years (Voronov, 1980). The other investigator Argunov (1987), on the basis of various geological–geomorphological data, assumes that, since the time of Dioskuria's destruction (2,500 years ago), the sea engulfed the land strip up to 200-m wide. Dioskuria was built during the Phanagorian regression when the land advanced on the sea. The seaward part of the city was submerged in the period of subsequent transgression.

Another example of the effect of ocean level fluctuation can be found on Java. It is possible that disastrous changes in the coastline of Java (Indonesia) in the 10th–11th centuries caused the collapse of the old Mataram kingdom and the devastation of a huge temple system – Borobudur – in central Java. Among the causes of this calamity, the catastrophic eruption of volcano Merapi situated close to Borobudur is generally mentioned. However, not all geologists, who believe that the age of this volcanic eruption is older, agree with the above claim (Voûte, 1999).

Caesar Voute and his colleagues (geologists and archaeologists) think that the cause of collapse is different (Voûte, 1999). According to them, in the 10th–11th centuries, sedimentation processes in river mouths on the northern coast of Java resulted in shallowing of the offshore part of the sea, displacement of the coastline and destruction of two important sea ports (separated thus from the sea). As a result of these events, trade ties with Central Java were broken off, pilgrimage to Borobudur stopped, the most important route of the south sea branch of the Silk Road (from Japan via north Java to the coast of India) changed on the whole, and the centres of social and religious activity of the population in the central and north areas of Java moved to the east of the island.

Changes in the coastline in the north of island, due to post-glacier eustatic fluctuations in World Ocean level and the sedimentation processes in river mouths as mentioned in Voûte (1999), maybe an important link in disastrous natural changes effecting, along with social and religious factors, the collapse of the kingdom and devastation of Borobudur. The discharge, through tectonic crevices, of gas and water emissions harmful for human health that are mentioned by the authors, among other links to disastrous natural changes, seems to be less probable.

2.1.6 Lake level changes

Lake shores have long since attracted people, due to their freshwater, rich soil, and the nearness of underground water. Lakes were convenient as transport waterways for trade. Siting of settlements and cities at lake shores was advantageous for fishing, agriculture, and for military purposes. Subsistence in many regions of the world depended on fluctuations in lake level, migration of the coastline, and disappearance of lakes (Mokrynin and Ploskih, 1988). One such region is the coast of lake Lobnor. It is shown in modern maps as a lake that temporarily dries up, located in the eastern part of the Takla Makan Desert. Nowadays, this lake, located in an extensive depression 50,000 km^2 in area, has become dry. Its bottom, as is clearly seen in space pictures, is occupied by solonchaks. The lake started drying up in the

1960s–1970s owing to the fact that large water reservoirs were built on the river Tarim flowing into it.

The water from the reservoirs is completely used for irrigation. In the lower reaches of the Tarim and Kurun-Darya, in their joint delta, there were found the remains of two ancient oases Lulan and Miran, which do not exist now. The great Silk Road passed along the shore of lake Lobnor. Excavations have shown that Lulan and Miran pre-date Christ. Anyway, the oases degenerated many centuries ago, which is likely to be explained by worsening of the natural situation, and mainly by migration of the coast line of Lobnor. This lake has attracted and surprised investigators for a long time. Almost everyone who visited this area calculated the shape of Lobnor differently to predecessors, sometimes in different places at the bottom of a huge basin with the same name. This provoked disputes and reproaches, resulting in inaccurate mapping of the lake shore. At present, changes in the outlines of the lake, its extent and drying up are related to the river flowing into it (Songqiao and Xuncheng, 1984). They changed their bed direction and sometimes flowed into smaller basins Kara-Kosha and Taitem, and not to Lobnor. So, Lobnor became much smaller and even dried up.

The size of the lake plane depended on river run-off: in the years of water abundance the lake overflowed, in dry years it decreased. It has been established that between the 1st century BC and 4th century AD (from ancient Chinese historical sources) the lake overflowed extraordinarily. Its area was then as great as 5,000 km^2. It is at this time that the Lulan and Miran oases appeared on its shore and flourished. Approximately in AD 330, the Tarim and Konka turned in the direction of the Kara-Kosha Depression. The result of this was drying up of the extensive Lobnor reservoir, as well as of the Tarim and Konka Delta. Freshwater disappeared, the land became saline. All this made farming impossible, and the Lulan and Miran oases, covered with moving sands, were ruined.

Such phenomena as lakes drying up and, as a result, the appearance of an ecological situation that was unfavourable for the coastal population are also observed in other deserts in China. The traces of anthropogenic activities in the prehistoric period and early historical time have been found in the north-west part of the Alashan Desert in north China. Here, in the semi-sandy and semi-rocky desert Badein-Jaran, there is practically no water, except for two small lakes. It has turned out that they are part of an extensive lake reservoir that existed in the past (Walker *et al.*, 1987). Its flat clay bottom and its outlines on the whole were successfully studied with the help of space pictures. The space pictures looking beneath the modern sand dunes. Two dry valleys 'flow' into these two remaining small lakes (the ruins of Zimiao City are on one of the valleys). The drying up of the lake reservoir most probably forced the ancient settlers to leave its shore (Songqiao and Xuncheng, 1984).

Lakes located in the area of desert and semi-desert dried up temporarily. Some time later, they were again filled and overflowed with water. This occurred slowly, generations of people changed at their shores for this time period.

In the periods of lake level decrease, people naturally moved closer to the shifting coastline of the drying reservoir. Both settlements and farming land turned out to be on the site of the former reservoir bottom. In periods of lake transgressions and rise of lake level, the reservoir advanced slowly but progressively on settlements and farming land located around it, which were gradually submerged. Examples of such natural effects on human activities in the historical past can be found in various regions of the world. The Caspian Sea region is highly indicative of these.

The Caspian Sea (lake) broke away from the Black Sea about 6,000 years ago. Subsequently, its level changed constantly, but irrespective of fluctuations in World Ocean level. That is why we consider the Caspian Sea region as a large lake. Many traces of human activities were found in its shallow water – ancient graves, fortresses, settlements, and even cities. The remains of fortresses Naryn-Kale and Derbent were discovered not far from modern Derbent under water at a depth of 2 to 7 m; among them were parts of fortress walls situated 300 m from the coast, quarries, and the ruins of piers. These structures remained from a series of defence works of the fortress Naryn-Kale built here in the 5th–6th centuries and the Derbent fortress built in the 7th century (Razumov and Khasin, 1978). The ruins of the ancient temple of fire-worshippers found in the Bay of Baku are well known. According to estimates made by hydrologist B. A. Appolov, since the time of its construction (1235) the sea level has become 3 m higher at this place (Razumov and Khasin, 1978). The Caspian Sea level rise in the 8th–9th centuries also led to the destruction of the Iranian port Abaskum.

It is unclear why one of the major Caspian states – the Khazar Kaganate – perished. According to L. N. Gumilev (Blavatsky, 1976), much of the land, settlements, and cities of this state suffered greatly from the rapid rise in Caspian Sea level that took place in the 10th–11th centuries AD. It is likely (according to L. N. Gumilev) to be one of the basic causes of the decline of the Khazar state and its destruction as a result of the flooding of its capital Itil, located near the Volga mouth.

Lake level fluctuations often had a disastrous effect on the life of the population and the lake coasts not only in arid, but also humid areas of the Earth. Some lakes became dried up and gradually bogged up, which could not but affect those settlements that turned out to be on the swamp margin and not on the lake coast. In other cases, the settlements and farming lands were flooded as a result of lake transgression.

Lots of cities and settlements situated on lake coasts had to change their location due to changes in the coastline, drying up of reservoir, or, vice versa, the increase in water volume. It is probable that the cessation of anthropogenic activities at the coast of Lake Titicaca was caused by natural water events, like reservoir level fluctuations. Here, in Bolivia and Peru, on the slopes of mountains surrounding the lake, have been found old plots of land and once cultivated fields, but neglected long ago. They were seen in aerial photographs as a combination of rectangles (Erickson, 1988; Lambrick, 1963). These fields were located on artificially terraced slopes of the mountains, to the south of the lake. They are situated at a distance of 15 km from the coast. The length of old fields reached 200 m, and the width 5–15 m. Canals approached some of them. Excavations have shown that artificial terraces started being built about 1000 BC. Farming on this territory prospered between AD 400–AD 1200. The ruins of Tiahuanaco were also found here. The city, located in the high mountain steppes, existed for over 2,000 years – from 1000 BC to AD 1200. It was suddenly deserted by people.

There is no doubt that the existence of this ancient city – the centre of Andes civilization – was closely related with agricultural activities on the mountain slopes. Judging from the excavation results, farming on artificial terraces stopped about AD 1200 (i.e. at the same time as life in Tiahuanaco). It is possible that the demise of the Andes civilization was caused by the rise in water level in Titicaca. The traces of a more recent rise in its level are found on the same slopes of mountains surrounding the lake. These lake clay deposits are observed at the same level as the location of the artificial terraces. Flooding of the low mountain slopes could have resulted in the destruction of settlements and cessation of agricultural activities (Erickson, 1988; Kolata, 1987).

Like any lake, Lake Titicaca experienced not only level rises, but also falls in its level. We mentioned above the possible effects of such an increase in the lake's level. The other extreme situation, which could lead to disaster, is a considerable decrease in reservoir level, disrupting its coast, and then rising up again once more. There are data (although not numerous) that such situations also took place in this area. The legends of the Aymara Indians, living along the coast of Lake Titicaca in Bolivia, give information about an ancient city that lies at its bottom. These sorts of data bring home the importance of understanding how changes in the natural situation can impact on the lives of ordinary people. According to Toynbee (1946), in some cases, changes in the natural situation, particularly natural disasters, awakened the 'sleeping energy' of the population and promoted the origin of some civilizations, in particular the Sumerian, Egyptian, and Minoan ones. This was how the Sumerians, Egyptians, and Minoans rose to the 'challenge' of nature. Thus, the increased dryness of the climate and desertification of the location made people migrate to the Nile Delta, to the bogged-up valleys of the Tigris and Euphrates. This was not a mere population migration, but a relocation that resulted in an explosive appearance of new civilizations. According to Toynbee, they appeared to be due to special psychological features of the population at a certain historical moment. Nevertheless, civilizations would have not appeared without the primary natural impetus. In most cases, there were no such population 'responses' to the 'challenges' of nature.

Natural events in the historical past, particularly disasters, sharply changed the ecological situation, which became unfavourable for the population. Mass migrations and deaths of large numbers of people (Table 2.4) are related to natural forces. Millions of people died due to natural events over several thousands of years. The material damage is also great: hundreds of cites and thousands of settlements vanished from the Earth. Entire states were under the threat of destruction, which was realized in some cases.

It is essential that humankind learn the lessons of experience in order to overcome the consequences of natural events, by studying the regularities of these events and population response to them. Of particular importance are the studies of the frequency of natural events and the cyclicity of their development. Geographer

Table 2.4. Largest-scale natural disasters in the history of mankind (100,000 or more dead) (data from different sources are given in parens in the right-hand column).

Year	Country (location)	Disaster	Number of deaths (1,000)
893	Armenia	Earthquake	100
1099	England, Holland	Cyclone, flood	100
1164	Germany	Cyclone, flood	100
1200	Holland (Frisland)	Cyclone, flood	100
1212	Holland	Cyclone, flood	306
1290	China (Jili)	Earthquake	100
Mid-14th century	Europe, Asia	Plague epidemic	25,000
1421	Holland (Frisland)	Cyclone, flood	100
1556	China (Shensi)	Earthquake	830
1642	China	Flood	300
1665	England (London)	Plague epidemic	100
1731	China (Peking)	Earthquake	100
1737	India	Earthquake	300
1876	India (Bay of Bengal)	Cyclone	100
1881	Vietnam (Haifon)	Cyclone	300
1887	Brazil (NE part)	Drought	2,000
1887	China (Great Chinese Plain)	Flood	1,000 (2,000)
1898	India (Bombay)	Plague epidemic	12,500
1908	Italy (Messina)	Earthquake	120 (42–82; 150)
1920	China (Gansu)	Earthquake	180 (200)
1923	Japan (Tokyo,Yokogama)	Earthquake	140
1931	China (Yangtze River)	Flood	140
1951	China (north part) (according to Western experts)	Drought	30,000
1970	Bangladesh) (Ganges River Delta	Cyclone, flood	300 (265–500)
1976	China (Tianshan)	Earthquake	242 (to 700)
1991	Bangladesh	Cyclone, flood	139

E. Huntington (1907) was one of the first who detected the relationship between the deterioration of natural conditions and the destruction of ancient settlements and cities. As a result of investigations, he concluded that there was a relationship between the prosperity and decline of some centres of ancient population and climatic pulses, like the alternation of arid and humid periods. Some scientists in the USSR, such as A. V. Shnitnikov and E. V. Maksimov *et al.* (Maksimov, 1990; Shnitnikov, 1961), determined and specified the duration of natural rhythms of different order. At the same time, establishing the relationship between the rhythm of various natural processes and the prosperity and decline of ancient population centres requires further investigations to be made by geographers and archaeologists.

In many cases, these investigations are difficult to make due to indistinct rhythms, scanty information on different sides about the subsistence of cities and settlements, including the interrelations between ancient civilization centres and the nature.

These mankind has accumulated a considerable experience of studying natural events that are recorded in legends, literary monuments, historical documents, maps, and scientific works. Underestimation and imperfect understanding of the essence of such natural events, their development, and rhythm led to repeated 'conflicts' between man and nature in the areas in which he settled. This experience of tragic, repeated, ecological calamities is instructive and requires further study, systematization, and interpretation.

2.2 ANTHROPOGENIC ENVIRONMENTAL DISASTERS IN THE PAST

2.2.1 Causes and basic types of disaster

The opinion prevails in modern science that natural calamities and slow changes in the natural environment, as well as wars and other social events were the principal, if not the only, causes that determined the destruction of cities and the decline of ancient civilizations. Although the significance of natural factors is indisputable in this respect, the importance of social events is also unquestionable.

At the same time, in connection with the intense ecological situation in many regions of the planet, including Russia, a question now arises increasingly frequently: 'Could it be that ecological disturbances caused by man also played an important role in the historical past in the decline of earlier developed lands, cities, civilizations?' After all, in the past, man must have had an impact on nature, while using it. In particular, anthropogenic impacts on landscape were not to a certain extent, only external, superficial changes. Far more serious, adverse effects in the change of natural situation are due to indirect impacts of human activities.

The ruins of cities, remains of ancient engineering structures, traces of farming are indicative of the considerable human impact on the environment in the historical past. The case in point is the human impact on natural events rather than the extent to which man tried to master nature. It is becoming increasingly clear that man was the cause (as he is at present) in the past of many unfavourable changes in the natural environment. Moreover, as a result of these anthropogenic impacts, in some cases the natural situation deteriorated so much that they led to disaster and decline of some cities. There is a viewpoint advocating that whole civilizations perished in such a way.

So, other than natural calamities, it may also be that man could produce a natural situation disastrous for himself. If this is so, then the importance of the anthropogenic factor in affecting nature should be recognized in the development of civilization prosperity and decline.

Therefore, investigations into the anthropogenic impact on nature in the past is justified and significant. However, such studies have not been made to any great extent so far due to the interdisciplinary nature of the problem. Archaeologists and, partly, ethnographers, are the ones who mostly engaged in this problem, but specialists in the environment deal with it to a smaller extent. Concepts developed by archaeologists and geographers respecting the relationships between man and nature differ widely. Many archaeologists believe that the prevailing role in the prosperity and decline of civilizations is played by social factors, while geographers think that these are natural factors.

Modern interpretations of well-known scientific data have enabled some investigators to conclude that, in the past, ecological crises occurred that were due to human activities. It seems natural that cities, the centres of population, and their environs were most vulnerable at times of threat to the ecological situation. Cities, undoubtedly, were also sources of environmental pollution (to smaller extent that of the air, to greater extent the reservoirs) in past times. Considerable problems arose when disposing of sewage and garbage. For example, near Rome there was a huge dump that formed a hill 35 m in height and 850 m in circumference (Blavatsky, 1976). It is not surprising that epidemics (including plague and cholera) were frequent and resulted in depopulation of whole cities. However, human impact on the environment was actually even larger in scale. It extended beyond the city borders to the surrounding territories, where it related to farming.

Of particular interest were the deliberate military actions aimed at worsening the natural situation. Such an aim was most often achieved by destroying the means of irrigation.

In the 13th century, for example, Genghis Khan acted in such a way. In Turkestan and Western Asia, everywhere his troops invaded irrigation plants were destroyed. The lands, devoid of water, soon dried out. The warriors of Genghis Khan in 1221 destroyed, in particular, complex irrigation structures in Horesm. This provoked the Amu Darya to change flow to the west. It turned from the Aral Sea, into which it flowed, to the earlier dry Sarykamysh Depression (Bartold, 1902). As a result, an extensive agricultural district was desertified in the Aral Sea area.

Unfavourable phenomena in the environment and critical situations had different causes. Among these worth mentioning are: soil degradation, particularly soil salinization; sand movements caused by man; degradation of vegetation cover and, finally, deterioration of climatic conditions provoked by anthropogenic activities. In all the above cases, change in the natural situation occurred as a result of the increased impact of man on the landscape. It also occurred for other reasons of a more general nature, in particular: incompetence in using hydrotechnical plants, underestimation of the limits in the development of land (e.g. as a result of exhausting water resources), and, finally, the change in types of farming. Let us examine, in this connection, examples of developing land in different regions of the earth.

2.2.2 Soil degradation

A strong human impact was exerted on the soil in the past. Soil erosion developed as a result of incorrect (without consideration for location, relief, and rock-washout degree) incompetent farming. As present studies have shown, the process of soil erosion caused by man in the ancient Mediterranean Sea area was 10 times higher than that caused by nature (on adjacent plots) (Godbrecht, 1987). It is not surprising that some authors believe land erosion to be one of the basic causes of farming decline and civilization destruction in the Mediterranean Sea area (Eberhard, 1982; Fantechi and Margaris, 1986; Godbrecht, 1987). Innovations in agriculture and an increase in the use of farming equipment on land were the impetus for further development. At this time, iron plows started to be used and the plowing area expands and moves from plains to hill and mountain slopes. At the same time, pastures increase in size, forests are cut down, and mining of mineral resources and use of building materials increase.

The results of incompetent overbreeding of cattle undoubtedly affected the soil considerably. The ancestors of the present Greeks probably did not even imagine that in future goats would destroy arboreal vegetation not only in Greece, but also in many countries of South Europe.

The destruction of vegetation and over-grazing by goats is not the only cause of soil degradation and the subsequent development of the ecological crisis in Greece and the Middle East. A combination of factors, most likely, contributed to the formation of numerous ravines, facilitating the processes of plane washout, degradation of soil cover, and an abrupt decrease in soil fertility.

Deterioration in the ecological situation in the Mediterranean region astonished the contemporaries and, in particular, the ancient Greek philosopher Plato. He wrote in *Critius*: '. . . at those times the area not yet damaged had high multi-hill mountains and plains which are now called rocky, and at that time were covered with soft soil, and abundant forests in mountains' (Plato, 1971). According to Van Andel's studies (Diamond, 1986), in Greece and the Peloponnese there were at least four periods of prosperity and decline of settlements, erosion, and devastation of lands between 2000 BC and the Middle Ages.

The processes of desertification in the countries of the Mediterranean Sea coast occurred not only because of the human impact. It is of no doubt that natural factors played an important role in this case, which is justly indicated by some authors (Bondarev, 1987). In the second half of the 1st millennium BC, in the Mediterranean Sea countries a relatively more humid period started that lasted for about 1,000 years. A great amount of rainfall resulted in the increase in soil erosion. The processes transforming Mediterranean Sea ecosystems, which should be called natural–anthropogenic, produced wide and peculiar changes in the appearance of landscapes. Partly due to cutting down forests, desertification facilitated further deforestation of the location. Karst-forming phenomena increased under their influence in Greece and modern Yugoslavia (Fantechi and Margaris, 1986).

Such phenomena of ecosystem degradation were observed not only in the north of Mediterranean Sea area, but also in the east, in West Negev (Palestine). This is an area of semi-desert with a small amount of rainfall and, on the whole, is unfavourable for subsistence. It looks the same nowadays. However, 1,500 years ago, life was thriving. This is proved by the remains of farming and traces of cultivated lands discovered by archaeologists (Williams, 1973). The Byzantines started developing land here during the Justinian dynasty (about AD 550). Rich loess soils facilitated the development of farming. Water was collected in tanks with the aid of dykes built of clay and partitioning the valleys. Later on, the former fields, not sown for a long time, declined and were divided by ravines everywhere. The area of the desertified farmland together with buildings was about 5,000 m^2.

In the middle of 5th century AD , the area under consideration was subject to the invasion of nomads. The newcomers were engaged not only in cattle breeding, but also farming. However, they could not adapt to the surrounding landscape. As a consequence of excessive cattle grazing, the vegetation cover, already scant, completely disappeared. As a result of this, the soil became bare and ravines formed quickly into loesses. A period of intensive land erosion started that led to soil washout, location division, and oasis destruction.

Some investigators consider the destruction of soil cover as one possible cause of the collapse of the Mayan civilization, which was situated earlier in the area adjoining the borders of the present Mexico, Belize, and Guatemala (Turner, 1974). The soil could have been impoverished due to cultivating single crops, or degraded by erosion. The state declined in the 7th–8th centuries AD.

In some regions of the arid zone, land salinization was principal cause of degradation of agriculture, economy, and the decline of ancient centres of civilization. Mesopotamia is one such region. Extensive tracts of lands with evidence of ancient irrigation are located in this area. They are partly covered by sand, solonchaks are abundant, and ground deflation is observed. According to Adams (1980), oases disappeared because of the disturbance of the water–saline regime of the soil, leading to its salinization. This was caused by imperfect irrigation systems. Local investigators (Andrianov, 1978; Gumilev, 1989; Ivanitsky, 1987; Mrost, 1987a, b) are of the same opinion. They assume that soil salinization was the consequence of unrestrained wide-scale irrigation without drainage. Salts rose to the surface as a result of excessive watering of the soil under when the ground was weakly permeable to water.

An additional cause of agrosystem destruction was the natural shift of the Euphrates, which led to the irrigated land drying up. This shift was partly anthropogenic in its cause: the silting of many canals led to accumulation of sediments, which contributed to river displacement. Assumptions (Mrost, 1987a, b) are quite correct that the agrosystem crisis was caused not only by ecological factors, but also the imperfect technology of irrigation and reconstruction of the hydrosystem. It was also closely related with social and economic causes (wars, destruction, epidemics), with a tendency toward introducing large-scale forms of irrigated farming – a kind of 'farming expansion'. The failure of farms and decline of civilizations on the whole in Mesopotamia caused by joint ecological and social effect happened more than once: especially in the 21st century BC and between the 3rd and 2nd centuries BC.

There is an assumption that the Bronze Age Central Asian cities Altyn-Depe and Namazga-Depe, major centres of ancient farming oases, declined for similar reasons (Masson, 1966). They prospered between the second half of the 3rd millennium and the early 2nd millennium BC. Intensive irrigated farming without consideration for the location led to exhaustion and salinization of soils, and the destruction of oases.

2.2.3 Desertification

One cause of ecological disturbance leading to destruction of farming oases situated in deserts was the advance of moving sand, which started moving for a number of reasons. In nature, they are held in place by vegetation, though scant and rare. Sand, devoid of protective cover due to farm animals trampling down the vegetation, gradually 'came to life' and started moving. The advance of sand was one of the causes of the destruction of many ancient cities, the ruins of which are now found half-covered by sand.

Many ruins of ancient cities have been discovered in the desert of an autonomous region of China – Inner Mongolia (Hou-Ren-zhi, 1985). These cities were built along the so-called Silk Road – the oldest route that stretched from Peking to Rome. The desert has predominated for a very long time in this area (North China). These conditions are highly unfavourable for human subsistence, and settlements along this route are very rare now. However, the ruins found here show traces of a thriving population. Imagine the surprise of European travellers who visited this region in the 19th–20th centuries: they saw majestic ruins in the depths of the desert. Subsequent excavations showed that some time ago this area was flourishing with cities and farm oases located along this stretch of the Silk Road.

One of the largest cities was Hara-Hoto built in AD 1035. In 1272, the route of Italian traveller Marco Polo passed through it; he saw this administrative centre at the peak of its prosperity. Russian traveller P. K. Kozlov turned out to be in these parts in the early 19th century. Hara-Hoto also struck him. However, this was a different Hara-Hoto – a dead city surrounded by sands.

According to Chinese specialists, active development of the Takla-Makan Desert by man started as long ago as the early Neolithic Age (Chao Sungchiao, 1984). At that time, irrigated farming was practised at outlying areas of the desert and along the valleys of rivers crossing it. Numerous settlements appeared here. The traveller Chang Chin, who visited the region in the 2nd century BC, saw a lot of cities. According to historical data, in the period from 86 up to 74 BC over 500,000 farms existed around the cities Luntay and Yuli in the lower reaches of the Tarim. There are only ruins and moving sands here nowadays.

The ruins among the hills, yardangs, and barkhans cut out by the wind have also been discovered at the site of the ancient state Lulan on the west coast of Lake Lobnor. Many cities along the Silk Road appeared before Christ. Among them is Yu-Yan (Hou-Ren-zhi, 1985) situated 15 km to the east of Hara-Hoto. It existed as long ago as 102 BC (i.e. more than 2,000 years ago). Yu-Yan has long been dead and buried in the sand, like other cities built along the Silk Road.

During military operations lots of oases and cities were doubtless ruined, including Hara-Hoto. At a later time, they were revived. It should be emphasized that nature also did not spare the cities along the Silk Road in North China. Its changes in the historical past led to a worsening situation for the population. A small lake is located on the outskirts of the Hara-Hoto and Yu-Yan ruins. Traces of a much larger reservoir can be seen here. At the time of these cities prospered, the lake area exceeded 1,000 km^2. Due to climate drying, the rivers that supplied this lake with water dried up, and its area reduced. At the same time, the poplar forests perished. The deterioration in climatic conditions (i.e. the decrease in water supply to the location) was undoubtedly one of the causes of oasis destruction and city decline in the region, resulting in the general advance of desert. Investigations by Chinese historians and geographers have shown in recent years that it was not only newcomers–nomads and natural calamities that were responsible for the disappearance of oases (Hou-Ren-zhi, 1985). The inhabitants of these places contributed to desertification of their lands as well.

Pastures were trampled down as a result of overgrazing by cattle. Pulling up bushes and trees for firewood bared the sands. In such sands, water went deep down, river beds became shallow, and extensive areas around Hara-Hoto and the other cities in this stretch of the Silk Road dried up. The canals built for irrigating the fields failed to bring life-giving water. The sands started moving and covered fields, canals, and settlements. Studies by Chinese specialists have shown that some of these settlements and cities were not surrounded by sands during their prosperity. Sands only covered clay river deposits. However, the settlers destroyed the thin layer of vegetation on top of the of river sands, resulting in sand encroaching from the river banks onto the farming land.

Thus, the flourishing of oases and settlements, located along the Silk Road in North China, was subject not only to alternation of arid and humid climatic periods, but also the change in river-bed direction and water supply of the area. The inhabitants themselves contributed to the deterioration of the natural situation, which was particularly vulnerable in dry years. Nature could not endure the additional anthropogenic stress. Ecological disasters resulted and life in the oases stopped completely.

2.2.4 Forest decline and soil degradation

Forest cover is an important indicater of how ecosystems are affected by human activity. Considerable changes in landscapes occurred for a long time as a result of forest burning, with the aim of subsequent use of lands for farming. Fires deliberately started often ran out of control. Natural vegetation changed at such places. Sparse forests and savannas appeared where once there were dense forests. Some investigators believe the greater part of modern savannas, particularly in India and Africa, are of artificial origin (Balandin and Bondarev, 1988), since savanna naturally appears at sites where forests have been burned for thousands of years.

We have already mentioned that the destruction of natural landscapes in the Mediterranean Sea area, beginning about 500 BC, was partly due to cutting down the forest. In a number of regions of the world, the degradation of ecosystems started with the large-scale cutting down of forests (Eckholm, 1976). Many traces of ancient civilizations have been found in the forested areas of the earth, in tropical, equatorial forests (Eidt, 1987). The sudden resettlement of people who left cities cannot only be explained by social problems. This very picture, without the destruction of cities and settlements, is observed, for instance, when studying the location where the Maya people lived at one time (Doyel, 1979; Adams, 1980, 1985). Many of their cities situated on the territory of modern Guatemala, Honduras, and Mexico, which once, particularly between AD 400 and AD 800, prospered, were, probably, left by the population. Some investigators suggest that the Maya people were short of food, which was difficult to get here, at a certain period of their development.

Jungles had to be cleared for farming around the cities. Soil fertility was lost due to long-time use of the same lands. The inhabitants were unlikely to move deeper into jungles and cultivate new plots far from the cities. So, a decision was taken, as a result, to leave populated sites. This took place approximately in the 9th–early 10th centuries AD. This is at that time that the cities of the central part of Maya empire became deserted. At the same time, the population in its provinces started increasing, in particular on the Yucatán Peninsula. This is one of the suggested causes of civilization breakdown in ancient Meso-America. All these anthropogenic changes in the ecosystems of Meso-America were not the only or even the major factor explaining the decline of ancient civilization. At the same time, important social events – such as major population uprisings, population migration – occurred due to demographic causes. Approximately at AD 900, deterioration in climatic conditions is finally observed. Most probably, Sanders and his colleagues (1979) are right to assume that the collapse of civilization was the result of a combination of the above factors, including soil exhaustion. Such a viewpoint is also convincing because the authors considered the civilization as closely related to the dynamics of the ecosystem. This break in their close links to the Meso-American ecosystems, both natural and social, which were closely interrelated, resulted in their degradation.

A similar ecological crisis, one of the causes of which was the destruction of forests, happened in the historical past on Easter Island. According to Diamond (1986), the decline of the civilization was due to an ecological crisis, and not to separate social or natural causes. This island is famous for huge stone statues of incomprehensible design and origin. They number about 800 and weigh as much as 85 tons. In addition to statues, about 245 stone platforms were found. The present, very small in number, local population knows practically nothing about the origin of the stone sculptures. The stone idols are silent witnesses to the rise of the Easter Island civilization. As is well known, approximately in AD 400, Polynesians reached the shore of the island. By 1500, they had erected the mysterious stone structures. By that time, the population of the country was 7,000 and they were prospering. Then almost suddenly, the population reduced, the economic basis was disturbed, and stone idols were thrown to the ground. What happened?

According to Diamond (1986), in the process of Island development, the settlers gradually removed the forests. Hardly anything was left of not only the palm groves, but also the shrubs that once covered the island as a continuous green carpet. The removal of forests caused land erosion, wash-out of the richest upper layer of soil, and ravine formation, resulting in a decrease in crop yield. The population stopped making fishing boats (canoes) because of the absence of wood, and so fishery reduced. Since important natural resources were exhausted or became inaccessible to the population, the economy started to decline. The ecological crisis led to strife, war, cannibalism, and population degradation.

2.2.5 Depletion of water resources

One of the causes of declining oases in deserts was the distinctive features of self-development of these farming centres situated along river valleys. The critical ecological situation could arise due to lack of water. Water resources were exhausted owing to an increase in the area of irrigated land oases. For the oases located in river valleys which started in mountains and moved to the deserts of inter-mountain depressions, this occurs usually according to the following scheme. At the beginning of the appearance of settlements at the river banks and the irrigation of the surrounding lands by its water, the growth in farming centres is observed, as is their expansion along river valleys up and down the river flow. However, later on, in the lower reaches of the river, water resources were exhausted and oases started degrading. Such events occurred, most likely, in one of the regions of the Kara Kum Desert, which extends around the deltas of Murgab and Tedjen.

At the present time, oases are located and life seethes in the deltas of these rivers, which flow far into the desert. In the desert on either side, including the region situated northward of the oases right into the depth of the desert, at different times were found the remains of ancient settlements that disappeared long ago (Kes *et al.*, 1980). Thus, archaeologists from the Institute of Archaeology(Russian Academy of Science) aerial photographs the traces of ancient, now dry beds of the Murgab River (Central Asia). They assumed that canals and settlements could exist between the now inactive and barely noticeable ground channels. Indeed, they were discovered here in 1981 (Kohl, 1981). A series of structures (towers, buildings) built in the Bronze Age was found. They are likely to be the remains of the ancient state of Margush (now covered by sands), which is mentioned in the inscription cut on the Bekhistun Rock in Iran dated at about 500 BC. Such finds were, as a rule, random. They can be explained by the fact that in the past the natural situation in these places was more favourable for farming and building permanent settlements.

The space pictures of Kara Kum Desert in the area located to the north-west and north of the modern deltas of the rivers Murgab and Tedjen show some features of relief that are helpful for reconstruction of the geographical situation of the historical past. The US archaeologist Kohl (1981) was the first to observe this. In particular, the dry and long-abandoned beds of Tedjen and Murgab are seen in the pictures as thin, winding curves. It is possible that, even now, somewhere along their bottom the water streams flows under the layer of sands. This is why on the surface, above the beds that are buried under the sand, more plants grow than on the surrounding area without water, as a result of which we can see the traces of the beds in the pictures.

Another feature of this landscape is the presence of a lot of takyr-like surfaces – clay deposits left behind after floods. Together with buried beds, they indicate the areas of deltas of past rivers, which flowed tens of kilometres farther into the depths of desert than at present. It is important to emphasize that the external boundary of ancient deltas, visible in the pictures, corresponds to the maximum boundary of the extent of ancient settlement traces discovered by archaeologists at different times. This proves the assumption about the undoubted dependence of prosperity or decline of farming oases on the degree of irrigation of this area and, moreover, on the use of water by the population. The interaction between nature and man occurred in the following way. Man first settled in the upper reaches of the rivers and gradually colonizing along the beds of the Tedjen and Murgab which carried life-giving water, finallymoved into the depth of desert. Thus, extensive farming oases with large tracts of lands irrigated by water from canals and many settlements appeared not only in the upper, but also in the middle and lower reaches of the rivers.

As oases and the demand for water use for irrigation grew, the lower reaches of rivers dried up. Therefore, oases in the lower reaches of the two rivers perished, covered by sediments. The deserts of the Tarim Basin were crossed by green belts of oases in the historical past. They were located along rivers. This is confirmed by the investigations made in the south of the Tarim Depression in the area of a modern oasis, Minfeng (Chao Sungchiao, 1984a, b). It is situated in the valley of the Kerija River. To the north of the modern oasis, there are ruins of the ancient oasis with the same name. It was also located on the banks of the river. The valley is now lifeless. It is clearly seen from the space observations. We can see in the space picture how suddenly the valley vegetation disappears when moving to the north, into the depths of desert. The ruins of ancient Minfeng are surrounded by sands. Major oases also existed at a distance of 120 km to the north of modern Minfeng, in the delta of the Nija River. According to historical data, a state appeared there at the time of the Tang Dynasty (AD 618–907). Nowadays, according to the data available, huge barkhans up to 35 m in height can be found at the site of the former oasis. Dry poplar forests have been preserved here as well. A combination of natural and anthropogenic factors was the possible cause of the death of the oases under study (Chao Sungchiao, 1984). The climate was probably slightly more humid than at present. So, a smaller amount of water was taken for irrigating the lands in the upper and middle reaches of rivers flowing to the north into the depths of deserts. This amount of water was also enough for the dwellers of oases situated in the lower reaches. When the climate became drier, water became insufficient even for the upper reaches of rivers. At that time, oases in the lower reaches of rivers died.

The decline of farming oases in river valleys and deserts doubtless occurred as a consequence of the gradual extension of oases areas and, as a result of this, there was a water shortage. The depletion of water resources could also arise against the background of the development of unfavourable climatic processes and the increase of climate aridity. Some current oases in the Tarim Depression are likely to perish in such a way (i.e. a combination of anthropogenic and natural causes).

2.2.6 Destruction of ecosystems due to human activity

The fall of some ancient civilizations has yet to be explained. Viewpoints have already been presented in this chapter stating that opposing factors were often responsible for the destruction of civilization. There are no data now available to prove the prevailing effect of any of the factors. This concerns, in particular, some river states, the welfare of which was determined to a considerable extent by the wide use of water resources.

The great ancient states, which appeared in the valleys of the Nile in Egypt, the Euphrates and Tigris in Mesopotamia, the Hwang Ho in China, and the Indus in India, were subject to the effect of nature and the consequences of their interference with nature. The river states that prospered between 3000 and 2000 BC, as some investigators believe (Pracad *et al.*, 1987), declined due to inadequate knowledge of water engineering and the water system. While it was supposed earlier that such major centres of the Harappian civilization as Harappa and Mohenjo-daro (Hindustan Peninsula) were destroyed by invaders – the Aryan conquerors – now there are many other factors to consider. Prior to the invasion of Aryans from the north, the Harappian state had serious 'conflicts' with nature. It underwent repeated river floods, as well as the salinization of irrigated lands. The internal social problems provoked by these calamities could on their own lead to the decline of a formerly powerful state. A little-investigated page in the history of man's attempt to develop nature on the African Continent is the gradual destruction and death of oases in the Mediterranean part of North Africa. The vast areas of desert and semi-desert were developed by the Romans. Cities appeared here that were surrounded by agricultural lands. The Romans succeeded in adapting to nature and to the arid climatic conditions. They created irrigation systems and water reservoirs. A vast territory in this region became a flourishing area due to irrigated farming. Its prosperity ended in the 7th–8th centuries AD. There are different opinions on the destruction of oases in North Africa. One explains their destruction by an increase in aridity, drying up of the climate and the soil of cultivated lands. According to another view, the life in oases came to a standstill because of the invasion of foreign conquerors (Marphey, 1951).

According to data available, there were no dramatic climatic changes in the period of oases decline. As for the invaders, they really had settled in North Africa. First, it was Byzantines, then Arabs. However, neither were engaged in the systematic, deliberate destruction of the hydrotechnical systems built by the Romans. According to Marphey (1951), the decline of civilization in North Africa was due to a change in human economic activities. The main blow that was inflicted on nature by the Arabs was the basic type of economic activity – cattle raising. The Roman irrigation systems, unnecessary to the Arabs, fell into decay. The stagnant reservoirs and ponds became breeding grounds for malaria. The unwatered soil dried up and any remaining vegetation was trampled down by cattle. Not only the plowed lands, but also pastures started degrading in such a way. The oases became neglected. The largest oasis of North Africa – Karga – was deserted.

Thus, according to Marphey, the destruction of North African oases was closely related to change of economic structure. This change was drastic. Such a view is likely to be close to the truth. Indeed, before the Arabs came, between AD 395 and AD 638, this region was inhabited by the Byzantines who replaced the Romans. They maintained the irrigation systems and used them for farming. In their time, the oases were not neglected. The Arabs were mainly interested in cattle breeding. Each Arab family kept from 15 to 50 head of cattle. Almost one million people had several million sheep and goats. So, the coming of nomads with different economic structures to North Africa was the principal factor for desertification of a vast territory. As a consequence, North Africa gradually depopulated, as was observed from the 9th up to the 18th centuries.

The decline of human subsistence in oases, located in deserts, occurred for different reasons. The cases were well-known when the oases died out because of a decrease in the level of farm development. However, this took place, sometimes, under the exactly opposite circumstances, i.e. when the lower level of farm development was replaced by the higher one. It was just the case with Oasis Dakleh, the largest oasis (about 2,000 km^2 in area) in East Sahara. It is situated 600 km to the south-west of Cairo.

As archaeological investigations have shown, past life in Oasis Dakleh depended mainly on climatic conditions, on the alternation of humid and arid periods (Holett and Grzymski, 1985/1986). In humid periods, the oasis attracted people. Giraffes, elephants, zebras, and antelopes inhabited it and the adjacent area (drawings of which were found on rocks). The oasis attracted people to this location as long ago as the Neolithic Age. The remains of primitive irrigation structures have also been preserved, which were built about 3000 BC (Holett and Grzymski, 1985/1986). They are indicative of just how developed agriculture was on this territory. However, earlier between 4000 and 3000 BC, the climate dried up slightly. Many water sources ran dry in the oasis, and the landscape was subject to desertification. Despite this fact, the Egyptians who lived here got water with the help of different devices and maintained the agriculture, and thus life in the oasis did not die.

The romans came to the oasis in the first few years AD to replace the Egyptians. They possessed better hydraulic engineering skills and built deep wells and aqueducts. However, the new equipment did not enable the Romans to link up with the surrounding landscape. Some 400 years later, they had exhausted the soil, and the oasis was abandoned right up to the present time. Nowadays we can see the numerous Roman buildings, water conduits, and wells half-filled by sands. A sand–solonchak desert has predominated for many centuries at the site of the oasis. The oasis most likely died as a result of water exhaustion and excessive use of underground water, which led to soil salinization and sand movement.

According to some investigators, mainly Bryson and Murray (1977), a number of social disasters occurred in the historical past that were due to human impact on local climatic conditions. This impact was not deliberate and took place as people went about their daily business. It resulted, first, in some changes in the natural situation, which caused a deterioration in the local climate. And this, in turn, contributed to an even greater change in the environment, to which the population was unaccustomed, making the life of people impossible. This led to civilization decline. Bryson and his associates adhere to such a scheme, to understand some twists in history that cannot be explained in other way. According to them, desertification developed intensively as a consequence of human activity in some regions of the Earth. It was followed, as is well known, by a change in albedo (the reflective characteristic of the landscape surface).

The albedo increase brought about a change in a number of physical processes such as a decrease in the amount of clouds and precipitation. As a result of this, the natural situation became more arid and less favourable for the population. This very situation, caused by human activities, happened in the area of the ancient cities Harappa and Mohenjo-daro, the centres of the Harappian State. The decline of Harappa, in particular, started as long ago as 1700 BC. As a consequence of unsystematic farming mainly overgrazing, the natural vegetation was either trampled or eaten, and the unanchored soil layer was broken and blown about by the wind. The location where food products were produced for the inhabitants of Harappa became a huge dust bowl. Though, later on, farming in this territory was renewed, dust storms occurred in subsequent years as well. By AD 1100, the location again became a desert and was left for good.

In a similar way, in the historical past, extensive areas in North Africa, in the north Sahara (Bryson and Murray, 1977), were laid waste by anthropogenic desertification. This took place after 1900 BC. Settlements and cities were abandoned and farming lands were buried under sand dunes that started moving. The unfavourable ecological situation was not, of course, the result of any single cause (e.g. soil exhaustion or sand movement). It is most likely that a series of interrelated causes, a chain of events, took part in its development, as happens at the present time.

2.3 THE POSSIBILITIES OF USING EXPERIENCE IN OVERCOMING THE CONSEQUENCES OF ECOLOGICAL DISASTERS

These results do not give a categorical answer to the true causes of ecological disasters in the past. The decline of cities and entire civilizations is explained by different investigators in different ways. Sometimes, it is not clear whether it is a natural or anthropogenic factor that causes deterioration of the natural situation. These questions still cannot be clarified conclusively: there are not enough known factors. However, this was not the goal of our book. The main object of analysis and comparison of different viewpoints on the causes of ecological disasters is recognizing the importance of anthropogenic factors in their development.

Unfavourable ecological events and disasters in the past were basically related, not to pollution of the natural environment as occurs now, but to its transformations. The degradation and exhaustion of soil should undoubtedly be considered as the basic transformation. In this connection, the understanding of interrelations between man and soil seems to be very important, as suggested by Hyams (1952) who arrived at the conclusion that if man considers himself to be and behave as a part of the 'soil community', then the results of the interactions between man and soil are satisfactory. If man forgets about his 'kinship' with the soil, this can lead, as it did in the past, to ecological disasters. It should be remembered, however, that ecological disasters occur when man forgets that he is an integral part of one or another ecosystem.

The lessons we have learned about interrelations between man and nature are very helpful, since we have entered a great spell of large-scale transformations of nature, which are fraught with various consequences. It turns out that man encountered many ecological disasters in the historical past that are similar to those of the present day. One of the most instructive lessons can be learned from analysis of the ecological situation that formed in the Aral Sea region in the past.

It is well known from archaeological sources that major farming states existed in the Aral Sea Region that later disappeared. Their decline is generally explained by natural disasters or historical causes (wars or social unrest). In this case, the possibility of ecological disaster as a result of human disturbance of the natural balance is almost never discussed. The Khorezmian state was one of the states that developed in the Aral Sea region. This state, which appeared in the 6th–5th centuries BC disappeared long ago (Andrianov, 1969; Tolstov, 1948). The existence of the Khorezm Oasis, its prosperity and growth were due to the development of a high level of irrigated farming. A dense network of canals allowed Khorezmians to reclaim extensive territories from the surrounding desert. Similarly, on the basis of irrigated farming, other states in the Aral Sea region grew and prospered, in particular in the lower reaches of the Syr Darya. The ruins of cities, covered by sand dust, land-reclamation canals buried under sediments, saline lands and even solonchaks, can now be seen at these places.

Many features of human activity in the abandoned oases of the Amu Darya and Syr Darya deltas are particularly clearly seen from aircraft. Among these seemingly surprising (after all, this is a desert) traces of life, it is the canal lines that catch the eye first of all. Mostly covered by sands, they appear through them due to the growth of vegetation – hygrophytes. We can even see at some sites the street layout, the outlines of the foundations of structures wiped out long ago, but still noticeable from the air. The remains of buildings, fortresses, and mausoleums are often found. Hundreds of city-fortresses and settlements have been discovered in the deltas of the Amu Darya and Syr Darya (Andrianov, 1969).

Some investigators consider climate change as the cause of destruction of some ancient civilizations in Central Asia. For example, Sinitsyn (1949) explains the decline of most farming oases, in particular Sogda and Baktria, by a progressive 'drying up' of Asia. Shnitnikov (1961) has a different viewpoint. He believes that climate was also the cause of reduction in the area of irrigated lands in the Khorezmian state, which took place in the first thousand years AD and resulted in the destruction of its cities. However, unlike Sinitsyn (and other supporters of such a viewpoint), Shnitnikov denies the progressive 'drying up' of Asia. He proved the existence of periodical alternation between humid and dry epochs, both long-lived and short-lived. In one dry epoch in the first thousand years AD, the amount of water in the lower reaches of Amu Darya decreased, which resulted, he believes, in the destruction of the Khorezmian Oasis.

Some investigators attribute the decline of Ancient Khorezm and other civilizations in Central Asia to social causes, mainly wars, as well as their consequences

(Tosi, 1986). Another cause of the decline of the Aral Sea region civilizations is possible and is underestimated by many people. This is a change in ecological conditions, necessary for life subsistence, as a consequence of the human impact on nature. Different investigation results confirm the probability of such an ecological change in the Aral Sea region in the past. According to calculations made by Kvasov (1976), the water level in the Aral Sea in the historical past decreased at times of intensive irrigation and increased when irrigation systems were no longer used.

According to the same calculations, the area of irrigated lands in the historical past reached modern size. According to new data (*The Aral*..., 1991), the lands under ancient irrigation in the lower reaches of the Amu Darya and Syr Darya exceeded in area modern ones by a factor of more than 3; in the period from the 3rd century BC to the 3rd century AD, their area was 3.5–3.8 million ha. This means that the Aral Sea became shallow due to excessive use of water for irrigation. A similar process occurs at the present time as well. It should be remembered that even the small (by only 4 m in 1975) modern decrease in the Aral Sea level damaged natural processes in the Amu Darya and Syr Darya deltas. In particular, the level of groundwater decreased, the tugai forests started drying up, and soils were blown away. All these processes were intensified in the late 1980s, at a time of maximum development of irrigation and an increase in the area of cultivated land. The water level in the Aral Sea dropped 14 m. The cause of this is well known: excessive use of water from rivers flowing to the Aral Sea (see Section 5.6).

Undoubtedly, whole series of events, unfavourable for human subsistence, resulted in further destruction of oases in the Aral Sea in the far past. It is not easy (though it is possible) to prove they happened now, many years later. It has been proved, for example, that more than half the area of the Khorezm oasis was affected by salinization (Kes, 1981; Kes *et al.*, 1980). This evidence shows that an ecological disaster could affect the farming oases in the Aral Sea region not only at a time of social unrest, but also during prosperity at a time of maximum development. Degradation of the present oases occurs in just this way.

Ecological disasters most likely always occurred owing to inefficient agricultural and, in particular, water management practices in a large region. Agricultural expansion not only in the Amu Darya and Syr Darya deltas, but in their middle and, possibly, lower reaches (in the whole Aral Sea basin) led to disturbance of the natural balance and undermined the economic basis of ancient states. This lesson is mainly aimed at the reorientation of the modern economy in the Aral Sea basin.

Special archaeological and geographical studies are now being made in China in the outskirts of the Takla Makan Desert, along the Tarim River. Intensive agricultural development of the location is currently taking place there at the present time and has caused degradation of the landscape, with the death of poplar forests, meadows, and movement of sand (Chao Sungchiao, 1984a, b). Chinese specialists are studying how the population adapted to natural conditions in the past, when both anthropogenic and natural factors) led to the degradation of oases and cities along the Silk Road.

Interesting, practical inferences from the ecological lessons of land development in Lower Mesopotamia in the historical past have been made by Mrost (1987a, b).

According to him, the present state of irrigation in this region still corresponds, to a considerable extent, to the irrational scheme operated in ancient times. At present, as in the past, governments demand agricultural expansion and the traditional system of fallow irrigation dominates. We have already mentioned that, as a result of such an approach to irrigation, salinization and finally degradation of irrigated land occur. To avoid undesirable ecological consequences, we must give up the system of fallow irrigation and develop technologically sound irrigated agrosystems that are designed to reduce or eliminate salinization. This should put a stop to the growth of anthropogenic desertification.

At the present time, as a result of studies made by many specialists, an increasing number of unfavourable ecological events and even disasters in the remote past, due to inefficient human economic activities, have been discovered. The role of such events in the history of human society has been underestimated. Unfavourable ecological events were mainly related to natural factors, apart from wars and civil strife. No doubt, the role of both is great. However, more information has now been accumulated on the ecological consequences caused by inefficient economic policies. Disasters can befall a city or a whole state not only at a time of decline but also at a time of prosperity. At the same time, it would be premature to blame the destruction of cities and even civilizations on inefficient economic policies in all cases. What we can emphasize, though, is the joint effect of deteriorating natural conditions and human errors in economic development.

We should also emphasize the major role of rhythmic processes in the interrelations between man and nature in the historical past. Evidence about the general frequency of natural and social events (their cyclicity) can be drawn from the prehistorical past. This is convincingly shown in the book by the well-known French philosopher Mirchi Eliade *Space and History* (1987). Cyclicity is most clearly seen in critical ecological situations with natural causes (e.g. climate cyclic fluctuations). At the same time, the rhythmic processes are observed in the self-development of social institutes. Such well-known thinkers as Schpengler (1923) and Toynbee (1946) believed in periodical surges and declines in the development of societies. In connection with this are undoubtedly the studies made by Gumilev (1989) on ethnoses, communities of people inseparably linked with the natural situation and in many respects governed by the laws of nature, and thus by natural rhythm.

Unfortunately, it is difficult to reveal these natural rhythmic events, their coordination with public events, and specific features of nature development by man in the historical past. On the one hand, this relates to the difficulties in recognizing natural rhythms and their boundaries, because they overlap each other. On the other hand, this is explained by insufficient historical–archaeological evidence. In view of the above, the studies of Maksimov (1990) are interesting. These deal with investigating the internal structure of elements in the 1,850-year natural cycle: the comparison of changes in the natural situation with the prosperity and decline of ancient civilizations. The ecological experience of mankind, gained in past historical ages, enriches our ideas about the possible critical situations mankind could find himself in. Figure 2.4 shows the areas of overflow during historical

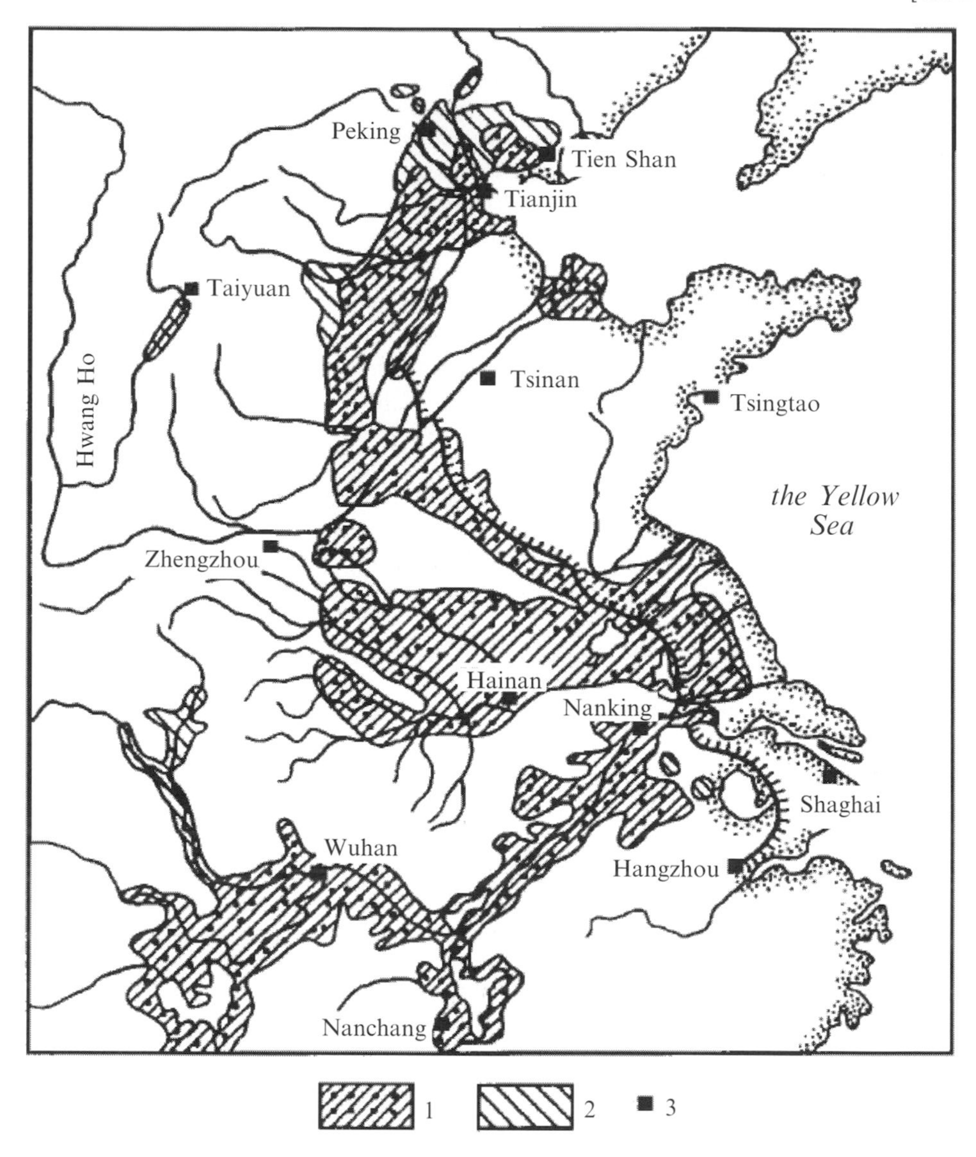

Figure 2.4. The areas of overflow in the eastern part of China during historical floods. 1 – disastrous floods, 2 – high floods with rare frequency, 3 – cities (cities with population > 1 million) (from Muranov, 1957).

floods in East China. The ecological lessons of the past are useful estimate the modern state of the environment, and for working out the best policy for dealing with it in the future.

Historical experience in overcoming the consequences of various natural disasters is also useful. Thus, for example, the consequences of the disastrous earthquake in Armenia in 1988 (see Chapter 3) would have been less serious if historical information had not been underestimated by governments and builders. Major destructive earthquakes in Armenia have occurred throughout its history. The traces of earthquakes having intensities of over 8 on the Richter scale are seen, for example, on the walls of the Urartian fortress – Erebuni – built 2,500 years ago at the place of modern Yerevan (Armenia) (Nikonov and Aslanyan, 1991).

The ancient capital of Armenia – Dvin – located 30 km to the south-east of Yerevan was destroyed more than once: in 854 and 869 (when tens of thousands of people died), in 893 (about 100,000 died). The intensity of the shocks in the latter case was 9–10 on the Richter scale. Yerevan was severely damaged in 1679 (shock intensity was up to 10).

On the basis of historical evidence (as well as the recent Spitak earthquake), the seismic danger to Armenia has been estimated anew with the risk possibility of shocks of intensity 9 over almost all the territory (prior to 1988 the larger part of the republic belonged to the zone of seismicity with intensity 8, the smaller part, in the north-east, to that with intensity 7) (Nikonov and Aslanyan, 1991). Shocks with intensity up to 10 are possible in the area of Yerevan and the atomic power station. Because of the high seismicity of Armenia and, particularly, of the area where the atomic power station is located requires a new approach to building, in general, and to the wisdom of constructing atomic power stations, in particular.

2.4 BIBLIOGRAPHY

Adams, R. E. (1965) *Land behind Baghdad* (211 pp.). Chicago University Press, Chicago.

Adams, R. E. (1980) Swamps, canals and the locations of ancient Mayan cities. *Antiquity* **544**(212), 206–14.

Adams, R. E. (1985) Ancient Maya canals/Grids and lattices in Mayan cities. *Archaeology* **35**(6), 28–35.

Andrianov, B. V. (1969) *Ancient Irrigating Systems in the Aral Sea Area* (220 pp.). USSR Academy of Sciences, Moscow.

Andrianov, B. V. (1978) *Agriculture of Our Ancestors* (168 pp.). USSR Academy of Sciences, Moscow.

Aprodov, V. A. (1976) *Millennia of East Magrib* (131 pp.). Mysl, Moscow.

The Aral Sea Crisis (1991) (220 pp.). USSR Academy of Science, Moscow.

Argunov, M. V. (1987) *The Antique Navigation Directions for the Black Sea* (157 pp.). USSR Academy of Science, Moscow.

Balandin, R. K., and Bondarev, L. G. (1988) *Nature and Civilization* (393 pp.). Nauka, Moscow.

Bartold, V. (1902) *Information on the Aral Sea and the Lower Reaches of Amu Darya from Ancient Times to the 17th Century* (120 pp.). Russian Geographical Society.

Blavatsky, V. D. (1976) *Nature and Antique Society* (79 pp.). USSR Academy of Science, Moscow.

Bolt, B. A. , Horn, W. L., MacDonald, G. A., and Scott, R. F. (1978) *Geological Natural Forces* (translated from English, 440 pp.). Progress, Moscow.

Bondarev, L. G. (1987) Denudation-accumulative processes in the countries of antique Mediterranean Sea area. *MSU Bulletin Geogr. Ser.* **6**, 27–32.
Bryson, R. A. and Murray, T. J. (1977) *Climates of Hunger* (192 pp.). University of Wisconsin Press, Madison.
Chao Sungchiao (1984a) Analysis of desert terrain in China using Landsat imagery. *Deserts and Arid Lands* (pp. 115–132). Elsevier, The Hague.
Chao Sungchiao (1984b) The sandy deserts and the Gobi of China. *Deserts and Arid Lands* (pp. 95–113). Elsevier, The Hague.
Cleator, P. E. (1973) *Underwater Archaeology* (224 pp.). Methuen, London.
Cornell, J. (1976) *Lost and Forgotten Peoples* (230 pp.). Bantam books, New York.
Delgado, J. P. (1998) *Encyclopedia of Underwater and Maritime Archaeology* (135 pp.). Yale University Press, New Haven, CT.
Diamond, J. M. (1986) The environmental myth. *Nature* **324**(6092), 19–20.
Doyel, L. (1979) *Flights to the Past* (translated from English, 296 pp.). USSR Academy of Science, Moscow.
Eberhard, J. E. (1982) Unveiling the Sahara's hidden Face. *Sci. News* **121**(26), 419–20.
Eckholm, E. P. (1976) Losing ground. *Environmental Stress and World Food Prospects* (224 pp.). Norton, New York.
Eidt, R. C. (1987) Prehistoric lowland maya environment. *Geogr. Rev.* **77**(4), 481–4.
Eliade, M. (1987) *Space and History* (translated from French, 290 pp.). University of Moscow, Moscow.
Erickson, C. L. (1988) Raised field agriculture in the lake Titicaca. *Expedition* **30**(3), 8–16.
Eydoux, H.-P. (1967) *A la recherche les mondes perdus les grand découvertes archéologiques* (276 pp.). Editions des Hespérides, Paris.
Evans, A. (1921–1935) *The Palace of Minos at Knossos* (4 vols). London.
Falkon-Barker, T. (1967) *1600 Years Under Water* (translated from French, 164 pp.). Progress, Moscow.
Fantechi, R. and Margaris, N. (eds) (1986) *Desertification in Europe* (311 pp.). McGraw-Hill, Boston.
Flemming, N. (1959) Underwater adventure in Apollonia. *Geogr. Mag.* **31**(10), 497–508.
Frezer, D. D. (1986) *Folklore in the Old Testament* (translated from English, 512 pp.). Progress, Moscow.
Fytikas, N., Kolios, N., and Vougioukalakis, G., (1990) Post-Minoan volcanic activity of the Santorini volcano: Volcanic hazard and risk, forecasting possibilities. In D. A. Hardy *et al.* (eds) *Thera and the Aegean World* (Vol. III, pp. 183–198). London, Thera Foundation.
Ginko, S. S. (1963) *Disasters on River Banks* (129 pp.). Gidrometeoizdat, Leningrad.
Godbrecht, G. (1987) Irrigation through our history – problems and solutions. *Water for the Future* (pp. 3–17). Balkema, Rotterdam.
Grigoryev, Al. A. (1986) *Historical Lessons of Interaction between Man and Nature* (32 pp.). Znaniye, Leningrad.
Grigoryev, Al. A. (1991) *Ecological Lessons of the Historical Past and the Present* (240 pp.). Nauka, Leningrad.
Gulyaev, V. (1981) Destruction of Chalchuapa and Seren: Natural disasters in the ancient maya history. *Science and Life* **4**, 129–33.
Gumilev, L. N. (1966) *Discovery of Khazaria* (362 pp.). Mysl, Moscow.
Gumilev, L. N. (1972) Climate changes and nomads migration. *Nature* **4**, 19–32.
Gumilev, L. N. (1989) *Ethnogeny and the Earth's Biosphere* (496 pp.). Nauka, Leningrad.
Helms, S. W. (1981) *Jawa, Lost City of the Block Desert* (270 pp.). Methuen, London.

Herodot. (1972) *History in Nine Books* (translated from Greek, 600 pp.). USSR Academy of Science, Leningrad.

Holett, A. and Grzymski, K. A. (1985/1986) Surveying the sands. *Rotunda* **18**(3), 20–29.

Hou Ren-zhi (1985) Ancient city ruins in the desert of the inner Mongolia Autonomous Region of China. *J. Hist. Geogr.* **11**(3), 241–52.

Huntington, E. (1907) *The Pulse of Asia* (312 pp.). John Wiley and Sons, New York.

Huntington, E. (1922) *Civilization and Climate* (307 pp.). Yale University Press, New Haven, CT.

Hyams, E. (1952) *Soil and Civilization* (270 pp.). Methuen, London.

Ilyinskaya, L. S. (1988) *Legends and Archeology: Ancient Mediterranean Region* (176 pp.). USSR Academy of Science, Moscow.

Ivanitsky, G. (1987) Why did Babylon perish? *Science and Life* **9**, 46–53.

Ivanov, V. V. (1989) Water, saline, land. *Ways to the Unknown* (Coll. 21, pp. 57–68). Progress, Moscow.

Jaquet, G. (1988) Images d'un Sahara Fertile. *Archeologia* **239**, 34–41.

Kapo-Rey, R. (1958) *French Sahara* (translated from French, 496 pp.). Progress, Moscow.

Karbonyin, L. (1985) The Earth' surface descent – a disastrous phenomenon of the global scale. *Nature and Resources (Paris, UNESCO)* **21**(1), 2–11.

Kes, A. S. (1981) Anthropogenic impact on the formation of the Amu Darya alluvial-delta plain relief. *The Ancient Khoresm Culture and Art* (pp. 72–80). USSR Academy of Science, Moscow.

Kes, A. S., Andrianov, B. V., and Itina, M. A. (1980) The dynamics of hydrographic network the Aral Sea level change. *The Fluctuations of the Aral–Caspian Region Moisture in the Holocene* (pp. 31–3). USSR Academy of Science, Moscow.

Kohl, P. L. (1981) The Namazga civilization: An overview. *The Bronze Age Civilizations of Central Asia* (pp. 21–32). McGraw-Hill, New York.

Kolata, A. L. (1987) Tiwanaka and its hinterland. *Archaeology* **40**(1), 36–41.

Kondratyev, K. Ya. (1985) *Volcanoes and Climate* (Science and Technology Results, Meteorol. and Climatol. Ser. Vol. 13, 204 pp.). VINITI, Moscow.

Kondratyev, K. Ya. and Galindo, I. (1997) *Volcanic Activity and Climate* (382 pp.). A Deepak, Hampton, VA.

Kvasov, D. D. (1976) The causes of Uzboy run-off stop and the Aral Sea problems. *Problems of Desert Development* **6**, 24–9.

Lamb, H. H. (1981) *Climate: Present, Past and Future* (712 pp.). Methuen, London.

Lambrick, H. T. (1963) The Indus flood plain and the Indus civilization. *Geogr. J.* **133**(4), 483–94.

Lot, A. (1973) *Searching for Tassilin–Adjer Frescos* (translated from French, 111 pp.). Art, Leningrad.

McIntosch, S., McIntosch, R. J., and Jenne-Jeno (1986) An ancient african city. *Archaeology* **33**(1), 41–5.

Maksimov, E. V. (1990) The 1850-year Shnitnikov rhythm and civilization waves. *Geoecology: Global Problems* (pp. 37–42). Russian Geographical Society, Leningrad.

Manning, S. (1989) A new age for Minoan Crete. *New Scientist* **121**(1651), 60–3.

Marinatos, D. (1973) *Die Ausgrabungen auf Thera und ihre Probleme* (32 pp.). Osterr. Akad. der Wiss., Vienna.

Marphey, R. (1951) The decline of North Africa since the Roman occupation: Climatic or human. *Ann. Assoc.Amer. Geogr.* **41**2), 116–37.

Masson, V. M. (1966) *The Country of One Thousand Cities* (231 pp.). USSR Academy of Science, Moscow.

Metcafe, S. E. (1987) Historical data and climatic change in Mexico – A review. *Geogr. J.* **153**(2), 211–22.
Mokrynin, V. P. Ploskih, V. M. (1988) *Sunken Cities* (192 pp.). USSR Academy of Sciences, Frunze, Ukraine.
Mrost, A. Yu. (1987a) Desertification processes in the Iraq Republic and measures for struggling against them. Abstract of Thesis for Cand. of Geogr. Science, Ashkhabat, 25 pp.
Mrost, A. Yu. (1987b) Map of desertification danger in the lower Mesopotamia. *Problems of Desert Development* **4**, 37–45.
Muranov, A. P. (1957) *Hwang Ho River* (88 pp.).Gidrometeoizdat, Leningrad.
Muranov, A. P. (1959a) *Yangtze River* (124 pp.). Gidrometeoizdat, Leningrad.
Muranov, A. P. (1959b) *Euphrates and Tigris Rivers* (140 pp.). Gidrometeoizdat, Leningrad.
Nikiforov, L. G. (1975) The post-glacial increase in ocean level and its importance for developing sea coasts. *The Fluctuations in World Ocean Level and the Problems of Marine Geomorphology* (pp. 12–40). USSR Academy of Sciences, Moscow.
Nikonov, A. A. and Aslanyan, I. I. (1991) Erevan earthquake in 1679. *Nature* **10**, 91–5.
Plato (1971) *Critius* (Collected works in three volumes, Vol. 3, 311 pp.). Mysl, Moscow.
Plumet, P. (1987) Les Vikinges amériques. *La Recherche* **18**(192), 1160–7.
Possehl, G. L. (ed.) (1987) Harrapan civilization: A contemporary perspective. *In the World of Science* (Vol. 10, p. 211). Progress, Moscow.
Pracad, T., Kumaz, B. S., and Kumaz, S. (1987) Water resources development in India – its central role in the past and crucial significance for the future. *Water for the Future* (pp. 211–14). Balkema, Rotterdam.
Razumov, G. A. amd Khasin, M. F. (1978) *Sinking Cities* (200 pp.). Mysl, Moscow.
Russel, R. J. (1956) Environmental changes through forces independent of man. *Man's Role in Changing the Face of the Earth* (pp. 453–72). University of Chicago Press, Chicago.
Sanders, W. T., Parsons, J. R., and Santley, R. S. (1979) The basin in Mexico. *Ecological Processes in the Evolution of Civilization* (pp. 453–72). Academic Press, New York.
Schpengler, O. (1923) *The Decline of Europe* (translated from German, Vol. 1. 468 pp.). L. D. Frenkel, Moscow.
Sheets, P. D. (1981) Volcanoes and the Maya. *Natur. Hist.* **90**(8), 32–42.
Shervashidze, L. A. (1967) *A Story of the City Taken by Waves* (65 pp.). Russian Geogr. Soc., Sukhumi (Abhazia).
Shilik, K. K. (1984) Localization of ancient Akkra as an example of comprehensive analysis in historical–geographical studies. Abstracts of papers for the *Conference 'Combined Methods in Studying History from the Oldest Times to Our Days'* (pp. 108–11) Moscow University Press, Moscow.
Shnitnikov, A. V. (1961) The variability of total moisture of the northern hemisphere continents. *Transactions of Geogr. Soc. New Series* **16**, 271.
Shuttle imaging radar discovers hidden features in Sahara (1983) *COSPAR Inform. Bull.* **96**, 66–8.
Sinitsyn, V. M. (1949) Tectonic factor in the change of Central Asia climate. *Bull. of Moscow Soc. of Nature Testers Geol. Dept.* **29**, 31–2.
Songqiao, Z. and Xuncheng, X. (1984) Evolution of the Lop Desert and the Lop Nor. *Geogr. J.* **156**(3), 311–21.
Steensberg, A. (1951) Archaelogical dating of the climatic change in north Europe about AD 1300. *Nature* **168**(4277), 672–4.
Street, M. (1984) Un Pompéi de l'âge glaciare. *La Recherche* **17**(176), 534–7.
Thompson, J. O. (1953) *The History of Old Geography* (translated from English, 592 pp.). Foreign Literature Press, Moscow.

Tolstov, S. P. (1948) *Following in the Tracks of the Ancient Khoresm Civilization* (211 pp.). USSR Academy of Science, Moscow.

Tosi, M. (1986) From Shahri-Sohte to Mohendjo-Daro. *Science and Mankind* (pp. 57–62). Progress, Moscow.

Toynbee, A. J. (1946) *A Study of History* (Abridgement of Vols 1–6 by D. Somervell, 320 pp.). Oxford University Press, London.

Turner, H. B. L. (1974) Prehistoric intensive agriculture in the Mayan lowlands. *Science* **185**(4146), 118–24.

Trifonov, V. G. and El Hair, Yu. (1988) Biblical legend in the eyes of geologists. *Nature* **8**, 34–5.

Voûte, C. (1999) Environmental geology, archaeology and history – what happened in Central Java (Indonesia) in the 10th–11th centuries AD./ *Cogeoenvironment Newsletter* **6**, 8–39.

Vozgrin, V. E. (1987) Greenland of Norsemen. *Problems of History* **2**, 184–7.

Voronov, Yu. N. (1980) *Dioskuriada – Sebastopolis* (120 pp.). USSR Academy of Science, Moscow.

Walker, A. S. and Wilson, D. K. (1987) The Badain Jaran desert: Remote sensing investigation. *Geogr. J.* **152**(2), 205–10.

Waltham, T. (1982) *Disasters: Furious Earth* (translated from English, 224 pp.). Nedra, Leningrad.

Williams, D. P. (1973) Environment archaeology in the West Negev. *Nature* **243**(5399), 501–33.

3

Natural disasters and their consequences

3.1 METEOROLOGICAL HAZARDS

3.1.1 Introduction

The borderline between two millennia is marked by the growing impacts of natural calamities on human life and economic activity. Mass media reports on such phenomena as destructive earthquakes, severe floods, tropical storms, volcanic eruptions, and so on have become almost daily occurrences. In this connection, the issues of great importance consist in understanding the reasons for such a situation. In particular, it is important to understand whether natural elemental phenomena occur more frequently than in the past, and whether they are anthropogenic. We shall discuss these issues using meteorological calamities and hazards, as examples.

Kunkel *et al.* (1999) analyse results of recent works on trends during the past century in social and economic impacts (direct economic losses, fatalities) in the USA from extreme weather conditions and climate. Obtained results testify that, in recent years, economic damage has increased considerably and reached its maximum in 1990. This damage is associated with property and harvest losses due to floods, hurricanes, thunderstorms, and winter storms. On the other hand, the frequency of occurrence or intensity of most weather extremes did not show trends increasing with time. Therefore, increase in damage was determined mainly by specific features of social and economic development. They are the following: growing population in coastal areas and large cities; developing insufficiently stable economic infrastructures; lifestyle and demographic changes giving rise to increasing vulnerability to natural disasters. All these hypothetical circumstances need to be analysed, though.

Data obtained over the past several decades reveal the decreasing frequency of severe hurricanes and insignificant changes in such phenomena as thunderstorms, hail, tornadoes, droughts, and extreme temperature. Contradictory, though, are flood data. From some data, it follows that precipitation has increased, while according to other data there was no trend of considerable river run-off (the connection between precipitation change and devastating floods needs to be further analysed). Rather contradictory is the information on fatalities as well. For example, only insignificant changes in the number of killed due to lightning and extremely low temperatures were found. Fatalities due to tornadoes and hurricanes have decreased with time, but probably the number of tornadoes and hurricanes has increased due to extreme temperature elevations. In this connection, it was important to improve the system of warning the population about hazardous weather phenomena (probable mortality growth due to high temperature was caused by specific conditions of living in cities).

Extreme weather and climate impacts are characterized by considerable regional variability in the territory of the USA. For example, harvest losses due to hail increased with time in the territory of the Great Plains, but decreased in the Midwest. While winter storms in the area of the eastern coast have become more frequent, in the basins of the Great Lakes the contrary situation has occurred. As a whole, related to the USA, observational data agree with the assessment presented in the second report by the Intergovernmental Panel of experts on problems of Climate Changes (Houghton *et al.*, 1996).

Related to the USA, data on individual observations make it possible to formulate the following (hypothetical, to some extent) assessments. They are as follows:

(1) Destruction and fatalities in the last 25 years due to floods appear to be more considerable than in the previous 65 years, and frequency of heavy precipitation in particular has increased.
(2) Gradual increase in the number of fatalities due to hurricanes takes place, although there is no data on the increasing frequency or intensity of hurricanes.
(3) Three major consequences are true for storms due to convection (thunderstorms, hail, tornadoes):

    (a) there have been no trends in them observed on the scale of entire country;
    (b) regional trends of different sign have been observed;
    (c) increasing losses have taken place.

(4) In the last 10–15 years, the devastating impact of winter storms in the east of the USA has both increased and decreased in the area of the Great Lakes.
(5) Although the drought of 1988 was accompanied by considerable economic damage, trends to droughts were not discovered in the USA.
(6) Recent heat waves have given rise to considerable fatalities among the population, but there are no data on the growth of their frequency. The most severe phenomena of this kind took place in 1930. There are neither any data on increasing frequency of cold waves.

For a number of reasons, during the last 20 years, application ranges of data and information about climate were rapidly expanding in domains of economics sensitive to climatic change such as agriculture, water resources, and so on (the difference between data and information is determined by the fact that, in the latter case, synthesis and data interpretation aiming at solving certain problems are meant: Changnon and Kunkel, 1999). One of the reasons mentioned was that in the period 1970–80, there were a series of natural calamities such as the devastating drought in the Sahel area, severe winters in the USA (at the end of 1970), heat waves, and considerable harvest losses (in 1980 and 1983). The second reason was radically enhanced capabilities to use computers for processing and analysing observational data. The third circumstance is connected with the emergence of readily available and inexpensive means for transmission and dissemination of information (in particular, the Internet). Fourth, very important was the successful development of numerical modelling methods for weather and climate change impacts on crop yield, river run-off, diseases of plants, probability of frosts, irrigation cycle, etc. The fifth important factor was the sufficiently clear understanding achieved at the onset of the 1980s of user's needs for meteorological information (primarily, concerning agriculture). One more (the sixth) important point was deepening the general understanding of the role of different information on the environment due to increasing sensitivity of economic infrastructures to ecodynamics and population growth in ecologically dangerous regions. Emergence of new centres of information for users has become the seventh factor favourable for wider use of information about weather and climate. Finally, the eighth factor was the considerable progress in improving climate forecasts.

In the context of the above-mentioned circumstances, Changnon and Kunkel (1999) discuss particular features of the use of data and information about climate for solution of problems related to agriculture and water resources. In the first of these areas, the main users are farmers, governmental institutions, agricultural business, and the scientific society. And, in the area of water resources, there are governmental institutions of all levels, consulting engineering companies, the scientific society, and private companies providing water users' needs. The main directions in the use of information about climate are:

(1) risk assessment;
(2) operational use;
(3) various assessments.

Human life and the existence of ecosystems depend essentially on such anomalous phenomena as shower-type precipitation, droughts, temperature extremes, and so on. In this connection, Huth *et al.* (2000) analyse observational data for eight meteorological stations in the southern part of Moravia (Czech Republic) and the results of climate numerical modelling (including both 'reference' climate and conditions of doubled CO_2 concentration). They characterize events of formation of heat waves and drought outbreaks. They consider surface air temperatures (SAT) computed by means of general atmospheric circulation (GAC) model ECHAM-3 for spatial grid points related to the area under consideration.

Analysed 'reference' integration data have shown that heat waves computed beforehand were too long (when compared to the ones observed), were formed later, and were characterized by higher SAT values. As for the quantity of computed heat waves, it was underestimated (June–July) or overestimated (August). Drought outbreaks computed beforehand were too long and their seasonal variations did not correspond to the actual one. The relation between computed heat waves and drought outbreaks with circulation conditions in the middle troposphere appeared to be weaker than actual one.

Since atmospheric circulation numerical modelling data are reliable enough, it is believed that possible reasons for the mentioned discrepancies were either overestimated intensity of air mass transformation and local processes, or underestimated advection. Numerical modelling has shown that, under $2 \times CO_2$ conditions, heat waves become more frequent. Increase in SAT by 4.5°C caused by doubled CO_2 concentration leads to a fivefold increase in the frequency of days with 'tropical' weather and considerable enhancement of extreme heat waves. Obtained results testify that the following is necessary:

(1) reliable validation of GAC models before their application for assessment of anthropogenic impacts;
(2) application of methods of 'scaling' or empirical statistics when using large-scale GAC models for local assessments.

The coast of the Gulf of Mexico in the USA is highly vulnerable to the devastating impact of hurricanes. For example, more than 8,000 persons became victims of the hurricane that battered Galveston (Texas) in 1990. In the context of further efforts to improve forecasts of hurricanes, Maloney and Hartmann (2000) analyse the process of tropical cyclone activity modulation in the Gulf of Mexico and in the western part of Caribbean Sea due to Madden–Julian oscillations (MJO) (tropical intraseasonal atmospheric circulation variations). The main feature of MJOs consists in oscillations of convection and wind fields in tropics on time scales of about 30–60 days.

Tropical convection accompanied by wind anomalies develops over the Indian Ocean and shifts eastwards in the direction of the Pacific Ocean. MJOs give rise to alternating periods of west and east wind anomalies in the eastern sector of the Pacific Ocean. Exactly these seasonal wind variations determine considerable changes in tropical cyclones in the Gulf of Mexico. As an indicator of MJO variability, the principal component (PC) of the first empirical orthogonal function (EOF) has been used. It was calculated from zonal wind data at the level of 850 hPa.

Obtained results testify that, when MJO anomalies occurred in the field of west winds in the lower troposphere, formation of hurricanes in the considered region are four times more probable than in the case of east winds. As a whole, MJO-related variations in tropical wind and precipitation fields are comparable as regards their value (for the area under consideration) with annual variability due to El Niño/ Southern Oscillation (ENSO). Due account for MJO must undoubtedly facilitate improvement of hurricane forecast.

As already mentioned, a severe tropical hurricane hit Galveston (Texas) on 9 September 1900 that led to more than 6,000 fatalities (approximately 20% of the population of the town). Currently, a hurricane of this kind is classified as a storm of the 4th category, characterized by steady wind speed within 133–155 miles an hour. In the last 150 years, the coast of the USA located along the Gulf of Mexico and the Atlantic Ocean was exposed to such severe hurricanes a total of 10 times (Travis, 2000). Even more rare are hurricanes of the 5th category, they only occurred twice in the 20th century.

Forecast of severe hurricanes is considerably complicated by the fact that in the past such hurricanes were rare. A feasible way to make forecasts more accurate is to use indirect information about hurricanes in ancient times within the line of scientific research called paleotempestology. One of the sources of information for paleotempestological analysis are stratified sandy deposits of coastal lakes and swamps formed under the action of floods due to hurricanes.

Analysed in this way, sedimentary cores 1–10-m long made it possible to study hurricane evolution over the last 5,000 years. Moreover, the obtained results offer possibilities of forecasting hurricane evolution under global warming conditions. Core data time referencing was performed from radiocarbon data. Cores from 16 lakes and swamps located along the coast of the Gulf of Mexico have been analysed. The obtained data revealed interdecadal variations of the number and intensity of hurricanes and variability on a 1,000-year scale, too.

In this context, the last 1,000 years were comparatively calm. Devastating hurricane events were four to five times more frequent in the period 1,000–3,500 years ago than in the last 1,000 years. It was found that, in the last 3,500 years, the coast of the Gulf of Mexico was exposed to hurricanes of 4th and 5th categories once in every 300 years. Thus, the hurricane that hit the town of Galveston in 1900 has 0.3% probability of recurring within 1 year. Information of this kind is very important for insurance companies. Historic documents proved to be a very important source of information about typhoons in China in the last 1,000 years. Certain data on hurricanes can be obtained from the results of chemical analysis of corals, data on plant pollen, and the isotope composition of precipitation.

It is well known that frequency and intensity of extreme changes of weather and climate exert tremendous effects on society and the environment (see any of the entries in the Bibliography). In this connection, an international workshop on the problem of indices and indicators of extreme climatic phenomena took place on 3–6 June 1997 in Asheville, North Carolina (Karl *et al.*, 1999a). The general purpose of the meeting was to analyse available data on time dynamics of extreme phenomena and their intensity (mainly to reveal whether climate has become more variable and (or) extreme). The specific tasks of approximately 100 experts (including representatives of insurance companies) from 23 countries consisted in answering the following questions:

(1) What is it necessary to do to augment the observational database and to improve its quality for monitoring and analysis of the consequences of extreme weather phenomena?
(2) Is it possible, in this connection, to substantiate relevant priorities, as well as indices and indicators of extreme weather and climate phenomena?
(3) What are the difficulties in solving these problems?

As reported by IPCC-1996 (Intergovernmental Panel of experts on problems of Climate Changes) (Houghton *et al.*, 1996): 'By and large, observation data for the 20th century have not revealed global-scale variability of extreme weather and climate phenomena. On the other hand, one should keep in mind incompleteness and insufficient reliability of observation data. On regional scales there are certain evidences for variability of a number of indicators of extreme climatic conditions. In some cases such data testify to growing variability, and in other cases they testify that it is decreasing.' For instance, there is no information testifying that there is growth in frequency (or intensity) of droughts or precipitation (not considering local situations, including the Sahel, where there were precipitation decay trends).

Although (starting in 1988), from satellite observation data, growth was noted in the number of extremely intensive extra-tropical cyclones in the North Atlantic, activity of tropical cyclones over the Atlantic Ocean in the last few decades attenuated (except for 1995). It was observed that the number of situations with extremely low temperatures decreased, but, as for extremely high-temperature events, no changes were observed. Daily temperature variability in the middle latitudes of the Northern Hemisphere became weaker.

On the other hand, global warming numerical modelling data suggest that in these conditions the following should take place (Houghton *et al.*, 1996; Karl *et al.*, 1999b). First, extremely high-temperature events will become more frequent, and the number of winter days with very low temperature will become smaller (in particular, the number of days with light frosts). Second, according to many models, precipitation intensity will increase, but, from others, it follows that the probability of dry weather and duration of dry weather periods (the number of successive days without precipitation) will increase. Third, since we should expect intensification of the global water cycle, this situation will result in more frequent hydrological anomalies (droughts and floods) in some regions and similar attenuated anomalies in others.

On the basis of the nine-layer model of general atmospheric circulation on the grid of 4° latitude × 5° longitude developed by Goddard Space Flight Center (USA) for the purposes of numerical modelling of the climate, Druyan *et al.* (1999) discuss two approaches to estimation of 'greenhouse' warming impacts on the frequency of tropical cyclones (TCs) in the Northern Hemisphere in summer. A tropical cyclone is defined as a storm under steady wind speed of more than 17 m/s, followed by a deepening minimum of atmospheric pressure.

According to IPCC-1996 (Houghton *et al.*, 1996), it is still unclear whether global climate warming will result in changes in frequency, location of initiation, time of formation, and average and maximum intensity of TC.

Climatic parameters averaged over the period of July–September were estimated by Druyan. *et al.* (1999) for prescribed current ($1 \times CO_2$) and doubled ($2 \times CO_2$) carbon dioxide concentration. Prescribed values of climatic parameters were used to compute seasonal genesis parameter of TC (SGP) defined by six components. They are the following: heat energy of the ocean (proportional to the difference between sea surface temperature (SST) and 26°C); humid stability (proportional to the vertical lapse rate of the equivalent potential temperature in the layer of ocean surface – 500 hPa); relative humidity in the layer of 700–500 hPa; vertical wind shear; near-surface wind velocity vortex – Coriolis parameter. Each of these components determines conditions required for TC formation, since SGP >0 only when all the components are positive.

Numerical modelling data for the cases of $1 \times CO_2$ and $2 \times CO_2$ were compared. It was shown that in the second case the number of TCs originating in the western sector of the North Atlantic/Gulf of Mexico must increase by 50%, and that over the Pacific Ocean in the Northern Hemisphere by 100–200%. Growth of this kind is mainly caused by increase in SST, but it might be overestimated. It seems likely that certain correction of SGP may be required since, for instance, SGP distribution computed from observation data for 1982–1994 is insufficiently reliable to simulate actual distribution of TC frequency of occurrence.

The second of the mentioned approaches is based on revealing easterly waves in the eastern sector of the North Atlantic by means of spectral analysis of the vortex field and the temporal trend of wind from computation data for conditions of $1 \times CO_2$ and $2CO_2$. Numerical modelling data of this kind indicate the shift in trajectories of eastern waves over Western Africa southward in the case of $2 \times CO_2$, and considerable growth in their average amplitude when crossing the African coast and further propagation in the eastern sector of the Atlantic Ocean along 14°S.

Assessments of the time dynamics of storms under global-warming conditions proved to be rather contradictory (this holds true especially for extremely intense storms). Computations have shown that the formation of TCs depends not only on the temperature of the oceanic surface, but also on a number of other factors (Karl *et al.*, 1999b).

The decreased activity of TCs was typical for the Atlantic Ocean Basin in the period 1970–87. Indicators for this relative calm were the fall in the number of severe hurricanes and landslides caused by them along the eastern coast of the USA, as well as a general decrease in hurricane activity in the Caribbean region. Short-term restoration of hurricane activity in 1988 and 1989 could have been symptomatic of a return to the high-level activity that was typical between 1920 and 1960, but by 1991–4 hurricane activity attenuated again. This was partly due to the long-term (1990–5) El Niño phenomenon that came to an end at the beginning of 1995. And, by the end of the same year, one of the most active hurricane seasons throughout the whole observation time was evidenced. Since then (except for 1997) hurricane activity has still been high (considerably higher than average).

Goldenberg *et al.* (2001) undertook an analysis of hurricane data in order to understand whether it is possible to consider this trend as long-term (a similar suggestion was made earlier by some specialists). Considerable annual and decadal variability in TC activity is typical of the North Atlantic Basin (NAB) (including the North Atlantic, Caribbean Sea, and Gulf of Mexico). This is particularly true for especially high-intensity cyclones that arise as a result of long-wave formation in the atmosphere propagating from Africa to the tropical North Atlantic.

Analysis of data relating to 1995–2000 has led to the conclusion that this period was marked by increased hurricane activity in the NAB compared to the preceding period 1971–94. In the last 6 years throughout the whole NAB, there was evidence of doubled hurricane activity and 2.5-fold growth in the number of severe hurricanes (wind speed higher than 50 m/s), as well as a fivefold increase in the number of hurricanes in the Caribbean Basin. The reasons for such a situation were the simultaneous rise in sea surface temperature (SST) in the North Atlantic and fall in vertical wind shear. Since changes of this kind occur on multi-decadal timescales, we believe that the currently observed high level of hurricane activity will be the same for the next 10–40 years. This demands thoroughly-planned measures to minimize the consequences for people and the economy.

Among the most well-defined regions of low-frequency variability in SST are the areas of generation of currents along the western boundaries of land in the Southern Hemisphere and of similar currents in the Indian, Pacific, and Atlantic Oceans. SST variability in these areas is important as a factor in track formation and intensity of extratropical cyclones and frontal systems connected with them. This is true because of the location of these areas in the zone of west–east transport in midlatitudes and advection of warm waters originated in the tropics to higher latitudes.

To analyse the regular features of this, Reason and Murray (2001) performed atmospheric circulation numerical modelling using the spectral (R21) model developed at Melbourne University. The main computational task was to find how subdecadal/multidecadal timescale variations (this is especially true for storm-tracks) respond to the positive SST anomaly in midlatitudes of the Southern Hemisphere that occur on almost hemisphere-size scales. Numerical modelling performed earlier for the Northern Hemisphere discovered that an intrinsic feature in the areas of the Pacific and Atlantic Oceans is the existence of storm-track modulation in midlatitudes caused by SST anomalies. New computations related to the Southern Hemisphere have demonstrated a similar response is evidenced in this case (on subdecadal/multidecadal scales) by formation of a large-scale trough in midlatitudes in the course of ridge formation (in the pressure field) in high latitudes of the Southern Hemisphere.

Similar atmospheric baroclinicity variations are accompanied by strengthening cyclonic systems in the belt 40°–50°S, and weakening in higher latitudes. According to results obtained, during the second decade of the integration interval, weakening response of atmospheric circulation to SST takes place in mid and highlatitudes of the Southern Hemisphere. However, it does not occur during the first decade. The temporal dynamics of SST anomalies impact on storm-track evolution in mid-latitudes of the Southern Hemisphere are similar to these obtained earlier for the Northern Hemisphere.

The main conclusion is that a steady interaction between the atmosphere and sea never occurs. As a consequence, formation of self-maintaining long-term oscillations in atmospheric circulation is possible in midlatitudes. The following features of atmospheric circulation in the Southern Hemisphere have not been considered. These are such essential features as ENSO and the Antarctic Circumpolar Wave, in the field of SST, as well as variations in near-surface atmospheric pressure, wind shear, and sea ice cover extent. An important task of further developments is to analyse possible interrelation between this variability and considered subdecadal/multidecadal variations.

Anxiety as a result of ongoing global climate change dictates the necessity to analyse possible trends in extreme meteorological and climatic events. In this connection, Changnon and Changnon (2001) consider data from 86 meteorological stations located in the USA over the period 1896–1995 in order to reveal fluctuations and long-term trends in thunderstorm frequency (TF) in the course of the preceding century. Short-term (<10 years) TF fluctuations at neighbouring stations quite often turned to be different because of the influence of local factors. This made it difficult to interpret data on spatial TF distribution. However, on timescales of about 20 years and more, the influence of large-scale atmospheric circulation was evidenced rather distinctly as a factor of spatial TF distribution.

With 20-year periods, it has become possible to identify six types of TF distribution and 12 typical regions in the territory of the USA. One of these types occurred in the Midwest, and another in the south of the USA, was characterized by maximum thunderstorm activity in the period 1916–35 followed by a fall in 1976–95. The distinctive feature of the second type of TF, whose maximum fell in the same years, was that its minimum came earlier in 1956–75. The distinctive feature of the third TF distribution in the Midwest and in the south was its maximum in the middle of the century (1936–55) and subsequent minimum in the period 1976–95. In the case of the fourth type, TF maximum fell also in 1936–55, but its minimum took place in 1896–1915 mainly in the areas of the northern plains and Rocky Mountains. The main feature of the fifth TF type was its maximum in 1956–75 recorded at stations located in the following four regions: central part of the Great Plains, the south-west, north of the Great Lakes, and the south-east. A characteristic feature of spatial–temporal variability under conditions of the sixth TF type is the gradual rise of thunderstorm activity throughout the whole century with the maximum in 1976–95 embracing a vast area from the Pacific seaboard of the north-west to the southern part of the Great Plains; the central part of the Rocky Mountains is located in the centre of this area.

Temporal TF variation averaged over the entire territory of the USA reached its maximum in the middle of the century. Considerable discrepancies between averaged TF variation with time over the entire territory of the USA and over individual regions suggest the necessity to allow for regional features. Determination of the linear trend over 100 years has also revealed the necessity to allow for regional specificity. A considerable positive trend is typical for the major part of the western two-thirds of the territory of the USA, but there was no such trend in the northern plains or in the Midwest, and in the major part of the eastern territory there was a negative TF trend. A positive trend has also been recorded in the area of the southern plains where the devastating consequences of the storms are the most significant, which is especially important in view of the growing population in the region. Regularities in spatial–temporal variability of thunderstorms are similar to relevant regularities of hail.

All these results testify to the need to improve observation systems and numerical modelling methods. For the latter purpose, we make use of the framework of the CLIVAR programme to detect the causes of climate change, implemented within the framework of the WCRP (World Climate Research Programme) and under support of the WMO (World Meteorological Organization). The programme is aimed at the following:

(1) database augmentation by invoking paleoclimatic information and numerical modelling data;
(2) production of more reliable estimates of variability in natural and anthropogenic impacts on the climate for sufficiently reliable recognition and quantitative estimation of the latter.

In this context, a special role is played by substantiation of a long series of global and regional indices and indicators of extreme changes in climate on the basis of observation data. To obtain these data, considerable contributions are made by implementing the GCOS (Global Climate Observing System) programme and satellite observations in general (Changnon and Changnon, 1998).

As noted by Karl and Easterling (1999), annually, tens of thousands of people become victims of climatic catastrophes and material damage reaches tens of billions of dollars in 1980, the insurance industry started to quote exponential growth of damage with time). For example, more than 140,000 people became victims of severe floods in Bangladesh in 1991, and the floods in 1970 were even more serious (approximately 300,000 people lost their lives). In the context of the grave socio-economical consequences of natural catastrophes, the following four fundamental questions need to be answered:

(1) Are any changes detected in the frequency of events associated with extreme variability of weather and climate?
(2) Are these changes unusual when compared with normal variability?
(3) Are there any data on the anthropogenic genesis of these catastrophes?
(4) What are the priority problems to be solved in order to decrease uncertainty in existing estimates?

Answers to these questions have been considerably hampered by insufficient attention, until recently, to the study of trends and long-term variability in the weather and climate. An important contribution was made by implementation of the GCOS programme toward reducing the consequences of natural calamities. The major obstacle is incomplete and inadequate observation data, preventing, in particular, reliable estimation of anthropogenic factors' impacts on the dynamics of meteorological catastrophes. If, for example, in some regions, an increase was observed in the frequency of extreme events, in other regions, decrease in their frequency took place.

Available data on variability in surface air temperature (SAT), precipitation and storms are discussed by Karl and Easterling (1999) in the context of formation of extreme changes in weather and climate. Insufficient data do not allow us to make a conclusion on increasing frequency and (or) intensity of extreme meteorological phenomena on a global scale.

It follows from numerical modelling data that anthropogenically-induced ('greenhouse') global warming of the climate must be accompanied by intensification of water cycle so analysis of pertinent observation data is of interest. It showed the following: evaporation in the tropics intensified; the amount of convective clouds and cirrus clouds associated with them increased; the amount of clouds over dry land increased, which favoured a decrease in daily temperature amplitude and a decline in evaporation for water basins on dry land; precipitation on dry land in middle and high latitudes increased, which favoured intensification of evaporation and river run-off; water content in the atmosphere in North America, China, and tropical regions increased; precipitation in many regions of the Northern Hemisphere intensified; extratropical storms became more severe.

Many of the observed features are in accord with the general conclusion concerning water cycle enhancement in the course of global climate warming, but sufficiently reliable conclusions can only be made on the basis of better data. Further elaboration should be focused on meeting the following objectives:

(1) working out and implementing more efficient ways for international observation data exchange;
(2) a search for different data, making it possible to extend the length of observation series under sufficiently high temporal resolution;
(3) increasing homogeneity of available observation series;
(4) more efficient use of satellite observation data;
(5) elaboration of more reliable monitoring methods for the dynamics of such local phenomena as tornado, hail, lightning, and wind storms;
(6) more thorough analysis of available information on meteorological catastrophes.

As mentioned earlier, an important aspect of the problem of antropogenically-induced global warming that attracts considerable attention, not only by experts but also the mass media, is the presumed enhancement of extreme weather phenomena (tropical storms, wind storms, and surges, etc.) under global warming conditions (Schmith *et al.*, 1998). According to data on numerical climate modelling, at a doubled CO_2 concentration, intensification and shift of Atlantic storm tracks to the east must take place (i.e. amplifying storms in the north-east Atlantic/north-west Europe). Yet, it is not clear what was the cause of the observed enhancement of storms, since changes in storm conditions are connected with the three-dimensional dynamics of the atmosphere, and can easily be caused by internal variability in the atmosphere–ocean system.

Schmith *et al.* (1998) analyse new observation data on winter storms in the north-east Atlantic for the period of 1975–95, being the results of long-term, more frequent (several times a day) observations of atmospheric pressure at sea level of eight stations (instrumental observations only started in 1950). Solution of such a problem may appear trivial at first glance, since a quite voluminous observation database is available for the region under consideration. The quality of these data is, however, poor (because of changes that took place in observation methods and processing of observation data), which makes the use of inhomogeneous observation series to reliably detect weak trends dangerous.

In order to eliminate the inhomogeneity of the series (and related false trends and discontinuities), Schmith *et al.* (1998) apply the method of filtering out high-frequency components of variability.

Analysis of obtained results revealed the trend toward amplification in storms in the recent 20–30 years. But annual and decadal variations dominates the contribution to variability in storms. Variability of this kind proved to be statistically connected with low–frequency circulation variations, characterizing the atmospheric pressure near the surface in the North Atlantic averaged over the winter period. Causal relationships of this kind are very important for analysis of the impacts of external disturbances on the atmosphere.

3.1.2 Storms

Apart from such extreme meteorological effects on society as droughts, floods, and heat waves, an important role is also played by storms. Storms, for which there is no strict definition, include extreme meteorological phenomena such as tropical and extratropical cyclones, thunderstorms, and so on, often involving victims and (quite often) enormous material damage.

Hellin and Haigh (1999) prepared a summary (Table 3.1) characterizing the most severe hurricanes in the Western Hemisphere for the entire period of available observations.

According to Nicholls *et al.* (1998), for the period of reliable satellite observations in the Australia region (to the south of the equator, 105–60°E),

Table 3.1. The most devastating storms in the Western Hemisphere.

Hurricane	Dates	Exposed region	Number of victims
The Great Hurricane	10–16 October 1780	Martinique and Barbados	22,000
Mitch	21 October–4 November 1958	Honduras, Nicaragua	11,000+
Galveston, Texas	8 September 1900	Galveston Island	8,000
Fifi	14–19 September 1974	Honduras	8,000
Dominican Republic	1–6 September 1930	Dominican Republic	8,000
Flora	30 September–8 October 1963	Haiti, Cuba	7,200
Martinique Island	6 September 1776	Point-Petre Bay	>6,000

in 1969–70 decrease took place in the number of TCs, but the number of severe TCs increased slightly. For a number of reasons, available estimates of the frequency and intensity of extreme phenomena (and, all the more, their forecast) are far from being adequate. The reasons for this are mainly limited observation data, which lack the necessary space and time resolution, as well as data completeness and reliability.

As was noted by Trenberth and Owen (1999), the growth in observation data completeness and reliability in recent years has resulted in the illusion of increasing frequency of extreme phenomena. For example, according to available statistical data, annual stable growth in the number of tornadoes took place in the USA, though, in fact, this was no more than a growth in observation completeness. That's why it is of fundamental importance to increase the reliability of databases. In view of great difficulties in obtaining long and homogeneous observation series for such mesoscale phenomena as thunderstorms, hail, tornadoes, etc. (and, in particular, insufficient reliability of wind observation data), it was recommended to retrieve the geostrophic wind field from more reliable data on the near-surface atmospheric pressure. For this purpose, corresponding indices should be introduced (demand for wind vector data determines the necessity to take into account both parameters). When it is possible, it is expedient to use data on pressure at the sea level from two stations. For instance, in the case of North Atlantic oscillations (NAO), determined by normalized pressure difference in Lisbon (Portugal) and Stykkiskholmur (Iceland), large-scale zonal and meridional wind indices are computed from instantaneous values of pressure at a typical space scale of about 1,000 km. It should be taken into consideration, however, that average wind speed is only one of the factors determining storm-induced destruction (e.g. data on turbulence and atmospheric gusts are very important). Investigation of tropical and extratropical cyclones is of special importance.

3.1.2.1 Tropical cyclones

TCs are defined as convection-induced mesoscale formations with a warm core over the ocean with the radius of the maximum wind speed area of about 10–100 km. Tropical coasts of the Atlantic, Pacific, and Indian Oceans are TCs at regular intervals (Figures 3.1). Sometimes, enormous numbers of victims (as, e.g., in Bangladesh in 1970, when almost 300,000 lost their lives) and massive material damage (US$26 billion (e.g. Hurricane Andrew in Florida in 1992) are involved.

As for the possible influence of climatic change on TCs, it is still rather uncertain (Trenberth and Owen, 1999). Although, for instance, under conditions of potential global warming, oceanic surface temperature must increase and, consequently, TC formation processes must be more intensive. Simultaneous changes in stability and atmospheric circulation must exert a contrary effect. Unfortunately, available general atmospheric circulation models do not have sufficient space and time resolution to simulate relevant processes. From observation data during El Niño periods, the number of TCs as a rule decreased over the Atlantic Ocean in the far west part of the Pacific Ocean and in Australia, but increased in the central region of the Pacific Ocean.

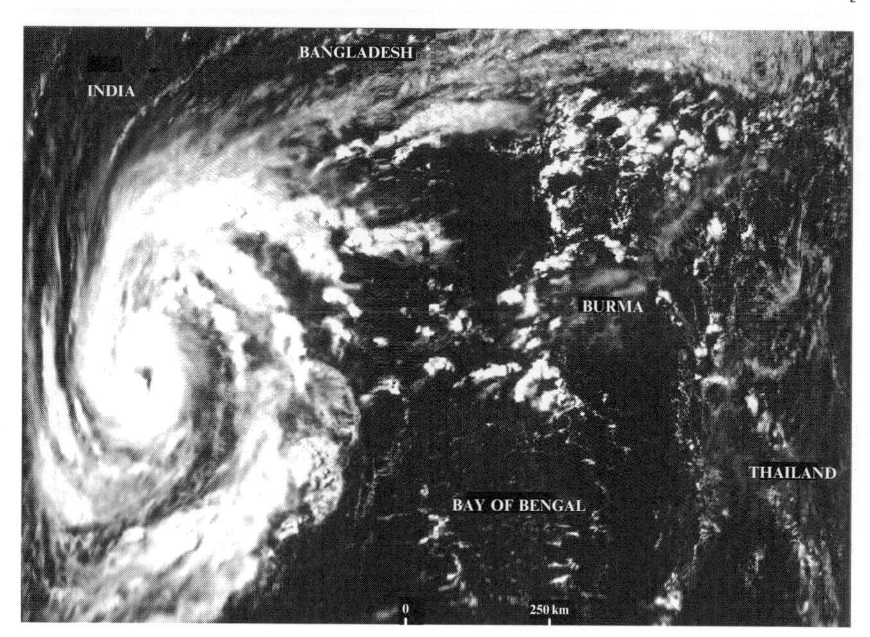

Figure 3.1. NOAA (AVHRR) satellite image of the tropical cyclone detected on 26 April 1981 over the Indian Ocean 3 days before it reached the coast in the area of Bangladesh (as a result 139,000 persons were killed and 10 million persons were displaced) (Grigoryev and Kondratyev, 2000a).

For many years, estimates of TC effects on agriculture and buildings have prevailed. One of the indicators of relevant impacts are threshold values of maximum (relatively stable) wind speed. For TCs ('devastating storms') in the Atlantic Ocean Basin and in some other regions, such parameters are used as the number of storms (at wind speed >18 m/s), number of hurricanes (>33 m/s), and severe hurricanes, and the number of days with devastating storms. It is expedient to use these characteristics in all regions, as well as the following values: radius of maximum wind speed zones for wind speeds reaching 18 m/s and 33 m/s; characteristics of spatial patterns of TCs; time of occurrence and duration of TCs coming to dry land. Diverse meteorological information such as air temperature, dew point, sea-level pressure, and precipitation are also important.

3.1.2.2 Extra-tropical cyclones

Extratropical cyclones (ETC) are low-pressure baroclinic systems that are formed in midlatitudes and, in many cases, cause considerable damage in winter. In this context, information on the wind observed (or on geostrophic wind retrieval from pressure field), air temperature, and precipitation are most important. TCs not only result in damage, but can also exert a positive effect by providing (when necessary) additional water resources. Coastal storms, of course, cause most damage.

Since 1874, data on ETC trajectories have been published on a regular basis.

These data, together with information on atmospheric pressure, are used for quantitative characterization of the frequency, trajectory and duration of ETCs for different threshold values of their intensity (the solution to this problem is further complicated due to scarce observation data in the Southern Hemisphere).

The principal feature of such mesoscale phenomena as thunderstorms is their quasi-occasional space and time variability. Tornadoes and thunderstorms, accompanied by lightning and hail, dust storms, and other dangerous phenomena are sometimes responsible for victims among the population and material losses (noted, however, that in the USA only 5–10% of thunderstorms and 1–25% of hail precipitation cause considerable damage).

Because of inadequate, routine observation data (as regards their space and time resolution), the main sources of information about these extreme phenomena are satellite observations (Grigoryev and Kondratyev, 1996; Kondratyev, 1999; Ghazi *et al.*, 1998; NESDIS, 1998; Oike and Yamada, 1993; *Putting NASA's . . .*, 1998; Zebrowski, 1997). The hazard index for such phenomena is the number of days (absolute or relative) when damage occurs.

3.1.2.3 Small-scale and mesoscale anomalies

In the USA, severe (sometimes catastrophic) weather phenomena occur every year on typical space scales from less than 1 km^2 to large regions of the country (Skinner *et al.*, 1999). Events of this kind (extreme temperature anomalies and various storms) cause victims and considerable material loss. The inadequacy of observation data does not allow us to estimate (or forecast) the frequency of occurrence and scales of catastrophic situations. This situation still has not changed despite the fact that, in the USA, the Cooperative Observation Network (CON) for monitoring of natural hazard phenomena, in the regions where they occur most frequently, has been functioning since 1980 through its dense network.

Currently, CON comprises about 8,000 stations but regular observations on precipitation, thunderstorms, hail, and extreme temperature are only carried out by about 1,000 stations. Some 200 measurements (points) have been taken daily on atmospheric pressure and wind speed since 1990, which makes it possible to have some information on the dynamics of anomalous weather conditions for the last 10 years. At the same time, CON data have a space and time resolution that is inadequate to obtain information of such small and mesoscale events as atmospheric gusts, hail, and tornadoes. So, data of this kind do not contain the necessary information on severe storms. It is important to remember that only 5–10% of all thunderstorms, 10–25% of hail precipitation, and 30–40% of all tornadoes result in severe after-effects.

Point observations do not allow us to define specific features about the spatial distribution of anomalous phenomena or assess the danger of damage to specific objects (e.g., wheat or maize in the early stage of growth; the city of Denver or Chicago, and so on). Better harvests and property insurance practices must rest on analysis of combined information about anomalous meteorological conditions and related economic damage.

In this connection, Skinner *et al.* (1999) discussed two particular situations:

(1) determination of damage to property, taking account of weather anomalies starting from 1949;
(2) advantages of combined analysis of meteorological information and terms of insurance.

Since 1949, insurance companies have defined as 'catastrophic' situations when damage reaches US$1 million or more. In 1983, this threshold was raised to the level of US$5 million because of inflation.

Computation of annual per capita normalized damage (taking account of a number of factors) on the territory of the USA for the period 1950–95 did not show a distinct trend, but showed several peaks, especially in the second half of 1990. Data on the number of catastrophes that caused damage of more than US$100 million were analysed for different regions of the USA for the period 1948–94. It was shown that in the south and west of the country, where population grew rapidly, considerable increase in the number of catastrophes of this kind was observed. Elaboration of a number of indicators characterizing weather changes and their effects on agriculture or construction made it possible to substantiate a more reliable insurance strategy.

3.1.3 Regional meteorological hazards

It is quite natural that dangerous weather phenomena are region-specific. Let's consider some illustrations of this kind.

3.1.3.1 Atlantic Ocean

According to the IPCC-1996 report (Houghton *et al.*, 1996), expected 'greenhouse' climate warming may be accompanied by an increase in SST in the tropics and enhancement of precipitation due to intensification of the ITCZ (Inter-Tropical Convergence Zone). TCs receive latent heat from the tropical ocean, and this heat is then 'released' into the upper troposphere, stimulating the formation of storms. So, concern arises with regard to potential effect of increasing SST that will manifest itself as more frequent formation and intensification of hurricanes, typhoons, and heavy TCs (considerable impact, however, may be caused by a number of other factors of variability in the tropical atmosphere).

These premises have not yet been confirmed by observational data. Landsea *et al.* (1999) undertake an analysis of long-term variations in TCs in the Atlantic Ocean Basin (north Atlantic, Gulf of Mexico, Caribbean Sea) using the available observational data. In doing this, they took into account the disastrous after-effects of TCs (in 1995, material damage in the USA due to TCs amounted to about US$5 billion).

The hurricane season usually embraces the period from July to November. TCs are classified into three categories:

(1) tropical storms (TS), maximum stable (for 1 min) wind speed within the limits of 18–32 m/s being their indicator;
(2) hurricanes, wind speed not less than 33 m/s;
(3) heavy hurricanes (HH), wind speed not less than 50 m/s.

Landsea *et al.* (1999) process observation data on hurricanes that invaded the USA territory during the period 1950–90 (some of the data refer to the entire century). Coastal ground-level stations and meteorological survey airplanes obtained these data.

Statistical data on hurricanes (their inter-annual trends and long-term variability) and losses caused by them are also processed by Landsea *et al.* (1999). It was shown that long-term variability and only weak linear trends are typical for the region under consideration. For example, it was discovered that HHs occurred very often in the period 1940–60 but became considerably more rare in the period from 1970 up to beginning of 1990. In 1995 and 1996, a return (unexplained so far) to high HH activity took place.

Similar variations are typical for more prolonged hurricanes, as well. Earlier, they were more prolonged (about 25–40 days per year), but became less prolonged (10–25 days per year) in the last few decades. Variability of this kind was not homogenous over the entire North Atlantic Basin, however. For instance, increase in activity of subtropical hurricanes was observed northward of 23.5°N from 1970 up to the beginning of 1990, but similar, as regards magnitude, decrease in activity was observed southward of 23.5°N.

It is possible that statistics about hurricanes have been much affected by the considerably more complete information obtained in recent decades by means of geostationary satellites. One point is clearly established, however; inhabitants of the countries of the Caribbean area and the eastern coast of the USA suffered fewer severe hurricanes in the last 20 years than the previous 20.

Analysis of observational data for the entire 20th century, relating to the eastern coast of the USA and the Gulf of Mexico, has led to the conclusion that the relatively calm period of the last two decades is similar to conditions of the first 2.5 decades of the 20th century. Analysis of dependence on factors such as sea-level pressure in the Caribbean Sea and zonal wind at the level of 200 hPa, quasi-biannual variations in the stratosphere, ENSO phenomenon, precipitation in the west Sahel and SST in the northern Atlantic has revealed, in all cases, the existence of a certain correlation between frequency, intensity, and duration of hurricanes. For instance, ENSO dynamics can serve as an indicator for destruction caused by TCs. The major portion of long-term variability in hurricanes may be due to the effect of the so-called multi-year Atlantic mode – a specific spacial structure of empirical orthogonal functions (EOF), computed from SST observation data.

The existence of this correlation creates the basis for a statistical forecast of long-term hurricane dynamics. As for filtering out the anthropogenically-induced contribution to this variability, solution of this problem will be extremely difficult, for many decades ahead, because of high natural variability and inadequacy of numerical models.

Figure 3.2. Meteorological satellite image of a series of hurricanes (tropical cyclones) over the Atlantic Ocean moving in the direction of the Gulf of Mexico. (One of a group of hurricanes that battered the coasts of the USA between June to November 1995 – the second highest hurricane activity in the history of the country) (Hewitt, 1997).

As Saunders and Harris (1997) note, after earthquakes, TCs are the next biggest threat to people's lives and property. In the USA, the total damage caused by hurricanes in the last 50 years reached, on average, US$2 billion per year. In this context, are of great importance (data on year-to-year variability in the number of hurricanes illustrated by the anomalous season of 1995). Figure 3.2 illustrates a series of TCs over the Pacific Ocean.

Since 1886, when it was decided to keep records of hurricanes in the Atlantic Ocean region on a regular basis, only twice did the number of observed tropical hurricanes and tornadoes exceed the number of these phenomena in 1995 (Table 3.2). In the period 1956–95, as regards activity, all indices given in the table for 1995 exceeded average conditions by approximately 3σ and more ($\sigma =$ mean square deviation) for the last 50 years.

Note that the criterion for a tropical storm (hurricane) is a wind speed of 17 m/s (32 m/s) for 1 min. TCs are defined as 24 hours of tropical storm (hurricane), broken down into defined by four 6-hour intervals, during every-one of which the tropical storm (hurricane) wind speed continues.

Only once in the 50 years under consideration were the indices for 1995 exceeded (for every one of the indices): 21 tropical storms (1933), 12 hurricanes (1969), 136 days with tropical storms, and 72 days with hurricanes (1993). The maximum activity of tropical storms and hurricanes in the north Atlantic falls in

Table 3.2. Activity of tropical hurricanes and tornadoes in the north Atlantic in 1995, compared to three previous years and average conditions for 50 years (1946–95).

	1995	1994	1993	1992	1946–95, average 6σ
Tropical storms	19	7	8	6	9.4–63.2
Hurricanes	11	3	4	4	5.8–62.3
Number of days with:					
Tropical storms	121	28	30	38	47.8–622.2
Hurricanes	62	7	10	16	24.0–614.3

August–September–October (ASO), when (as statistical data show) 85% of storms occur.

In 1995, in the ASO period, 84% of storms occurred. The specific feature of that year, however, was the major region of storm formation (viz. 10–20°N, 20–60°W), where 93% of 'anomalous storms' formed (i.e. the number of storms in excess of their average number for the last 50 years) and in the ASO period 91% of all storms formed. This region embraces almost 60% of the latitudinal belt of 10–20°N, extending from the western coast of Africa to Central America. We designate this belt as the 'region of development' of TCs in the north Atlantic.

The preceding studies revealed five factors of interannual variability in TCs in the ASO period. They are the following:

(1) vertical wind shear in the troposphere (SHEAR);
(2) ENSO;
(3) quasi-biannual oscillations (QBO) in the stratosphere;
(4) monsoon precipitation in the western part of the Sahel;
(5) atmospheric pressure anomaly in Caribbean basin (SLPA).

Most likely, the ENSO-induced wind shear effect is the most significant. Earlier, it was found that small values of the latter (less than 5 m/s between altitudes of 1 and 12 km) are favourable for TC development, while high wind shear values retard development of a vertically coherent vortex. Regression analysis performed, in this context, for the role of SST growth in the 'region of development' has led to the conclusion on the dominating role of positive SST anomaly as the reason for the high frequency of storms. It turns out that SST in the 'region of development' was the highest since 1865.

Table 3.3 illustrates regression analysis data that clearly reflect the dominating contribution made by SST anomalies. Here $\Delta(N\text{TS})$ and $\Delta(N\text{H})$ are changes (%) in the number of tropical storms (NTS) and hurricanes (NH) formed within the 'region of development' in 1995 that can be explained on the basis of linear regression and the anomaly of 1995.

As can be seen from Table 3.3, all the factors under consideration (except for the Sahel) had a beneficial effect on formation of anomalous activity of storms in 1995,

Table 3.3. Influence of different factors caused by interannual variability in frequency of tropical cyclones in the 'region of development' in 1995.

Influencing factor	ASO, average $\pm\sigma$	1995 Anomaly (σ) 1979–95	$\Delta(NTS)$ (%)	$\Delta(NH)$ (%)
SST	$27.45 \pm 029°C$	2.00	59 ± 46	75 ± 58
SHEAR	11.93 ± 40 m/s	-1.48	24 ± 38	32 ± 48
		1951–95		
SST	$27.38 \pm 127°C$	2.41	41 ± 26	61 ± 34
ENSO	$24.76 \pm 084°C$	-0.79	8 ± 13	9 ± 17
QBO	$0.00 \pm 1{,}395$ m/s	0.79	12 ± 15	17 ± 19
Sahel	0.00 ± 068	-0.12	-2 ± 10	-3 ± 14
SLPA	0.00 ± 038 hPa	-2.18	-25 ± 2.5	34 ± 33

but the dominant role of OST is without doubt. It is important to realize, however, that, in all cases, a high level of errors is involved. TCs form within the boundaries of the 'region of development' under the influence of easterly waves coming from North Africa. These waves are troughs in the lower troposphere characterized by intensive convection and move westward through the Atlantic Ocean in the 10–20°S band of the trade winds. Annually, about 60 waves of this kind cross this region between May and November (every 3–4 days, on average).

The number of easterly waves that transformed into TCs within the limits of the 'region of development' in 1995 was higher than in other years. Latent and sensible heat transport from the ocean are the sources of energy for TCs. Within the range 26–29°C, development of TCs is very sensitive to even small changes in SST, and most probably depends not only on the local SST but also on SST in remote regions (through heat and moisture advection by trade winds).

Wilson (1997) made a remark regarding an article by Landsea *et al.* (*Geophys. Rev. Lett.*, 1996, **23**, 1657–70) where a conclusion was made on decline in frequency of heavy hurricanes in the period 1944–95 from observational data for a series of TCs in the Atlantic Ocean, including heavy hurricanes (of 3rd, 4th, and 5th categories on Saffir–Simpson's scale) and weaker cyclones (hurricanes of the 1st and 2nd categories, tropical and subtropical storms). This decline was interpreted as evidence of the existence of a long-term (5-decade) linear decline trend in frequency of hurricanes.

This remark consists in statistical substantiation of the fact that, as stated earlier, a 'sudden decline' was observed, but not a linear trend. Consideration of all TCs for 50 years (their number was 507), weaker cyclones (394), and HHs (113) showed that, only in the latter case, can data characterizing the decline trend be considered as statistically significant.

Breaking an entire 50-year interval down to two subintervals (1944–69 and 1975–95) clearly showed that this decline should be interpreted not as a linear

Table 3.4. Statistical data on Atlantic tropical cyclones (hurricanes) of three categories in 1944–64 and 1965–95. Here: y = average frequency of cyclones (per year); S_y = mean-square deviation; and n = total number of cyclones.

	All cyclones		Weaker cyclones		Heavy hurricanes	
	1944–69	1970–95	1944–64	1970–95	1944–69	1970–95
y	9,922.87	9.58	7,232.37	7.92	2.69	1.65
S_y	258	3.16	188	2.50	1.69	1.13
n		249		206	70	43

trend but as a step-like decrease in the frequency of IIIIs from one interval to another. This is evidenced by the data in Table 3.4.

The frequency of occurrence histogram, constructed for the entire period 1944–95, revealed that for all cyclones the histogram is bimodal. It reflects the superposition of a broad frequency distribution of the number of weaker cyclones and distribution of HHs that were characterized by a clear-cut maximum. Table 3.4 shows that the average frequency of HHs decreased from 2.7 per year (at MSD of 1.7) in the period 1944–69 to 1.7 per year (at MSD of 1.1) in the period 1970–95.

Although an undeniable decline in the number of HHs was observed within each of the intervals, no statistically significant trend was revealed. This casts some doubt about the existence of a linear trend in the data over the 50 years.

An attempt to explain the sudden drop in frequency of occurrence of HHs in 1965 came up with no satisfactory answers, as was the case for 1995 when a sharp increase in the number of TCs (including HHs and weaker cyclones) took place.

3.1.3.2 North and Central Europe

Heino *et al.* (1999) analyse meteorological observation data in northern (Finland, Norway, Sweden) and central (Czech Republic, Germany, Poland, Switzerland) Europe. His aim was to reveal the characteristic features of space and time variability in anomalous manifestations of the climate, paying special attention to the 'tails' in frequency distributions of occurrence of different phenomena. The results refer mainly to the period 1901–95. They show that in that period the following was observed: the increase in minimum SAT was slightly stronger than the growth in maximum SAT that caused decay in the daily amplitude of SAT; decline in the number of days with slight frosts; absence of distinct changes in extreme daily precipitation totals; absence of changes in the number of days with precipitation of more than 10 mm; absence of a long-term trend toward intensification of high winds; decline in frequency of thunderstorms and hail (especially regarding the Czech Republic in last 50 years).

The high daily frequency of 'extraordinary' precipitation was typical for the climate of northern countries in 1990 and after 1980. Consideration given to proxy palaeoclimatic data for Switzerland revealed the existence of profound inter-decadal variation and a unique climate (in the context of palaeoclimatic variation) in the period 1986–95. Important future aims are to augment the databases to be analysed (using not just SAT and precipitation data), and to study numerical modelling potentials that simulate climate and its extreme variability.

3.1.3.3 Russian Federation

Analysis of the global database on climate for the last century, contained in the IPCC-1996 report, revealed the complex and inhomogeneous space and time structure of changes in global climate, so much so that it is unclear whether climate became more variable and (or) extreme in the 20th century.

From data on mean monthly SAT values and precipitation from 455 meteorological stations in the former USSR from 1886 to the present time, Gruza *et al.* (1999) analyse the regional variability in climate on the scale of the entire Russian Federation and especially its western part – 'Russian territory free from permafrost' (RTFPF). To characterize climate variability, two aggregated indices are also used: the extreme climate index (CEI-3) and index of the greenhouse effect on climate (GCRI-3) computed from SAT, precipitation, and drought index data. Values of the Bagrov climate anomaly index (CAI) are also computed.

The results testify that in the 20th century, as a whole, SAT increased over the RTFPF (0.88°C/100 years), which manifested itself most distinctly in winter (1.3°C/100 years), frequency of droughts increased as well, but precipitation decreased. Changes of this kind are characterized, however, by nonhomogeneity of spatial distribution (climate warming manifested itself most significantly in the latitude band 50–55°N). Corresponding linear trends (period 1901–95 was considered) describe only a small part of variability in climatic parameters. Variations on the scale of decades turned out to be the largest.

Relative values of a CEI index, computed for the territory with extreme anomalies of climate (with frequency $\leqslant 10\%$), are specifically dependent on annual mean SAT, precipitation total, and frequency of droughts. CEI values increase with increase in SAT, decrease with increase in precipitation, and only slightly increase with increase in frequency of droughts. In the case of an aggregated CEI index, computed using three parameters, only a slight increase is observed (explaining only 1% of variability), also manifesting itself in the case of the GCRI-3 index. It shows to a certain extent that numerical modelling data on 'greenhouse'-induced changes in climate correspond to observational data (especially for the last 50 years).

Preliminary analysis of the data from 41 stations has led to a conclusion that, at the majority of stations, decline in the number of days with extraordinarily low SAT values were observed both in winter and in summer. The number of days with intensive precipitation on the ETR (European Territory of Russia) in summer increased. Observed climatic changes are too weak to reject, with confidence, the hypothesis that they represent natural changes in climatic parameters in the context of the stationary climate concept. However, reality and the importance of observed changes in climate are beyond question. Further more-detailed analysis of observational data is required.

3.1.3.4 China

The climate of China is subject to south-Asian monsoons (Zhai *et al.*, 1999). In the cold (winter) period (October–March), the climate is mainly cold and dry, and major extreme changes in climate are manifested as cold waves, high winds, and low temperature. In summer, the precipitation belt moves gradually from the south to the north. High variability of precipitation in summer is the most specific feature of the south-Asian monsoon climate.

Analysis of SAT and precipitation for the period 1951–95 (data from 369 meteorological stations) has revealed considerable growth in mean minimum SAT in China, particularly in the last 40 years. It manifested itself most considerably in winter in the north of China. At the same time, a decrease in cold wave activity throughout the country and a considerable decline in the number of cold days in the north of China took place.

Although no statistically important trend in mean maximum SAT was observed for the country as a whole, a distinct decrease in mean maximum temperature in summer was noticed in the east of China, where the number of hot days decreased. While the trends of maximum SAT were mainly negative, contrary regularity was observed in the case of minimum temperature.

Throughout China (data from 361 meteorological stations) a statistically important decrease (as compared to mean climatic values) was observed in the number of days with high positive anomalies of precipitation totals. On the contrary, a trend toward growth was detected in the case of positive anomalies in precipitation intensity (ratio of precipitation and number of days with precipitation) for the last 45 years. A distinct trend toward precipitation intensity growth was observed in the west of north-western China, but in the north of the country the climate was getting drier (considerable decrease in its share of territory with excess annual precipitation total). An increase in the number of very intensive precipitation events was also noticed.

3.1.3.5 Australia and New Zealand

From observational data in Australia and New Zealand during the 20th century, Plummer *et al.* (1999) analyse such extreme changes in climate as floods, droughts, heat waves, frosts, and high winds that exert considerable (sometimes disastrous) effects on people's lives and economic activity.

Data for Australia showed that:

(1) those areas of Australia where extreme humidification (aridity) conditions occur, increased slightly (decreased) since 1910;
(2) within the same period, certain growth in precipitation intensity was observed in some regions;
(3) after 1961, frequency of occurrence of warm days and nights increased and that of extremely cold days and nights decreased;
(4) since the end of 1960, a growth in total number of TCs was observed, which could be explained mainly by the influence of ENSO, while the number of intensive TCs reduced;
(5) since the middle of 1960, the activity of extratropical cyclones over the major part of midlatitudinal waters southward from Australia became weaker, but, with further increase in latitude (in south-west direction), activity of this kind became stronger.

From observational data in New Zealand, it follows that:

(1) due to climate warming in the period 1941–90 the additional number of days when temperature dropped below zero decreased by approximately 10, and the additional number of days when temperature was higher than 30°C increased by 2 days;
(2) in the period 1951–80, the frequency of moderate and severe droughts decreased (compared to the preceding 30-year period);
(3) changes in temperature extremes and frequency of droughts were the result of changes in atmospheric circulation in the region.

3.1.4 Air temperature and precipitation anomalies

3.1.4.1 Surface air temperature (SAT)

Extreme changes in SAT mainly manifest themselves as positive and negative anomalies (usually, hot summers and cold winters) and give rise to considerable socio-economic impacts (Evangelista *et al.*, 1995). In the UK, for instance, the cold winter of 1947 and dry hot summer of 1976 caused serious economic damage, and in Scotland, at the end of December 1995, a cold 'wave' caused a lot of damage to buildings.

According to numerical modelling data, amplification of SAT extreme values must take place under 'global warming' conditions. Taken together, this testifies that adequate SAT monitoring (as well as monitoring of the phenomena influencing the SAT field such as ENSO and the north Atlantic Oscillation) and substantiation of representative indices of extreme changes in SAT are required:

- BI-1 – the share (diurnal) of intervals with minimum, medium and maximum SAT. Estimates of this kind may be applied to longer periods of time (month, season, year).

- BI-2 – it is recommended to use shares of monthly, seasonal, and annual SAT anomalies.
- BI-3 – the amplitude of the diurnal course of SAT on dry land.
- BI-4 – the 'interval to interval difference', determined as the difference in SAT between two adjacent intervals of time. Note that BI-4, as well as the other indices, are to be mapped.
- BI-5 – characterizes frosty conditions. It is determined as a fraction of time (in the course of a given month) when minimum SAT is lower than 0°C or lower than other threshold value in the interval 2–5°C. Otherwise, it is best to assess the severity of frosts by the length of frost periods from the dates of the first and last SAT minimum (below 0°C).
- BI-6 – introduced to characterize duration of heat (HWDI) or cold 'waves' (CWDI). Corresponding threshold SAT values must be locally-specific.
- BI-7 – heat stress index. There are several options for this index, which is computed not only for SAT data, but also data on air humidity and wind speed.

It is recommended to use the following indices as key ones (KI) (Folland *et al.*, 1999):

- KI-1 – based on BI-2 and characterizing mean seasonal and annual SAT maximums;
- KI-2 – determined by the change in annual mean-global amplitude of the current diurnal temperature course (for the last 20 years) as compared to the period 1951–70, when global warming had not yet manifested itself.
- KI-3 – identical to BI-5.

Easterling *et al.* (1999) considered observational data for Argentina, Mexico, and the USA that can be used for analysis of extreme changes in SAT and precipitation (considerable attention was paid to assessments of the quality and homogeneity of observational data). Two indices were applied to observational data on diurnal, monthly, seasonal, and annual SAT values and precipitation: (1) extreme changes in climate for Canada and (2) analysis of dangerous SAT drops in Florida and their effect on citrus plants.

Both the GCRI and the CEI were used. The second of these indices takes account of the following factors:

(1) Maximum SAT that is considerably higher than normal;
(2) minimum SAT that is lower or higher than normal;
(3) the doubled relative territory of the USA where an extraordinarily large contribution to total precipitation was made by extremely high (more than 50.8 mm) diurnal precipitation;

(4) situations when the number of days with precipitation or without precipitation was higher than normal.

High variability in CEI (since 1910) on a decadal scale is typical of American conditions. Since 1970, CEI values considerably exceeded mean values, testifying that areas of the USA where extreme changes of climate were observed increased by approximately 1.5%. It was found that there are at least two ways to assess susceptibility to extreme changes in climate. They estimate the frequency of events when certain threshold values of meteorological parameters are exceeded during a fixed time interval or low probability of situations for a limited time interval.

Easterling *et al.* (1999) consider examples of variability in the parameters that characterize extreme variability most adequately (i.e. maximum SAT in winter and summer, early and late snow cover melting, most intensive rain and hail, synoptic-scale high wind, severe droughts). Data on the frequency and intensity of cold waves in the USA are discussed. The necessity to have long and homogenous observational data series is emphasized.

The usual widely available databases on climate contain information on mean monthly values or the totals, and, in recent years, these archives have been augmented by data on mean monthly minimum and maximum values SAT. Information (more limited, however) on daily values was also obtained. Observational series of this kind in central England embrace the period from 1772 up to the present time.

Using available data, Jones *et al.* (1999) analyse the following:

(1) mean monthly and daily SAT extremes and atmospheric pressure from the data for central England (CET);
(2) global database on mean monthly SST for the grid of 5° latitude by 35° longitude.

Analysis of the 225-year CET data series (mean monthly SAT values are available since 1659) did not reveal considerable growth in SAT in very hot days at the present time, compared to earlier data, but considerable decrease in frequency of very cold days was detected. Growth of SAT in the last two decades was mainly caused by decline in the number of cold days.

According to CET data on frequency of days with unusual SAT values, during the last two centuries as a whole, a gradual decrease in frequency of unusual SAT values was observed. For the period 1881–1997 in Great Britain, there were no distinct changes in the number of annual heavy storms, either. Under certain atmospheric circulation conditions and pressure field structure in the area of the British Isles, long-term variations took place. Changes in monthly global temperature fields, since 1951, show that the growth in SAT in the last 10–15 years was accompanied by decrease (increase) in areas with extremely low (high) temperature (i.e. change in spatial structure of the global temperature field).

3.1.4.2 Precipitation

Intensive precipitation, characterized by high space and time variability, every year is a threat to people's lives and the economy. Inadequacy of available observational systems and data, and absence of a well-developed and generally approved system of hazard precipitation indicators are closely connected with these problems.

In this connection, Nicholls *et al.* (1999) substantiate recommendations on introducing a sufficiently universal minimum number of indicators to characterize extreme precipitation. Three indicators are of primary importance. They are the following:

(1) variability in frequency of days when precipitation falls 95% of the time or more;
(2) share of such precipitation in seasonal and annual precipitation totals;
(3) share of extreme precipitation (estimated as 5% of extreme precipitation) that characterizes periods of drought and overhumidification for a country or a region.

To estimate these indicators, it is necessary to use both routine direct observational data and data obtained by means of satellite remote sensing. Since observational data are fragmentary, information on daily precipitation totals should be used as the basis. The number of days without precipitation, with hail and snow, diurnal mean precipitation, maximum duration of drought periods, and so on may serve as additional indicators.

Analysis of precipitation observational data for different regions of the world, performed by Groisman *et al.* (1999), reveals various trends. In many countries (e.g., Russia, Norway, Sweden, Canada), trends toward increase of precipitation were observed. They were more distinct in winter (about 10–15% for 100 years) than in summer (about 5% for 100 years), but comparable in magnitude because of specific features in the annual course of precipitation in extratropical regions of the Northern Hemisphere. In Poland and Austria, a century-long trend in precipitation took place mainly during the warm part of year. In the major part of Norway, the annual sum of precipitation increased by 8–14% during 100 years, but intensification of precipitation in summer was less significant (5–10%) and was observed mainly in the northern part of the country.

Of special interest are the data on precipitation variation in North America, Australia, and Japan. They show a growth in extreme and very heavy precipitation. Groisman *et al.* (1999) suggest a very simple statistical model for variability in daily precipitation, based on parametrization of variability by means of gamma-distribution (GD) applied to the summer period of the year (June–August in the Northern Hemisphere and December–February in the Southern Hemisphere). Using GD, precipitation total P may be presented as follows:

$$P(\eta, \lambda, x) = \text{const}\,(\lambda, x)x^{\eta-1}\exp(-\lambda x)$$

and soil drying out, gave rise to development of dust storms and soil deflation over a huge area of the steppes and forest steppes of the country. Suffice is to say that 10 million ha of only sand particles, unfit for agricultural activities, were the net result.

In Russia, much damage is caused to agriculture by dust storms not only in summer but in winter as well, especially in winters without snow. In 1960, so-called black storms developed in the south of the European part of the former USSR, covering more than 1 million km^2. Dried-out soils were easily blown and transported together with snow (Nalivkin, 1969).

During the 1950s–1960s, a 'dust bowl' formed over northern Kazakhstan, similar to the American one. A huge dust storm developed over the former virgin, uncultivated lands due to careless development and, in particular, excessive ploughing by machinery that was too heavy for the soil structure.

Between 1954 and 1968, the area of ploughed land here increased from 6.6 to 22.0 million ha. As a result of erosion (deflation) of the soil, the loss of humus in this 'dust bowl' represented 10–25 to 30–35% for 40 years (Belgibayev, 1993). This dust storm (anthropogenically-caused) led to devastation of the vast granary of the country, a regional-scale catastrophe.

Dust storms cause damage not only to agriculture, but to transportation as well, impeding movement. In the USA, where dust storms are customary in western states, between the cities of Phoenix and Tucson there are special traffic signs installed on highways, warning against possible emergencies as a result of dust storms.

There are records of crashes of airplanes caught in dust storms. For example, in 1975 not far from the city of Dakar (West Africa), a plane was caught in a dust storm and crashed. Onboard this plane there was a delegation from Mauritania headed by the Prime Minister. The plane crashed in the sea at a site that, due to frequent export of dust from deserts, acquired the name 'sea of gloom' among sailors. A similar catastrophe took place in Mauritania in 1995 (94 people lost their lives).

Catastrophic events provoked by dust storms quite often happen in big cities. A dust hurricane that carried about 170,000 tons of dust in December 1985 passed over Ashkhabad (the capital of Turkmenistan). In darkness and in violent turbulence, roofs were torn away, overhead wires collapsed, trees were uprooted, traffic was brought to a standstill because of sand and dustflows, and electric power and gas supples were cut off. In June 1986, a similar phenomenon occurred in Delhi (India). Dust storm with atmospheric gusts up to 110 km/hour caused damage to buildings, plantations, gave rise to fires, and several blocks lost their power supply.

There are other consequences of dust storms. For example, diseases can break out caused by pathogenic microbes transported large distances, or suffocation due to dust. In 1895, 20% of cattle in eastern Colorado (USA) suffocated.

Consideration given to dust storms as catastrophes or disasters for people must take into account their coupled nature. As already pointed out, in arid and subarid regions, dust storms are one of the interdependent components of the triad 'droughts–desertification–dust storms'.

The catastrophic consequences of dust storms and droughts, and the desertification that accompanies them, need different protective measures for their prevention and forecast. Dust storm prediction is often based on the forecast of meteorological conditions under which they develop. For instance, correct forecast of long-term dry weather makes it possible to predict possible intensification of dust storms, as well.

Study into the cyclic character of hydrothermal conditions (i.e. alternation of dry and humid periods) is promising for prediction of dust storms. In historical times, periods of dust storm intensification and attenuation are clearly seen. The same is true for wind–erosion activity. In particular, periods of drought and dust storm development relate to 11- and 22-year cycles of solar activity.

There are some achievements in the development of forecasts for catastrophic dust storms. For instance, specialists from Alma-Ata (Kazakhstan) developed forecasting methods for local dust storms in Kazakhstan and Central Asia based on analysis of the development of atmospheric fronts, zones of storms, and other phenomena. Application of satellite observation results are essential in the study and forecast of regional and planetary dustiness of the atmosphere (which in its turn leads to climate cooling, crop failure, starvation), environmental dynamics, and large-scale dust clouds, revealing dust storm centres and specific features of their behaviour.

3.2.2 Satellite observations of dust storms

3.2.2.1 Types of dust contamination of the atmosphere and their satellite images

Dust contamination of the atmosphere of different types, which may be classified according to their origin (depending on dust source), can be monitored from space. Dust rises into the air above the surface, carrying loess-type deposits and sands, over saline soils. At the same time, although dust outbreaks from various sources differ in their composition, it is rather difficult to find sharp distinctions between them in satellite images.

If dust pollution is considered as a whole, it is identified by a complex of brightness and structural indications that make it possible to distinguish them from other natural (or anthropogenic) formations. They differ from normal clouds mainly by their reduced brightness.

When dust pollution is located over cloud cover, the overall albedo of normal and dust clouds is lower than that of normal (water) clouds, which can be seen in the images. Moreover, the dust flow's dust mantle tones down the usually sharp contours of clouds. Like smoke, dust pollution in many cases can hardly be distinguished by their brightness (tone) from cirrus. Regarding brightness, dust and smoke atmospheric pollution are very similar. However, the latter are more widespread in wetter air masses and transfer to water clouds with higher albedo more often.

Dust formations surveyed in narrow spectral ranges are, like smoke, the most bright in the range 0.5–0.6 μm. However, in the presence of thick haze, dust pollution is more clearly displayed in the interval 0.6–0.7 μm. Moreover, dust atmospheric pollution surveyed over water in the range 0.5–0.6 μm cannot be distinguished from water pollution, because of the dredging of rivers. Simultaneous surveying in two intervals (e.g. 0.5–0.6 and 0.8–1.1 μm) makes it possible to differentiate them. In the images taken in the second range (in the near-infrared range) water pollution is not reflected, and only atmospheric pollution is seen. On many occasions, comparison of the images obtained in these intervals made it possible to distinguish dust storms from pollution of water area of Aral Sea as a result of dredging (Kondratyev *et al.*, 1979, 1983).

Infrared images (in the range 10.5–12 μm) allow identification of dust pollution thanks to temperature contrast between a cloud of dust and the background surface. Dust storms being colder than the surface of the Earth, are lighter in the images. Cloudiness (to be more precise, its upper surface) in images of this kind, on the contrary, is colder and consequently lighter. During the day, dust formation is more clearly displayed over dry land than over the sea because temperature contrasts in the pollution–land system are higher than those of the pollution–sea system (Kondratyev *et al.*, 1983a,b). It is especially important that this effect takes place while surveying the dust storm at night, when it is extremely difficult to trace. Contrasts in the dust cloud – dry land system are smaller at night than during the day because dry land is considerably cooled at night. As a result of application of these images, it is possible to trace the development of dust storms in the West Sahara at night. Dry smoke pollution in the atmosphere does not produce such an effect. It is not only cloudiness and smoke pollution, which can be close in brightness to dust storm images, that make it difficult to decode. Dust storms are spread in arid and subarid regions where the underlying surface is characterized by rather high albedo. In such cases, in particular when a dust storm is developing over extensive saline soils or lightly coloured sands free of plant cover, it is very difficult to distinguish it from the background. It is possible, though, to trace the dust flow if it is sufficiently extensive and traverses underlying surfaces of other kinds with lower albedo. For instance, the intense dust outbreak, observed from space in the area of Mesopotamia, passed for some distance over lightly coloured saline soils and a watershed valley with scarce vegetation (Grigoryev and Lipatov, 1974). Dust streams can be traced where they pass over coloured dark surfaces (i.e. over individual lakes and oases in river valleys).

Similarly, clouds of dust over the Sahara Desert can be seen in satellite images. Against the lightly coloured background of the desert, dark-coloured elevations, isolated hills, can be seen. Dust streams are clearly seen over the latter.

It is especially useful, for this purpose, to compare satellite images of one and the same region obtained in the course of development of dust outbreaks and in the periods when the atmosphere is clean and undisturbed.

Like smoke, dust pollution of the atmosphere differs in a number of cases from normal clouds by structural indicators. Images of dust outbreaks are characterized by longer and less pronounced contours at boundaries. Dust storms of a flow type are quite often characterized by a multi-stream flow that determines band-type structure. Dust flows of this multi-stream type cannot be mistaken for cloud formations.

Large, structure-free, 'shapeless' clouds of dust cannot be distinguished from

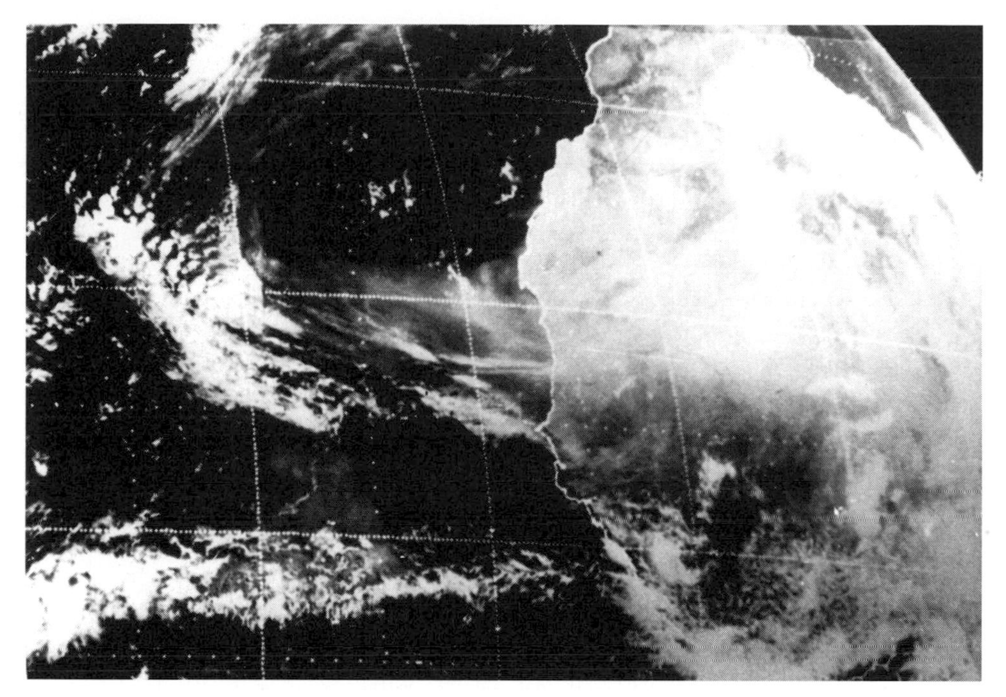

Figure 3.3. Dust outbreaks over the Atlantic as seen in TV images in the visible spectrum range, from the geostationary GMS-1 satellite on 30 July 1974 at 12:30 GMT (Kondratyev *et al.*, 1983b).

similar dust formations. Multi-stream flows of dust have no analogues (as regards their images) among smoke pollution and can easily be distinguished from them.

Although dust pollution in the atmosphere of any size is always oriented, smoke pollution quite often (in the case of weak air flows) does not have a distinct orientation. This feature is one of the most important indicators for differentiating these pollutions from one another.

Often, dust pollutant transport is accentuated by the orientation of normal clouds when both natural formations are involved in motion by one and the same air flow. This phenomenon is often observed over the Atlantic where dust pollutants and clouds combine in a vortex-type motion caused by development of the Azores anticyclone (Figure 3.3).

When decoding dust outbreaks and smoke pollution in the atmosphere, the geographical location of either of them is usually taken into consideration. The absence of sources of smoke pollution in the atmosphere (with rare exception) in African deserts makes it possible to decode dust outbreaks with confidence. Consideration of location can only be applied to decoding 'local' atmospheric pollution that is observed within the boundaries of one and the same natural zone and does not migrate far beyond its boundaries.

Dust pollution in the atmosphere, crossing natural zones, arrives in regions where air mass characteristics may differ sharply. Specks of dust become nuclei of water vapour condensation, and dust formations partly transform into water clouds. Clouds of this type, from which yellow–brown dust sell along with rain, were detected, for instance, in meteorological satellite images over Italy in April 1977 (Prodi *et al.*, 1979).

A similar phenomenon was also observed in the course of saline dust transport from the coast of the Caspian Sea to Povolzhje (Volga River area) (Grigoryev and Lipatov, 1982). Dust outbreaks proper were seen here in images over the Caspian Sea and Low Povolzhje. In the images of Middle Povolzhje, only normal clouds could be seen, from which dust along with rain fell.

Only in a few cases have dust clouds of different origin shown distinctive-enough features to identify them in the images. For example, colouring of dust formation is important. Pink shades, which are often typical of cloudiness images over Africa in colour pictures (Nalivkin, 1969), are evidence that iron-accumulated reddish dust, abundant in a number of regions of the West Sahara and not common for other regions of the Earth, rise to the atmosphere. Dust clouds developing in the course of winter and spring storms in the Ukraine and carrying coloured particles of black humus are seen in the images as a dark grey formation (Semakin and Nazirov, 1977). Sometimes, dust clouds of different genesis differ in their structural features. For instance, dust formations over loess valleys or valleys built up with loess-type deposits are characterized by almost shapeless contours and the absence of sharp boundaries. This is due to wide spread, easily blown-away deposits that begin to raise dust instantaneously over a large area. Very different are the images of narrow out stretched dust outbreaks that originate from small depressions with saline soil or occur over isolated areas with very saline soil. But, as a whole, structural peculiarities are mainly evidence of the nature of circulation flows, and, secondarily, of the nature of the region of occurrence of the dust storm, and they are only indirect evidence of the type of transported deposit.

3.2.2.2 Transport of dust pollutants

In the pre-satellite period of the study of Earth data on intense dust outbreaks, long-range transport of dust clouds, dust transport trajectories, and, finally, dust deposition were very scant. And this is not surprising. To trace the motion of a dust cloud to dozens or even several hundreds of kilometres from the centre of its origin by means of on-land observations was impossible. Ship observations and, in the 20th century, aircraft observations could be equally episodic. It was especially difficult to trace the transport of dust clouds over difficult access, sparsely populated regions of the Earth (i.e. over deserts and oceans).

Nevertheless, unusual cases of dust depositions in the regions where no dust storms occur, have long been known (e.g. in forested areas, over water areas). It was equally difficult to judge the sizes of observed dust clouds.

Satellite observations mainly made it possible to obtain new data on the sizes of dust outbreaks. It was found that intense dust outbreaks can spread several dozens to several thousands of kilometres. Small (local) dust clouds several dozens kilometres long were also discovered by satellites, though infrequently. For example, in

September 1973, in the image from the *Skylab* orbital space station near the coast of Venezuela (Bodechtel and Gierloff-Emden, 1969), a dust flow was detected. It had formed above a sandy coast in a band 50-km wide and stretched out to sea for more than 70 km.

In January 1973, the *Landsat* satellite obtained images of six small dust outbreaks in the Mojave Desert (Nakata *et al.*, 1976). Dust clouds stretched 15 to 75 km.

Dust clouds several hundreds of kilometres long (regional) are more often traced by satellite.

Dust outbreaks in the area of the Aral Sea have been observed from satellite more than once in recent years up to 200–400-km long. Dust clouds of similar range were observed from the manned spacecraft *Gemini-12* in 1966 in the south of Iran and over the Gulf of Oman (Otterman, 1978): five bands 375–400-km long could be seen.

More considerable, up to 500–800-km long were dust outbreaks over Mesopotamia. Equally far (almost 900 km) moved the dust cloud that originated on 16 June 1971 over the north-eastern regions of Sudan, stretching toward the Red Sea.

Dust formations, stretching over subcontinental areas from 1,000 to 3,000 km, have been traced many times from satellites. For example, on 5 April 1969 a dust cloud formed over the Chinese deserts, stretching 1,200 km; it was detected by the *ESSA-9* satellite (Ing, 1972). An even larger cloud, detected on 24 February 1977 by the *NOAA-5* satellite, originated in the west of the USA. It extended from Albuquerque (New Mexico) 2,400 km eastward and went beyond the boundaries of the continent. Maximum width of the dust outbreak reached 1,600 km (Dust storm ..., 1997).

Finally, the formation of unique continental-scale dust outbreaks extending 3,000–5,000 km have been recorded. Such a dust outbreak was the one that stretched from the Sahara to the Atlantic on 30 July 1974. It was about 3,000 km long and its width near the coast was more than 400 km. It extended over the ocean to almost 1,500 km. Another dust cloud, also formed in the West Sahara, and detected on 4 July 1969 in images taken by *ATC-1* (Carlson and Prospero, 1972), stretched an even greater distance, this time it embraced the whole ocean. Its length was 4,300 km and its width was between 800 and 1,200 km.

One of the largest dust clouds, about 5,000 km long, formed over the Atlantic Ocean on 4 July 1974.

Depending on the length of dust outbreaks (from dozens to hundreds of kilometres), their area changes as well. It can vary from hundreds and thousands to hundred thousands and even millions of square kilometres. For example, the areas of the dust clouds over the Aral Sea (22 May 1975), Mesopotamia (17 July 1970), and the Atlantic (4 July 1969) amounted to about 25,000 km^2, 150,000 km^2 and 4,000,000 km^2, respectively. The largest outbreak of dust from Sahara (its area was about 7,000,000 km^2) was observed from the *GMS-1* satellite on 30 July 1974 (Kondratyev *et al.*, 1983b).

Satellite images are the most reliable and plausible and, in many cases, they are the only source of information on trajectories of dust cloud motion. For instance, it was established that dust storms, originated as a rule in the north-eastern coast of Aral while a cold front passes over this region, more often move over the sea. As the dust cloud moves toward the south-west, dust falls out in the Amu Darya Delta resulting in salinification of fertile soils.

Analysis of the trajectories of intense dust outbreaks observed in 1975–9 in Aral regions has shown the following. The main direction of dust cloud motion (in about 50–60% of events) was south-west, to oases in the Amu Darya Delta, and in 30–40% of events it was west-south-west to the plateau of Ustyurt. In other cases (less than 10%), dust was carried away toward the south or south-east.

The trajectory of dust cloud motion often passes over different natural zones, including those where dust storms do not occur. Such was the case, in particular, when a huge dust cloud moved. It originated on 24 February 1977 in valleys in the state of New Mexico. In the course of its eastward migration from the dry steppes where it originated, it crossed steppes and near-ocean damp forests in the south-east of the USA.

Dust cloud motion trajectories often run from continent to continent. Satellite observations made it clear that dust from deserts in North Africa is carried away by cyclones (over the Mediterranean Sea) to the European coast, in particular the shores of Italy (Grigoryev and Lipatov, 1974; Prodi *et al.*, 1979). From the Sudan, dust is carried by monsoons over the Red Sea to Asia (e.g. on 16 June 1971).

Huge masses of dust carried away by the trade winds from west Saharan deserts across the Atlantic to the shores of North, Central, and South America. In particular, on 4 July 1969, the *ATC-1* satellite recorded that a huge dust cloud, which originated in the Sahara, had reached Haiti. American scientists were the first to discover ultra-long-range dust transport of this kind. They investigated samples of reddish atmospheric dust that fell near Barbados. Judging from its composition, it could only have originated in the Sahara. Analysis of satellite images has proven the existence of such dust transport by large-scale anticyclone vortexes (Prospero and Nees, 1977).

However, dust clouds, originated in the Sahara and carried away to the Atlantic, can move northward as well towards Europe. Such was, for example, the trajectory of dust cloud motion on 4 July 1974. In this way, dust from the Sahara reaches Spain, and sometimes England. Another way of dust transport to Europe is from North Africa over the Mediterranean Sea.

In winter, trajectories of dust transport from Africa to the Atlantic and the zone of dust outbreak propagation shifts northward to 30°N, and in summer southward to 10°S. This is due to migration of pressure fields and their seasonal variations. In particular, in winter, subtropical high pressure area shifts southward, as does the dust transport zone.

Satellite observations identified the areas of atmospheric dust outbreak transport (Figure 3.4).

For Asia, it was found that most often dust clouds associated with north-west flows occur over the Mesopotamian lowland. Dust rises from land between the Tigris and Euphrates Rivers, then it is carried away in the direction of the Persian Gulf up to 600–700 km away. Sometimes, the dust flows deviate southward and penetrate

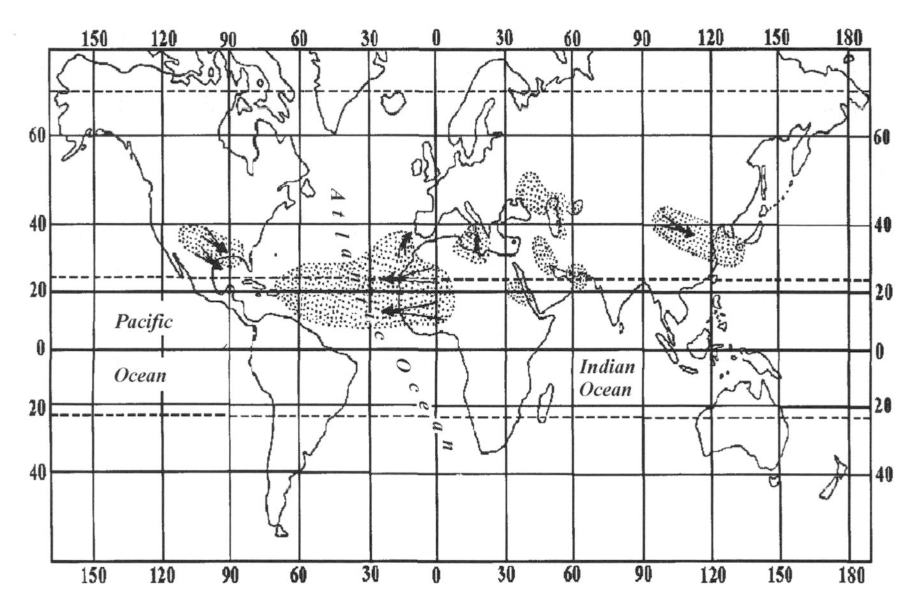

Figure 3.4. Dust cloud transport in the atmosphere of the Northern Hemisphere built up from satellite data (Kondratyev *et al.*, 1979).

deep into the Arabian Peninsula. In particular, on 12 August 1968, a dust cloud moved from the area between the two rivers (north-west of Baghdad) southward, a distance of about 800 km. The zone of intense dust outbreak diffusion in this region exceeds 500,000 km^2.

Dust-induced atmospheric turbidity over southern areas of Iran and Pakistan and the adjacent water areas of the Gulf of Oman and the Arabian Sea is very rare. In 1966, the manned space craft *Gemini* successfully took a picture of a dust outbreak in this region (Wobber, 1972). Judging from the photograph, dust was rather different in concentration in different parts of the zone of its diffusion, and covered about 200,000 km^2.

A very large atmospheric dust outbreak-induced turbidity zone is located over Mongolia and China, but satellite survey outline its boundaries are still scant. A report by Ing (1972) enables us to judge the dust diffusion range in this region. It discusses the large dust storm in the central part of China in April 1969, whose motion was traced for several days by satellite. Intense dust outbreaks were observed in the zone that extends about 1,200–2,400 km over China and, farther, over the Yellow Sea and East Siberian Sea up to Kyushu Island (800 km more). The area of the whole zone of migration of this dust cloud, according to our estimates, amounted to about 1,500,000 km^2. This figure does not account for dust diffused sideways from the above-mentioned zone. Its presence was recorded by a number of meteorological stations, as can be seen from the maps in Ing (1972), at a distance of several hundreds of kilometres northward, especially near the coast. This dust cloud was characterized by considerably lower concentration and, therefore, it was not seen in the pictures. For the same reason, dust diffusion over water has only been traced up to a distance of 800 km. At the same time, we know the dust cloud was transported farther eastward to the Pacific Ocean. It was observed by ships in the area 11 days after the dust storm occurred in the deserts of China at a distance of 2,500 km from the shore (Ing, 1972).

In Africa and over adjacent areas of water, satellite surveys have revealed the different regions of dust cloud transport. Dust clouds originating in East Sudan migrate towards the Red Sea. The area of this zone of atmospheric pollution can reach 200,000 km^2. A considerably larger zone of atmospheric dust pollution transport is situated in North Libya and area of water adjacent to it. Dust flows from the African shore reach the coast of Italy and Crete. The area of this zone exceeds 700 km^2.

The largest zone of dust cloud transport in the atmosphere starts in West Africa, and extends across the Atlantic Ocean to the shores of North, Central and South America. Dust motion in this zone followed by formation of the so-called Saharan aerosol layer in a vast subtropical belt was studied in detail in the course of the Atlantic Tropical Experiment (ATEP). It was carried out in 1971 within the framework of the Program of Investigations of Global Atmospheric Processes (PIGAP) (Kondratyev *et al.*, 1976). This program used both satellite images of the Earth (polar and geostationary orbit meteorological satellites) and observational data from land, ships and aircraft. The area of this zone is about 15 million km^2.

In North America, individual dust pollution events were observed over the Mojave Desert. During the dry year of 1977, heavy dust outbreaks were observed many times over central states of the USA. Dust clouds moved towards the Gulf of Mexico and the Atlantic Ocean. One of these dust clouds formed on 23 February 1977 according to *GMS* satellite survey and reached Bermuda on 24 February (Kondratyev *et al.*, 1983a, b). So, in North America as well, the zone of dust cloud transport is very large.

In the European part of Russia, Kazakhstan, and Middle Asia (within the limits of the former USSR), five regions of intense dust cloud transport are identified, three of which merge and overlap each other.

The smallest is located in Turkmenistan on the south-eastern coast of the Caspian Sea and in the adjacent water area where dust is transported. Heavy atmospheric pollution with dust is rare but, when it does happen, the area observed from space can reach 10,000 km^2. The largest atmospheric pollution zone (covering about 800,000 km^2) is located in the Ukraine; sometimes, it embraces northward regions as well. The zone of dust pollution in the area of the Salskaya Steppe in Russia practically merges with it. The area of both zones exceeds 1million km^2. A vast zone of dust flow transport in European Russia and Kazakhstan occupies territory extending, in the form of curved band, along the Volga, from the north of the Caspian Sea (in Kazakhstan) to Kazan in Russia. In its southern part, this zone (the water area of the Caspian) is identified by heavy dust clouds observed from satellites (Figure 3.5). Boundaries of the rest of the zone were established on the basis of land-based information on dust storm impacts on the landscape and by analysing

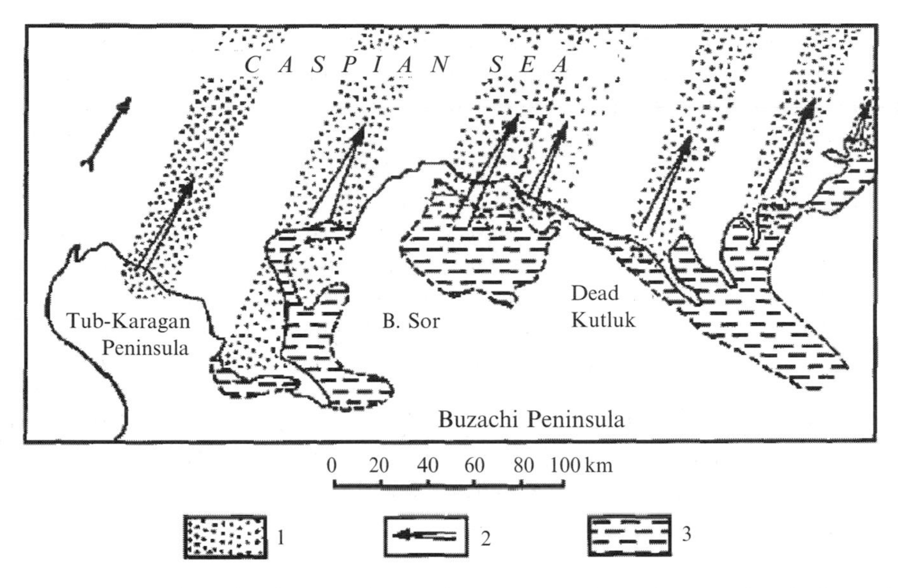

Figure 3.5. Heavy dust clouds from the north-east coast of the Caspian Sea, built up from analysis of photographs taken by the manned spacecraft *Soyuz-12* and TV images from *Meteor-Priroda-25* satellite, that recorded heavy dust outbreaks on 22 August 1976 (Kondratyev *et al.*, 1983a). 1 – dust flows (predominantly, saline) in the atmosphere; 2 – directions of dust cloud transport; 3 – dust outbreak locations (all from considerably saline soils).

satellite images. The dust comes from the north-western and (or) north-eastern coasts of the Caspian Sea. The area of the zone is about 400,000 km^2. In the west, the zone merges with the two preceding ones.

Heavy and frequent atmospheric pollution due to dust arises in Middle Asia in the area of the Aral Sea. Dust is transported over the major part of the sea and its environs reaching the Amu Darya Delta. Heavy dust outbreaks of Aral origin were observed from surveys deep in the Ustyurt Plateau, at a distance of several hundred kilometres from the Aral Sea. The area of this atmospheric pollution zone is not less than 150,000 km^2 (Figure 3.6).

3.2.2.3 Locations where large amounts of dust are blown away

The studies of locations (hearths) of heavy dust outbreaks only started recently, because of the necessity to study them in connection with investigations of meso- and macro-scale dust formations traced by satellite. Our notions about dust formations will be incomplete if we do not answer the following questions: Where do heavy dust outbreaks originate? Why and how do they break out?

Most hearths can be discovered from the photographs taken by the meteorological satellites that detected the dust outbreaks. Small-scale pictures of this kind can determine (by the initial part of the plume) the approximate location of

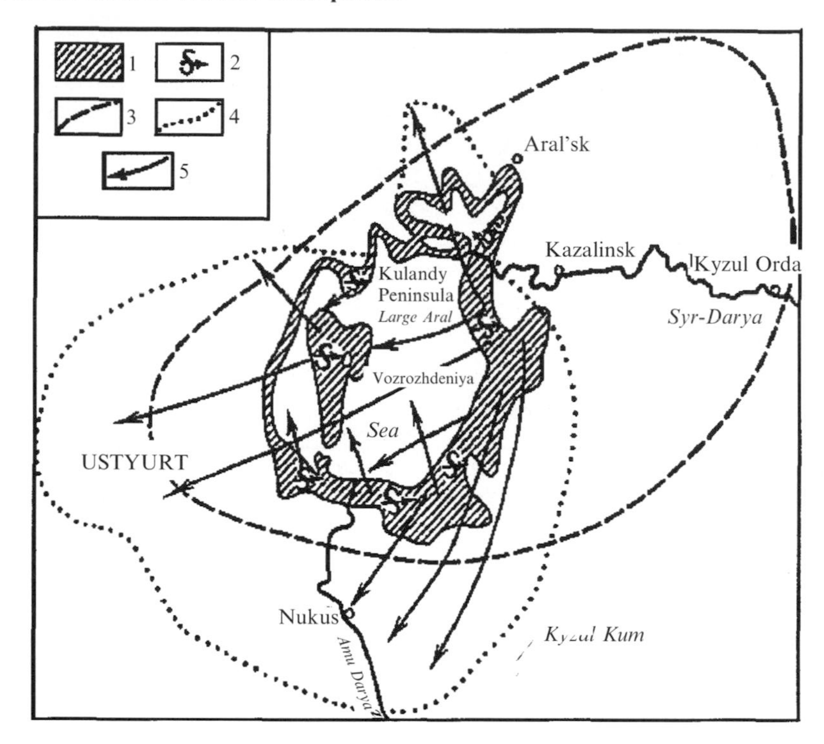

Figure 3.6. The centres and ranges of dust outbreaks in areas near the Aral Sea, builtup from satellite survey materials (Kondratyev *et al.*, 1983a).

a dust hearth. More detailed information on the composition of deposits (ground soils), fixing them with the help of vegetation, type of relief, specific features of anthropogenic development, and so on can be obtained under combined use of more detailed satellite images and data from known cartographic and literary sources.

The characteristics of some of the hearths can be found in a number of publications (Grigoryev and Zhogova, 1992; Grigoryev *et al.*, 1976a, b; Grigoryev and Lipatov, 1979; Grigoryev and Lipatov, 1982; Kondratyev *et al.*, 1983a). Depending on their location, all known hearths can be classified into three types:

(1) hearths located in lake and sea shores (e.g. in the Aral Sea);
(2) hearths located in depressions between mountains (e.g. in the south of Iran);
(3) hearths of flat interfluve locations (Salskaya Steppe).

Hearths of heavy dust outbreaks also differ by the type of loose deposits that develop at their surfaces.

Hearths of heavy dust outbreaks differ by the intensity of the outbreaks, and frequency and duration of their activity. Available information obtained from satellite and land-based observation data shows that, in the course of a dust storm, more than 1 million tons of matter can be transported. For example, the hearth of dust storms in the north-east coast of the Aral Sea has been known to emit about 1.7 million tons of matter (Grigoryev *et al.*, 1976a, b). About 5 million tons of matter were transported on 13 July 1970 from a larger hearth located in Salskaya Steppe (Grigoryev and Kondratyev, 1981a, b).

At the present time, approximate estimates are made for aggregate dust outbreaks per year. A hearth near the Aral Sea in 1990 emitted about 90 million tons of matter into the atmosphere (Grigoryev and Zhogova, 1992). From the hearths in the West Sahara about 100 million tons of dust are carried away during four summer months (Carlson, 1975). Modern data show that West Sahara is the largest area of dust outbreaks on Earth.

Dust storms differ in their duration. It varies from several hours to several days. For instance, a dust storm that originated near the Aral Sea on 22 May 1979 lasted about 10 hours, while hearths in West Africa can pollute the atmosphere for 4 days.

Finally, hearths differ in the frequency of heavy dust outbreaks. In 1975 near the Aral Sea, nine dust outbreaks were recorded by satellite survey. In the West Sahara, hearths of heavy dust outbreaks, during the four most 'active' summer months, can release dust as many as 16 times (Carlson, 1975).

Dust outbreak hearths are mainly of natural origin (Kondratyev *et al.*, 1983a, b). At the same time, there are known hearths that are to a greater or lesser extent due to anthropogenic activity. For example, all the hearths of local dust storms in the Mojave Desert, detected by the *Landsat-1* satellite, are of anthropogenic origin (Nakata *et al.*, 1976). They occur because of disturbance of plant and soil cover of different natural surfaces, as a result of road construction, off-road driving, building of settlements, hydraulic and road improvement work, and agricultural activity. In recent decades in different regions of the Earth, the process of anthropogenic desertification of territory has intensified considerably. Its after-effects, associated with changes in the physical properties of the surface (albedo, radiating capacity) are clearly seen in satellite images.

The largest destruction of plant–soil cover, on the scale of an entire region, took place in Africa in the savannah zones of the Sahel. Heavy dust outbreaks originate here and are detected by satellite images; for example, the hearths located south-west of the Air Plateau are undoubtedly caused by the loose surface layer due to heavy grazing.

Satellite images show (Muchlbaerger and Wilmarth, 1977) that annual burning of grass also favours the intensification of dust storms in the Sahel. It should be mentioned here that, as satellite observations of smoke from artificial fires indicate, the process does more harm than good. These fires are carried out at the end of the rainy season directly before the onset of the dry season. As a result, the soil, not held together by grass, becomes loose and tired. Formation of one of the most active hearths of dust outbreaks in 1975 near the Aral Sea was also caused by human activity. The sharp lowering of Aral Sea level, which was solely due to taking excessive amounts of water from the inflowing Amu-Darya and Syr-Darya Rivers (Grigoryev and Lipatov, 1979), was responsible for it. Dust outbreaks comprising-mainly saline sulphates and chlorides were to some extent unexpected by many scientists who did not foresee this development in the region. Emissions of salts into the atmosphere and their transport hundreds of kilometres away has been going on since 1975 and, as satellite observations show, has become more intense with time.

In order to estimate the frequency, transport, and consequences of dust outbreaks, it is necessary to study their hearths. At present, only a few major hearths of dust storms in the Northern Hemisphere have been outlined and briefly described (Kondratyev *et al.*, 1979).

3.2.2.4 The dynamics of dust pollution of the atmosphere

The dynamics of heavy dust outbreaks is so far poorly known. Satellite images (mainly taken by meteorological geostationary satellites), obtained with high periodicity, are used for the study of the rate of dust cloud motion, frequency of occurrence, and specific features of their motion.

The motion of dust flows, trajectories, and speed of motion (i.e. the dynamic features of dust clouds) are closely related to meteorological conditions. For example, heavy dust outbreaks near the Aral Sea usually take place when a cold front passes over this area (Grigoryev and Lipatov, 1979), especially when the cold front reaches the north-east coast of the Aral Sea. Where many hearths are located.

A similar storm situation, periodically occurring over the north-eastern coast of the Caspian Sea, determines the volume of dust in this region and its transport in the atmosphere.

Dust clouds that form in the West Sahara are rather different in configuration and size. Quite often satellite observations discover comparatively small (by Saharan standards) almost linearly-extended dust clouds several hundred kilometres long. As a rule, they move south-west under the influence of the north-east trade winds and are 'discharged' in the 'sea of gloom'. A dust cloud of this kind, up to 1,300 km long and 200 to 500 km wide was observed on 29 July 1974 in the centre of the Sahara on pictures taken by the *GMS-1* satellite (Kondratyev *et al.* 1976; 1983a, b).

However, as analysis of these pictures shows, dust transport in the atmosphere from the West Sahara may behave differently. Dust clouds from the West Sahara are often transported over the Atlantic by large anticyclones.

Super long-range transport of this kind to the coasts of North, Central, and South America occurs when strong stable north-east winds arise in the lower troposphere. A situation of this kind comes into being when the Azores cyclone reaches the African coast, accompanied by formation of pressure lapse rates at high altitudes (Kondratyev *et al.*, 1976; 1983a, b). As a result, the trade wind over the South Sahara becomes stronger, as does the harmattan.

The pulse-type character of dust emissions from the West Sahara to the Atlantic, discovered by satellite observations, can be explained, according to Carlson and Prospero (1972), in the following way. The cyclic character of atmospheric disturbance over Africa gives rise to changes in pressure lapse rate and, consequently, wind speeds at ground level, which determines periodic dust elevation and formation of dust clouds.

Dust outbreak transport rates from the West Sahara recorded by geostationary satellite survey appear rather different. For example, the dust cloud over the Atlantic from 4 p.m. on 29 July to 7 p.m. on 30 September 1976 moved at a speed of about 36 km/hour. At a much greater speed (108 km/hour), another dust outbreak took place on 29 July from one of the hearths in deserts of the West Sahara. This dust cloud travelled a distance of 5° latitude in 5 hours (Kondratyev *et al.*, 1983a, b). Bearing in mind that the aerological network in this region is rather sparse, data obtained on dust cloud speed are very important. They reflect the speed of the north-east trade wind at altitudes of 1–3 km. Using satellite images, the average speed of dust cloud transport through the Atlantic has been estimated, as well. It turns out to be much lower than the values given above (15–25 km/hour).

In different regions of the Earth, heavy dust outbreaks develop in different seasons, which is determined by specific features of climate at one or another dust storm centre. In the West Sahara, for instance, two maximums in development of dust outbreaks are observed – those in January and in July. In the Aral Sea region, heavy dust outbreaks occur six to nine times a year during two periods (i.e. from the end of March to the beginning of August and from the beginning of August to the beginning of September).

The persistence of heavy dust pollution in the atmosphere, judging from satellite survey data, vary from several hours to several days. Heavy dust outbreaks near the Aral Sea (from satellite observations) usually begin and finish in 1 day, though they can last 2–3 days. A dust cloud that appeared in July 1970 in Libya was observed by satellite for 24 hours.

Dust outbreaks from the West Sahara, moving westward, after about 6–8 days (sometimes a bit more) reach the coast of America. A huge dust cloud was observed from the *ATC-3* satellite for 9 days, from 26 June to 4 July 1969, and approached Haiti (Figure 3.7). Judging by satellite observational data (*ATC-3*) and land-based observations, a similar dust cloud, that migrated from Africa to the coast of North Florida, existed even longer – from 1 to 14 July 1970 (Carlson and Prospero, 1972).

Many specific features of the macrostructure of dust outbreaks, concerning the type of air flow or the type of underlying surface (sometimes to both), are reflected on satellite images. They make it possible for us to understand the dynamics of dust motion in the atmosphere. It is impossible to judge the structure of dust clouds from the Earth, since the field of view is very narrow. It is much easier to do so from aircraft at a height of 3–8 km above the dust outbreak, but these observations can only be episodic.

Analysis of satellite images makes it clear that many flow-type dust storms are characterized by a stream-like macrostructure. A dust flow, spreading over dozens or hundreds of kilometres, is sometimes not a single whole, but is broken up into several streams. These stream-shaped dust clouds are repeatedly observed from satellites, in particular over Mesopotamia and the Aral Sea (Grigoryev and Zhogova, 1992; Grigoryev and Kondratyev, 1980; 1981a, b) (Figure 3.8 and 3.9).

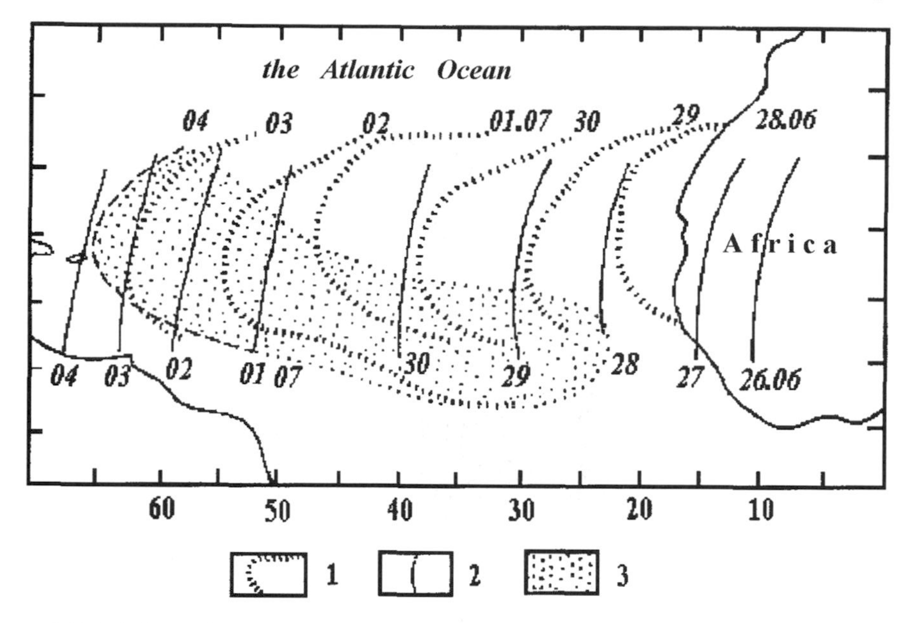

Figure 3.7. Motion of the dust cloud that developed in deserts of the West Sahara, over the Atlantic on 26 June–4 July 1969 from satellite *ATC-3* survey (Carlson and Prospero, 1972). 1 – front edge of the cloud; 2 – wave axis; 3 – dust cloud range as of 4 July.

The macrostructural features of dust clouds are to a considerable extent determined by air streams that carry the dust away. For example, dust clouds that move from the West Sahara to the Atlantic quite often bend towards the West European coast while moving westward. Satellite survey, in these cases shows the giant bow-shaped structure of the dust cloud. For example, on 30 July 1974, a dust formation extended a distance of 4,500 km and was 500 to 700 km wide. The bow-shaped structure of the cloud was due to the fact that dust from the Sahara was transported in a giant Azores anticyclone.

In contrast to satellite images of comparatively low resolution obtained by meteorological satellites, pictures taken from manned orbital stations make it possible to see the mesostructure of dust clouds. For instance, the *Skylab* manned orbital station, in September 1973, photographed a dust storm over the Caribbean Sea. The dust flow that originated in the narrow littoral band of Paraguay Peninsula was composed of multiple small (several kilometres long) bent streams. This phenomenon reflects the turbulent character of dust motion.

Only a few attempts have been made to determine the composition of aerosols in dust clouds; for example, using multi-range spectral data obtained by CZCS, taking into account changes in the distribution and characteristics of aerosols, and data for visible range obtained by the *Meteosat* satellite (Moulain *et al.*, 1994). In dust clouds over the Mediterranean Sea, sulphates, sea salt, and desert dust have been identified.

Figure 3.8. Dust outbreak over the Aral Sea of anthropogenic origin. Near-infrared image (0.8–1.1 μm) obtained by the *Meteor* satellite on 22 May 1975 (Grigoryev and Lipatov, 1979).

3.2.2.5 Zones of dust deposition

Satellite observations of dust outbreaks enable us to identify not only the zones of dust deposition but also their boundaries. Dust cloud deposition can have unfavourable (sometimes even critical) consequences for human beings.

In some cases, satellite observations enable us to trace the areas where deflation and accumulation of eolian material develop simultaneously. This phenomenon was, for example, recorded from images taken by the meteorological satellite *Space* during heavy dust storms, directly after they passed over the south of the European territory of the USSR in 1969 (Semakin and Nazirov, 1977). Hurricane winds at a speed up to 30–40 m/s first passed through the west of the Ukraine, and then embraced vast areas of adjacent regions (i.e. Krasnodar, Rostov, and Volgograd, the areas at the centre of a chernozem region). Apart from snowstorms and snowfall, dust storms arose in the above regions. This was facilitated by extensive and widely developed easily removed ground soil comprising loesses and loessal deposits.

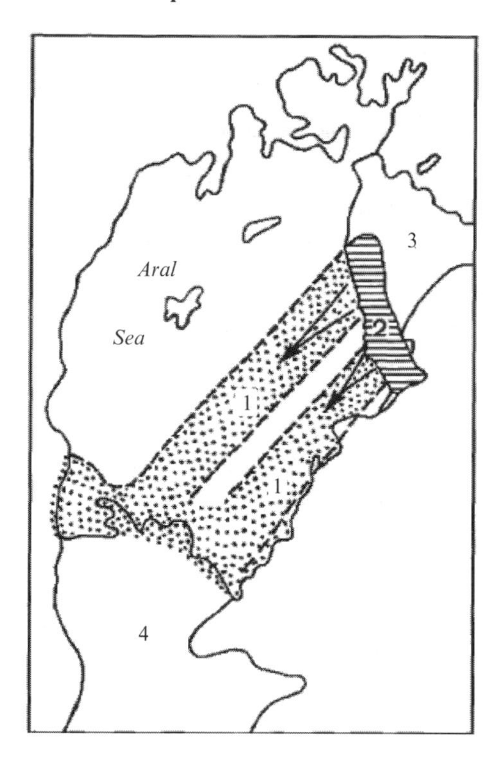

Figure 3.9. Image decoding diagram: 1 – dust outbreak streams; 2 – centre of the dust storm on a dried area. It was formed as a result of Aral Sea level decline due to excessive withdrawal of water from the rivers that flow into it; 3, 4 – Syr Darya and Amu Darya Deltas (Grigoryev and Lipatov, 1979).

The wind first blew away the snow, then lifted the dust into the air, which under conditions of dry and sparse vegetation is very easy. In the images obtained in the course of this dust storm, from 9 February until it finished on 22 February, an anomalous black spot extending from the Black and Azov Seas in a north-east direction was clearly seen. It should be labelled anomalous because at this time of year the above-mentioned territory is seen on satellite images as very light, which is due to extensive snow cover.

The total area of the zone made up about 700,000 km^2. It included both snow-free territories a result of snow blowing away, and areas of snow covered with dust that changed the albedo of this territory (e.g. ice on the Azov Sea was completely dust-bound). It was not only snow that was blown away, but the upper layer of soil – the layer of fertile chernozem. Soil deflation in more southern regions was quasi-synchronous with accumulation of dust carried by wind in more northern regions bringing with them such eolian forms as hillocks, dust banks, snow-and-dust barriers across roads, and so on. Although satellite images and, especially, overview images, do not make it possible, of course, for us to see these and other details of dust storm impacts on the underlying surface, they do present the possibility of determining its aggregate effect (i.e. to reveal the region of intense soil deflation and dust accumulation).

Long-range dust transport, whose estimation has been made fully possible by the use of satellite data, may undoubtedly as well exert other meso and macroscale effects on the landscape. They may, in particular, manifest themselves not only in the mechanical movement of the surface layer of the soil from one region to another, but also in the change in saline composition of the soil. As was found from analysis of satellite images and land-based investigations, in a number of regions dust hearths are on intensely saline ground; due to this, they are saturated with saline particles. Salt transported during a dust storm deposits in the band of dust cloud motion, giving rise to soil salinization.

Depletion of vegetation on the Ustyurt Plateau and especially in the delta of the Amu Darya River correlates with transport of dust–salt clouds originating in thc eastern coast of Aral (for more details, see Chapter 5).

Satellite monitoring of dust–salt storm propagation from their hearths in the northern Caspian area has made it possible to specify the causes of some accidents. For example, the failure of power lines in large cities of Lower and Central Privolzhskaya in a vast band from Astrakhan to Kazan, due to heavy deposition of salts on insulators (Grigoryev and Lipatov, 1982).

Analysis of satellite images has also made it possible to understand the origin of the deposited salt (Figure 3.10). For example, in the images are seen dust storms that arose from hearths off the northern coast of the Caspian Sea. Within the limits of these hearths, highly saline deposits are developed. Saline dust, determinable from the images, reaches the Volga Delta. Northward of the delta, saline dust flows are not traced from satellite images, because as they move farther away, saline dust concentration decreases. However, dust transport takes place far beyond the lower reaches of the Volga.

This is also indicated both by the orientation of the dust streams in the images and the fact that all damage to power lines only occurs under south-east winds, blowing, in particular, from the north-eastern coast of the sea. Saline particles transported by the airflow up to Kazan were active condensation nuclei. They moved over Povolzhskaya and mixed with clouds of air masses from the Caspian Sea, and were deposited at the same time as precipitation. It should be mentioned that the number of power line accidents increase during precipitation and drizzle under prevailing south-east winds (Grigoryev and Lipatov, 1982).

The last example relates to the problem of dust storm impact on humans and on economic activity (salinization of soils, deposition of salts on power lines). This problem is urgent. Impacts (e.g. transport of pathogenic microbes with the dust, roads buried under dust, etc.) are widely known (Grigoryev and Kondratyev, 1981a, b; Idso, 1976). It is only with satellite images that it has become possible to study impacts of this kind at the macroregional scale, as follows from the examples presented above.

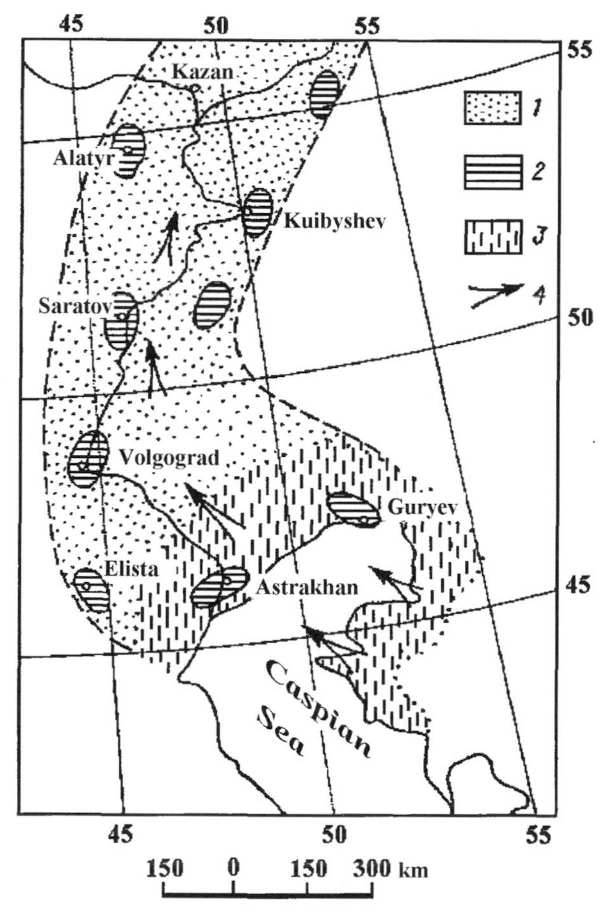

Figure 3.10. Atmospheric dust pollutant deposition (mainly saline) in Privolzhskaya, built up using satellite and land-based data (Grigoryev and Lipatov, 1982). 1 – zone of dust deposition; 2 – regions of mass-scale dust deposition; 3 – centres (hearths) of saline dust outbreak; 4 – direction of dust transport.

3.3 HYDROLOGICAL HAZARDS

3.3.1 River floods

River floods are among the most dangerous natural calamities (Table 3.5).

Diverse factors give rise to floods. More often, they occur as a result of heavy precipitation, snow melting that feeds rivers, and, finally, due to cyclones and typhoons, especially in coastal and marine regions. Sometimes, floods are caused by inefficient drainage in flat, low, easily flooded plains, as well as by simultaneous peak floods at many inflows into river systems. Quite often, flooding of a territory is associated with anthropogenic activity: construction of channels, water storage basins, cutting down of forests, etc.

Table 3.5. The largest river floods of the 20th century by number of victims (1,000 persons and more).

Date	Country, location	Number of victims (1,000 persons)
1911	China: Yangtze	100
1931	China: Yangtze	140
1933	China: Hwang Ho	18
1939	China: Tientsin	1
1951	China: Manchuria	5
1953	North Europe	2
1954	Iran: Kazvin	2
1955	Pakistan and India	1.7
1963	Italy: Belluno	2
1965	India: Bengal, Assam, Bihar	2
1968	USA: New Jersey	1
1973	Pakistan: Indus River	1
1978	India: Ganges River, Jumna River	2 (1.2)
1980	India: Orissa, West Bengal	1
1981	China: Yangtze	1.5 (1.3)
1982	India: Orissa	1
1982	Peru	2.5
1983	India	1.6
1987	Bangladesh	1
1991	Afghanistan	5

Floods on large rivers bring the highest threat to population, because their flood plains of large rivers are the places where there are lots of settlements and cities. In Asia, about 20 million persons are affected by floods annually (Smith, 1993; Smith, 1997).

Floods in different regions of the Earth are often caused by different factors. They may be snow melting, glacier melting, intensification of rain, etc. Although periods when such situations develop are known in many cases, their ranges cannot be always foreseen. It is important to remember, mainly concerning developing countries especially poor ones, that catastrophes on flood plains are to a great extent connected with overpopulation, with the reluctance of people, on the eve of calamity, to be evacuated.

The data on floods and their consequences by continents and countries is interesting. This information is based on statistical data (United Nations Environment Programme, 1991) on major floods (1,000 or more victims and (or) 1,000 or more evacuated persons and (or) considerable material damage) for the period 1981–9 (Table 3.6).

Altogether, there were 177 major floods in this period (United Nations Environment Programme, 1991). Most occurred in Asia (85), followed by South America (34), and an approximately equal number in Africa (16), Europe (14), Central and

Table 3.6. The most serious river floods for the period 1981–9, causing large numbers of homeless people (more than 10,000) and deaths (more than 100,000) (from United Nations Environment Programme; 1991).

Country	Year	Number of homeless people (1,000 persons)	Number of evacuated people (1,000 persons)	Number of deaths (1,000 persons)
China	1981	1,130		
India	1981	500		
China	1982	210		
Spain	1982			225
Ecuador	1983			700
Korea	1984	200		
Brazil	1985			600
India	1985	400		
Colombia	1986		250	
Bangladesh	1986	1,000		6,000
China	1986			4,000
China	1986			5,000
India	1986	650		
India	1986	150		
Sri Lanka	1986		300	554
Bangladesh	1987			3,000
Bangladesh	1987			600
China	1987			900
China	1987			200
China	1987			1,000
India	1987			1,000
India	1987			Several million
Nepal	1987			351
Sudan	1988			1,500
Argentina	1988			4,500
Bangladesh	1988	28,000		45,000
China	1988			280
China	1988			22,000
China	1988	110		
China	1988			2,800
China	1988			300
Iran	1988			150
Thailand	1988			1,000
Tanzania	1989			300
Malawi	1989			137
China	1989	100		
Sri Lanka	1989	250		

North America (16). The smallest number of floods were recorded in Australia and Oceania (8) and in the former USSR (4).

The number of victims approximately correlates with the number of floods in large regions of the world. The highest number of victims was in Asia (20,313) and South America (4,658), followed by Africa (779), North and Central America (479), Europe (261), former USSR (189), and Australia and Oceania (25). Altogether, 26,754 persons lost their lives between 1981 and 1989 (United Nations Environment Programme, 1991).

The comparatively small number of victims in the given period (compared to the entire century) is explained by the fact that there were very few major floods between 1981 and 1989.

The number of those affected by the floods (65 million persons) was immeasurably greater than those who died. More than 48 million persons were affected in Asia, more than 5.2 million persons in South America, and about 1,977 million persons in Africa. The numbers were considerably smaller in Europe (261), the former USSR (189), and Australia (25).

Among countries with the most victims (between 1981 and 1989), China, India, and Bangladesh stand out (6,203, 5,857 and 4,929 persons, respectively) (Table 3.7).

In the countries of Europe, North and Central America, and Africa, the number killed due to floods, with rare exceptions, does not exceed 100 persons per country and comprises: Europe (Italy, France, and Spain) 50, 48, and 105 persons, respectively; North and Central America (Guatemala, Haiti, and USA), 94, 82, and 81 persons, respectively; and Africa (Sudan, Mozambique, and South Africa) 90, 100, and 425 persons, respectively.

As a whole, distribution of the number of victims due to floods in many respects reflects specific features of the natural environment of a country. In China, India, and Bangladesh, as well as in South America, are found some of the largest rivers of the world whose floodplains are the most densely populated (Yangtze, Hwang Ho, Ganges, Indus, Brahmaputra, Amazon). It is precisely here that the largest, in terms of number of victims, floods occur (Table 3.8).

At the same time, the absence of victims in major floods in a number of European countries such as, for example, in 1981 and 1983 in France and in 1981 in England (Table 3.9) is striking. This is undoubtedly explained by less extensive large-scale floods and, of course, by the large amounts of money spent in these countries on flood-prevention measures.

The extent of flood consequences is magnified by their frequency. Only in four countries of the world have there been 10 floods or more in a given year. These are China, India, Brazil, and Indonesia (25, 15, 11, and 10 floods, respectively). In four countries (Bangladesh, Peru, the USA and Iran), there were six floods. And in three countries (France, Philippines, and the former USSR), there were four floods.

Tables 3.7 and 3.8 show that the total number of killed depends not so much on the number of floods, but on the size of individual, disastrous ones. Relevant US data provide the most sound evidence of it.

Every year about 20,000 persons lose their lives and about 75 million people are displaced or suffer damage caused by floods (Smith, 1997). Annual global material

Table 3.7. The number of victims in countries with the highest number of floods in the period 1981–9 (from United Nations Environment Programme, 1991).

Country	Number of floods	Number of victims
China	25	6,203
India	15	5,857
Brazil	11	1,160
Indoncsia	10	329
USA	6	81
Iran	6	582
Peru	6	2,107
Bangladesh	6	4,929

Table 3.8. The largest river floods for the period 1981–9 with number of victims higher than 1,000 (from United Nations Environment Programme, 1991).

Country	Year	Number killed
China	1981	1,500
India	1981	3,000
Peru	1982	2,500
India	1983	1,600
Bangladesh	1987	2,055
India	1987	1,200
Bangladesh	1988	2,000
India	1988	1,000
China	1989	2,500

damage due to floods equals US$2 billion (Rao, 1996). Among the countries that experience most material damage, due to flooding of their territory, are India and Bangladesh.

In India, every year more than 1,500 persons die because of floods. Moreover, during the floods 8 million ha of land are often submerged, including 4 million ha of sown areas, and damage is caused to 1.2 million houses. Total material damage is estimated as US$300 million. In the years of especially severe floods, the scale is considerably (sometimes, threefold) higher. In 1988, for instance, damage due to flooding exceeded US$1.5 million, 10.2 million ha of sown areas were flooded, and 2.3 million houses were destroyed (Rao, 1996).

The size of river floods and related calamities in Bangladesh were no less considerable. In the 1960s, the total damage due to floods for 10 years totalled US$10 billion. Damage was caused to 100 million ha of agricultural land and 1.2 million houses.

Table 3.9. The largest floods in the world between 1981 and 1989 with material damage exceeding US$25 million from United Nations Environment Programme, 1991).

Country	Years	Damage (in US$ million)	Number killed
France	1981	166	–
Australia	1981	230	–
France	1982	152	18
England	1982	174	–
France	1983	604	–
India	1984	500	741
USA	1986	25	–
Iran	1986	1,500	230
Italy	1987	1,000	50
Switzerland	1987	40	1
Iran	1988	150	–
Spain	1989	375	12

In developed countries, the major damage caused by floods is material (deaths are usually not very high, (Table 3.9). For example, in 1993 in the USA, the most severe flood in the entire history of the country occurred (floods of this kind occur once in 500 years).

Water from the Mississippi and Missouri Rivers, and their inflows that burst their banks, flooded parts of nine states (along the 800-km length of the Missouri). The huge calamity did not result, however, in the death a single person because the 54,000 inhabitants of the immediate vicinity were quickly evacuated. Material losses were high at US$15–US$20 billion. In that year, crop production in the country dropped by one-third because more than 4 million ha of agricultural land were flooded and the farmers were made homeless (more than 50,000 houses were destroyed) (Smith, 1997).

The damage caused by river floods as well as the number of victims in the 20th century increased with time. It is mainly characteristic of developing and, especially, poor countries. The increase in the number of victims is associated with population growth. At the same time, the same trend is observed in developed countries, as well. For example, in Australia insured material losses increased ten times in 1985, compared with 1953 (Smith, 1997). In the USA, the growth of material losses due to floods was more distinct than that in the number of people killed in floods. The reasons for the growth in material damage (even though investment in measures to minimize the scale of calamities was huge) are connected with development of flood plains.

Despite the wealth of experience amassed by people in different countries of the world who came across disastrous floods and the progress achieved in measures to reduce their impacts (in China this experience exceeds 5,000 years), the consequences are just as serious. In some countries, they actually become even more serious. This is explained by the growing human activity in flood plains. The advantages of life near big rivers obviate the risk of building cities on their banks or increasing the areas of agricultural production (mainly crop lands). As before, man continues foolishly to confront that most dangerous element – water – by permanently expanding the ranges of his activity over river banks.

Floods often occur because of lack of information on environmental conditions in the upper reaches of rivers, in the areas where they are fed. These areas are often located many hundred of kilometres or even thousands of kilometres away from the area where floods occur; sometimes, they are located in the territories of neighbouring countries. When severe floods occur, then, it is not surprising that engineering constructions cannot withstand them (this is especially clearly seen in China, despite a wealth of experience in constructing flood-protection projects).

It is very difficult to overcome the catastrophic situations caused by floods. There are only two ways of cushioning the effects of floods. The first way involves construction of various hydraulic structures (embankments, water storage basins, channels, dams) that enable us to have some control over the flooded river. A wealth of experience of this kind has been gained in China. In lowlands, through which the Yangtze and Hwang Ho Rivers flow, there are plenty of structures protecting settlements, cities, and agricultural lands from floods.

The second way is to try and reduce the consequences of floods. It includes the following:

- zoning by extent of risk;
- arrangement of observations on environmental conditions with the aim of warning people about the development of dangerous phenomena;
- taking measures on evacuation of people;
- planning rational locations for human activities (settlements, roads, cities, enterprises), agricultural lands, including types of crop that can to some extent withstand long-term inundation.

The current tactics and strategy used to overcome the consequences of floods have changed. This is mainly true of the majority of developed counties. Earlier, floods were tackled by physical methods (construction of dams, etc.). In other words, the consequences of the calamity were mainly associated with the natural calamity itself, its intensity and size. In recent decades a lot more attention is paid to the social component of the calamity, while earlier this was almost ignored. To overcome the consequences of floods, it is expedient to have in place an adequate social policy, and one principally regarding the distribution of population and industries.

There are no comparative data available for the risk of flooding in different countries. However, there are some examples typical in this respect. In New South Wales, about 200 cities are located near riverbanks that are subject to floods. But, as a whole, in Australia (as in England), the risk of severe floods is not considerable: less than 2% of these countries occupy territories that flood only once in 100 years. In the USA, such risk is considerably higher: one-sixth of all urban lands in the country is located on territory subject to flooding once in 100 years. So, 10% of the population of the USA is subject to the risk of flood (Smith, 1997).

Of all the countries in the world, Bangladesh is the one where the risk of river floods is highest. During floods (especially at the time of monsoon rains), 20–30% of its territory is flooded. This is not surprising, because 80% of its territory is a low alluvial plain. Floods from the Brahmaputra, one of the rivers that crosses this plain, are constant threats to the lives of 127 million persons living in 22,000 villages located along its banks (Rao, 1996).

Having said this, it follows that creation of maps of flood risk zones is very important. In the USA, maps are prepared for zones of risk, due to floods that happen once in 20, 100, and even 500 years (the absence of such maps resulted in the especially grave consequences of Mississippi River floods in 1973 and 1993).

A number of measures are required in order to give valuable forecasts about impending floods and to reduce the damage caused by them to a minimum. Among them are the following:

- monitoring cloudiness and meteorological conditions (not only in the territory of the country but beyond its boundaries);
- tracking snow cover, especially in mountains, glacier melting;
- land use management;
- zoning of the territory according to extent of risk;
- construction of hydraulic projects (dams, channels);
- organization of a population warning system and timely evacuation of people.

3.3.2 Satellite observations of river floods

Satellite observations enable us to quickly identify flooded areas, and keep watch on their development. They were efficiently used, in particular, in 1973 in the USA, during development of the second largest flood by the Mississippi River and its inflows, caused by rainfall in the upper reaches of the river (Figure 3.11). From the results of high-water mapping using satellite images taken in the course of the flood (floods of this size occur once every 100 years), an unknown zone of flood risk was discovered, being a rarely flooded high-water bed (Rango and Anderson, 1974).

Images taken from *Landsat* (MSS) satellite, *Landsat* (TM), and *MOS* (MESSR) between 1972 and 1993 have been used to keep watch on Brahmaputra stream-bed changes in Bangladesh accompanied and resultant catastrophic floods (in 1988 waters of the river flooded two-thirds of the territory of the country) (Noorbergen, 1993).

Infrared images (in the near and medium infrared spectral range of the spectrum) are especially useful to distinguish areas of land inundated by flooded rivers and areas unaffected by high water. In these images, water is clearly differentiated from vegetation, and dry areas from moist ones. Although cloudiness impedes infrared images, they appear to be useful even several days after the flood is over for estimation of the damage within the territory subjected to the flood (these areas with elevated moisture of soils and grounds are clearly seen in infrared images). Decoding flooded regions is made easier by the use of colour (pseudo-colour) images

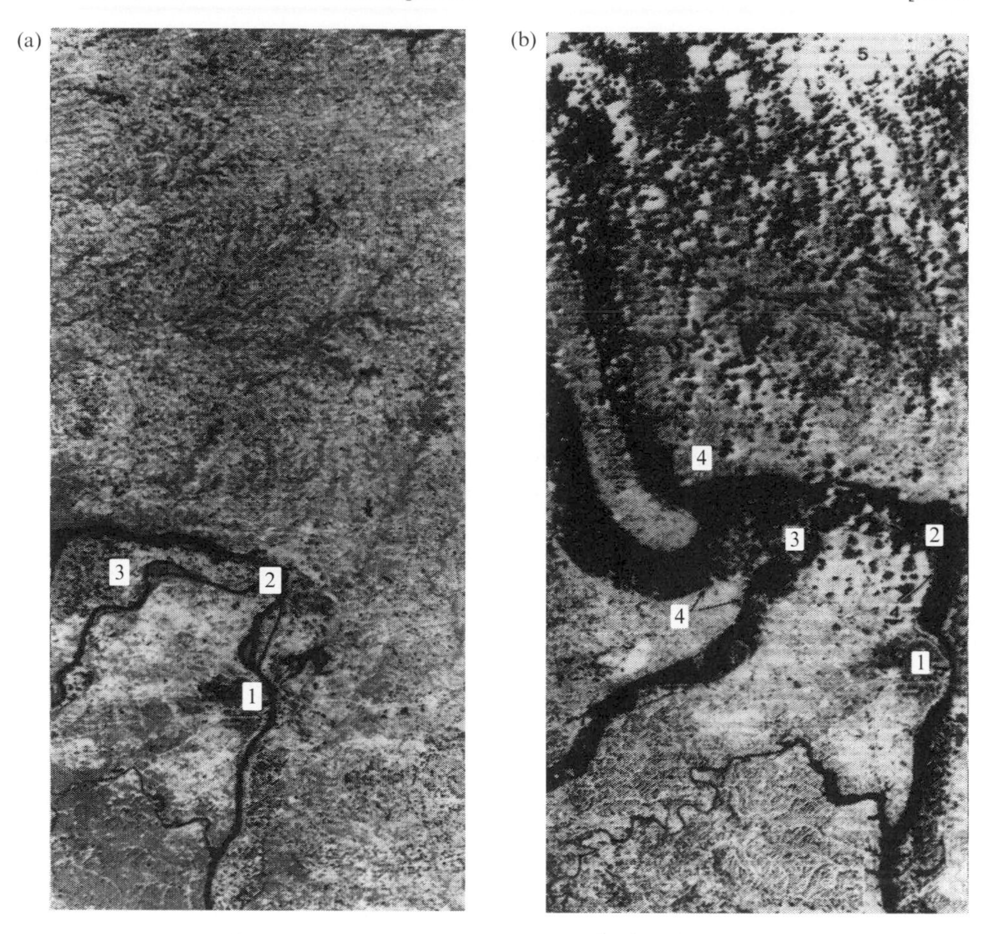

Figure 3.11. Disastrous flood on the Mississippi and Missouri Rivers in spring of 1973. a, b – satellite images taken by *Landsat* (0.8–1.1 μm spectral range) on 28 August 1972; (a) before and (b) after the flood on 31 March 1973. The river burst its banks (2), in places considerably (4), flooded agricultural lands (3), including the outskirts of Saint Louis (1) (Rango and Anderson, 1974).

obtained in the visible and near-infrared spectral regions. They allow clearer distinction of different gradations of landscape subject to flood.

In France, colour images taken by the *SPOT* satellite were used for estimation of catastrophic floods in 1994 in the valleys of Sarthe and Maine, the Men River in the Camargue, and the town of Angers (in the *départements* of Men and Loire). Photographs taken before the area was flooded and during the flood identify those regions at highest risk of damage. Furthermore, on the basis of these data a project to build a dam on the Loire River was prepared with the aim of minimizing any future flood (Dafait, 1994).

Satellite radar images are very prospective for and used to monitor floods. Images from the *Cosmos-1500* satellite were used, for instance, to track Amur

River floods. In doing so, the main parameters that contribute to inundated areas were identified and monitored (Pichugin, 1985).

The most fruitful means of monitoring is simultaneous use of satellite and aircraft images (principally those obtained by radar with synthetic aperture that gives information under all weather conditions). In this way, using *NOAA* and *Landsat* satellites in 1991 in China, monitoring of floods on the Yangtze River and its inflows was successfully carried out (Grigoryev and Kondratyev, 1993a; Cao Shuhu, 1991). Since then, satellite and aircraft observations have become a key part of flood survey in this country.

Satellite images can be successfully used to map inundation zones and, more importantly, they can do so for large territories and very quickly. In India, hydrological maps at a scale of 1:50,000 are made from satellite images every week. They enable us to estimate the size of a flood, tendencies in its development, and form the basis for taking short and long-term measures. Satellite observations of this kind were carried out from the Indian satellite *IRS-1A* to track floods in the basin of the Brahmaputra River (Grigoryev and Kondratyev, 1993a, b).

Satellite images were used to map inundated areas at the time of the catastrophic Mississippi River flood in the Midwest in 1993. This flood was the largest in the history of the USA (as estimated, it occurs only once in 500 years). Heavy rainfall in the Midwest caused it. This flood resulted in inundation of the suburbs and outskirts of hundreds of towns and vast agricultural areas (covering several million acres). Using satellite images, the zone of coastal inundation, which extended many hundreds of kilometres, was mapped (Bottorff and Mayney, 1994) within a few days.

Landsat satellite images were the main source of information on inundation of the shores. They were compared with the images of the same territory obtained in 1989 in the river under normal (flood-free) conditions.

Maps for different inundated regions were prepared within 24–28 hours and were presented to all the institutions concerned. Radar images taken from the *ERS-1* satellite were also used since the objects to be photographed were obscured by periodic and frequent cloudiness. These images, close in date (approximately 4 days) to the images taken from *Landsat* (TM), were used in combination with the satellite images. The obtained data made it possible to operationally track the course of the flood, to organize evacuation of people from dangerous places, and to outline the best way of normalizing the situation (Gardner, 1994).

Satellite images are useful to estimate the consequences of major floods. In China, after the next major flood in summer 1991 that caused damage to some regions in the basin of Yangtze and Hwang Ho Rivers, data from the Thematic Mapper (TM) were used. The images made it possible to determine the boundaries of the flood, soil moisture after the flood was over, and specific features of vegetation growth that determine future yield (Changda Dai *et al.*, 1992).

In the course of decoding the satellite images, natural and anthropogenic formations in inundated and adjacent territories were classified by their spectral indices. Using images obtained from the *Landsat* (TM) satellite in the area of flood development in China in 1991 (in the basins of the Yangtze and Hwang Ho Rivers), classification of natural and anthropogenic formations was performed. In doing this, areas that suffered from the flood (three classes of damage), areas free from flood, and residential territories were identified (Changda Dai *et al.*, 1992).

Using satellite images of the Brahmaputra River for several years (separately for each years), classification of the four categories of natural formation (i.e. water, vegetation, sand, and soil) was performed (Grigoryev and Kondratyev, 1993a, b). It was used to determine the sequence of changes in riverbed migration. These data, covering a large territory but obtained in a short period of time, are useful for hydrological modelling and further engineering works.

For the purpose of decoding inundated areas, it is especially efficient to use methods of digital classification of natural and anthropogenic formations in accordance with their spectral features. By doing so, we can successfully differentiate a wide class of objects in the flooded area that are important for estimation of damage caused by the flood. Among them are areas under water, areas where water begins to regress, areas with inundated agricultural crops, and so on. Information of this kind from different satellites obtained using digital methods of satellite image processing may be presented in the form of maps at a scale of 1:50,000 and higher.

Application of images from different satellites enables us to manipulate them with the aim of overcoming the lack of information on the inundated area at times of cloudiness. Satellite survey in the microwave range assists monitoring the course of river flood development through the clouds.

The high periodicity in obtaining satellite images in a flood period allows us to monitor its development and, thus (together with other data), to plan population evacuation and control measures at this critical time.

Satellite information currently plays an important or even the key role in the flood risk management programs of a number of countries (e.g. the USA, France, India, and China). This information is used for assessment of the situation in vast territories (sometimes, entire river basins) estimation of the extent of risk in individual parts of them, and finally, working out the best strategy to combat river floods and a risk management strategy.

On the basis of the different specific features of flood development and the land inundation pattern obtained from satellite images, it is possible to determine the areas of flood risk for different river basins and individual parts of them. For example, in India, satellite images were used for determination of flood risk in the basins of a number of rivers, including the Ganges (Rao, 1996). This pointed the area of flood risk at a high water period. In China, satellite images were similarly used to forecast inundation in the lower reaches of the Hwang Ho River (Grigoryev and Kondratyev, 1993a, b; Jighan Jiang and Shuhu Cao, 1990).

At the time of serious flood development in the USA in 1993, the Federal Prompt Reaction Agency used Global Positioning System (GPS) and geo-information technologies (GIS) for prompt assessment of the situation. Satellite images served as the most important source of data for this assessment. This made it possible for a group of engineers and environmental specialists to prepare a series of maps for inundated regions within several days after the start of the flood.

GPS was used for a more precise tie-in with co-ordinates, and GIS was used for mapping and displaying the maps in interactive mode. The obtained data served as the basis for operational assessment of the situation and helped the administration to take the necessary measures, as well as for information and warning of the population (Bottorff and Mayney, 1994).

Satellite images may be used not only for monitoring river floods, but for their prediction, as well. For this purpose, data on spatial–temporal variations of precipitation in river basins are required. In this context, data on soil moisture obtained by means of microwave survey are useful. Satellites may be used to obtain information on soil moisture within a river basin days before the flood.

Furthermore, flood forecasts may involve data on the weather-forming atmospheric phenomena that determine precipitation. These phenomena have long since been successfully identified by means of satellite images.

Especially heavy (storm) rainfall may be estimated by analysis of simultaneous images in the visible and infrared spectral ranges (as well as together with data on temperature and albedo of clouds). Satellite images in the visible and infrared spectral ranges coupled with data from microwave survey and land-based radiolocation enable us to forecast precipitation reliably enough.

Using satellite images, it is possible to obtain information on snow cover too, including information on its unusual development in mountains, which to a large extent determines the evolution of disastrous floods on rivers that originate in the mountains. All these data enable us to build and apply models that forecast expected rise of water in rivers.

One of the most important components of flood forecasts is information on cloudiness, development of cyclones, rainfall, and snow melting in mountains.

Bangladesh is one of the most vulnerable countries in the world to risk of floods and TCs. In the years of disastrous floods (e.g. in 1988), half of the country was inundated. In Bangladesh, a land-based meteorological station is installed that receives satellite images from *NOAA* and *GMS* at frequent intervals. By means of a specially-developed method, the station can predict the area that will be affected by a cyclone (tracked by satellite images) at least 24 hours before the event (Choudhury, 1980).

In Africa, in order to give prompt warning about Blue Nile floods, an operative system for application of *Meteosat* satellite images has been created. These images are used to determine mean daily rainfall amount in the river basin. Estimated was precipitation from 'cold' clouds with brightness temperature below a certain threshold. Information of this kind is was used as input data when forecasting heavy rainfall-induced high waters. The forecasting model applied is part of a more general hydrological model used for flood forecasts.

Such a system (which also uses thermal infrared images) is only valid for regions with convective precipitation. This model can be used for practical purposes and since 1992 has been successfully applied to management of river water handling facilities on the Blue Nile (El Amir El Nur *et al.*, 1933).

In India, for the purpose of high water forecast, considerable attention is paid to assessment of the snow-melting situation in the Himalayas. From satellite data, mainly from *Landsat*, diverse data on snow cover are obtained. Principal indicators of snow melting are the sizes of snow cover, snow depth, and its water equivalent. In particular, *Landsat* satellite-derived images for the period 1975–89 of neighbouring Nepal were analysed. For assessment of high waters, data on water flow and temperature are also used. On the basis of all this information, regression equations for the correlation 'area covered with snow – water flow' and ' temperature – water flow' are constructed (Thapa,1993).

Forecasts of river floods using satellite images are important when planning measures for risk reduction, including related hydraulic works, communication services, information, and evacuation of population.

3.4 GEOLOGICAL DISASTERS

3.4.1 Earthquakes

3.4.1.1 Specific features and consequences of disastrous earthquakes

Earthquakes are one of the most terrible natural phenomena. Occurrence of the most disastrous earthquakes is presented in Figure 3.12. Together with TCs and floods caused by them, as well as river floods, they bring about both massive material damage and high numbers of victims. Figure 3.13 shows destruction and fires in San Francisco after an earthquake. In the period from 1900 up to 1989, 1.2 million people perished from earthquakes (United Nations Environment Programme, 1991). The USSR was among the first six countries on whose territory 82% of all earthquake-related deaths occurred. Russia is among the 35 countries where there is a probability of catastrophic earthquakes (with magnitudes of 6–8 is the Richter scale).

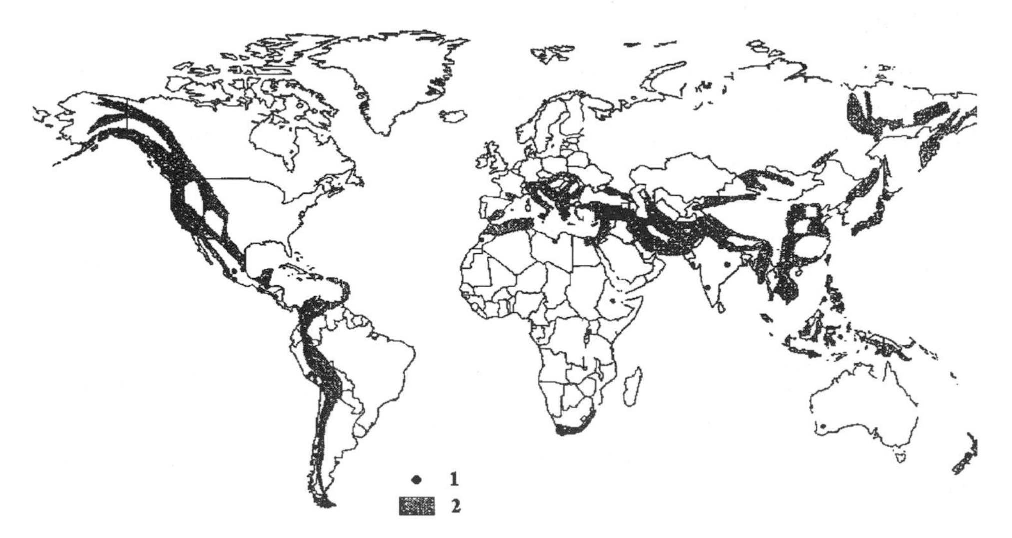

Figure 3.12. Global pattern of major severe earthquakes, 1950–79 (Hewitt, 1997). 1 – centres of individual earthquakes; 2 – major active orogenic zones.

Figure 3.13. San-Francisco, on the San Andreas Fault in the Cordilleras, 5 hours after the catastrophic earthquake on 18 April 1906, enveloped in the flames of fires that broke out in almost 500 quarters of the city (Grigoryev and Kondratyev, 2001).

For example, the earthquake in Neftegorsk (Sakhalin Island) in 1995, and those in Turkey and Greece in 1999.

The global temporal distress pattern due to earthquakes is spasmodic. The number of persons killed in the 1980s as a result of earthquakes was about 57,500. In 1999, in just 1 year, the number of victims amounted to 52,000. It should be mentioned, however, that the annual number of victims for the period 1977–90 was even less, namely 30,000. At the same time, the great Chinese earthquake in 1976 exceeded all the records of the 20th century – 693,000 persons (Anon, 1991).

For the period 1980–90, throughout the world there were 170 severe earthquakes that involved 43 countries, some earthquakes (neighbouring ones) happened simultaneously (United Nations Environment Programme, 1991). The most devastating earthquakes (by number of victims) took place in Iran (1990 – 50,000 perished), the former USSR (1988 – 25,000 perished), the simultaneous earthquake in Turkey (25,000 perished), and in Mexico (1985 – 10,000 perished). Although they were of different intensity (7.3, 7.6, and 8,1 on the Richter scale, respectively), they resulted not only in large number of victims, but also very high material damage (more than US$25 million).

Let's consider the specific features of earthquake occurrence and their consequences in large regions of the world (Asia, Australia and Oceania, North and Central America, South America, Europe, former USSR) and in a few countries in the period 1980–90. The highest number of victims were in Asia (86,212) and the former USSR (25,392), and third place belongs to North and Central America (11,313). In Africa, Europe, and South America, the number of victims were almost of the same order of magnitude (5,070, 3,253, and 2,506), and the smallest number were in Australia and Oceania (459 persons).

The ratio of number of victims to population number for different regions can be significant (for 1990):

(1) Asia $3{,}108{,}475 \times 10^3$;
(2) Africa $647{,}518 \times 10^3$;
(3) Europe $497{,}740 \times 10^3$;
(4) North America $425{,}682 \times 10^3$;
(5) South America $296{,}687 \times 10^3$;
(6) former USSR $28{,}790 \times 10^3$;
(7) Australia and Oceania $26{,}476 \times 10^3$.

The most disastrous earthquakes for population occurred in the former USSR and Asia, followed by North America, Australia and Oceania, South America, Africa, and Europe. As the last two regions were the less dangerous, the number of victims compared with population number was lower than in other regions.

In Asia, there were 78 earthquakes in 12 different countries. They were the most frequent in China (16), Indonesia (13), and Iran (11). As for the number of deaths, the first place belonged to Iran (51,564), followed by Turkey (25,020); in the latter, there were five earthquakes.

Six disastrous earthquakes were recorded in the former USSR; in Australia and Oceania there were 16 earthquakes, and the number of victims in the former USSR was many times greater. In North and Central America the number of disastrous earthquakes was the greatest in the USA (17), but the highest number of victims (10,001) were recorded in Mexico. There were 31 earthquakes in these seven countries (USA, Guatemala, and Mexico head the list with 17, 4, and 4 earthquakes, respectively).

In South America, there were 19 disastrous earthquakes in seven countries. The list is headed by Peru (six earthquakes), and the highest number of victims (1,350) were recorded in Colombia, where two earthquakes occurred.

In Europe, frequent earthquakes were recorded in Italy (nine times) and Yugoslavia (six times), less frequent earthquakes were recorded in Greece (three times), Albania and Romania (twice in both countries), and very rare earthquakes were recorded in Portugal, Bulgaria, and Belgium (once in each). Most victims were recorded in Italy (3,111).

In Africa four out of five disastrous earthquakes happened in Algeria, where the greatest number of victims (5,070) were recorded, as well.

Overall, the distinct relationship between the number of victims and disastrous earthquakes was not reported for the 10–year period (although in Europe and Africa there is such relation, in other regions it is absent or not so distinct). Sudden severe earthquakes (e.g. in Mexico, Turkey), easily upset a relationship of this type. Evidence of this can be found in Table 3.10 containing data on all earthquakes causing more than 1,000 deaths (United Nations Environmental Programme, 1991).

As is well known, every earthquake is accompanied by material damage (Caviedes, 1991). For the period 1980–90, of all the countries where disastrous earthquakes happened, 11 had material damage that was especially high (more

Table 3.10. The most disastrous earthquakes for the period 1980–90 and the number of deaths and displaced people.

Country	Number of disastrous shocks	Number of people displaced	Year	Richter magnitude	Number of deaths (1,000)
Iran	13	54,573	1981	6.9	3
			1981	7.3	1.5
			1990	7.3	50
Former USSR	7	25,392	1988	7.6	25
Turkey	4	25,030	1983	6.9	1.4
			1988	7.6	25
Mexico	4	10,067	1984	8.1	10
Algeria	3	5,031	1980	7.4	5
Italy	9	3,133	1980	6.9	3.1
Yemen (Democratic Republic)	1	2,800	1982	6.0	2.8
Nepal	2	1,150	1988	6.5	1
Salvador	3	1,123	1986	5.4	1.1
Ecuador	2	1,008	1987	6.9	1

than US$25 million) (Table 3.11) (United Nations Environmental Programme, 1991).

Table 3.11 shows that the greatest material damage and, at the same time, smallest number of victims in the course of disastrous shocks were recorded in the USA, Italy, and Belgium (United Nations Environment Programme, 1991). It is characteristic especially of the USA where, in 1980–90, there were 17 disastrous shocks and none of them resulted in more than 1,000 deaths, but four shocks were accompanied by very high material damage (more than US$25 million). This reflects in the very high efficiency of measures taken in the USA to protect people during earthquakes.

A specific feature of earthquakes is the fact that all victims and wounded arise, not as a result of the ground shocks themselves, but as a result of destruction of different structures. Although currently available data testify that, sometimes, earthquakes may be caused by engineering structures (e.g. large water storage basins), underground nuclear explosions, and mountain mining, nevertheless, this is by and large a natural phenomenon. However, the destructive effect is closely connected with anthropogenic factors (building cities in a seismic region, inadequate seismic stability of buildings, and so on).

The catastrophes in Armenia (1988) and in California (1989) (Table 3.12) (United Nations Environment Programme, 1991) may serve an impressive examples of effects caused by earthquakes similar in strength. Despite the former being weaker than the second one (their magnitudes were 6.9 and 7.1, respectively), the consequences of underground shocks in both populated areas were exactly opposite.

Table 3.11. The most disastrous earthquakes for the period 1980–90 with material damage higher than US$25 million.

Country	Year	Richter magnitude	Number of victims
Yemen (Democratic Republic)	1982	6.0	2,800
Italy	1982	4.5	0
USA	1983	6.5	0
Japan	1983	7.8	104
Turkey	1983	6.9	1,400
Belgium	1983	4.7	2
USA	1984	6.1	0
Italy	1984	5.6	0
Mexico	1985	8.1	10,000
USA	1987	5.9	8
Burma	1988	6.1	730
China	1988	6.1	730
Turkey	1988	7.6	25,000
Former USSR	1988	7.6	25,000
USA	1989	7.1	67
Iran	1990	7.3	50,000

In Armenia, the number of dead totalled 25,000 (according to alternative estimates, it was up to 50,000), and injured or displaced totalled 500,000, while in California 13 people died (i.e. approximately 4,000 times less). The main damage in California was caused to property and totalled US$7 billion.

However, in Armenia, the material damage was even greater – US$14 billion, which is easily explained: destruction was so considerable because, contrary to California, investments in preparations to combat such a catastrophe were inadequate. In particular, data on the seismic potential of the territory were underestimated, different structures were insufficiently earthquake-proof, the people had not been shown what to do during earthquakes, and experience of similar catastrophes in this region was not taken into consideration.

How can we avoid the catastrophic consequences of earthquakes? The most important thing that should be done is to avoid large-scale construction in seismic zones. We should learn from historical experiences (e.g. in Ashkhabad, Turkmenistan and Spitak, Armenia) and build cities in safer places. Nevertheless, one-tenth of our planet's population is concentrated in seismic zones. Of the 100 largest cities of the world, 70% of them may be exposed to underground shocks up to 6 on the Mercalli scale once in 50 years (Figure 3.14). In 25% of these cities, shocks of 8 are quite possible (Degg, 1992; Degg, 1993).

In this context, seismic zoning of the territory and fulfilment of comprehensive observations on seismic activity are also very important, as are mass-scale observations connected with unusual precursors of earthquakes (animal

Table 3.12. Consequences of earthquakes for the period 1988–90.

Date	Country	Number of victims (including displaced)	Damage (US$ million)	Richter magnitude
21 August 1988	India, Nepal	>1,052	25	6.7
6 November 1988	China (Yunnan Province)	730		7.6
7 December 1988	Armenia	25,000 (50,000)	14,000	6.5
23 January 1989	Tajikistan	<284	25	6.0
1 August 1989	Indonesia (Irian Jice Province)	120		5.6
18 October 1989	USA (San Francisco)	63	7,000 (1,000)	7.1
19 October 1989	China (Shani Province)	29		6.1
28 December 1989	Australia (Newcastle)	10		5.5
26 April 1990	China (Kinghei Province)	>126		6.9
29 May 1990	NE Peru	>200		6.2
21 June 1990	Iran (Dkhilan Province)	50,000	8,000	7.4
16 July 1990	Philippines (Luzon Island)	>1,600	>370	7.7
13 December 1990	Italy (East Sicily)	>20	500	4.7

behaviour, etc.) noted by people. A combination of scientific and people's observations formed the basis for correct forecast of the largest earthquake in China (1975) with a magnitude of 7.3 in the area of Haisheng. Evacuation of citizens (which took place twice, second time 5 hours and 36 minutes before the shock) helped to reduce the consequences of the event.

It is clear that a complex of all known measures made it possible to reduce sharply (even by many times) the consequences of underground shocks in California in 1989. On the contrary, failure to understand the importance of such a comprehensive approach (as in Armenia) can aggravated many fold the ruins caused by the earthquake in 1988.

3.4.1.2 Satellite-derived information on earthquakes

As indicated by the data presented above, earthquakes are among the most difficult phenomena to predict. Satellite observations (Grigoryev, 1975; Dusmukhamedov, 1989; Ishakov *et al.*, 1990) can obtain some useful information about earthquakes.

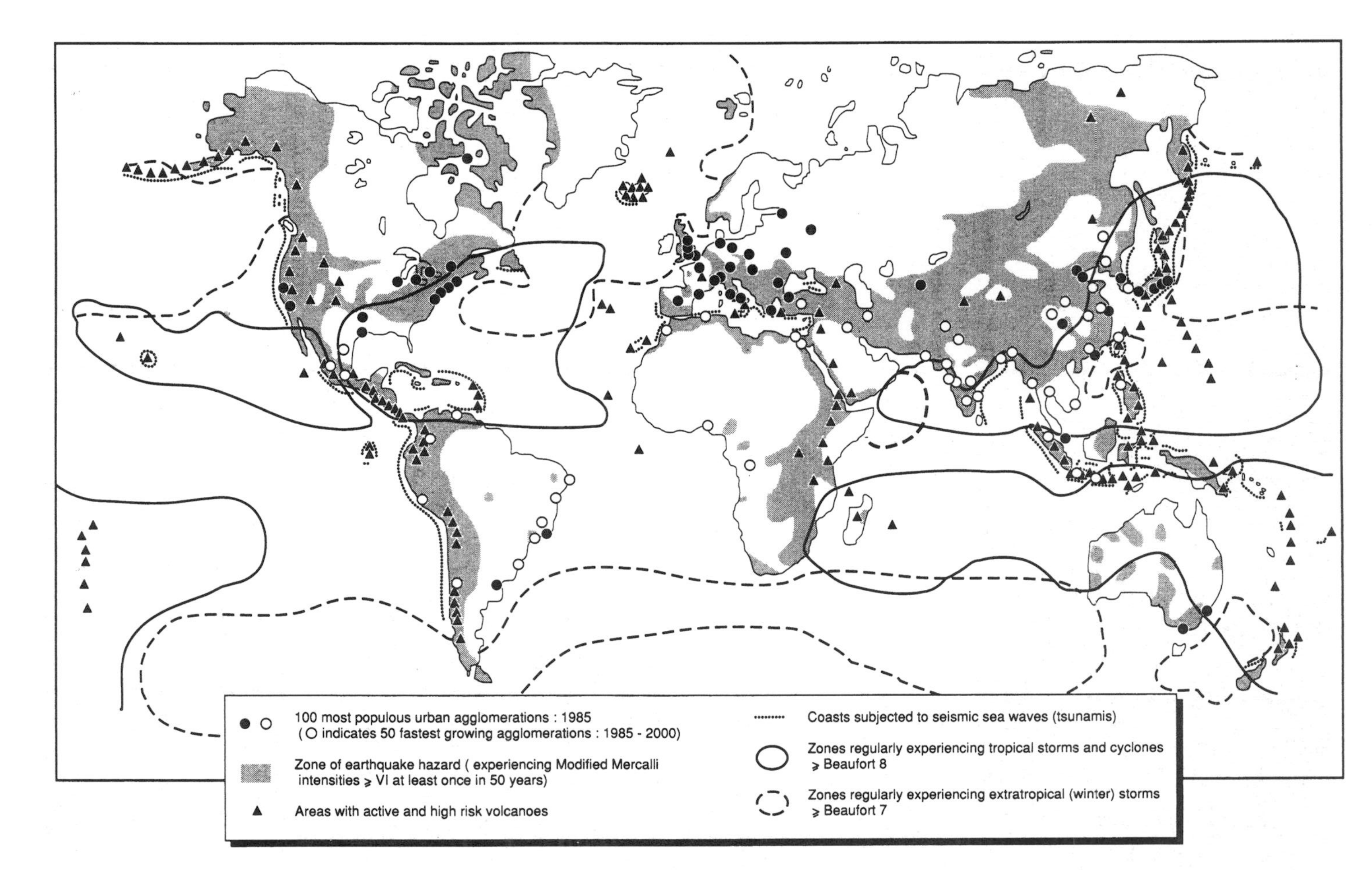

Figure 3.14. Risk to the largest 100 cities due to earthquakes, volcanic eruptions, tsunami, storms and cyclones (Smith, 1997).

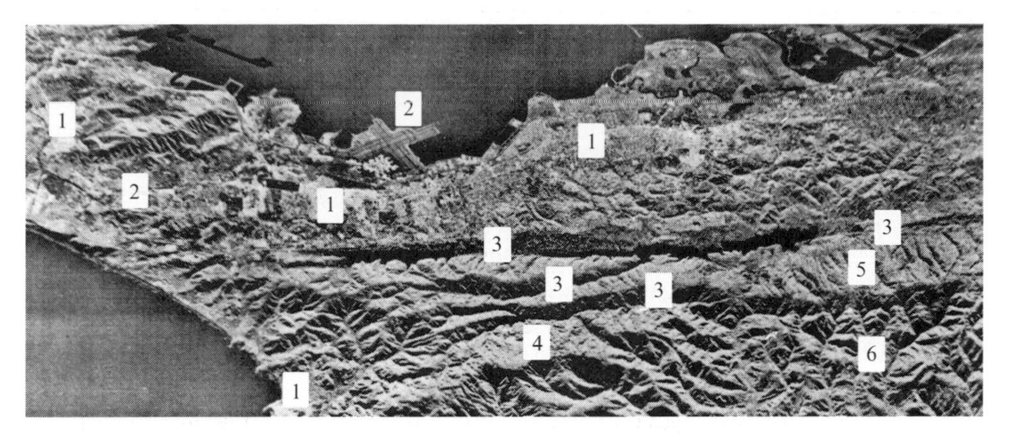

Figure 3.15. High-altitude aerial photograph of the San-Andreas Fault and its branches (3) that cut through the mountain ridges (4, 5, 6) near the Bay of California. Fracture displacements causing earthquakes threatening San Francisco (1 – city housing area, 2 – airport) (Grigoryev and Kondratyev, 2001).

They can be used to reveal and map active tectonic disturbances. Such investigations (from the beginning of the 1970s) opened up the opportunity to establish a correlation between centres of earthquakes and active breaks in various seismic regions of the world, including California, Alaska, and the Caucasus (Figure 3.15). Satellite observations in China proved to be rather promising. In China, ground shocks annually result in the death of more people (54%) than as a result of other natural calamities. This figure exceeds the mean world number of 51%, because this region is characterized by high seismic activity and, moreover, is densely populated. Satellite images of a number of regions of China made it possible not only to reveal many active tectonic disturbances, but also to establish high correlation with the orientation of groups of seismic centres for earthquakes depicted on the newest geological map of 1991. (Jin Zhang and Qi Tuan, 1990).

Satellite images proved to be useful for assessment of the tie-up between tectonic structure and seismic activity in the area of Rachinsk (South Georgia) in 1991. Satellite images of this earthquake location area were used to prepare 11 maps of specific length of lineaments (faults) determining the field of tectonic splitting of mountain rocks at a depth interval of 2–20 km. In doing this, two submeridian zones with low tectonic splitting values and five running blocks were identified. It turned out that vertical and lateral zoning of the field of tectonic splitting of mountain rocks correlated quite well with seismic observation data of the epicentre area (Nechayev *et al.*, 1993).

In the Mackenzie Mountains in Canada (Northwest territories) where a number of strong earthquakes (with Richter magnitudes of 6.5, 6.9, and 6.2, respectively) occurred in recent years (5 October 1985, 23 December 1985, 25 March 1988), satellite images also proved to be useful. They helped to find the correlation between specific features of geomorphological and geological structures and manifestation of seismic activity in this region. It was found that the region of epicentres of underground shocks and aftershocks coincide with the crossing point of a submeridian lineament and sub-latitudinal zone (Moon *et al.*, 1990).

The study performed by Russian scientists of satellite images of an area affected by strong earthquakes in 1976 and 1984 near Gazli in Turkmenition enabled us to take a new glance at the epicentre area of these underground shocks (Gololobov *et al.*, 1994). In so doing, seismoactive and seismogenic breaks were revealed, zones of seismic dislocations were outlined, and certain changes in the landscape were determined. Seismic activity in the region is assessed by two methods. The first analy as the planned change of characteristics of the geoindicator. Among these characteristics, faults (lineaments) are the most physiognomic and dynamic ones, and are easily seen in satellite images.

The second method is a composition coding of multi-zonal satellite image data. The study of tectonic activity evidence and consequences of the two earthquakes near Gazli made it possible to determine the centres of foreshocks and aftershocks between 1979 and 1984. Their location was marked by a concentric distribution of geo-indicators and fields of composition codes of multi-zonal satellite images. What's more, scientists believe that such characteristics of the distribution of the mentioned formations determined from the state of the Earth surface for 1972 and 1975 may be considered foreshocks of the earthquakes in 1976 and 1986, respectively (Afanasyev and Gololobov, 1994).

All these studies of areas prone to active earth shocks and destructive earthquakes (in China, Turkmenistan, and Tajikistan) are based on the high information content of different-type satellite images. The latter not only fix the fracture disturbances themselves that cause earth shocks, but how they change after earthquakes, as well as various relief disturbances. What's more, satellite images fix certain changes in the landscape that occur before earthquakes. The point is that a day before earthquakes, changes take place in fluids circulating over fractures, reaching the earth surface or near-surface mountain rock layers, and influencing surface deposits and plant cover (their reflectivity and thermal features are fixed by remote observations).

The study of other phenomena such as precursers to earthquakes is of great interest. Among them are the gaseous emanations rising to the surface along fracture disturbances. It was noted that, sometimes (in the course of their activation), they lead to formation of clouds that stretch over mountain rock fractures, and are fixed in satellite images (Morozova, 1993). Of special interest is the emergence of specific clouds over active fractures in earthquake regions. For example, it was discovered that, sometimes, cloud rings (like smoke rings) are generated several days before earth shocks over seismic regions. Such objects were fixed in satellite images of Middle Asia in the area of Gazli on the eve of the two strong earthquakes in April and May 1976 (Shultz, 1978).

Analysis of satellite images taken by *NOAA-11* and *12* satellites enabled the discovery of cloud lineaments that formed above many fractures in the course of earthquakes whose epicentres were located near the fractures (Morozova, 1993). Such cloud lineaments were discovered in the course of earthquakes in Armenia in

December 1988. They were observed over the north and east Anatolian fracture and the North Armenian fractures, as well as over territories located quite far from the epicentre of the earthquake (in the square limited by the following co-ordinates: 30° 47°N; 29°–45°E).

Cloud lineaments were also discovered during the earthquake series in March–April 1992 in Turkey. They were, in particular, tiedup with the east and north Anatolian fractures. Development of the anomaly of cloud cover over the latter had already begun a day before (19 April) the earthquake. Huge masses of clouds began to dissipate directly in the area of the fracture.

The cloud cover anomalies seen in satellite images (as banks of clouds or linear bands of dissipated cloudiness) are linked to specific features of energy activity during seismic shocks (negative or positive accelerations due to gravity).

Although it is impossible to predict earthquakes by means of cloud lineaments, an indicator of this kind is useful for revealing tectonically active areas.

Thermal anomalies of the earth surface, fixed in infrared satellite images, sometimes serve as an important indicator of earthquakes. For example, in China, thermal anomalies have been discovered in the areas what major earthquakes form. One such earthquake took place in November 1986 in Yunnan province and two more occurred in October 1989 in Shensi province (in the region of Datong). These temperature anomalies of the earth surface, developed on the eve of the earthquakes, were discovered both from satellite observational data and land-based data (Ma Zonjin and Gao Xianglin, 1991).

Simultaneous development of thermal anomalies (within the timescale of 2 to 10 days) and clusters of frontal cloud fields (occurring 1–2 days before the event) may be precurser events to strong earthquakes. Phenomena of this kind were seen many times in *Meteor* and *NOAA* satellite images directly before strong earthquakes in a number of regions of Central Asia.

A continuous series (100–250 days) of infrared night-time images taken by the *NOAA* satellite (about 10,000 images) were analysed for seismic regions in Central Asia, as well as for Iran and Egypt. A correlation between thermal anomaly occurrence, fractures, and seismic activity phenomena (for a period of 10 years) was found (Tronin, 1996).

Satellite survey is also useful for assessment of the consequences of destructive earthquakes and for determination of the zone and extent of disturbance caused both to natural landscapes and technological structures.

The high resolution (about 10 m) of images obtained from the *SPOT* satellite makes it possible to use them to discover tectonic displacements of about 1 m or less. By displaying images from the *SPOT* satellite obtained on 27 July 1991 and 25 July 1992, details of mountain rock displacements caused by the earthquake of 28 June 1992 in the region of Landers, California were revealed (Crippen and Blom, 1992).

Details of the earth surface structure in the zone of fracture such as shears, overturned blocks, and tension fractures have been fixed. As a result, the direction and magnitude of horizontal shifts along the fracture over the whole width of the zone of displacements can be estimated. It should be emphasized that radar satellite images and interferograms are very promising in the study of tectonic motions (to an accuracy of a few cm). For example, from data obtained by *ERS-1* a shift of mountain rocks by 10–12 cm was discovered after the earthquake on 17 May 1991 in California (*New Views*, 1995).

Infrared images from the *IRS* (*LISS II*) satellite were analysed for one of the regions of the Himalayas with the aim of understanding the changes in relief and landscape as a whole that are due to earthquakes. Photographs taken before and after the earthquake (26 October) made it possible to clearly identify these changes. This opens up the possibility of using these images to obtain real-time information (i.e. at the moment of the catastrophe). By doing so, it is possible to outline to actual locations of the most destructive earthquakes, which, in turn, helps to activate the relevant activities to help the population (Gupta and Joshi, 1989; Gupta *et al.*, 1994).

Satellite images together with air photos and land-based observations were used to estimate the consequences of the Khisorsk earthquake on 23 January 1989 in Tajikistan that had a magnitude of 7–7.5 on the Richter scale (Ishakov *et al.*, 1990). They showed the sequence of events necessary for the occurrence of earthquake consequences. From satellite photos obtained before the earthquake possible consequences were assessed by means of decoding lineaments that determine specific features of relief and its future changes in the course of the earthquake. This is true, in particular, regarding the development of landslides and mudflows. The latter, as well as rock displacement, gave rise to massive destruction of populated areas, roads, and channels.

Satellite information is useful for identification of dangerous areas in case seismic shocks are repeated in highly seismic regions, and for planning measures on minimization of the seismic hazard.

Satellite images are useful for simultaneous (combined) assessment of several interrelated geological risk phenomena. Thus, they have been used to assess geological risk events of three types: volcanic, seismic, and landslides in south Peru (in the area of the Kolka River) in the Andes (Heamen-Rodrigo *et al.*, 1993). Large structural forms, displacements, and fractures that determine the general geological activity in the region were seen in the images obtained by *Landsat* (MSS). More detailed images from the *SPOT* satellite proved to be useful for more detailed mapping of the territory. They were used to map lithostratigraphic features, zones of volcanic rock location, and finally landslide phenomena.

3.4.1.3 The electromagnetic precursory phenomena

The recent experimental results on electromagnetic (EM) precursory signals detected in Japan, Greece, the USA, etc. and the complex field measurements carried out in Japan of ULF magnetic field variations, acoustic emissions (AE), and the most recent theoretical models forwarded for explanation of EM precursory effects are summarized in a recent review article (Varotsos, 2001) and are presented below. We should clarify that the study of EM signals was inititated in the 1980s by Varotsos and co-workers who reported the detection of Seismic Electronic Signals (SESs). The recent developments in the SES studies are summarized below. We should clarify that the study of EM signals was initiated in the 1980s by Varosos and co-workers who reported the detection of Seismic Electric Signals (SESs). The recent developments in SES studies are summarized below.

3.4.1.3.1 Summary of the recent theoretical study into Seismic Electric Signals

SES physical properties (which have been reported on empirical grounds by Varotsos and Alexopoulos, 1984a, b; 1991) seem to be natural consequences of the fact that the current-emitting dipole source always lies in the vicinity of a highly conductive path.

About 20 years ago, Varotsos and Alexopoulos proposed that the Solid State Physics aspects relating to the thermodynamics of defects imply the emission of electric currents from the focal area before an earthquake. We can simplify the picture by saying that they *theoretically* discovered that a rock under heavy pressure emits electric currents even before it is fractured. This universal physical phenomenon should be obeyed by the rocks in the focal area of an impending earthquake and, thus, by recording these currents we can obtain information concerning the magnitude, place, and time of the earthquake before it happens. So, all that remained was to prove that such an earthquake prediction method (Uyeda, 1996) can be applied to actual conditions. This was the reason that the VAN network was set up, in Greece, by Varotsos, Alexopoulos, and Nomicos (the acronym VAN comes from the initials of these investigators), which has been operating for the last 20 years (Varotsos and Lazaridou, 1991; Varotsos *et al.*, 1993; Varotsos *et al.*, 1996a).

Types of telluric anomaly observed before earthquakes

SESs are 'abnormal' variations in currents that flow on the Earth's surface (telluric currents) before an earthquake occurs. When we say 'abnormal', we mean that SESs are signals that have nothing to do with the diurnal variation in telluric currents, magnetic storms, currents due to atmospheric phenomena like lightning, or even the anthropogenic (artificial) noise that also exists in such measurements. They are measured by deploying a couple of electrodes (called measuring dipoles), at a distance L apart, in the Earth and near its surface and measuring the potential difference between them. For the actual application of SESs in earthquake prediction, we should (Varotsos and Lazaridou, 1991, Varotsos *et al.*, 1993) deploy many such dipoles with different directions and lengths because, as we mentioned above, there are many 'normal' variations in telluric currents that have to be identified and discriminated against SESs.

SESs are not the only telluric precursors that are observed before earthquakes by the VAN method (Varotsos and Alexopoulos, 1984a, b; Uyeda *et al.*, 2000). When we study the recordings of these dipoles that have been deployed in a sensitive site for some years, we can easily distinguish a few variations in telluric currents before a major earthquake. Basically, there are three types:

(a) Gradual Variation of the Electric Field (GVEF), which is a gradual change in the electric field (which causes telluric currents) that is observed a few weeks before a major earthquake. GVEF is a very slow variation that takes around a few weeks to reach its maximum amplitude, which is about one order of magnitude greater than SESs. Its main shortcoming, as a reliable earthquake precursor, is the fact that it is not always observed before an earthquake. For example, in two catastrophic earthquakes in Greece, the 1988 Killini-Vartholomio (M = 6) earthquake and the 1986 Kalamata (M = 6.4) earthquake, GVEFs were not recorded even close to the epicentre dipoles, even though SESs were recorded at remote stations.

(b) SESs, which that are observed a few days to several weeks before the earthquake. A detailed study of the properties of SESs can provide information for the time, place, and magnitude of the impending earthquake. SESs may have various types (e.g. signals that look like brushes or hats in the recordings). In general, Varotsos and Lazaridou (1991) propose the following two main types of SES:

   (1) Single SESs (i.e. we recorded an isolated SES on 25 April 1985 at the Assiros [ASS] station). That SES preceded the 30 April 1985 (M = 6.4) earthquake that occurred near Volos. For a single SES, the time lag Δt between the SES emission and the occurrence of the earthquake lies between 7 hours and 11 days.

   (2) Series of SESs or, as it is commonly known SES activity. Such a case corresponds to the SESs recorded on 31 August 1988 at the Ioannina (IOA) station that was one of the three SES sequences (the other two were recorded on 29 September 1988, and 3 October 1988) that preceded the disastrous Killini-Vartholomio earthquakes on 22 September and 15 and 16 October 1988. The last one seriously damaged Vartholomio, but casualties were fortunately avoided, because on 5 October 1988 a public warning was issued by H. Tazief, Minister of France for Natural Disasters, on the basis of scientific evidence provided him by VAN. Figure 3.16 depicts the time correlation between this series of SESs and the actual earthquakes. The detailed properties of SESs will be studied extensively below.

(c) Electric pulses which are expected to be emitted from the seismogenic source at the time of occurrence of the earthquake, travel faster than the seismic wave produced at the same time. This is observed a few seconds or minutes before the earthquake, depending on the distance from the epicentre. This phenomenon is not that different from lightning, which we see before we hear the thunder in the case of thunderstorms. However, the solid Earth crust is conductive and, in contrast to the atmosphere, it significantly attenuates this pulse, making it undetectable at large distances (Varotsos *et al.*, 1997; 2000b). Clearly, such a pulse, even if observed before an earthquake, does not give us enough time to take measures against it, because the time lag before the earthquake is too small.

The likely time sequence of the three types of telluric (electrical) precursor is the following:

(1) GVEFs – a few to several weeks before the earthquake.

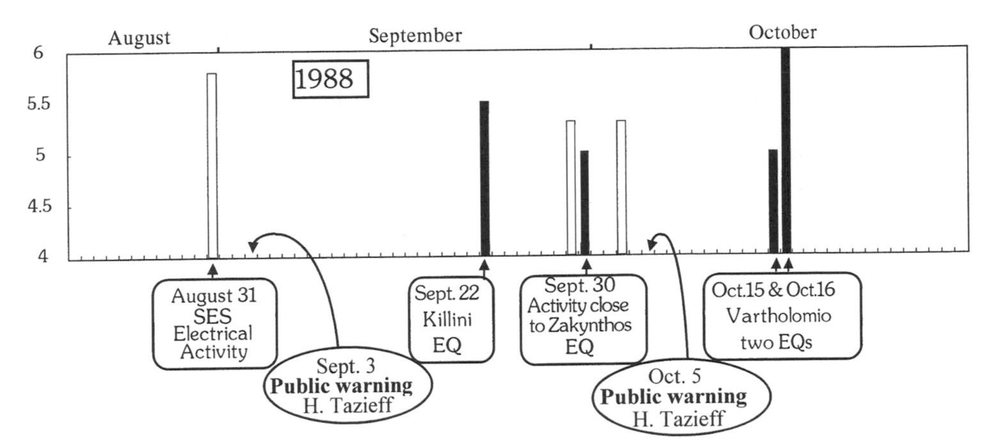

Figure 3.16. The time chart for correlation between SES activity (open lines) and the disastrous Killini-Vartholomio earthquakes (solid line) of 1988 (from Varotsos *et al.*, 1996a, b).

(2) SESs – a few days to a few weeks before the earthquake.
(3) Electric pulses – a few minutes before the earthquake.

However, depending on the type of SES, the above lead times can change and we can have SES activity several weeks before the earthquake.

Relation between SES amplitude and the magnitude of the impending earthquake
Varotsos and Lazaridou (1991) and Varotsos *et al.* (1993) study a variety of SESs that were recorded by the VAN network in Greece at many different observation stations and concluded the relation between the SES amplitude $\Delta V/L$ (with ΔV in mV and L in metres) and the magnitude M of the impending earthquake is (for a given sensitive station and a given epicentral area, and hence, for a given epicentral distance from the station):

$$\log(\Delta V/L) = (0.34 - 0.37)M + b$$

where b is a constant that depends on the observation station and has to be found experimentally, by relating the SES of the station to the magnitude of the respective earthquakes. The value of the proportionality constant 0.34–0.37 is roughly the same for all observation stations and is related to the universal physical phenomenon we mentioned in the introduction.

Relation between the observation station and the epicentre of the impending earthquake
The 20 years of experimentation is Greece has shown (Uyeda *et al.*, 2000; Varotsos *et al.*, 1996a) that SESs are not observed everywhere on the surface of the Earth. Those sites that are sensitive to SESs are called SES-sensitive sites. Moreover, it has been shown that an SES-sensitive site is sensitive only to SESs from some specific focal area (or areas), which are not always close to them. A map identifying those focal area(s) (SESs that are sensed by a site) is called the *selectivity map* of the site. In other words, if an SES is observed at one station then there is a certain area (not

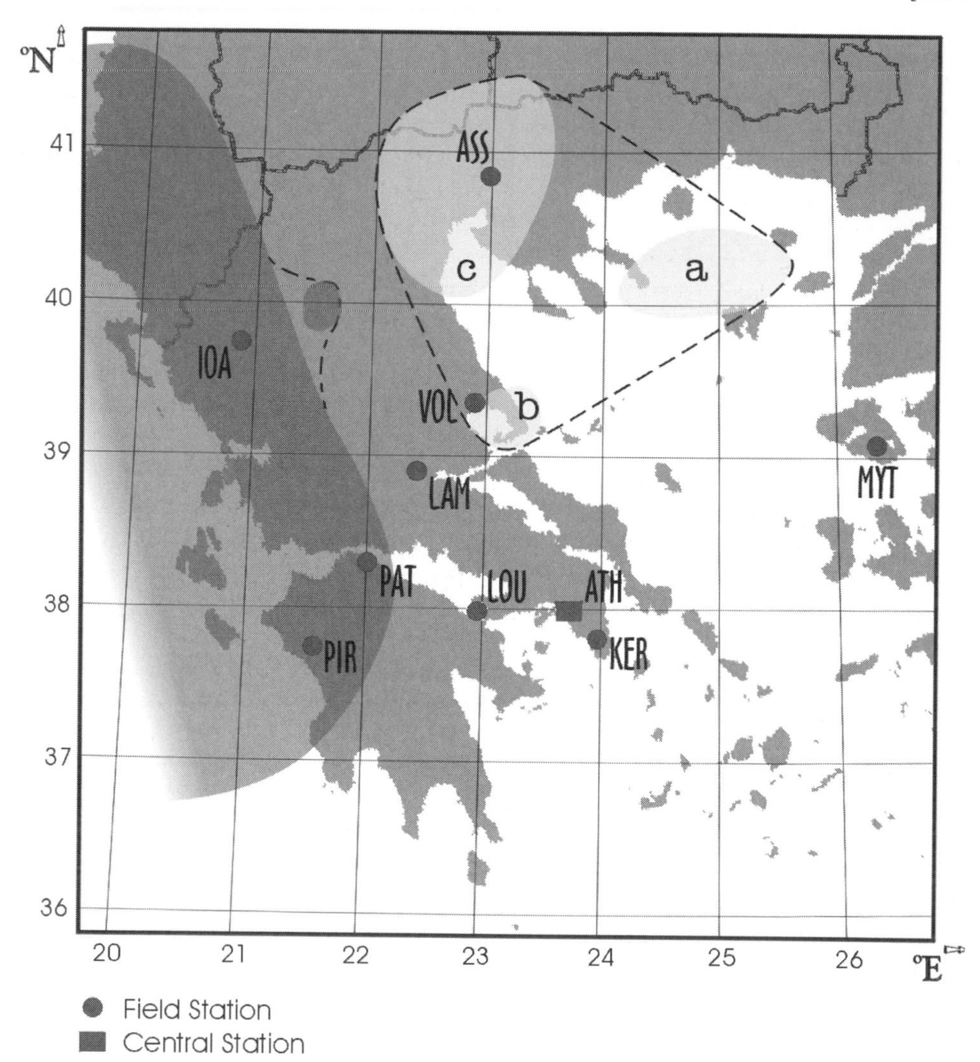

Figure 3.17. The solid dots show the sites of the stations that are telemetrically connected to Athens (ATH). The line-shaded areas correspond to the selectivity map of ASS, i.e., the seismic areas from which SES are collected at station ASS; the dotted line, that surrounds them, indicates that these three are probably interconnected to a larger area (Varotsos *et al.* [1996a]). The large dotted area (to the left) is the selectivity map of IOA (until May 13, 1995) (Varotsos *et al.* [1996b]).

necessarily close to the station) where the epicentre of the impending earthquake lies. The possible epicentral areas can be found using the selectivity map for the station. Figure 3.17 depicts the selectivity map for the IOA and ASS stations, as well as the operating stations of the VAN network in Greece now. The *selectivity phenomenon* is that an observation station collects SESs from many different and, sometimes, distant locations of the epicentre and is the key phenomenon for understanding the physics of the propagation of SESs from the epicentre to the observation station.

In order to determine the epicentre using the present non-dense configuration of the VAN network in Greece, we have to rely on the selectivity map of the station that recorded the SESs (Varotsos *et al.*, 1996a). Then, by considering the SES polarity and the exact value of the ratio of $(\Delta V/L)$ EW for dipoles directed in the east–west direction over $(\Delta V/L)$ NS for dipoles directed in the north–south direction, we can select, as a candidate epicentre region(s), those regions of the selectivity map that, from earlier experience, give SESs with characteristics consistent with those of the given SES.

Physical models for the generation of SES

We mentioned in the introduction that there is a universal physical phenomenon according to which a rock under heavy pressure emits electric currents even before it is fractured. The theoretical discovery of this possibility was made by Varotsos and Alexopoulos (1984a, b), when they studied the thermodynamics of point defects in crystalline solids, and triggered the practical application of this phenomenon for earthquake prediction according to the VAN method.

The currents that are produced this way are called *piezo-stimulated currents* and their origin is closely related with the existence of point defects in crystalline ionic solids (CIS). A CIS is a solid in which the ions (charged atoms) that constitute the solid occupy certain positions in an 'infinite' lattice. In every solid, apart from the ions of the solid, there are also a few foreign ions, which are called impurities (e.g. Ca^{+2} [calcium] ions in a NaCl [sodium cloride] crystal). They usually occupy positions in the same lattice or reside in an interstice between the positions of the lattice, in the latter case they called interstitials. At a given temperature T, a few atoms of the solid, both native and impurities, abandon their positions leaving vacancies in the lattice. Since the ions are charged and the vacancies are also charged in the sense that they represent the absence of a charged particle, electric dipole moments are formed in the crystalline ionic solid. The stress (pressure) that is applied to a CIS forces these electric dipole moments to change their orientation, thus producing an electric current. Moreover, when stress reaches a critical value this (re)orientation is enhanced and a transient current is emitted. CISs are very common in the Earth's crust, and thus it is plausible to expect to measure on the surface of the Earth the transient current that is emitted when the stress reaches the critical value at the seismogenic area before the earthquake.

Apart from the above mechanism, which involves point defects, Slifkin (1996) suggested another *solid state generation mechanism* according to which the current measured by the VAN network may also be produced by the motion of segments of electrically charged dislocations upon a sudden variation in local stress at the seismogenic area. Dislocations are extended which can generally be described as slight changes in the lattice structure that involve many atoms. These dislocations are charged, near their ends, and mobile. So, when the stress is abruptly changed in the seismogenic area, we should expect to have a current due to their motion in addition to the piezo-stimulated currents mentioned previously.

It is interesting to know that the currents emitted by CISs when a stress is applied has actually been observed in *laboratory experiments* for both crystalline (Fishbach and Nowick, 1958) and rock (Hadjicontis and Mavromatou, 1996) samples. Thus, from this fact alone, it is physically expected to observe the similar phenomenon in the Earth's crust before an earthquake.

Finally, another physical mechanism for the generation of SESs, which involves the presence of fluids in the solids of the earth's crust is the *electrokinetic effect*, according to which the movement of water, which is supposed to exist near the seismogenic source, due to the increase of the stress there causes significant electric field changes that can produce an electric current flowing out of the epicentre before the earthquake. Other models (Gershenzon *et al.*, 1989) for explanation of SESs are based on the electrokinetic effect produced near the sensitive station.

Physical models for the propagation of SESs
The physical properties of the SESs:

- $f\,\text{SES} \gg 0.01–0.001$ Hz;
- $(\Delta V/L)$ short dipoles $\gg (\Delta V/L)$ long dipoles;
- $\log(\Delta V/L) \gg (0.34–0.37)M + b$;
- selectivity phenomenon.

These properties (mainly the last one) imply that currents produced by the heavily pressurized rock do not travel homogeneously in the Earth, because in this case we should observe SESs everywhere on the surface of the Earth; rather, they travel concentrated in some paths and then they emerge on the surface as these paths terminate close to it. What sort of paths can concentrate currents in the Earth's crust? Of course, highly conductive paths do exist in the Earth's crust such as graphite sediments or faults. Yes, faults are conductive with a fault resistivity of $\sim$1–10 Ωm (appreciably smaller than the mean resistivity of the host rock, which is of the order of 1,000 Ωm) and, at the same time, earthquakes always occur near faults. So, faults are also very close in Wm to the heavily pressured rocks that emit current. Indeed, the fault system of the Earth's crust can provide an effective 'highway' for the SES to travel from the epicentre to the observation station. The currents travel inside a channel of high conductivity, so they keep a high value although the distance may be long, and, as the channel terminates, they flow out of it giving rise to a detectable electric field in a highly resistive region, where the observation station is located. Thus, two points remain to be done:

(a) to study whether this model is quantitatively plausible and find the physical parameters that much affect the propagation of SESs, and
(b) use the above theoretical knowledge to locate points on the Earth's surface that are SES-sensitive observation stations.

The next section will present the theoretical results that we found concerning the propagation of SESs in the solid Earth crust and try to reveal the essential parameters that determine this phenomenon.

Results from models for propagation of SESs
The study of both numerical and analytical models for the propagation of SESs in the solid Earth crust reveals that the electric field, due to currents emitted in the focal area before an earthquake, is enhanced at certain areas of the Earth's surface due to the existence of a conductive path that 'connects' the epicentral area with the observation station. Note that such a path need not necessarily be connected, since (Varotsos *et al.*, 2000a) a partially disconnected path also works in a similar way when the disconnected parts are not far away from each other. Moreover, the distance between the epicentre and the observation station may be as high as hundreds of km, whereas in the intermediate region the SESs may be not detectable. The electric field that emerges on these sensitive areas varies as slowly as $1/d$ as a function of the distance from the 'edge' of the path (Sarlis *et al.*, 1999; Varotsos *et al.*, 2000a, b). Such a slow variation accounts for the experimentally observed spatial characteristic of SESs:

$$(\Delta V/L) \text{ short dipoles} \approx (\Delta V/L) \text{ long dipoles}$$

which is very useful for discrimination of SESs from 'normal' variations in telluric currents.

Further theoretical study of SESs
The theoretical study of SESs, although it explained the selectivity phenomenon and the spatial characteristics of SESs, has still to continue in order to provide us with the necessary knowledge to understand complementary physics questions such as: What happens to electric signals of higher frequency than SESs. Are SESs accompanied by detectable magnetic field variations in the case of large earthquakes ($M \gg 6.5$).

To answer these questions, research is underway and initial results imply that higher frequency signals (Varotsos *et al.* 1999a; 2000b) cannot propagate through the conductive Earth's crust due to their high attenuation. As for the magnetic field variations that accompany SESs they are too small to be detectable for earthquakes of $M \gg 5.0$ (Sarlis *et al.*, 1999) but their detectability in large earthquakes seems theoretically possible (Varotsos *et al.*, 1996a) and has actually been experimentally observed (Varotsos *et al.*, 1999b).

In conclusion, we can say that research into the physics of SESs should continue since the VAN method is a useful tool for earthquake prediction (see papers given at the RIKEN/NASDA symposium on *The Recent Aspects of Electromagnetic Variations Related With Earthquakes*, Tokyo, December 1999) that has already given successful warnings (Uyeda *et al.*, 2000; Lighthill, 1996a, b; Lighthill *et al.*, 1994) about several large earthquakes that took place within the areas covered by the selectivity maps of the operating stations.

- *Frequency domain (frequency range of SESs [$\leq$1 Hz]).* The frequency dependence of the electric field produced by a current electric dipole lying inside or very close to a highly conductive cylinder (σ), of infinite length, embedded in a significantly less conductive medium (σ'), was studied by Varotsos *et al.*, (2000b). The main conclusion was that, in a host medium containing highly conductive paths, the effective 'skin depth' ($\delta = \lambda/2\pi$ where λ denotes the wavelength) for the transmission is solely governed at low frequencies by the conductivity σ' of the outer medium (i.e. $\sqrt{2/\mu\sigma'\omega}$). This only gives $\delta \approx 50$–100 km, for $\sigma' = 10^{-3}$ S/m, for frequencies $f \leq 0.1$ Hz (Varotsos *et al.*, (1999a; 2000b).

- *Time domain (explanation of the minimum duration of SESs).* Varotsos *et al.*, (1999a), indicated that, in a full conductive volume with conductivity σ' (or resistivity $\rho' = 1/\sigma'$), the diffusion of the current from a delta function – in the time domain – current dipole leads to observation of a dispersed signal that has a duration of around $\mu\sigma' d^2/4$ when it is recorded at a remote point, at a distance d from the source. This case, at low frequencies, is similar to that in the presence of a conductive path (Varotsos *et al.*, 2000b), which is the case of SES propagation in the solid Earth crust. Varotsos *et al.* (2000b) studied this, but for a current dipole emitting source $Il \exp(-t/\tau)I(t)$ with a $Il = 1$ Am, where $I(t)$ is the Heaviside unit-step function. It has been demonstrated that irrespective of the short duration τ_e of the emitted signal, $f(t) = I(t)\exp(-t/\tau_e)$, the signal recorded at a distance d has a duration (larger than or equal with) $\tau_m \approx \mu\sigma' d^2$ due to diffusion. Taking $\sigma' = 10^{-4}$–10^{-3} S/m and $d \approx 100$ km, we find $\tau_m = 3$–30 sec (Varotsos *et al.*, 1999a).
- *Explanation of the selectivity effect.* A model was suggested long ago (e.g. Varotsos *et al.*, 1993), for explanation of the selectivity effect (Varotsos and Lazaridou, 1991). It was assumed that the current dipole emitting source lies near a channel of high conductivity σ embedded in a significantly less conductive medium σ'. The electric field is measured at a site lying on the surface layer close to the upper end of this channel and preferably close to a high-resistivity anomaly. Assuming reasonable values of the ratio σ/σ' and dipole sources of the order of 10^{-3} Akm (which are compatible with $M \sim 5$ according to solid state generation mechanisms), both analytical and numerical solutions (Varotsos *et al.*, 1998; 2000a; Sarlis *et al.*, 1999) find, at distances $d \approx 100$ km in the region close to the edge, detectable electric field values (i.e. ≈ 10 mV/km).
- *The non-detectability of the magnetic field only accompanies the SESs for earthquakes of $M \leq 5$–5.5.* Sarlis *et al.* (1999) using the aforementioned model found that, for dipole sources compatible with $M \sim 5$ and for the region above the upper end of the channel at epicentral distances $r \sim 100$ km, the electric field values are 5–10 mV/km while the calculated magnetic field values are $\sim 10^{-2}$ nT. This justifies the experimental observation (e.g. Varotsos and Alexopoulos, 1984a, b; Varotsos *et al.*, 1996a) that small-amplitude SESs are not accompanied by easily observable variations of the *horizontal* components of the magnetic field.
- *The detectability of magnetic field variations (accompaning the SESs) for earthquakes of $M \sim 6.5$ and explanation of their inclination.* Repeating the above calculation for stronger dipole sources compatible with $M = 6.5$, we find at

$r \sim 100$ km magnetic field values of the order of a few to several tenths of nT. Such magnetic field variations are actually detectable and agree with the experimental observations before the magnitude 6.6 Grevena-Kozani earthquake of 13 May 1995 (Varotsos *et al.*, 1996b, c; 1999b, c). As for the inclination of the magnetic field, the following main result was obtained (Sarlis and Varotsos, 2002): When the electric dipole has a fortuitous orientation and is located close to the centre of the conductive sheet, it was found that when the dipole is directed around the normal to the sheet in the approximate range $-50°$ to $50°$, the magnetic field close to the upper top of the channel has a dominant vertical component. The calculations were made at a depth 10 km. Considering that this depth is the most probable for shallow earthquakes, and the current aspects that the main stress axis (which is almost parallel to the current electric dipole associated with the SESs emission) is also directed around the normal to the fault in a range of angles smaller than the aforementioned one (see Scholz, 2000a, b), we may conclude that the magnetic field accompanying the SES has a dominant vertical component, which is detectable, as explained above.

3.4.1.3.2 Precursory ground-based ultra-low-frequency magnetic field variations

The main body of these observations was reported in the frequency range 0.01–10 Hz and the precursory time from a few hours or days to a few weeks.

3.4.1.3.2.1 Spitak (Armenia, M = 6.9, 8 December 1988) and Loma Prieta earthquakes (USA, M = 7.1, 17 October 1989)

In the Armenian earthquake, the three component magnetovariational observations, filtered in two frequency bands (0.005–1.0 Hz and 0.1–5.0 Hz) were made at a site (the Dusheti station) 130 km from the epicentre. The main result by Kopytenko and co-workers was that, before the main shock and before several strong aftershocks ($M \geq 5.0$), bursts (with duration of the order of 1 hour) were detected a few to several hours before each shock. The amplitude of the vertical component was comparable with those of the horizontal ones, unlike the polarization of natural geomagnetic pulsations. Beyond the bursts (some of which showed a pronounced frequency peak near $f \sim 0.3$–0.5 Hz), the average (day-to-day) intensity of the magnetic field (in the range 0.1–1 Hz) sharply increased just before the main shock and retained this high value (of the order of 0.1 nT) for about 1 month.

Note that the type and duration of bursts as well as their lead time is similar to that reported for the SESs in Greece. We should also recall (see Section 3.3) that, for $M \sim 6.5$ earthquakes, the SESs at epicentral distances $r \sim 100$ km are actually accompanied by magnetic field variations of about a few tenths of nT that are mainly recorded on the vertical component. Thus, there is a striking similarity between the aforementioned Russian observations before the Armenian earthquake and those repeatedly reported in Greece.

In the Loma Prieta earthquake, Fraser-Smith and co-workers, using induction coil magnetometers, reported observations in the frequency range 0.01–10 Hz. The main results could be summarized as follows: a broadband increase was detected in amplitude over 12 days before the earthquake; furthermore, 3 hours before the earthquake, a considerable increase (by a factor of around 5) in amplitude of the lower frequencies was observed.

A comparative study of these observations before the Spitak and Loma Prieta earthquakes was made by Molchanov *et al.* (1992). In spite of several differences, we may say that the general behaviour (e.g. frequency range, lead time of intensification) of these magnetic field variations is similar. It is important that this behaviour is compatible, as already mentioned, with that reported for the magnetic field variations associated with the SESs in Greece.

3.4.1.3.2.2 Guam earthquake ($M = 7.1$, 8 August 1993)

ULF magnetic field measurements have been carried out by a three-axis ring-core-type fluxgate magnetometer with sampling rate 1 s (and hence the highest analysable frequency is about 0.4 Hz) at the Guam Observatory located 65 km from the epicentre (12.98°N, 144.80°E). The data, as well as their analysis and/or plausible interpretation, have been presented in several papers (e.g. Hayakawa *et al.*, 1996a, b; Kawate *et al.*, 1998; Hayakawa *et al.*, 1999). They can be summarized as follows:

(a) ULF disturbances of a noise-like nature (with main frequency band about 0.02–0.05 Hz) with maximum intensity ~0.1 nT have been detected; the disturbances, which have been checked as not being associated with geomagnetic activity, also exhibit the property $B_z > B_h$. Thus, four essential characteristics of these disturbances (i.e. amplitude, $B_z/B_h > 1$, frequency range, and lead time) are similar to those associated with the SESs observed earlier in Greece.
(b) The temporal variation in B_z is similar to that for the Loma Prieta earthquake in the following respect: it shows a broad maximum 10 days to 2 weeks before the earthquake.
(c) When plotting the running mean of the ratio B_z/B_h over 5 days, it is found that this ratio was enhanced from the end of June and that a broad maximum was found to persist for about 1 month until the time of the earthquake.

3.4.1.3.2.3 Biak earthquake, Indonesia ($M = 8.0$, 17 February 1996)

The ULF magnetic field observations were carried out by a three-ring-type fluxgate magnetometer with 1-s sampling rate (and hence the higher analysable frequency is around 0.4 Hz) at Biak Observatory at a distance around 100 km from the epicentre (0.27°S, 136.54°E). These data were compared with similar data collected at a remote site (i.e. at Darwin, Australia about 1,200 km from Biak: Hayakawa, 2000b) and the following conclusions were drawn: The ULF activity at Biak was very much enhanced (compared with that at Darwin) about 1.5 months before the earthquake (in addition to a small peak 2 weeks before the earthquake). This enhancement reaches its maximum just before the earthquake. The difference in their perpendicular components, especially, became more enhanced. Furthermore, when studying the Biak data separately, it seems that the B_z component was also more enhanced before the earthquake.

Hayakawa (2000b) reported that these magnetic field variations were found to be enhanced in the frequency range from 5 mHz ($T = 200$ s) to 30 mHz (~30 s) with a maximum intensity of 0.1–0.3 nT. Thus, the characteristics (frequency range, amplitude, and lead time) are comparable with those associated with the SESs observations in Greece.

We also note that fractal analysis of the data for both earthquakes at Guam (Hayakawa *et al.*, 1999) and at Biak (Hayakawa, 2000b) was independently performed, which showed a power law behaviour of frequency spectrum $f^{-\beta}$; the power exponent was found to exhibit a temporal evolution and approach the critical value of unity about 1 month before these earthquakes.

3.4.1.3.3 Tentative explanation of precursory ULF magnetic field variations

Careful inspection of these ULF magnetic field variations reveals, as was the case for each earthquake separately, that their main characteristics (amplitude, frequency range, lead-time) are strikingly similar to those associated with SESs observed in Greece since 1981. Theoretical calculations have recently shown (see Sarlis and Varotsos, 2002) that, for $M \sim 6.5$ earthquake, the SESs at epicentral distances of $r \sim 100$ km are accompanied by magnetic field variations which (i) have an amplitude of the order of a few tenths of nT and (ii) have a prominent vertical component. The latter emerges as a natural consequence of the point that the emitting-current dipole source must have a significant component normal to the neighbouring fault. Therefore, the experimental fact $B_z > B_h$ seems to have well-founded theoretical justification.

Thus, looked at in this framework, the generation mechanisms for the observed precursory ULF magnetic field variations are those already suggested for SES generation. For the latter, more than 10 models have been forwarded to date. A review of these models and the relevant references can be found elsewhere (Varotsos, 2002). They can be classified in certain classes of mechanisms as follows: The first class of mechanisms considers the current solid state physics aspects that interrelate the emission of electric signals before fracture with the *imperfections* in the crystalline rocks; both point defects and linear defects have been suggested as responsible for SES generation. The second class of mechanisms uses the so-called streaming potential or electrokinetic effect (either in the earthquake preparation zone or close to the measuring site) produced by fluid flow in the crust, driven by pore pressure gradients related to deformation associated with the earthquake preparation. We consider two subclasses in this second class: either 'dilatancy' (or another type of 'overpressure') in the earthquake preparation zone (the fluid flow within the fault zone results in a self-potential distribution on the Earth's surface) or stress gradients at remote sites (with proper inhomogeneities). In the first subclass, the SES corresponds to the electric field originated from electrokinetic phenomena in the earthquake preparation zone, while the second is recorded at those sites that have favourable conditions (e.g. inhomogeneities) for 'local' generation of electrokinetic effects in response to precursory stress (strain) variations that have arisen from the (probably remote) earthquake preparation zone. The third class of mechanisms focuses on fluctuations in physical properties when the earthquake rupture is approached (if the latter is considered critical point). Finally, the fourth class uses other concepts of shallow crustal earthquake preparation (e.g. the magmatic mechanism).

3.4.1.3.4 Complex observations of ground-based ULF and Acoustic Emission (AE)

Acoustic emission (AE) is considered a rather reliable indicator of microfracturing. In an attempt to understand whether this process provides a general, generation mechanism for the detection of ULF magnetic field variations, SESs, and AEs (e.g. Molchanov and Hayakawa, 1995, 1998b; Molchanov, 1999), simultaneous AE and ULF measurements started 2 years ago at Matsuhiro (along with the installation of a ULF measurement network in the Kanto area).

The AE measurements were carried out in a tunnel at a depth of about 100 m (Gorbatikov *et al.*, 1999) using special receivers with magneto-elastic detectors the sensitivity of which increases as the cube of the frequency (i.e. $\propto f^3$. The latter is considered of primary importance because earlier measurements were not convincing due to very weak signals, with their amplitude spectrum sharply declining with frequency f; such an early example is the AE anomaly (in the range 800–1200 Hz) recorded 16 hours before the Spitak earthquake at an epicentral distance of around 80 km.

The AE results are summarized as follows (Hayakawa, 2000b): AE anomaly starts about 12 hours before an earthquake and decays in a comparable time period after it. The signal-to-noise ratio is higher for lower frequencies. It seems that there is a simple magnitude–distance model, in which the AE signal is assumed to be proportional to the square root of seismic energy and attenuates as $D^{-\alpha}$ with $\alpha \approx 1.33$. (This reminds us of approximate $1/r$ attenuation of the current density when an SES happens to be simultaneously recorded at a number of stations; see Varotsos and Alexopoulos, 1984b).

3.4.1.3.4.1 Power emitted from an accelerating body coupled to a wave; tentative explanation of the success of recent AE measurements

We are interested in the electromagnetic radiation emitted from an accelerating charge or acoustic radiation from an accelerating volume element. This radiation damping is reminiscent of problems in various branches of physics (e.g. radiation of phonons in solid state physics). It is therefore useful to clarify the way we obtain the power $P(t)$ emitted from a small accelerating element (with mass m_e and a typical linear dimension R) coupled to a wave, as well as the reaction force F_{rad} due to radiation. The power is proportional to the square of the acceleration but it is important to note that the force F_{rad} is proportional to the *time derivative* of the acceleration (and not to the acceleration, as is usually thought and incorrectly stated). These points could be understood as follows (Jackson, (1975)): Recalling Galilean invariance, $P(t)$ must be zero if the velocity dx/dt is constant, but becomes non-zero when the acceleration is non-zero; assuming symmetry under

$d^2x/dt^2 \rightarrow -d^2x/dt^2$ and performing a Taylor expansion of $P(t)$ in powers of the acceleration d^2x/dt^2, we obtain the relation (keeping the leading term only):

$$P(t) = m_e \tau \left(\frac{d^2x}{dt^2} \right)^2$$

where τ is a characteristic time proportional to R/c with c being the wave velocity. The reaction force F_{rad} is obtained by the requirement of energy conservation (i.e. the radiated power $P(t)$ should be equal to (minus) the work per unit time of F_{rad}) and hence:

$$\int_{t1}^{t2} dt P(t) = -\int_{t1}^{t2} dt F_{\text{rad}} dx/dt;$$

upon integrating the left hand side by parts (and disregarding the boundary term, which is zero for periodic motion) we find $F_{\text{rad}} = m_e \tau d^3x/dt^3$. This expression is important, in the following sense: For a given oscillatory amplitude, the damping due to radiation is proportional to ω^3 in contrast to the usual viscous damping which is $\sim\omega$. (This reflects that, above a certain frequency [around 0.5 Hz for the case of the acoustic radiation that controls friction] the radiation damping can drastically exceed the usual viscous term). This might provide the necessary basis for the explanation (see Varotsos, 2002) of the success of the aforementioned AE measurements, in which the detectors' sensitivity was proportional to f^3.

3.4.1.3.5 Precursory EM signals in the atmosphere and on satellites

3.4.1.3.5.1 Precursory atmospheric electromagnetic effects

Precursory phenomena when studying sub/ionospheric VLF/LF propagation. The method is based on the monitoring of the phase and amplitude of radio signals emitted from navigational transmitters and propagated inside the earth-ionosphere waveguide. When monitoring at a certain site and frequency (i.e., the transmitter frequency with the distance between transmitter-receiver fixed) the parameters of the observed signal are governed by the position of the reflection height; the latter depends on the value and the gradient of the electron density near the atmosphere-ionosphere boundary. This method has been used in detecting earthquake precursory phenomena (Gufeld *et al.*, 1994). In order for the seismogenic perturbations of the atmosphere and lower ionosphere to influence the VLF signal, it seems probable that the future epicentral zone should lie within the so-called Fresnel *zone* (which indicates the zone of the VLF signal sensitivity to perturbations in the medium during propagation). The latter is an elliptical area in which the sites of the transmitter and receiver are foci and the minor semi axis b is given by:

$$b = \lambda D/2(1 + dh^2/D^2)^{1/2}$$

where D is the distance of the two sites, λ the wavelength of the emitted signal and h denotes the height of the reflection point. A striking example reflects to the Kobe earthquake, that will be separately discussed below. Furthermore, during the last 10 years several papers analysed deviations from night-time monthly averages of signal phase and suggested that the phase differences increased in period by about one month to a few days before earthquakes.

Pulse-like signals in the VLF band. Such signals have been also reported by various workers (Oike and Yamada, 1993, Fujinawa and Takahashi, 1994) as precursory phenomena that usually appear a few days before large earthquakes. There are two main streams towards explaining the origin of these pulses: The first one is based on the concept of EM emissions from the seismogenic zone; this is supported by the laboratory measurements (Hadjicontis and Mavromatou, 1995; 1996) which undoubtedly demonstrates a strong EM emission before failure. The second stream is based on the concept that the emission originates in the atmosphere as a result of some modification in lightning activity. For example, there are suggestions (Hayakawa *et al.*, 2000b and references therein) that there is an increase of 'cloud-cloud' and 'cloud-ionosphere' strokes (which are *not* usually recorded by conventional lightning monitoring) due to seismo-associated gas release and decrease of the usual 'cloud-ground' strokes.

Beyond the aforementioned two classes of precursory phenomena (study of subionospheric VLF/LF propagation, pulse-like VLF signals), other seismo-associated electromagnetic effects have also been reported. Such reports can be found in the relevant monographs edited by Hayakawa and Fujinawa (1994) and by Hayakawa (1999).

3.4.1.3.5.2 The anomaly in the subionospheric VLF propagation observed before the Kobe earthquake

The receiver was located at Inubo (35°42′N, 140°52′E, near Tokyo) and the transmitter of the VLF Omega signal (10.2 kHz) at Tsushima (34°37′N, 129°27′E). Thus the distance D between the two sites is approximately 1043 km, the epicentre of the Kobe earthquake lies only 70 km from the main VLF signal path (i.e., the great circle path between transmitter and receiver). Considering also that the wavelength is 29.4 km and the height 80 km the aforementioned formula gives for the minor semiaxis of the Fresnel zone $b \approx 100$ km; thus plotting the Fresnel zone we see that the epicentre lies well inside this zone.

Hayakawa *et al.*, (1996c) (see also Molchanov *et al.*, 1998, Molchanov and Hayakawa, 1998a) reported a significant shift in the terminator times a few days before the Kobe earthquake. The terminator is defined as the time when the diurnal phase variation exhibits a minimum around sunrise and sunset; these minima are called morning and evening *terminator times* and are labelled t_m and t_e respectively. The sequential plot of diurnal variations during the period 3 January to 23 January 1995 (the Kobe earthquake occurred on the 17 January) at Inubo reveals that t_m shifts to earlier hours while t_e to later hours. This effect is also confirmed when analysing a much longer period and then it is clear that there is a significant propagation anomaly a few days before the earthquake. As also noticed by Hayakawa (2000a) this is likely to be associated with the earthquake, because it is not correlated with other phenomena like magnetic activity, solar activity, rainfall, etc. A probable interpretation of this anomaly was forwarded (Hayakawa *et al.*, 1996c, Molchanov *et al.*, 1998), stating that the lower ionosphere might have been lowered by a few kilometers probably connected to the intense radon emission observed before the earthquake. This explanation was reconsidered by Rodger *et al.* (1999) who applied more realistic propagation models. The latter indicated that the changes required in VLF reflection height to produce the observed effects are likely to be considerably larger (~4–11 km) than those (i.e., 1 to 2 km drop) suggested previously. Rodger *et al.* (1999) also state that if the reported terminator time changes were caused by alterations in the VLF reflection height associated in some manner with an imminent earthquake, these effects would be commensurate with the effects due to a solar flare. However, this would lead to changes in received amplitude (or phase) that would be significant at all times, and not just during the day/night transition. Thus, Rodger *et al.* (1999) argue that is not all clear that a simple height-lowering explanation for this effect is correct.

Motivated by the above result, Molchanov and Hayakawa (1998a) proceeded to a similar analysis (terminator times) of the 13 year data of the Inubo receiver for Tsushima Omega transmission. Studying a number of earthquakes with $M > 6.0$, they found that when the earthquake is shallow (depth less than 50 km) and located very close to the aforementioned great-circle-path, a significant shift in terminator times (and hence ionospheric perturbations) is actually observed. In addition they found that the data of terminator times have exhibited a modulation with periods of 5 days or 9–11 days. Such 'oscillations' had been observed in many atmospheric parameters, usually termed as atmospheric waves, planetary waves, atmospheric oscillations or Madden–Julian oscillations; they are believed to be normal mode Rosby waves, exhibiting planetary-scale westward moving features that represent the resonant states of the atmosphere (Hayakawa, 2000b). Thus this observed modulation implies that the atmospheric oscillations with such periods may play an important role in the coupling from the lithosphere to the ionosphere.

In summary, the main characteristics of the up to date observed VLF signals are (Hayakawa, 2000b):

(1) Such signals have been detected only for crustal earthquakes (while a relevant study for deep strong earthquakes led to negative conclusions). Such a detection is possible only when the epicentre is inside the Fresnel zone.
(2) Transient oscillations with a period 5–10 days appear a few days before a strong earthquake and disappear a few days or weeks after it. It seems that precursory changes appear only when resonant atmospheric oscillations also exist.

The models suggested for the explanation of precursory electromagnetic effects are summarized in Figure 3.18 and will be discussed later in Section 3.4.1.3.6.

3.4.1.3.5.3 Precursory signals claimed to be detected on satellites

We start with a general remark concerning the ionosphere: during the last decades, ground-based radiosounding techniques, as well as direct measurements of the ionosphere parameters on board satellites and rockets, revealed that the ionosphere has a

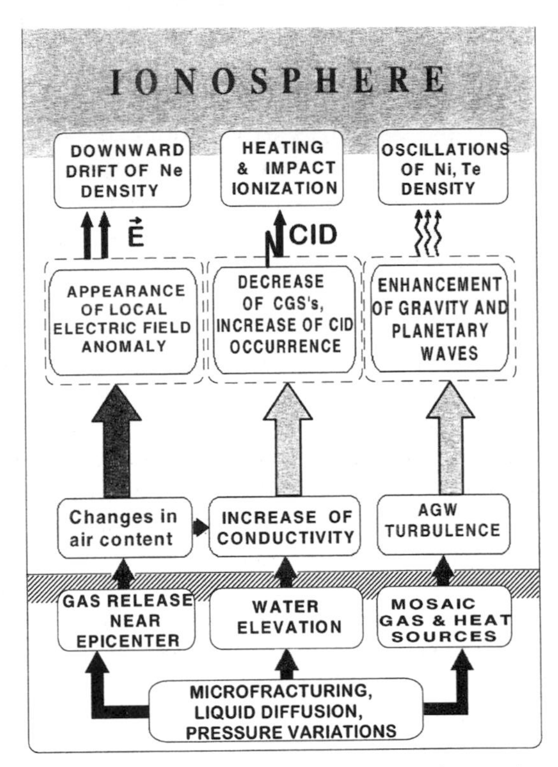

Figure 3.18. Schematic diagram of the three basic approaches followed to explain precursory electromagnetic effects (Hayakawa, 2000b)

strongly irregular structure with spatial scales of irregularities ranging from several centimeters to hundreds of kilometers. Several papers describing the experimental studies of the irregularities in the ionosphere as well as plausible mechanisms to explain them have been published (Kelley, 1989).

Before going to specific examples we note that, there are a lot of controversial reports concerning the statistical significance of signals observed by satellites. However, Hayakawa (2000a) recently reported that it seems likely that there may be some kind of perturbations in the ionosphere associated with earthquakes; beyond VLF subionospheric signals which are believed to be reflected from the lower D region of the ionosphere, precursory disturbances onto the upper ionosphere may also exist.

Examples of precursory variations. Satellite data reveal that both *small-scale* and *large-scale* variations in the ionospheric parameters occur a few to several days before strong earthquakes. Concerning the small-scale variations (e.g. 4–10 km along the orbit of *Cosmos-1809* satellite before the Spitak earthquake), inhomogeneities of electron density with scale $\ll$10 km arise in the upper ionosphere before an earthquake. It seems that they are localized in the geomagnetic force tube with the root on the projection of an epicentral zone onto the ionosphere E layer (Chmyrev *et al.*, 1999 and references therein). The disturbed region of these inhomogeneities has a horizontal size around 300–450 km. Such fluctuations of the plasma density ΔN (of the order of $\Delta N/N \geq 5\%$ and with characteristic spatial scales 4–10 km) are not typical for the midlatitude ionosphere under normal conditions. Concerning the large-scale variations, we mention, as an example, the significant increase (~50%) of the height of the F-layer and the considerable decrease (by a factor of 3) of the maximum value of electron density one day before the Spitak earthquake (Sorokin and Chmyrev, 1999a and references therein).

Beyond the above electron density fluctuations, other signals obtained from satellite data have been reported to preceed earthquakes (Sorokin and Chmyrev, 1999a and references therein): As a first example we mention the variations ~3 pT of the two horizontal components of the geomagnetic field in the frequency range 0.1–8 Hz together with a change $3 \sim 7$ mV/m of the vertical component of the d.c. electric field 15 min before a $M = 4.8$ earthquake. The second example refers to the Spitak earthquake: ELF/VLF measurements onboard the *Cosmos-1809* satellite over the Spitak region showed that seismogenic ELF emissions were generated in the zone <6 deg in longitude and 2–4 deg in latitude with respect to the earthquake focus; the measured intensities were about 10 pT at $f \sim 140$ Hz (in a frequency band of 25 Hz) and about 3 pT at $f \sim 145$ Hz (in a frequency band of 75 Hz). Furthermore, it seems that during the Spitak earthquake preparation stage the H^+ and He^+ ion densities in the ionosphere increased by a factor of ~3.

We also note that Afonin *et al.* (1999) found a reliable correlation between the global distribution of seismic activity and ion density variations. The maximum values of the normalized standard deviation of the ion density correlated with seismic activity are 10–15%; clear correlation was found only for daytime (10–16 LT) and altitude range from 500–700 km. In addition it was reported that the equatorial anomaly common at low latitude is strongly modified by the seismic activity (Hayakawa, 2000a).

Thermal IR satellite data. There are several old reports concerning observations of surface and near surface temperature changes prior to crustal earthquakes. Furthermore, some laboratory measurements have reported thermal changes due to stress fields. Thus, in view of these field and laboratory reports, it was found worthwhile to investigate (for the plausible existence of precursory variations) the satellite data of remote sensing in the IR spectrum (Tronin, 1999); a monitoring of the Earth's thermal field with a spatial resolution of 0.5–5.0 km and with a temperature resolution of 0.12–0.5°C is currently available. Such studies are currently carried out in China, Japan and the European Union (Hayakawa, 2000b and references therein).

The preliminary results in China show that there are thermal anomalies that appeared about 6–24 days before earthquakes and continued about the same time after earthquakes; the anomaly extends over a region around (700×50) km and has an amplitude around 3°C. In Japan, the thermal anomalies (with amplitude up to

6°C) seem to have smaller size and appear around 10 days before earthquakes (Hayakawa, 2000b). We should clarify however, that the nature of these (stable and non-stable) anomalies is not clear.

3.4.1.3.6 Theoretical models for the explanation of the precursory EM phenomena

Several models have been developed to explain the EM precursors at ionospheric heights. The most popular model assumes that EM signals are generated at a small depth under the Earth's surface and then propagate to the ionosphere through the layers of the Earth and the atmosphere. Another model assumes that the primary signals are generated under the Earth's surface (or in the air), but have an acoustic nature; they reach the ionosphere and produce EM perturbations due to the sound–plasma interaction. In addition to these models, that consider underground electrical or acoustic 'sources' respectively, the following mechanism was also suggested: radioactive gas emanation from the ground occurs above the earthquake preparation region; this produces an additional ionization of the air near the ground, thus leading to its conductivity growth from σ_0 to σ_1. Such a growth of the air conductivity by itself cannot give, of course, rise to electric fields and currents; however, this may occur if we recall that the gap between the Earth's surface and ionosphere forms a peculiar spherical capacitor charged to a significant voltage mainly due to the global thunderstorm activity. If gas emanation is assumed to be more or less homogeneous. The current lines bend in different directions, which leads to increase or decrease of current density in different regions of the ionosphere. The picture as a whole resembles the scattering of light rays after passing through a glass plate with distorted thickness; in such a case, focused and defocused light regions appear.

In summary, the explanation on the precursory EM phenomena has been attempted by the following three basic approaches, which are schematically illustrated in Figure 3.18: (i) changes in the tropospheric chemistry due to release of neutral and/or charged gas from the seismogenic zone, (ii) modification of lightning activity due to a change of the near-surface conductivity and (iii) intensification of acoustic-gravity turbulence, which will be shortly discussed below. We note that the following suggestion has also been made (Hayakawa, 2000b): these VLF transients are a triggered phenomenon in the critical state of the upper atmosphere and the specific resonant 5–10 days oscillation are just an indicator of such a nearby critical state.

3.4.1.3.6.1 Dissipative instability of acoustic-gravity waves in the ionosphere

Such instability has been suggested by Chmyrev *et al.* (1999). The argument goes as follows (see the chain of processes depicted in the left-hand side of Figure 3.19): In the low-frequency approximation, the ionosphere is considered as a continuous medium with tensor

Figure 3.19 illustrates the chain of processes forming the model of the ionospheric earthquake precursors (Sorokin and Chmyrev, 1999a) conductivity. In such a

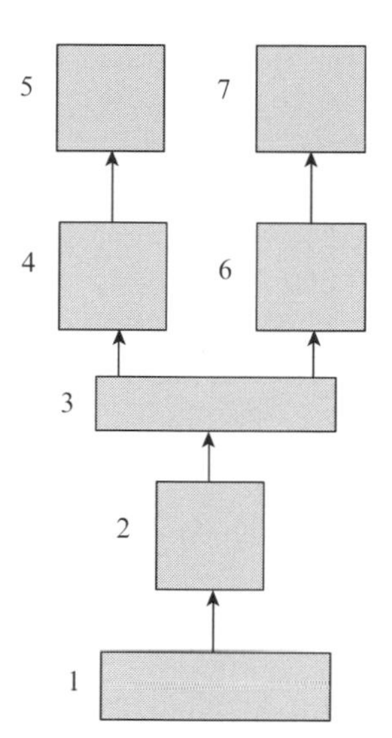

1. Injection of active substances into the atmosphere.
2. Modification of altitude profits of the electric conductivity in the atmosphere.
   Perturbation of the electric fields in the atmosphere.
3. Modification of conductivity and the electric field in the ionosphere.
   Perturbation of ionization in the lower ionosphere.
4. Dissipative instability of acoustic-gravity waves.
   Formation of the conductivity inhomogeneities in the ionosphere.
5. Field-aligned currents.
   Plasma density fluctuations, ULF/ELF-magnetic field oscillations.
6. Joule heating in the ionospheric *E*-layer.
   Formation of the vertical plasma motion.
7. Temperature increase and upward plasma transport.
   Ionospheric *F*2 layer deformation. Growth of the light ion density.

Figure 3.19. Schematic diagram illustrating the chain of processes forming the model of the ionospheric earthquake precursors (Sorokin and Chmyrev, 1999a).

medium, it is shown that a propagation of small acoustic oscillations is accompanied with a conductivity disturbance and hence with a perturbation of current. These disturbances, under certain conditions, have such a character that Joule heating of the distorted currents results in a growth of the acoustic wave amplitude. Thus, in contrast to earlier publications, Chmyrev *et al.* (1999) propose, as an energy source of the instability under consideration, the electromotive force of the external electric field. They suggest that the electric field energy transforms to the energy of acoustic oscillations without a change in the medium thermal balance.

Following Chmyrev *et al.* (1999), in the frame of the above model, the observed (in the satellite data) fluctuations ΔN_e of a plasma density result from spacecraft crossing the field-aligned plasma density irregularities. These irregularities arise from the formation of the field-aligned currents (with a transverse scale $\leq$10 km) excited in the lower atmosphere. A plausible mechanism generating a wave of field-aligned currents (shear Alfvén wave) is the aforementioned process of formation of the horizontal irregular structure of the ionosphere conductivity and its interaction with the ionospheric dc electric field. Such conductivity variations arise in a localised zone in the ionosphere over the earthquake preparation area. Thus, the observed precursory plasma density fluctuations, as well as the ULF/ELF magnetic field variations excited in the upper atmosphere, could be related to the physical processes resulting in intensification of acoustic gravity waves above the earthquake preparation zone.

Thus, Chmyrev *et al.* (1999) suggest that in the ionosphere the disipative instability of acoustic-gravity waves can arise. (The growth rate of this instability is proportional to the square of the external electric field.) This instability leads to a growth of waves with the frequencies of the order of the Brunt–Vaisala frequency $\omega_g(\sim 2 \times 10^{-2}\,\mathrm{s}^{-1})$; recall that:

$$\omega_g^2 = \frac{(\gamma - 1)g}{H}$$

where $\gamma \approx 1.4$ and that, in the simple model of a layered isothermal ionosphere, we consider an isothermal ($T_0 = \text{const}$) exponential irregular medium in which the neutral atmosphere density ρ_0 varies in the z-axis (directed vertically upward) as $\rho_0 \sim \exp(-zH)$ and $H = RT_0/g$. As a result, conductivity irregularities are excited in the lower ionosphere having a horizontal spatial scale l obeying the relation:

$$l = \lambda/2 = \eta v_g/\omega_g = \pi\alpha/\omega_g n(\omega_g)$$

λ is the wavelength of AGW, v_g the (minimum) phase velocity and $n(w_g)$ the refractive index. These conductivity irregularities (which move with velocities considerably *less* than the sound velocity) change the ionospheric electric field and the excitation of horizontal spatial structure of conductivity in the lower atmosphere results in formation of plasma layers stretched along the geomagnetic field. The transverse spatial scale of the layer coincides with the scale of conductivity irregularities. When a satellite crosses the field-aligned currents (carried out by electrons) and the related field-aligned plasma irregularities the oscillations of plasma density $\Delta N/N_0$ and ULF oscillations of the geomagnetic field are registered on the satellite with a period Δt determined from the relation:

$$\Delta t = l/v_s = \pi\alpha/v_s\omega_g n(\omega_g)$$

where v_s denotes the satellite velocity. Considering an acoustic velocity $\alpha = 3.10^4$ cm/s and taking $v_s \approx 10^6$ cm/s and $n \approx 1$–10 we find at $\omega_g = 2 \times 10^{-2}\,\mathrm{s}^{-1}$ the value $\Delta t \approx 0.3$–3 s which roughly agrees with the satellite data.

3.4.1.3.6.2 The heating model

Such a model was presented by Sorokin and Chmyrev (1999b) and the basic scheme is as follows (see the chain of processes depicted in the right hand-side of Figure 3.19): Above the zone of an impending earthquake a growth of the electric field (as well as conductivity growth) occurs in the ionosphere. Such a growth leads to an additional Joule heating of the lower ionosphere. This heat injection and the related increase of the gas temperature result in a vertical flux of neutral particles. This produces an elevation of the F-layer maximum, a decrease of the ion density in the maximum of this layer and a growth of the light ion density in the upper part of the ionosphere.

The starting points of the above model (i.e. growth of the electric field and the resulting Joule heating) are justified as follows: The growth of the electric field could be attributed to an intensification of aerosol and gas injection from the earthquake preparation zone, which can intensify the lower ionosphere ionisation. This increase of the electric field leads to intensification of ionospheric currents flowing in a thin conducting layer and a release of Joule heat in the disturbed region. The importance of such a release can be roughly estimated from the relation $q \sim \Sigma E^2$, where q stands for the thermal flux, Σ is the integral conductivity and E the horizontal electric field. Considering the values $\Sigma = 3 \times 10^{12}$–$3 \times 10^{13}$ cm/s, $E \approx 6$ mV/m we find $q = 0.1$–

1 erg/cm^{2}s; the significance of this value becomes evident if we compare it with the value $q_0 \approx$ a few erg/cm^{2}s, where q_0 denotes the thermal flux to the ionosphere driven by the absorption of the solar short wave variation ($\lambda < 1,026$ Å). Thus, it seems that the Joule heating of the ionosphere above the earthquake preparation zone supplies a considerable fraction of the total balance in the ionosphere and hence influences decisively the Ionosphere State.

A model that includes both: the acoustic-gravity waves instability and additional Joule heating. A physical model that includes both effects has been presented by Sorokin and Chmyrev (1999a). The chain of the relevant processes is depicted schematically in Figure 3.27. The injection of radioactive materials into the atmosphere and the increase of the ionospheric electric field before an earthquake are assumed to be the source of the processes. The growth of the electric field results in the dissipative instability of acoustic-gravity waves at the so-called Brunt–Vaisala frequency mentioned in Section 5. This instability is followed by formation of horizontal periodic irregularities in the ionospheric conductivity (Block 4). Interaction of these irregularities with the electric field is followed by upward radiation of the directed Alfvén wave; as a result, the layers of the field-aligned currents and plasma density stretched along the geomagnetic field arise. When a satellite crosses these layers, it registers the plasma density fluctuations and the ULF/ELF geomagnetic fluctuations (Block 5). The other effect of the electric field increase yields an additional source of Joule heating in the lower ionosphere (Block 6), which reflects a variation of the ionosphere thermal balance; a downward thermal flux is balanced by an upward mass transport and their balance determines the value of the flux and the ionosphere temperature (which are connected with the ionosphere electric field). In the presence of this vertical mass flux and of a dependence of temperature on the electric field, a deformation of the F-layer takes place; the calculations show that in the lower ionosphere a growth of atomic oxygen density occurs, which is followed by a growth of the light ion densities. We note that the model under discussion foresees a chain of processes that reflect the appearance of several precursory changes among which we have: an increase of the ionosphere electric field, an upward motion of the ionosphere F-layer, a decrease of the maximum value of the ion density in the F-layer and an increase of the light ion density in the top of the ionosphere above the earthquake preparation zone.

In summary, the aforementioned model of Sorokin and Chmyrev (1999a) reduces numerous effects on space plasma to a single cause: the increase of quasistationary electric field in the ionosphere. This field is controlled by the dynamics of lithosphere processes through variations of the lower atmosphere conductivity under the influence of the injection of active products into the atmosphere (by considering the atmosphere ionisation due to the radioactive radiation of these products).

3.4.1.3.7 *Spatio-temporal complexity aspects of Seismic Electric Signals. The concept of natural time*

Seismic Electric Signals (SESs) are, as mentioned in Section 3.4.1.3.1, low-frequency (≤ 1 Hz) changes in the electric field (E) of the Earth. They have been detected in

Greece and Japan and found to precede earthquakes with lead times ranging from several hours to a few months. Their analysis may well lead to estimation of the epicentral area. Varotsos *et al.* (2001a) recently showed that the spectral content of seismicity in areas that are candidates for earthquakes ultimately depends on the spectral content of SES activity, just before the occurrence of the main shock. The key point is that both spectra have to be calculated in a new time domain, termed 'natural time', starting from the instant of the SES recording. Thus, since the spectrum of the SES is known well in advance, continuous inspection of the spectrum of evolving seismic activity may lead to estimation of the time window of the impending main shock with an accuracy of around a few days.

The interrelation between the time evolution of seismic activity (measured from the start of SES recording) and the spectrum characteristics of SESs can only be achieved if we depart from conventional time t and think in terms of *natural time* χ (Varotsos *et al.*, 2001a, 2002). The latter serves as an index for the occurrence of an event (reduced by the total number of events, thus it is smaller than, or equal to, unity). Let us, therefore, denote by Q_k the duration of the kth transient pulse (single SES) of SES activity comprised of N pulses (Figure 3.20). The natural time χ is introduced by ascribing to this pulse the value $\chi_k = k/N$. By considering the evolution (χ_k, Q_k), we can define the continuous function $F(\omega)$ (this should not be confused with the discrete Fourier transform):

$$F(\omega) = \sum_{k=1}^{N} Q_k \exp\left(i\omega\frac{k}{N}\right)$$

where $\omega = 2\pi\phi$, and ϕ stands for the *natural frequency*. We normalize $F(\omega)$ by dividing it by $F(0)$:

$$\Phi(\omega) = \frac{\sum_{k=1}^{N} Q_k \exp\left(i\omega\frac{k}{N}\right)}{\sum_{k=1}^{N} Q_k} = \sum_{k=1}^{N} p_k \exp\left(i\omega\frac{k}{N}\right)$$

where $p_k = Q_k / \sum_{n=1}^{N} Q_n$. Thus, the quantities p_k describe the 'probability' of observing the transient at natural time χ_k. The normalized power spectrum $\Pi(\omega)$ can be obtained from:

$$\Pi(\omega) = |\Phi(\omega)|^2$$

For natural frequencies ϕ less than 0.5, $\Pi(\omega)$ or $\Pi(\phi)$ reduce to a characteristic function for the probability distribution p_k in the context of prabability theory. The procedure of reading a series of electric pulses in the natural time domain is depicted in Figure 3.20C. The evolution of seismic activity should be considered in the same framework by ascribing to the kth event, instead of Q_k, the corresponding seismic moment M_0 (Figure 3.20D). In what follows, we summarize the results of the applications of this procedure to the data about the four strongest earthquakes

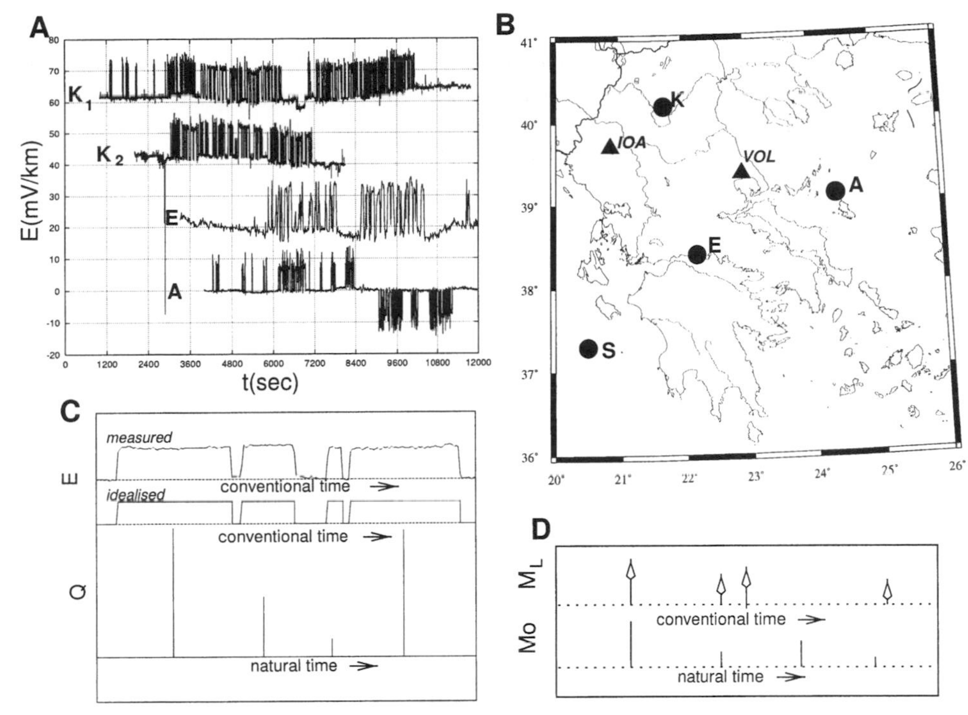

Figure 3.20. (A) SES activities recorded before the main shocks K, E, S and A given in Table 3.13; K_1 and K_2 refer to the two SES activities before the EQ labelled K. The upper two SES activities were recorded at IOA, while the lower two at VOL. (B) Map showing the EQ epicentres (cycles) and the sites (triangles) of the measuring SES stations. How a series of electric pulses (C) or a series of seismic events (D) can be read in 'natural time'. In both cases the time serves as an index of the occurrence of each event (reduced by the total number of events), while the amplitude is proportional to (C) the duration of each electric pulse and (D) to the seismic moment M_0 (Varotsos *et al.*, 2001a).

(EQs) (labelled K, E, S, and A, Table 3.13 and Figure 3.20B) that have occurred in Greece since 1988 (Varotsos *et al.*, 2001a).

We start with analysis of the SES activities. Once SES activity has been recorded, we can read it in the natural time domain and then proceed to analyse it. Figure 3.21 depicts, as an example, $\Pi(\phi)$ for SES activity, along with a number of artificial noises that have similar features to SESs. Inspection of this figure shows the following three facts: first, the curves practically fall into two different classes, labelled 'noises' and 'SES activity', respectively. This classification, which is theoretically understood (see below), provides a tool that distinguishes between noises and SESs. Such a distinction can be made alternatively (e.g. by ensuring that electric field variations precede associated variations in magnetic field by a lead time of the order of 1 sec: Varotsos *et al.*, 2001b). Second, Figure 3.21 reveals that, for *natural frequencies* smaller than 0.5, the corresponding $\Pi(\phi)$ values of SES activity scatter around the dotted curve, which has been estimated by theoretical considerations

Table 3.13. All EQs with MS(USGS) $\geq$ **6.0** within $N_{36.5}^{41.5}$ $E_{19.0}^{26.0}$ and relevant SES activity (Varotsos *et al.*, 2001a).

	Main earthquakes				SES activity			Regional considered
EQ label	Date	Time	Epicentre	MS(USGS)	Date	Time	Station	Coordinates
K	13 May 1995	08:47	40.2N–21.7E	6.6	18 and 19 April 1995	10:04	IOA	$N_{39.2}^{40.5}$ $E_{20.3}^{22.0}$
E	15 June 1995	00:15	38.4N–22.2E	6.5	30 April 1995	05:41	VOL	$N_{37.5}^{39.7}$ $E_{21.2}^{25.0}$
S	18 November 1997	13:07	37.3N–20.5E	6.4	3 October 1997	18:24	IOA	$N_{37.0}^{38.5}$ $E_{20.3}^{21.7}$
A	26 July 2001	00:21	39.1N–24.4E	6.6	17 March 2001	15:34	VOL	$N_{38.7}^{39.5}$ $E_{22.0}^{25.0}$

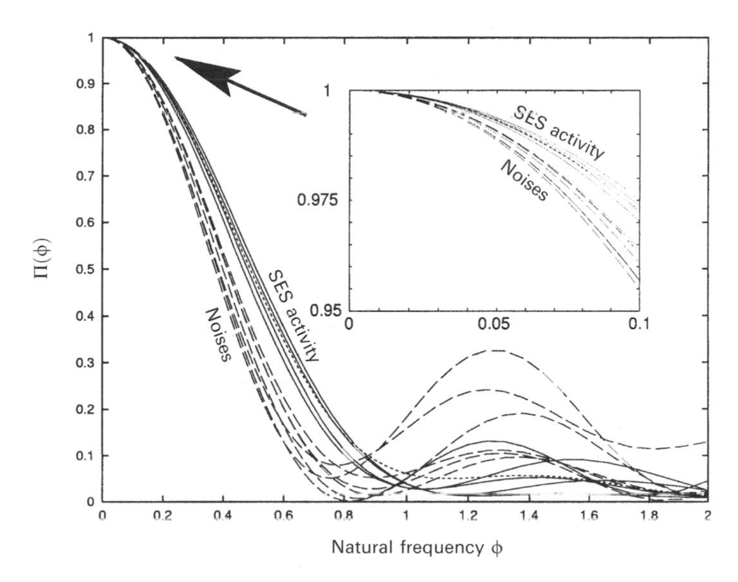

Figure 3.21. The normalized power spectra $\Pi(\phi)$ for SES activity (solid lines) related to the EQs labeled K, E, and A in the previous diagram (from top to bottom: K_1, A, E, K_2) along with those of a number of artificial noises (broken lines). The dotted curve corresponds to a theoretical estimation emerged from the theory of critical phenomena described in the text (Varotsos *et al.*, 2001a).

when approaching a critical point (see below). Third, the curves of SES activity cross at a point with a ϕ value very close to unity (i.e. $\phi \approx 1.05$). How this point is approached is studied by means of the so-called *β-function* (see Varotsos *et al.*, 2001a and references therein).

Let us now study how the natural spectra of SES activity interrelates with the evolution of subsequent seismic activity. The continuous lines in Figure 3.22 depict the normalized power spectra $\Pi(\phi)$ deducted from analysis of SES activity. In the same figure, we plot (broken lines) the corresponding quantity $\Pi(\phi)$ obtained from the seismic activity for each of the four EQs mentioned earlier, as it evolves *after* SES detection of each event (see below). Careful inspection of this figure shows that the broken lines fall on the continuous line *a few days* before the main shock – *at the most*. We emphasize that this only occurs if we consider the totality of SES activity (e.g. we must not omit the initial period of this activity). Despite this fact, in these four EQs, the corresponding lead times differ widely (from 3 weeks to 4.5 months; Table 3.13).

These suggestions of Varotsos *et al.* (2001a) open up the possibility of estimating the time of the impending event with an accuracy appreciably better than hitherto available.

Theoretical aspects of the concept of 'natural time'. In Figure 3.21, the normalized power spectra surves of SES activity lie above the $\Pi(\phi)$ curves of noises. An attempt toward understanding this classification was presented by Varotsos *et al.* (2001a). The behaviour of SES activity was explained by using aspects of the theory of critical phenomena.

The most useful quantity around $\omega = 0$ is the variance $\kappa_1 = \langle x^2 \rangle - \langle x \rangle^2$ of natural time distribution. This is so, because the various normalized power spectra that group together as ω or ϕ tend to 0 depending on their κ_1 values. The value of the variance κ_1 that reproduces noises is larger than 0.084, while for SES activity the κ_1 value is around 0.07.

According to the SES generation model of piezo-stimulated current (Varotsos and Alexopoulos, 1986) a (re)orientation of the electric dipoles occurs when approaching a critical pressure P_{cr} that obeys the condition:

$$\left(\frac{dP}{dt}\right)_T \frac{v^m}{kT} = -\frac{1}{\tau(P_{cr})}$$

where v^m is the migration volume, defined as:

$$v^{m,b} = \left(\frac{\partial g^m}{\partial P}\right)_T$$

g^m is the Gibbs migration energy and $\tau(P_{cr})$ the relaxation time for the (re)orientation process. The values of v^m associated with SES generation should exceed the mean atomic volume by orders of magnitude, and this entails that the relevant (re)orientation process should involve the motion of a large number of 'atoms' (see p. 404 of Varotsos and Alexopoulos, 1986). Recent experimental findings are consistent with such a 'cooperativity'. These experiments showed that, when approaching the glass transition, the emitted electrical signals have the form of a 'Random Telegraph Signal' (RTS). Analysis of RTS kinetics enables us to extract distributions of durations from individual RTS series that have two dominant levels. We recall that, for a thermally activated two-state process, the time durations t should be exponentially distributed with a probability density function (PDF): $(1/\tau)\exp(-t/\tau)$. Varotsos *et al.* (2001a) first considered the stochastic nature of the relaxation process. Such a stochastic analysis was based on the concept of clusters, the structural rearrangement of which develops in time. According to this analysis, the exponential relaxation of polarization is arrested at a random time variable η_i and the instantaneous orientation reached at this instant is 'frozen' at a velue $\exp(-\beta_i \eta_i)$, where $\beta_i = b =$ constant.

Assuming that η_i itself follows an experimental distribution, with a time constant $\tau_0 \ll \tau \approx \tau(P_{cr})$, an almost constant current would be expected for as long as this unit 'lives' (i.e. for a duration η_i). The RTS feature of SES activity might be understood in the following context: the duration Q of a pulse is just the sum of n such identical units $Q = \sum_{i=1}^{n} \eta_i$. Under this assumption, the duration Q_k of the kth pulse in SES activity follows the gamma distribution with a mean lifetime $n_k \tau_0$ and variance $n_k \tau_0^2$, where n_k is the number of exponential lifetime back-up units that act cooperatively. Varotsos *et al.* (2001a) consider that SES activity is emitted when approaching criticality in the focal area. If at the critical point n_k back-up units were

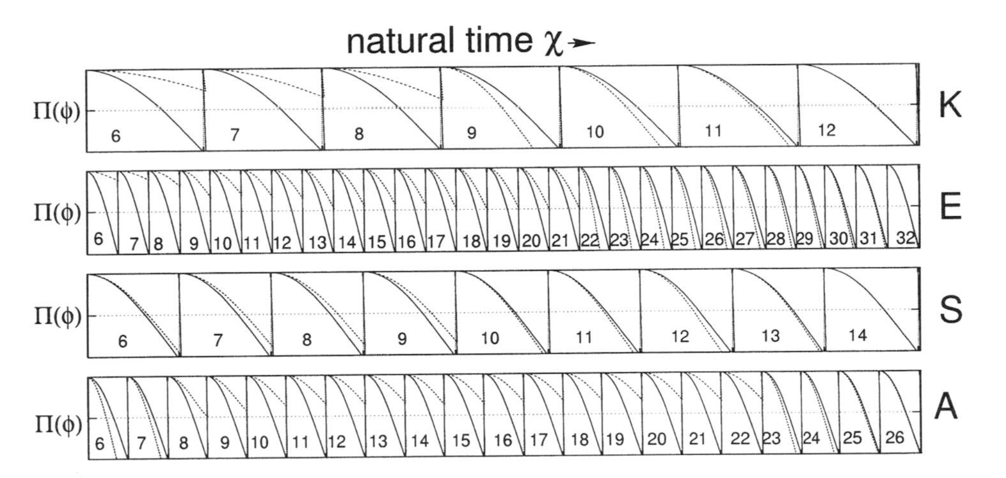

Figure 3.22. Time evolution of the normalized power spectra $\Pi(\phi)$, in the window $0 < \phi \leq 0.5$, of the seismic activities (broken line) along with those obtained for the SES activities (solid lines) for the cases of the strong EQs labeled K, E, S, and A. The numbers refer to last event considered in order to calculate the seismicity spectrum (and correspond to the event reported in the tables of the supplementary information). For the SES activities K_1, K_2 their average is used while for S the theoretical estimation (Figure 3.21) is plotted (Varotsos *et al.*, 2001a).

available at the kth current emission, then the average number of back-up units for the $k+1$ emission would be the same. This assumption is reminiscent of the fact that the reorientation of a spin, in the random-field Ising Hamiltonian, will cause on average one more spin to flip at the critical point. Under the above assumptions, Varotsos *et al.* (2001a) obtain for the normalized power spectrum the expression:

$$\Pi(\omega) = \frac{18}{5\omega^2} - \frac{6\cos\omega}{5\omega^2} - \frac{12\sin\omega}{5\omega^3}$$

After expanding around $\omega = 0$, we find:

$$\Pi(\omega) = 1 - 0.07\omega^2 + \cdots$$

which implies that:

$$\kappa_1 = \langle\chi^2\rangle - \langle\chi\rangle^2 = 0.07$$

The experimental results agree favourably, as seen in Figures 3.23b, c and 3.24 (for EQs and SES activity, respectively), with these theoretical estimations.

3.4.1.3.8 Plasmons

It is usually argued that electromagnetic waves that radiate from a seismic focus cannot reach the surface due to dissipation in the conductive Earth. Such arguments

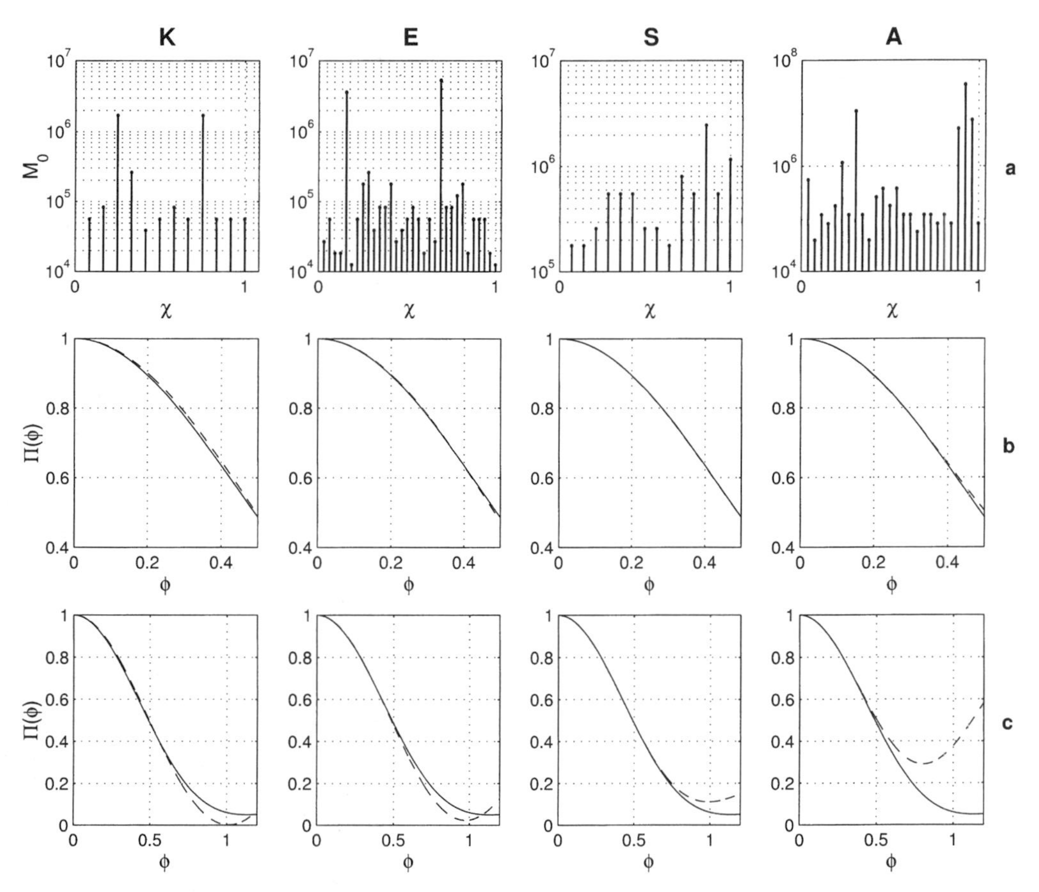

Figure 3.23. (a) How the EQs that preceded the main shocks K, E, S, and A (see tables A1 to A4 of Varotsos *et al.*, 2001a) are read in 'natural time'; (b), (c) comparison of the normalized power spectra $\Pi(\phi)$ for EQs in (a) (broken lines) with those predicted (solid lines). Note that (b) refers to the range $0 < \phi < 0.5$, while (c) to $0 < \phi \leq 1.2$ (Varotsos *et al.*, 2001a)

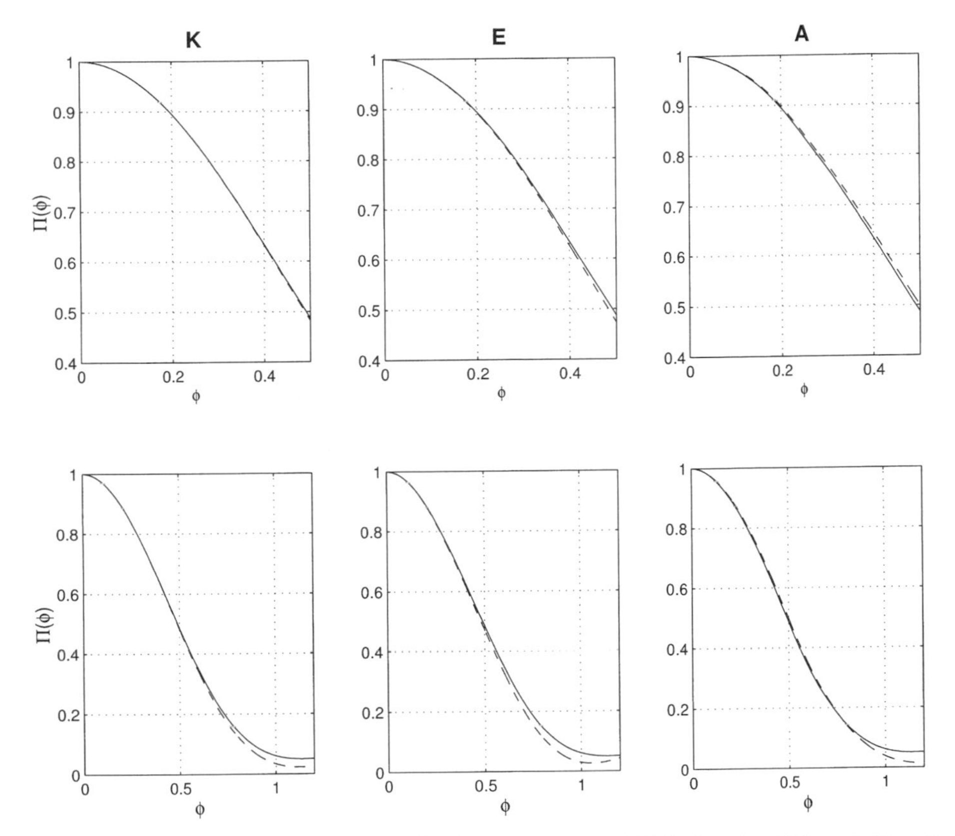

Figure 3.24. Comparison of the normalized power spectra $\Pi(\phi)$ for the SES activities that preceded the EQs labeled K, E, and A (broken lines) with those estimated from the theoretical aspects described in the text (solid lines). Upper: ϕ lying between 0 and 0.5; bottom: $0 < \phi \leq 1.2$ (Varotsos *et al.*, 2001a).

are only based on the electromagnetic theory for electric flux density $\vec{D}$. However, when writing the relation:

$$\vec{D} = \varepsilon(\omega)\vec{E} = 0 \qquad (\vec{D} = 0, \vec{E} \neq 0)$$

this is still satisfied if we consider the case of a *non-zero* electric field $\vec{E}$ when $\varepsilon(\omega) = 0$. The latter happens at a certain frequency ω_p associated with 'plasma waves' in the electron gas. Before explaining the physical background of this point, we first present current aspects that seem to defeat the universal validity of the conventional picture of a non-interacting electron gas.

3.4.1.3.8.1 Current aspects that defeat the Fermi liquid theory

For several decades, the study of the nature of the metallic and superconducting states of electrons in solids has been based on Fermi liquid theory, in which electrons in ordinary metals can be viewed as a non-interacting electron gas. In the frame of quantum physics, the metallic state of matter in Fermi liquids theory corresponds to a quantum gas of 'quasiparticles', which are like electrons in every way except that they no longer interact. The quasiparticles are fermions and have to obey the Pauli exclusion principle, which forbids any two fermions from being in the same state. This forces the electrons into states of quantum kinetic energy and these quantum motions completely wash away the interactions. In the presence of any residual attractive interaction, the quasiparticles form so-called Cooper pairs, which are bosons (and so do not obey the Pauli principle) and they immediately condense into a superconductor, foreseen in the Bardeen–Cooper–Schrieffer (BCS) theory.

Recent experiments, however (Mook *et al.*, 2000; Sharma *et al.*, 2000), provide evidence that the conventional theories of metals and superconductors are no longer universally valid and add further credibility to an alternative picture of electrons in solids known as 'dynamical stripes' (see Zaanen, 2000 and references therein). In stripes, the electrons gather together in complex, quantum fluctuating patterns and the electronic matter organizes itself into a topologically one-dimensional quantum texture.

Striped phases are inhomogeneous distributions of charge and spin and are suggested to play an important role in determining the properties of the high-T_c copper oxide superconductors (T_c is the superconducting transition temperature). In the simplest picture, the charge and spin can be thought of as being confined to separate linear regions in the crystal and thus resemble stripes (see Mook *et al.*, 2000 and references therein). The first experimental evidence (with neutron-scattering) on stripes in copper oxide superconductors was published by Tranquada *et al.* (1995), which suggested the following frozen version for the stripes (see the review by Zaanen, 1999): mobile charge carriers in the perovskite planes of cuprates are confined in narrow one-dimensional lines (*charge stripes*). These stripes are separated by insulating regions showing antiferromagnetic spin order. In other words (Zaanen, 2000), the static state of stripes is highly unusual and consists of antiferromagnetic (and presumably insulating) domains about a nanometre in width, separated by domain walls on which the charge carriers reside. In the static phase, these domain walls, or *stripes*, condense in a regular pattern (this phase arises from competition between quantum kinetic energy and electron–electron interactions on the microscopic scale). As for dynamical (as opposed to static) stripes, we simply note that this is the state that is believed to be associated with superconductivity, in which the system of stripes is disordered by long-wavelength quantum fluctuations. The popular view is shown in figure 1 by Zaanen (2000). We should emphasize, however, that in spite of the fact that we have gained a deeper insight into the electron states in solids in the past few years, the problem of high-T_c superconductivity should not be considered as being solved.

3.4.1.3.8.2 Theoretical background on plasmons

The zero of the dielectric function $\varepsilon(\omega) = 0$ is the condition that determines the *collective longitudinal 'oscillation'* (the relevant quantum $\hbar\omega_p$ is called a *plasmon*) of the conduction of an electron gas in a solid. Several textbooks report the relation (m is the mass of an electron and n is the electron density):

$$\omega_p = \left(\frac{ne^2}{\varepsilon_0 m} \right)^{1/2} \tag{3.1}$$

for the plasmon frequency. By inserting the *appropriate* value of n (under the restrictions of the simplified Jellium Model [MJ] that is usually applied) for a given metal, we can estimate the value of ω_p for which $\varepsilon(\omega) = 0$. This equation, however, does not mean that, if we insert *any* small value of n and hence find the corresponding small value of ω_p (say, $\omega_p = 10^6\,\text{sec}^{-1}$, frequency $\nu_p \sim 10^5\,\text{Hz}$), such a low frequency plasmon actually exists (i.e. so that $\varepsilon = 0$). This could be roughly attributed to the following physical reason: at high frequencies (e.g. 10^{15} Hz), the plasmons mainly refer to the 'oscillations' of electrons, while the 'cores' (positive ions) – which have characteristic oscillation frequencies 10^{12} 10^{13} Hz remain practically immobile and, then, equation (3.1) holds. On the other hand, at appreciably lower frequencies (i.e. when the motion of ions 'is accompanied' by that of electrons), equation (3.1) is *not* valid for plasmon oscillations. This will be shown more precisely below.

Let us define the plasma frequency as:

$$\omega_p^2 = \varepsilon_0^{-1} e^2 n/m \tag{3.2}$$

we find that the dielectric function is given by:

$$\varepsilon = 1 - \frac{\omega_p^2}{\omega} \frac{1}{1 + i/\omega t} \tag{3.3}$$

which for high frequencies only (i.e. $\omega\tau \gg 1$) becomes:

$$\varepsilon = 1 - \frac{\omega_p^2}{\omega^2} \tag{3.4}$$

Equation (3.3) is valid under the condition $\omega \gg v_F k$ (v = velocity, k = wave number, v_F refers to the Fermi velocity $v_F = \hbar k_F/m$). On the other hand, if $\omega \gg v_F k \gg v_F k_F$, we have a *static* ($\omega = 0$) dielectric constant:

$$\varepsilon(k, 0) = 1 + \frac{k_{T,F}^2}{k^2} \tag{3.5}$$

where

$$k_{T,F} = 1/l_c$$

which is the so-called screening constant and the subscript T, F refers to the Thomas–Fermi model. Equation (3.5) can be immediately proved if we assume that the potential φ (produced inside an MJ by an external charge q) has an exponential screening of the form:

$$\varphi \approx \exp\left(-\frac{r}{l_c}\right) \Big/ r$$

We now distinguish two cases.

(a) *Collective longitudinal oscillations for* $\omega \neq 0$ *and* $k \to 0$. In this limit, equation (3.3) becomes (disregarding the 'friction' term, i.e., $\tau^{-1} = 0$):

$$\varepsilon(\omega) = 1 - \frac{\omega_p^2}{\omega^2} - \frac{\omega_{pi}^2}{\omega^2} \tag{3.6}$$

where the term ω_p^2/ω^2 is due to the electrons and the last one to the positive ions (pi) (i.e. $\omega_p^2 = \varepsilon_0^{-1} e^2 n/m$ and $\omega_{pi}^2 = \varepsilon_0^{-1}(\varepsilon J)^2 n_i/m_\alpha$, where m is the mass of the electron, m_α the mass per atom, J the valence and n_i the concentration of ions). The condition:

$$\varepsilon(\omega) = 0 \tag{3.7}$$

gives for the frequency of the collective longitudinal oscillation:

$$\omega^2 = \omega_p^2 + \omega_{pi}^2 \approx \omega_p^2 \tag{3.8}$$

We draw attention to the fact that the latter approximation comes from:

$$\omega_{pi} \ll \omega_p \tag{3.9}$$

because

$$\omega_p^2 + \omega_{pi}^2 = \varepsilon_0^{-1} e^2 n(m^{-1} + Jn_\alpha^{-1}) \approx \varepsilon_0^{-1} \mathrm{e}^2 nm^{-1} \tag{3.10}$$

where $m^{-1} + Jm_\alpha^{-1}$ is the inverse of the 'reduced mass' m' (i.e. $(m')^{-1}$ as expected). This is the case for *plasma oscillation*, the quantum $\hbar\omega_p$ of which, as mentioned above, is called a *plasmon*. When the electrons are (homogeneously) displaced by x (regarding the positive ion background), we have a positive charge $Sx|e|n$ at the left side O and a negative charge $Sx|e|n$ at the right side L (S stands for the surface). These charges produce a field that, according to Gauss law, is equal to $E = \varepsilon_0^{-1} x|e|n$. Thus, the force $F = -|e|E = -\varepsilon_0^{-1} \mathrm{e}^2 nx$ is proportional to the displacement x with a constant of proportionality $\kappa = \varepsilon_0^{-1} \mathrm{e}^2 n$. Therefore, the oscillation frequency is:

$$(\kappa/m)^{1/2} = (\varepsilon_0^{-1} e^2 n/m)^{1/2} \equiv \omega_p$$

If we use the reduced mass $(m')^{-1} = m^{-1} + J(m_\alpha)^{-1}$, the more accurate formula Eq. (3.8) will result.

(b) *Collective longitudinal oscillation for* $\omega \to 0$ *when* $k \to 0$. We must be very careful here, because *for the electrons*: (i) if $\omega k \gg v_F$, we must use Eq. (3.3) and (ii) if $\omega/k \ll v_F$, we must use Eq. (3.5). As for the *ions*, the relation $1 - (\omega_{pi}^2/\omega^2)$ is valid when $\omega/k \gg v_i$, where $v_i \approx v_F(m/m_\alpha)^{3/4}$ (the exponent 3/4 results when we write $m_\alpha v_i^2 = \varepsilon_D$, where the Debye energy ε_D is $\varepsilon_D = \hbar\omega_D$, $\omega_D =$ Debye frequency). Thus, if the condition $v_i \ll \omega/k \ll v_F$ holds, the dielectric function becomes:

$$\varepsilon(\omega, k) = 1 + \frac{k_{T,F}^2}{k^2} - \frac{\omega_{pi}^2}{\omega^2} \tag{3.11}$$

where $\frac{k_{T,F}^2}{k^2}$ corresponds to the electrons and $-\frac{\omega_{pi}^2}{\omega^2}$ to the ions. In the limit $k \to 0$, the quantity $\varepsilon(k, \omega)$ given by Eq. (3.11) becomes zero (i.e. *collective oscillations*) when:

$$\omega = \frac{\omega_{pi}}{k_{T,F}} k \tag{3.12}$$

Such a longitudinal collective oscillation (i.e. with $\omega \sim k$) is just a *longitudinal sound wave* with velocity equal to that of the sound $c_0(\omega = c_0 k)$. Thus, Eq. (12) reflects:

$$c_0 = \omega_{pi}/k_{T,F} \tag{3.12}$$

which allows us to calculate $k_{T,F}$ as a function of the known quantities ω_{pi} and c_0. (Note that c_0 obeys the condition $v_i 0 \ll v_F$, under which Eq. (3.11) is valid.)

The basic difference between the two collective longitudinal oscillations in cases (a) and (b) is that the quantum in case (a) is a *plasmon* of the form $\hbar^2\omega_k^2 = \hbar\omega_p^2 + O(k^2)$ (i.e. it goes to a *non-zero* limit $\hbar\omega_p$ when $k \to 0$); on the other hand, the quantum in case (b) is a *phonon*, the energy $(\hbar\omega_k = \hbar c_0 k)$ of which tends to zero when $k \to 0$.

From a physical point of view, the basic difference between (a) and (b) is due to the *long*-range nature of electric forces. If the electrostatic force were of short range, there would be no 'returning force' (in the limit $k \to 0$). It can be shown, for example, that *if* the Coulomb potential were of short range (e.g. of the form $\exp(-br)/r$) then the quantum of *plasmon* would tend linearly to zero when $k \to 0$. Then, we could ask why the *phonon*, in which long-range electric forces are also involved, does not have non-zero energy (as the *plasmon* does) when $k \to 0$? The answer is as follows: *phonons* are the 'oscillations' of the positive ions (background 'core'); the fields produced by the oscillation of an ionic charge are 'screened immediately' by the electrons (which are more mobile) and, hence (due to this screening), long-range forces instantaneously become short range. Thus is the case of phonons $\omega \sim k$. On the other hand, *plasmons* are related to the oscillations of electrons; in this case, the ions, having appreciably larger mass, cannot follow the fast motion of the electrons and, hence, cannot 'screen' the fields of the latter (thus, the long-range nature of electrostatic forces remains and, hence, the plasmons do have *non-zero* frequency when $k \to 0$). All the above points disregard the *ionic nature* of the solid.

3.4.1.3.8.3 Plasmon model for the origin of EM precursors

Several laboratory experiments (e.g. Enomoto and Hashimoto, 1990; Enomoto *et al.*, 1994) have shown that exoelectrons and changed particles are emitted from rocks while microfracturing; this emission is of the order of 1 C/s for rocks of a few m^2.

Kamogawa and Ohtsuki (1999) suggest a model for EM precursory generation, which consists of the following stages (see Figure 3.25).

In the first stage, a huge number of exoelectrons are emitted from the stressed rocks around the focal area accompanying microfracturing (hence, before the final rupture).

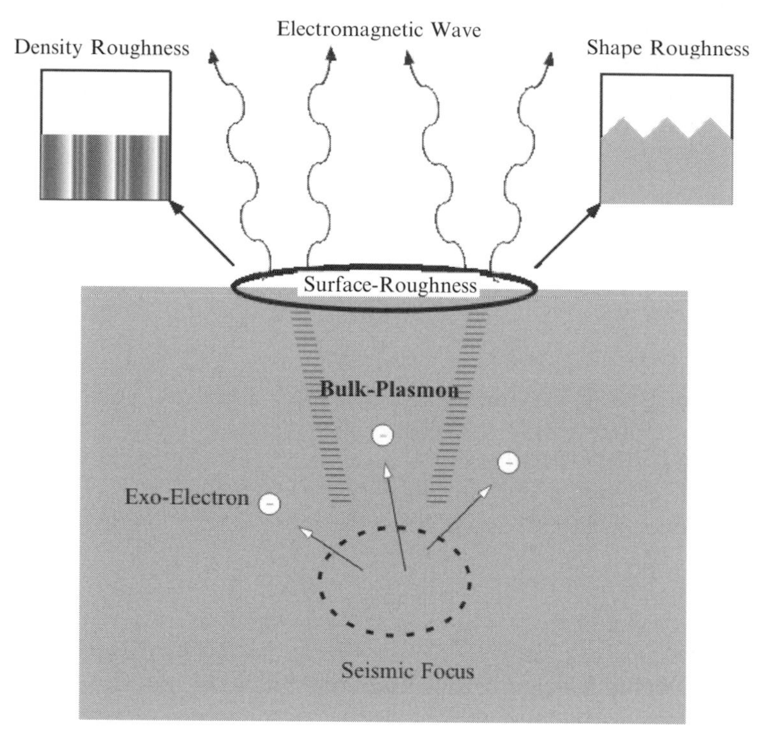

Figure 3.25. Schematic diagram of the plasmon model suggested by Kamogawa and Ohtsuki (1999) for the generation mechanism of precursory EM signals. Note that the bulk plasmons are excited by the exoelectrons or charged particles emitted from the stressed focal region (see the text) and change to electromagnetic waves by surface roughness (density roughness and shape roughness).

In the second stage, bulk plasmons are excited underground by the electrical shock of the emission of the exoelectrons from the rocks. Kamogawa and Ohtsuki assume that the electron density in rocks varies between 10^{15} and 10^{22} m^3, while in ordinary metals it is of the order of 10^{28}–10^{29} m^3. Inserting these values into Eq. (3.1), they found that the frequency ν_p is:

$$\nu_p \sim 10^7\text{–}10^9 \text{ Hz}$$

while in ordinary metals it is of the order of 10^{14} Hz.

In the third stage, the plasmon waves underground are transformed into EM waves in the air. Such a transformation is not possible if the Earth surface is smooth and the electron density homogeneous. In a theoretical study (Tanaka *et al.*, 1999), they investigate the transition probability from bulk plasmons to photons (EM waves) by using a hydrodynamic theory; they found that bulk plasmons can be changed to EM waves due to surface roughness (e.g. non-uniformity in surface topography and electron density).

As for the production mechanism of electrons, Tanaka *et al.* (1999) suggest the following: hydrogen gas is produced in faults and reaches the boundaries between rocks and aqueous electrolytes. At these boundaries, the hydrogen gas dissolves into the aqueous electrolyte leaving the electrons in the rock (i.e. the boundaries play a role similar to platinum in the hydrogen–oxygen fuel cell).

3.4.1.3.9 Main conclusions

(1) Ground-based ULF magnetic-field measurements: the main characteristics (e.g. amplitude, frequency range, lead time) of the precursory changes reported to date for large earthquakes (i.e. Spitak in 1988, Loma Prieta in 1989, Guam in 1993, Biak in 1996) are similar to those reported long ago by VAN for the magnetic field variations (a few to several tenths of nT) associated with SESs preceding earthquakes of $M \geq 6.5$ (for $M \sim 5.0$–5.5, such magnetic-field variations are not readily detectable). It seems, therefore, natural that SESs could have also been recorded before the earthquakes, had electric-field variations been simultaneously measured. These magnetic-field measurements can be explained by the mechanisms suggested to date for SESs generation.
(2) Complex ground-based ULF and AE measurements in Japan: the successful recording of the latter seems to be attributed to the quality of the sensors that have a sensitivity proportional to the cube of frequency. An understanding of the latter fact may be achieved when considering power emitted from an accelerating body coupled to a wave.
(3) Precursory EM effects observed in the atmosphere or on satellites: several tentative schemes have been suggested for their explanation that include changes in tropospheric chemistry due to a precursory release of gases from the seismogenic zone, modification of lightning activity due to a change in near-surface conductivity, intensification of acoustic-gravity turbulence (a triggered phenomenon in the critical state of the upper atmosphere), excitation of bulk plasmons due to exoelectron emission from the seismogenic zone, EM generation due to Earth's surface roughness, etc. However, among the various models suggested, despite having a reliable physical basis, none of them can be considered to be very promising, on the basis of current scientific knowledge.

3.4.2 Volcanic eruptions

3.4.2.1 Scales and consequences of volcanic disasters

By their scale and impact on ecosystems and mankind as a whole, volcanic eruptions rank considerably lower than earthquakes and floods, because they are rare events. But, on the other hand, they are among the most terrible and destructive natural disasters. In densely-populated regions, they may result in many victims (Table 3.14: Oppenheimer, 1996) and considerable material damage (Figures 3.26 and 3.27). In the 20th century, the eruptions of just two volcanoes (Mount Pelée in 1902,

Table 3.14. Volcanic calamities since 1000 AD resulting in more than 300 deaths.

			Ultimate cause of death				
Volcano	Country	Year	Pyroclastic flow	Lava flow	Flow of fragmental rocks	Death because of starvation	Tsunami
Merapi	Indonesia	1006	1,000	100,000			
Kelut	Indonesia	1586					
Vesuvius	Italy	1631			18,000*		
Etna	Sicily	1669			10,000*		
Merapi	Indonesia	1672	300*				
Awu	Indonesia	1711		3,200			
Oshima	Japan	1741					1,480
Cotopaxi	Ecuador	1741		1,000			
Makian	Indonesia	1760					
Papandayan	Indonesia	1772	2,960				
Laki	Iceland	1783				9,340	
Asama	Japan	1783	1,150				
Unzen	Japan	1792					15,190
Maion	Philippines	1814	1,200				
Tambora	Indonesia	1815	12,000				
Galunggung	Indonesia	1822		4,000		80,000	
Nevado del Ruiz	Colombia	1845		1,000			
Awu	Colombia	1846		3,000			
Cotopaxi	Ecuador	1877		1,000			
Krakatoa	Indonesia	1883					
Awu	Colombia	1892		1,530			
Sufrière	St Vincent and Grenadines	1902	1,560				
Mount Pelée	Martinique	1902	29,000				
Santa María	Guatemala	1902	6,000				
Taal	Guatemala	1911	1,330				
Kelut	Indonesia	1919		5,110			
Merapi	Indonesia	1951	1,300				
Lamington	New Guinea	1951	2,940				36,420
Khaibok	Philippines	1951	500				
Khaibok Agung	Indonesia	1963	1,900				
Mount St. Helens	USA	1980	60	>22,000			
El Chichón	Mexico	1982	>2,000				
Nevado del Ruiz	Colombia	1985					
Total			*65,140*	*53,900*	*28,000*	*89,340*	*53,090*

* Together with mud flows.

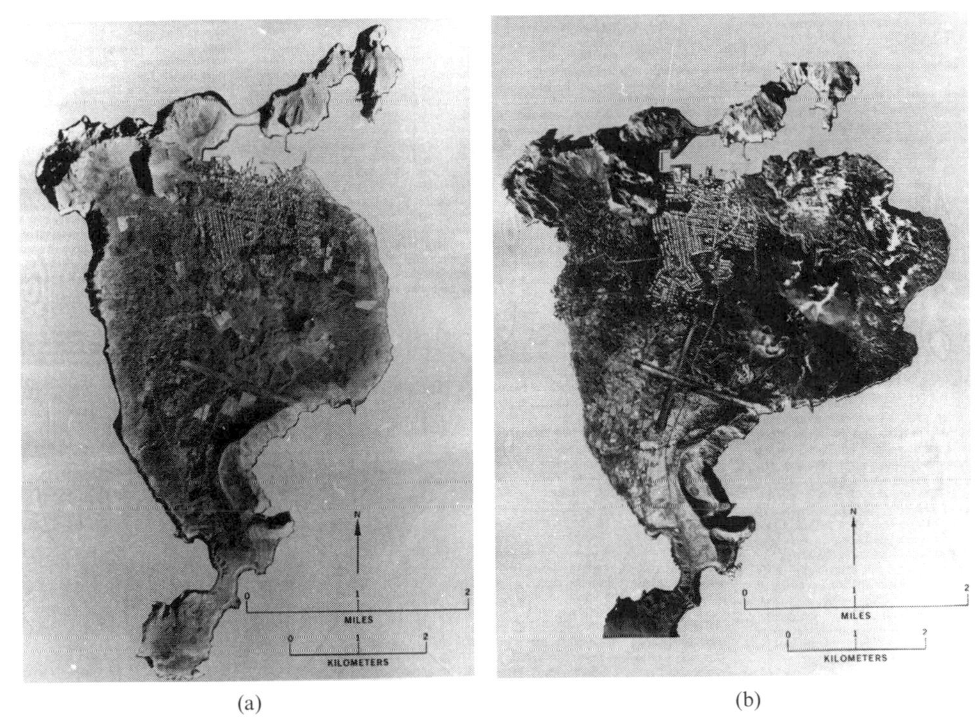

(a) (b)

Figure 3.26. Helgafell volcanic eruption (January 1973) resulted in an increase in the area of Heimaey Island (southern coast of Iceland) and partial destruction of the eastern part of Vestmannaeyar town: the island before (a) and after (b) the eruption (from Grigoryev and Kondratyev, 2001).

Martinique and Nevado del Ruiz in 1985, Colombia) resulted in 28,000 and 25,000 deaths, respectively.

Volcanic eruption data between 1900 and 1995 (Oppenheimer, 1996; Tiling, 1989) show that 76,960 persons were killed. Compared to other disasters, these figures are low. The same is true for the number of injured or displaced. For example, during the same period, the number of injured and displaced (based on catastrophes with more than 1,000 evacuated persons) totalled 1,310,000 persons.

All land volcanoes, both active and dormant, are well known, and this helps us determine related zones of risk for population. Although catastrophic volcanic eruptions are rare, they can be very destructive and, for small countries, they can destabilize life altogether. The volcanic eruptions in Mount Pelée and Nevado del Ruiz are the proof to this as concerns the number of victims. No less convincing are the data on the number of injured in the course of some volcanic eruptions (Table 3.14, and Tiling, 1989).

The eruption of Pinatubo in the Philippines (1991) resulted in 250,000 injured or displaced persons. The number of displaced persons was even greater as a result of the eruption of Golung in Indonesia (1982) and totalled 750,000 persons.

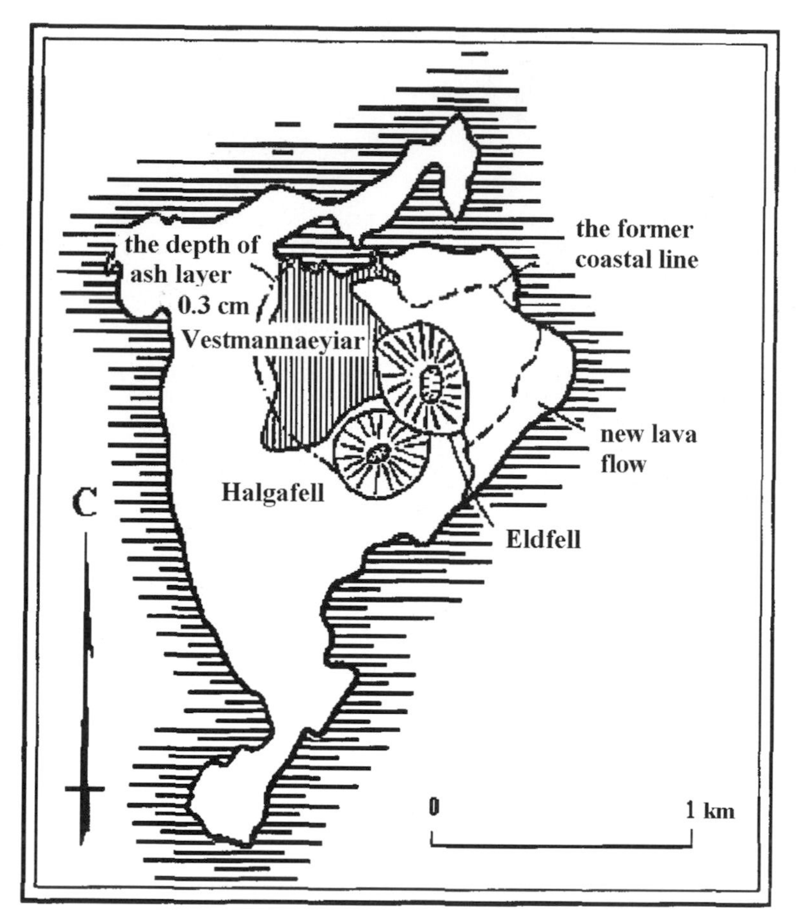

Figure 3.27. Map of Heimaey Island situated near the south coast of Iceland. Presented are volcanic cone and lava flows of the eruption in 1973, as well as location of the town and harbor of Vestmannaeyar. (Grigoryev and Kondratyev, 2001).

At the moment volcanoes and their global relation to earthquakes and tectonic active areas are undergoing intense study. Volcanoes are connected with continental rift zones, subduction zones, zones of oceanic slab submergence under the mantle and with deep convection cells (Oppenheimer, 1996).

Although volcanoes, especially those located in densely-populated areas, are being monitored, it is very difficult to predict their eruptions.

The consequences of volcanic eruptions can be both direct and indirect. They may be disastrous for people living in the vicinity. Large eruptions contribute to atmospheric pollution and the possible changes in climate associated with it, which may, hypothetically, be disastrous for people in other regions of the Earth (Kondratyev, 1985; Kondratyev and Galindo, 1997).

Like some other natural phenomena, volcanic eruptions sometimes have serious indirect consequences, apart from any initial damage caused. Often, earthquakes

Table 3.15. Volcanic eruptions resulting in more than 1,000 displaced persons.

Year	Location	Number of evacuated	Number of killed	Caused by
1980	Gamalama, Indonesia	40,000		
1980	Uluvan, Papua-New-Guinea	2,000		
1981	Paluweh, Indonesia	1,850		
1981	Semeru, Indonesia		373	Pyroclastic, and rock flows
1981	Maion, Phillipines		200	Mud and rock flows
1981	Galkonora, Indonesia	3,500		
1982	El Chichón, Mexico	10,000	1,879	
1982	Galunggung, Indonesia	750,000	68	Mud and rock flows
1983	Una Una, Indonesia	7,000		
1983	Gamalama, Indonesia	5,000		
1983	Miyake-jima, Japan	1,400		
1984	Merapi, Indonesia	1,000		
1984	Karangetang, Indonesia	20,000		
1984	Maion, Philippines	73,000		
1985	Sangeang Api, Indonesia	1,250		
1985	Nevado del Ruiz, Colombia		23,080	Mud and rock flow
1986	Nevado del Ruiz, Colombia	15,000		
1986	Nyos Lake, Cameroon		1,700	CO_2 gas emissions
1986	Oshima, Japan	12,200		
1987	Gunung Ranakah, Indonesia	4,200		
1988	Gamalama, Indonesia	3,500		
1988	Bandi Api, Indonesia	10,000	4	
1988	Makian, Indonesia	15,000		
1989	Longkuimen, Chile	2,000		
1989	Galeras, Colombia	2,000		
1989	Longkuimen, Chile	4,600		
1990	Kelut, Indonesia	60,000	32	Mud and rock flow
1990	Sabankaya, Peru	4,000		
1991	Pinatubo, Philippines	250,000	800	Mud and rock flow
1991	Panaya, Guatemala	1,500		Pyroclastic flows
1991	Lokon-Empung, Indonesia	1,000	1	
1991	Cerro-Negro, Nicaragua	28,000	2	Ash fall
1993	Maion, Philippines	57,000	75	Pyroclastic flows
1994	Rabaul, Papua-New Guinea	50,000	5	Ash fall
1994	Merapi, Indonesia	6,000	41	Pyroclastic flows
1994	Popocatepetl, Mexico	75,000		
1995	Fogo, Cape Verde Islands	1,050		
1995	Sufrière, Montserrat	5,000		

bring about other dangerous phenomena that are disastrous for the population (Table 3.15: Tiling, 1989). For example, in the course of Nevado del Ruiz volcanic eruption in Colombia in 1985, 25,000 persons were killed, not directly because of volcanic emissions, but because of mud and rock flows, due to melting of glaciers and snow at the top of the volcano.

Table 3.16. Victims of volcanic eruptions, 1600–1986.

Ultimate cause of death	1600–1899	1900–1986
Pyroclastic flows and avalanches	18,200 (9.8%)	36,800 (48.4%)
Mud and rock flows (lahars) and floods	8,300 (4.5%)	28,400 (37.4%)
Tephra and fall of impact bombs	8,000 (4.3%)	3,000 (4.0%)
Tsunami	43,600 (24.4%)	400 (0.5%)
Disease and starvation	92,100 (49.4%)	3,200 (4.2%)
Lava flows	900 (0.5%)	100 (0.1%)
Gas and acid rain		1,900 (2.5%)
Other or unknown	15,100 (8.1%)	2,200 (2.5%)
Volcanic total	186,200 (100%)	76,000 (100%)
Annual average	620	880

As for other dangerous phenomena (landslides, mud flows, avalanches) activated in the wake of a volcano, the zones of risk for the population get bigger. In the case of Nevado del Ruiz, the principal damage to the city of Armero, located several kilometres from the volcanic crater, were caused by mud and rock flows.

Decreasing the extent of calamities cased by volcanic eruptions can mainly be reached by using data about the areas around volcanoes, further determination of zones of risk and, taking this into account, settling people in other places. According to data for 1900 to 1986, in the course of volcanic eruptions 85.8% (65,200 persons) out of the total number of victims (76,000 persons) were killed by pyroclastic flows and avalanches, mud and rock flows (lahars) and floods (Tiling, 1989). The approximate routes to be followed by these flows (i.e. the probable zone of their propagation) can be determined and mapped beforehand.

To some extent, the consequences of disasters can be smoothed out by certain engineering structures such as dams, effluent channels for lava streams (built before the eruption and even bombardment of lava flows. In addition, in many countries of the world, there are special services for observation of active volcanoes. Attempts are also made to carry out regular satellite monitoring of volcanism.

Atmospheric pollution of volcanic origin is undoubtedly a subject of concern. As has been discovered, volcanic aerosols are formed in such great quantities that they can give rise to a global fall in temperature. For example, aerosols, associated in many respects with sulphur dioxide emissions, entered the atmosphere as a result of Pinatubo erupting in 1991 and were responsible for the global fall in temperature in the lower atmosphere by about 0.5°C in 1992 (Kondratyev, 1997).

Volcanic aerosols that enter the stratosphere have been proven to affect the ozone layer (Kondratyev and Varotsos, 2000). For example, in 1983, two years after Pinatubo erupted, the amount of global ozone reduced by 4% as compared with the preceding 12-year period. Despite the fact that there were no significant climatic catastrophes caused by volcanic eruptions in the 20th century, they do happen, as is evidenced by both history and geology. For instance, 73,500 years ago, the global ambient temperature decreased by 3–5°C as a result of the volcanic eruption of Toba on Sumatra. It was about this time that the area of glaciers increased and coincided with the period when the population of the Earth decreased as well (Oppenheimer, 1996).

All the known data on dangerous volcanoes made it possible for UNESCO, in 1983, to compile a list of the 89 most dangerous volcanoes (Bonneville and Gonze, 1992). The majority of these volcanoes are located in South-East Asia and on the western coast of the Pacific Ocean (42 volcanoes), in North and South America, on Caribbean Islands (40 volcanoes) and only 7 in Europe and Africa.

However, despite a supposed thorough assessment of volcanic danger, the list turns out to be incomplete. This is highlighted by exclusion of the largest volcanic eruptions of the 20th century (i.e. Nevado del Ruiz [1985], which caused a great many victims and widespread damage, and El Chichón [1982]). The latter was located far from residential areas, though it was staggering in the strength of its explosion and by its impact on the global atmosphere. Both volcanoes were not included in the list. This suggests that there are evident gaps in knowledge, not only on these two particular volcanoes but also on volcanoes as a whole. Successful forecasts of volcanic eruptions are further proof of lack of knowledge: in the 20th century, only three explosive volcanic eruptions were forecasted (Tolbachek in Kamchatka, Mount St Helens in North America and Sakurajima in Japan).

Thus, volcanoes are an extreme hazard to people, though they are mainly restricted to narrow areas. Notwithstanding this, the majority of active volcanoes (totalling some 600) are located in developing countries (where the consequences of eruptions are especially devastating and destructive), and many of them are located in densely populated regions.

3.4.2.2 Satellite information on volcanic eruptions

Observations carried out from various spacecraft enable us to obtain useful information on volcanic eruptions. A special role in the study of volcanoes belongs to satellite survey. The permanent operation of several satellites in outer space allows us to carry out 24-hour observations. Moreover, the entire surface of the Earth is at any one time in the satellite field of observations, which is very important, because many active volcanoes are located on desert islands or places not easily accessible for conventional observations.

Since 1964, many volcanic eruptions were discovered from IR photographs taken by satellites. These included, for instance, volcanic eruptions on Hawaii in 1964, Etna in 1967 and 1971, Surtsey in Iceland in 1966, Berenberg on Yan-Maiyen Island near Iceland in 1979 (Grigoryev, 1975), and many others.

Use of satellite images enables us to identify quickly the onset of volcanic eruptions by distinct 'smoke' emissions (Grigoryev and Lipatov, 1978; Kondratyev, 1993; Bonneville and Gonze, 1992; Colleau *et al.*, 1991; Evangelista *et al.*, 1995; Kondratyev and Galindo, 1997). It was in just this manner volcanic eruptions in the most remote and not easily accessible areas of the Earth, including the Arctic and Antarctic regions. For example, every so often a long, drawn-out volcanic train of smoke has been noticed in the area of Bennet Island (East Siberian Sea). Its origin is supposed to be associated with submarine volcanic activity.

One of the main volcanic eruption onset indicators is a smoke plume clearly identified in satellite images. On images obtained in the visible spectrum, it looks like a light-grey band drawn out in a certain direction, depending on wind direction. Smoke plumes are particularly clearly seen in photographs taken over populated territories, and over water areas as well. In these cases, images of smoke plumes and the background contrast especially well.

A smoke plume, as identified in satellite images, cannot be mistaken for an ordinary cloud, although its structure resembles an ordinary jet cloud and its hue is close to that of corresponding, ordinary cloud objects. The specific features of smoke plumes consist of the following: first, they originate from the known position of the volcano. Second, the plumes do not break down for a long time (unlike clouds). And, third, they maintain their initial start position (although their 'tails' can extend widely).

Smoke plumes from volcanoes can diffuse, as is evidenced by satellite observations, to many tens or even hundreds of kilometres. In cloudless periods, the trace of volcanic emissions is seen both in the visible and (although less clearly) in infrared images.

Sometimes, the trace of volcanic emissions is seen in infrared (IR) photographs even better than in images in the visible spectrum. This happens when a smoke plume is observed, not over the water surface with its low albedo, but over snowy surfaces and areas of high and middle cloudiness. In such a case, the smoke plume reaches the tropopause, and shows up because of its considerably lower temperature than the underlying snow cover or the thin layers of cloud. This is the reason the smoke plume can be clearly seen in IR photographs. A phenomenon of this kind was discovered, for example, while analysing visible and infrared images of Kluchevskaya Sopka, which erupted on 9 April 1974, obtained by the American meteorological satellite *NOAA* (Grigoryev and Lipatov, 1978). In the visible range image, the trace of volcanic emissions was not seen against brightly coloured snow cover and low cloudiness.

In the infrared photograph, the smoke plume, on the contrary is seen very clearly, because the snow cover and water surface turn out to be warmer here. The grey-coloured areas represent low and middle clouds that cover the entire western part of the Kamchatka Peninsula, starting from the central mountain ridge and extending from north-east to south-west. Against this background, considerably colder volcanic pollutant clouds, whose upper boundary reaches the tropopause, are seen as bright spots drawn out along an arc. The volcanic pollutant track extends farther away in the form of a narrow band 8–10 km long with blurred edges, being much darker than these spots. It has probably already reached the stratosphere.

Satellite data about the precise location of smoke plumes are important for flight servicing. For example, the lack of such information at the disposal of the pilots of two Boeing-747 planes in 1982 resulted in the two planes finding themselves in smoke shortly after Galung erupted (Java, Indonesia). Aerosols of volcanic origin stopped the engines of both planes and forced them to land (Bonneville and Gonze, 1992).

Another important volcanic eruption indicator in satellite photographs is the temperature anomaly associated with a sharp rise in temperature at the Earth's surface, especially near the volcano itself. A thermal anomaly of this kind is fixed in infrared satellite images. It can also be observed in day-light through a plume cloud, but it is most distinctly seen in night-time satellite images when there are sharp temperature contrasts at the Earth's surface (between the volcano and its environment). It was in just this way that an observation from *Nimbus-3* precisely identified the eruption of Berenberg in Yan-Maiyen Island in the Atlantic (not far from Iceland), which suddenly became active in September 1970.

The Berenberg volcano or, to be more exact, the volcano and the adjacent part of the coast and water body (covering an area of 200 km^2), were reflected in the night-time infrared image as an irregular-shaped very dark small spot. This spot contrasted against the dark background image of the Norwegian Sea. This image of the volcano can be explained by the considerable temperature variations of the surface water layer. The contrast between temperatures in the sea far from the island and that of the red-hot surface of volcanic formations (reaching several tens of degrees), as well as of water heated as a result of lava flows reaching the shoreline, is vividly clear (Grigoryev, 1975).

Thermal anomalies around many active volcanoes can be tracked in satellite images. They make it possible for us to ascertain volcanic activity, even through a thin cloud cover. Thermal anomalies of this kind, detected by using such means as the Thematic Mapper mounted on *Landsat* satellites, made it possible for us to monitor the volcanic activity of Erebus in the Antarctic and Laskar in Chile (Bonneville and Gonze, 1992; Oppenheimer, 1996). The appearance of thermal anomaly around the latter (with temperature of about 800°C) was noticed in July 1985, a year before this volcano erupted in September 1986. A thermal anomaly in Laskar (of about 800 1,000°C) was also discovered from satellite images in 1987. It is believed that a lava lake caused it (probably, such a lake gave rise to a similar anomaly in 1985 in the vicinity of Erebus in the Antarctic).

The onset of the explosive eruption of Mount St Helens (18 May 1980) was fixed on the basis of infrared spectrometric data from USAF satellites (*DMSP*). Observations of an aerosol-and-gas cloud generated a minute after the Earth shock, which triggered the eruption, was observed simultaneously from two satellites. This made it possible to carry out triangulation measurements and track consecutive stages of eruption development (Kondratyev, 1985; Rice, 1981). Satellite data about the length, direction, and temperature of volcanic trains can be used for estimation of eruption energy.

Photographs taken from space made it possible for us to track the newly awakened (in November 1992) Unzen volcano in Kyushu Island in Japan. This volcano suddenly became active after a 200-year break. The period of pronounced activity lasted for about 1 year. Currently, the volcano is under permanent observation carried out by means of *Landsat* (TM), *SPOT*, and *MOS-1* satellites. Satellite and aircraft images have proven to be useful in discovering changes in the

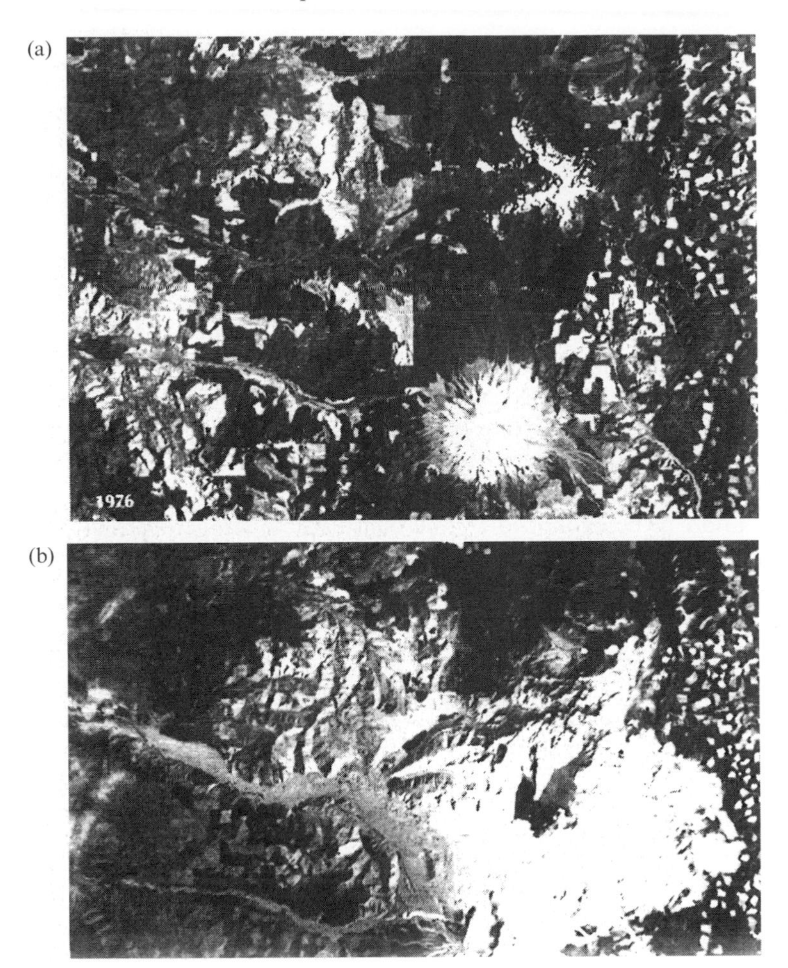

distribution of loose rocks and areas covered with ash, in order to determine the heat on volcanic slopes (Ohkura *et al.*, 1992).

By analysing satellite information, we can assess the consequences of volcanic eruptions, which is important for estimation of damage (e.g. desertification of farming lands). Satellite images allow us to track not only modern lava fields but ancient ones as well such as those on the slopes of Etna in Italy and Mauna Loa (in Hawaii).

Lava fields of different age can be identified in satellite images thanks to differences in the spectral parameters of these fields. In turn, these parameters are caused by a number of factors such as composition of lava, its extent, development of different types of plant cover, sometimes peculiarities in its micro-relief (Figure 3.28). Of course, precise information on the age of lava fields may be obtained from land-based data. Satellite images can also be used to reveal the ranges of

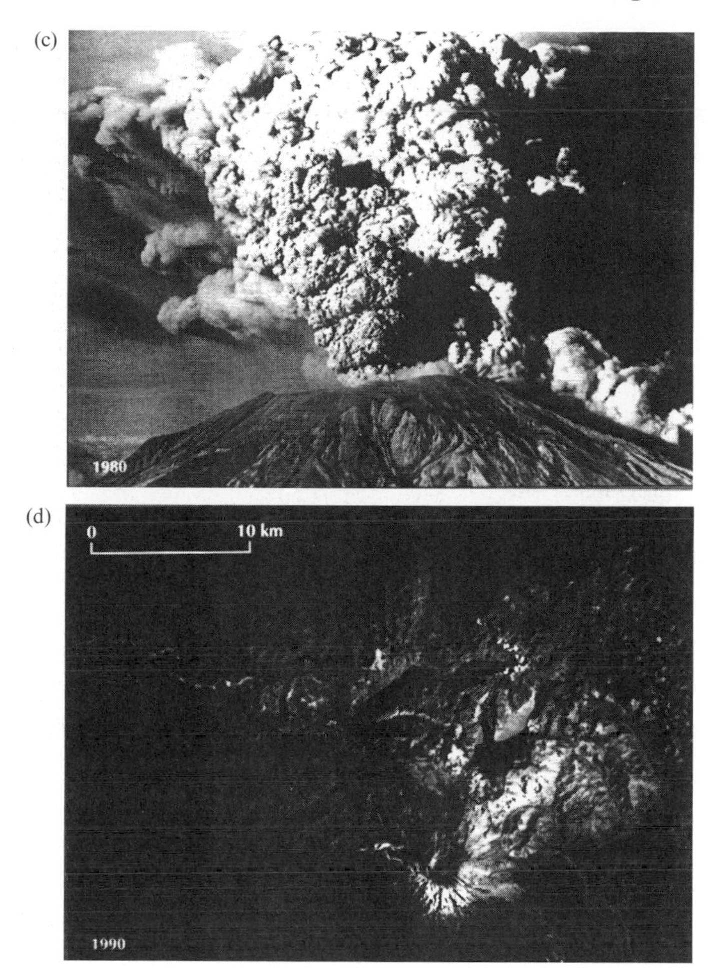

Figure 3.28. Eruption of Mount St Helens volcano (USA). Series of satellite images of the volcano (a, b, c) obtained by means of multi-spectrul scanner from *Landsat* satellite and on-land photograph of the eruption (c) performed before (a) in 1976, in the course of (c) in 1980, 4 days after the onset of the eruption (b) in 1980, and 10 years after the eruption (d) in 1990. Even on black-and-white photo (being coloured in the original) a vast zone of young volcanic objects is clearly seen (white spot); after 10 years (d) it is less pronounced because it is masked with vegetation. Primary volcano snow – ice cap (a) in 1990 did not form yet (c). (Grigoryev and Kondratyev, 2001)

ancient volcanic rocks, which is important for assessment of the vulnerable inhabited areas on the slopes of (Dircke Weltraumbild-Atlas, 1981).

High-resolution *SPOT* satellite-derived images were successfully used for recognition and mapping of the lahars (mud flows) that developed as a result of the Nevado del Ruiz volcanic eruption in Colombia on 13 November 1985 (Vandemenlebrouck *et al.*, 1993). This clearly identified the ranges of the most recent volcanic and mud emissions and, thereby, zones of risk.

Satellite observations have also proved to be useful for assessment of the risk of avalanche slides caused by hydrothermal volcanic activity. For example, *Landsat* images were used to assess the thermal activity of fumaroles on the slopes of the Mont-Bekr volcano (state of Washington), whose activity is associated with submarine lahar motion along its slopes (Frank, 1976).

Satellite observations are also able to locate those thermal anomalies that are indicators of future volcanic eruptions. For example, images obtained from *NOAA* (AVHRR) and *Landsat* (TM) in 1986 identified a number of anomalies in surface temperature fields on Etna. Some of these were associated with permanently active craters in the upper zone. The most considerable anomaly (200 × 300 m) correlated clearly with Etna eruption that occurred a week after observation on 3 October 1986 (Bonneville and Gonze, 1992).

Gathering operational information from and transmitted by satellites at special-purpose land-based automatic sensors helps to identify changes in the surface slopes and in the composition of gas emissions; this is very important for forecasting volcanic eruptions as well as earthquakes (which are quite often interrelated). A number of volcanoes are already being monitored by such an observation network.

SPOT images, obtained in 1990, proved to be useful for assessment of a dangerous volcanic zone around Nevado del Sabankal in Southern Peru. This volcano, which had been dormant for about 200 years, suddenly became active in May 1990. Eruption of this volcano was a clear threat to the 35,000 persons living on its outskirts and cultivating the lands in its valleys.

This volcano (stratovolcano) erupted on and off for 3 years and was recognized as dangerous because of its icecap, which did not melt for a long time despite the activity of the volcano.

Images from *SPOT*, land-based geological and geomorphologic survey data and especially information on tephritic and pyroclastic flow propagation made it possible for us to map dangerous zones. The following three zones of different activity were identified: (1) moderate activity, (2) isolated eruptions, and (3) probable principal stage of eruption (Thouret *et al.*, 1994).

At the present time, infrared sensors (SWIR) in the near-infrared range (1.55–1.75 μm and 2.08–2.35 μm) can be used to discover and measure thermal changes in a volcano, ranging from a few degrees Centigrade up to the high temperatures of magma of about 1,200°C (Rothery, 1992). From measurement data in two wavelength ranges, the size and temperature of radiation sources can be determined. These data are useful for determination of the temperature and thermal structure of lava flows.

Measurements of this kind of the thermal fields of active volcanoes were first made by means of an optical sensor (OPS) mounted on *JERS-1*, which was launched in 1992. Satellite images were taken in three wavelength ranges in the near-infrared spectral zone. In this way, both the lava cupola of Laskar in Chile and the lava lake in Ert-Ali in Ethiopia were studied. *Landsat* (TM) satellite images were also used for this purpose. Qualitative information of this kind is useful for development of a lava flow motion model.

3.4.3 Landslides

3.4.3.1 Occurrence and features of dangerous movements of mountain soils and rocks

There are several types of soil and rock movements on the slopes of mountains. They comprise landslips, mud flows, avalanches, taluses, and landslides. The most widespread, large scale, frequently occurring and therefore the most dangerous ones are landslides and mud flows.

As for number of victims, mass events are not among the most dangerous of natural calamities. For example, between 1945 and 1990, only 33 major landslides are known to have taken place in the world with a total number of victims not exceeding more than 50,000 (Jones, 1993). In the league table of natural disasters, landslides occupy 13th place. In many respects (compared with earthquakes, volcanic eruptions, and storms), this is due to the fact that it is easier to predict landslides and, to some extent, limit the damage they cause by protecting buildings and roads.

The true size of calamities caused by landslides is often masked by attribution of the damage they caused to phenomena such as earthquakes, volcanic eruptions and so on that give birth to them. Nevertheless, although individual mountain rock displacements are not considerable either in their size or consequence, when added together they constitute a true disaster. For example, in Hong Kong on 25 August 1972, several hundred landslides took place, and, in 1982, about 1,500, killing approximately 57 and 48 persons, respectively.

The mean annual number of victims due to small or major landslides varies in different countries of the world: in the USA, for example, it is 25 persons, in Indonesia 2,000 (Brabb, 1991). However, most damage associated with landslides is to settlements, agricultural land and, sometimes, the outskirts of cities.

The displacement of soil and rocks' occurs for a variety reasons such as earthquakes and volcanic eruptions. Sometimes, they are triggered by specific meteorological conditions (floods, heavy rainfall). The greatest damage is often caused by landslides and avalanches, rather than the primary disturbance that gave rise to them.

Anthropogenic activity can also be a factor in slope disturbance. Many rock and soil movements of caused by human activity are known. For instance, in Hong Kong (1966, 1972), Italy (1963), and Aberfan (1966), they resulted in the deaths of many people (64, 138, 2,600, and 144, respectively).

As we have already said, ther are factors that contribute to catastrophic displacements of rocks. The causes of landslides are very diverse. Sometimes, these are a seismic shock (1982 in Bolivia) or a volcanic eruption (1981 in Indonesia). In other cases, these are the heavy precipitation that falls in the course of typhoons or monsoons (1989 in Sri Lanka), slope erosion (1927 in China), and the like.

Among manmade causes of landslides (rarer than natural ones) the breaching of dams should be mentioned (1989 in the former USSR).

Although landscapes seem to be local in nature, and, moreover, can be controlled by special engineering constructions like specializing nets to catch falling rocks near roads, nevertheless their forecast is far from being correct. What is more important, although infrequent, the scale of these phenomena can be unforeseeably great and they the damage they cause devastating. One of the most disasterous landslides in the 20th century was the mud flow, volcanic in origin, that occurred on 13 November 1985 in Colombia (Thouret *et al.*, 1994). Nevado del Ruiz erupted causing its ice cap to melt at its top. Pyroclastic flows mixed with huge masses of melt water formed a vast mud flow.

Within 4 hours of the volcanic eruption, a lahar went through river valleys and moved a distance of 100 km through one of them. The volume of the material carried by the lahar ranged between value of $(82\text{–}102) \times 10^6\,\mathrm{m}^3$. The volcano's icecap was partly (25–30% of its volume) dislodged because of fractures that occurred due to seismic shocks. A part of the icecap (9% of itsvolume and 16% of the total area) disappeared altogether. The volume of displaced and melted snow, firn and ice made up $62 \times 10^6\,\mathrm{m}^3$, and that of relieved water made up $43 \times 10^6\,\mathrm{m}^3$.

Several settlements and almost the whole city of Armero situated 72 km from the volcano were buried under the mud mass. About 23,000 people lost their life. Mud mass spread over the area of 210,000 ha and covered farming lands (Figure 3.29).

The most destructive avalanche happened in 1970 in the Andes in Peru. A large chunk of the icecap was torn off the Nevados-Huaskaran Mountain, and together with snow and mountain rock fragments, the huge avalanche of $(5\text{–}10) \times 10^6\,\mathrm{m}^3$ swept down to the valley and totally ruined the city of Yungau, more than 2,000 people were killed.

The most considerable (regarding number of victims), displacement of mountain rocks in the 20th century occurred in China. That involved a displacement (an avalanche) of loesses as a consequence of an earthquake and led to the death of about 100,000 people (Brabb, 1991).

Landslides are undoubtedly the most widespread and hazardous phenomena among rock displacements on mountain slopes.

From 1981 to 1985, 86 major rock displacements occurred. They resulted in a number of deaths (at least 10 persons), evacuation of a great number of citizens, and material damage in each other of about US$1 million and more. Only three major landslides were not accompanied by deaths, and in another two cases the number of victims was less than 10. Only one-quarter of all rock displacements resulted in the death of 10 or more persons. Only once did the number of killed reach 1,000 persons (Table 3.17: Brabb, 1991).

Many of the landslides were accompanied by different related phenomena such as mud flows and floods, which aggravated matters. Rock movements have the capacity to destroy settlements, disrupt communication lines, and make people homeless. For instance, as a result of simultaneous rock displacements and floods in 1974 in Bangladesh about 50,000 people were made homeless. Many people were

Figure 3.29. Attempts to save one of the citizens of the city of Armero (Colombia) buried under the Earth flow induced by eruption of Nevado del Ruiz volcano on 13 November 1985 (Hewitt, 1997)

Table 3.17. The biggest landslide (1981–1989) as regards the number of victims (more than 250,000 persons killed).

Country	Year	Number of victims	Number of injured
Ecuador	1987	About 1000	6,000 injured
Indonesia	1981	500	3,300 evacuated
China	1989	800	
Colombia	1987	355	2,436 injured
Brazil	1988	277	95 missing
Brazil	1988	>250	
Sri Lanka	1985	305	250,000 homeless
Philippines	1985	300	300 missing
Peru	1983	232	
Nepal	1987	188	351,000 injured

Table 3.18. Catastrophic rock displacements in the 20th century.

Year	Country	Phenomenon	Number of killed and injured	Damage
1920	China	Avalanche of loess rock, caused by earthquake	About 100,000 persons	
1919	Indonesia	Mud-rock flow, caused by volcanic eruption	5,110 persons	
1970	Peru	Snow-and-ice avalanche from Nevado Huaskaran mountain	20,000 persons	
1985	Colombia	Mud-rock flow caused by Nevado-del Ruiz volcanic eruption	22–25,000 persons	US$212 million
1976	Guatemala	10,000 earthquake-induced rock displacements	200	500 houses destroyed
1982	Salvador	Rock displacements and floods	500 (25,000 homeless)	
1988	Turkey	Rock displacements	300	Several houses destroyed
1971	Turkey	Rock displacements	100	Villages ruined
1978	India	Rock displacements and floods	125	
1974	Bangladesh	Rock displacements and floods	39 (50,000 homeless)	
1976	Nepal	Rock displacement	150	
1983	Ecuador	Rock displacement	100	
1969	Turkey	Debris fall		US$1.5 million
1963	Italy	Mud and rock flow	3,000	
1967	Costa Rica	Mud and rock flow	20	US$3,5 million

made homeless as a result of a landslide in 1987 in the former USSR (16,000 peoplr) and, especially, in 1989 in Sri Lanka (250,000 persons). A huge landslide in Nepal in 1987 left 350,000 people without a roof over their heads.

In some cases, the prospect of rock displacement made it necessary to evacuate a great number of people (e.g. in Italy, in 1983, 3,000 persons were evacuated; in Indonesia, in 1981, 3,300 persons were evacuated; in Switzerland, in 1988, 2,000 persons).

The material damage caused by rock movements sometimes exceeds US$1 million. For example, damage due to a landslide in Turkey in 1969 totalled US$1.5 million; that due to two landslides in Switzerland in 1987 and 1988 was estimated as US$40 and US$7 million, respectively (in both cases there was one fatality). Damage due to mud rock flow in 1985 in Columbia reached US$212 million (Table 3.18).

From these examples, we can see that the cost of damage due to rock displacements differs considerably in different countries. In some of them, it does not generally exceed a few million dollars (in the Caribbean US$1.5m, in Denmark

US$1.4m, and in Ghana 0.5 US$1.0m). In others it totals dozens of millions (in South Africa US$20m) and even hundreds of millions (in Spain US$220m). The greatest material damage caused by landslides was in Japan and China (US$4m and US$5m, respectively). Although this phenomenon is neither the most dangerous nor of the largest scale, every year it leads to the death of thousands of people, and damage caused by it is measured in billions of dollars.

The geography of 33 of the most major and destructive rock displacements (Jones, 1993) is as follows. About one-third of them (12 events) were in South and Central America. One event was observed in each of North America, Africa, Australia and Oceania, and two in Europe, and the rest (16) were observed in Asia. Among the countries that found rock displacements to be most dangerous for their citizens, Peru and Nepal lead the league table (which is not surprising because they are both mountainous countries), followed by Japan and Indonesia (5,5,3, and 3 events, respectively).

Among the more than 100 countries that are commonly subject to calamities from rock displacements, of special note are the following: Russia, Kazakhstan, the USA, Canada, China, Japan, Ghana, South Africa, Australia, Sri Lanka, Thailand, Philippines, Malaysia, Hong Kong, Papua-New Guinea, Bulgaria, the Czech Republic, Slovakia, Poland, Denmark, Spain, Norway, Ecuador, Panama, Costa Rica (Brabb, 1991).

3.4.3.2 Satellite data on rock displacements on slopes

It is well known that satellite images can be applied to the study of the most frequently occurring dangerous phenomena – landslides and mud flows.

3.4.3.2.1 Landslides

Preliminary studies (laboratory and field studies) show the following. The spatial resolution capacity of the available remote-sensing systems makes it possible for us to determine, from satellite data, significant landslide features of at least 400 m in size (unless the contrast between the landslide event and the background is very high).

Because of its restriction images from *SPOT* and *Landsat* (TM) satellites can only be used for the cartography of landslides on a regional scale. To make more detailed maps, it is necessary to take more detailed air photographs. These would also present us with the opportunity to observe the stereoscopic effect, which is very useful for the assessment of landslides.

In practice, for the study of a landslide region, air photographs and satellite images of one and the same territory are used simultaneously, complementing each other.

Changes in relief and landscape induced by landslides were fixed for example, in the images from a *SPOT* satellite in the area of La Paz, in Bolivia (Colleau *et al.*, 1991). Simultaneously with in the pilot site, it was found that the most efficient use of images from a *SPOT* satellite was by complementing them with air photographs and various data on the area under investigation, and in particular, on the gradients of relief, slope exposure, and the hydrological network.

Note that in forested areas landslide relief is not easily scanned because it is masked with forests. In such areas, it is important to use data that are indicative of the role of forest cover whose features are recognized in satellite images. For example, for an assessment of how landslides develop in the forested region in the basin of the Iosina River in Secocu Island in Japan, *Landsat* (TM) satellite images were used. In so doing, the water-retentive soils and geological structure were considered the most important factors that determined how landslides develop (Samarkoon *et al.*, 1993).

Satellite images were used to identify the forest cover at times of different concentration of water for two periods: after prolonged rains and after prolonged drought. Then, the value of the standardized vegetation index (NDVI) was computed for two dates from data for two channels of the TM-3 sensor, NN 4 and 5. Finally, satellite images were used to identify regions of relatively high water content. They were then compared with the map of landslides. It was found that the majority of landslides were associated with these regions. Thus, these satellite data on vegetation made it possible for us to judge upon soil's water content and, moreover, they proved to be useful for other information about landslide events.

On the basis of satellite information maps of landslide threat can be made. One such map was made for the basins of the Bagirata and Bilandana Rivers in the Central Himalayas (Garg *et al.*, 1996). First, landslides were decoded from satellite images, and of the 35 landslides that were identified from land-based observations, 32 were verified by these satallite data. Then, on the basis of satellite information and data from the Geographical Information System (GIS) a 1:50,000 scale map of the landslide hazard with four gradations of its development was prepared. To make this map, maps of lineaments, land use and the geomorphological structure prepared from satellite photographs and GIS, as well as other maps (drainage, slopes, lithology) and *in situ* observations, were used.

For another Himalayan region (Garval), six gradations were identified in the map of landslide hazard on the basis of computer processing of the data obtained by means of decoding satellite images and GIS (Nagarajon *et al.*, 1996).

It should be mentioned that, although satellite images do not allow us to obtain (in contrast to air photographs) detailed information on the relief of landslide areas, the new RLSA interferometer method proved to be promising for very precise estimates of landslide displacements. For example, the rather insignificant displacements of individual landslides in France, of about 1.5 m/day, were discovered by means of RLSA-interferometer data. For the study, RLSA images obtained from *ERS* were used (Franeau and Achacte, 1935).

3.4.3.2.2 Mud flows

Satellite information has proved to be useful for acquisition of various data on mud flows and mud flow-prone regions. Overview satellite images enable us to identify the main centres of mud-flow origin, the path of mud-flow motion, as well as to make judgements about the geological and geographical features of the regions of mud-flow origin that might facilitate or even prevent their development. Of course, these data are not as detailed as those obtained from air photographs or, even more so, from *in situ* observations. Nevertheless, they enable us to monitor large and difficult-to-access territories at regular intervals.

Analysis of satellite images gave us a lot of information about the mud flow-prone areas in the Issyk Kul trough (Challaev, 1989). The images decoded mud flows, mud flow-made tracks and their subsequent entrenchment, and areas where mud flows had developed. By ascertaining the type of mud flow, it is possible to identify the place of its origin. Mud and rock flows develop in the upper reaches of mountain slopes, while alluvial mud flows develop in foothills. In the Issyk Kul trough, it was found (from satellite images) that there are 980 mud-flow basins. Of them, 90% are not less than $5\,km^2$, 8.6% from 5 to $50\,km^2$, and 1.4% of more than $50\,km^2$.

Another mud flow-prone region is located in the low mountains of Turkmenistan. In the last 100 years (from data as of 1992), mud flows have swept down from the Kopet Dagh Mountains more than 1,300 times.

From 1:200,000 scale satellite images magnified to the scale of 1:50,000, all the principal types of mud-flow area can be seen in the mountains of Turkmenistan: gravitational (taluses and placers), gravitational and infiltration (landslides), fluvial and erosive. Identification of the latter is especially important since the risk of mud-flow development induced by them is maximum: 80% of mud flows are generated in these areas (Osmanov, 1992).

Yet another mud flow-prone region is the Philippines. Images from *Landsat* TM were used to estimate the risk from mud flows during the rainy season in 1991–1993. The material of these mud flows emanated from the eruption of Pinatubo (Evangelista *et al.*, 1995).

To be able to forecast mud flows information about both present-day and ancient mud flows, especially wide-ranging ones, is very important. Currently, mud flows under certain conditions do follow, regarding size and direction of motion, the routes of certain ancient mud flows. Satellite images have revealed the routes and final resting places of many ancient mud flows around the lake of Issyk Kul. As evidenced by these data, mud-flow activity in many regions of the world in the past was considerably higher and, in particular, around Issyk Kul (subsequent to degradation in this glaciation region). Some, capes in the lake, including the one near the city of Cholpon-Ata (Challaev and Abdullaeva, 1989), are formed on deposits from ancient mud flows.

Maps giving information about the sources of mud flows, their routes, and the extent of the hazard they pose to the population have been prepared on the basis of decoded satellite images. From satellite survey data and other observations, a map of the mud-flow hazard for the Issyk Kul trough has been drawn (Challaev and Abdullaeva, 1989). Note the need to have information about mud flows in this region: in 100 years of observations, 395 mud flows have swept down from the mountains of Terskei Alatau ridge from the north and 213 mud flows have done so from the south. This map shows the three main routes of mud flows (debris and alluvial types) and the areas at risk of mud, flows (four categories of risk were identified). Seismic mud flows are rare here, although earthquakes in this region give rise to landslides and landslips, and thereby 'prepare' material for mud flows.

So, satellite images can be used to identify mud flows of different age, to map the boundaries of mud-flow basins where they are deposited, and to determine mud-flow routes. Together with other data (land-based and aircraft observations), satellite images are very useful tools for mud-flow forecasting.

Data on all types of rock movement on slopes that are obtained in isolation by satellite survey, by themselves are not informative enough. The efficiency of satellite images that look for rock movements on a massive scale is much higher, because not only are the above-mentioned phenomena observed but other features of the landscape that influence them as well. Among these natural features, of special significance is, the role of plant cover, which prevents any rock movements.

For instance, by means of satellite images the contours of zones prone to snow-slips and the routes of avalanches in the Himalayas have been mapped (Reuben *et al.*, 1993). Maximum attention, in doing this, was paid to decoding vegetation cover. This is due to the fact that thickly forested areas protect the surrounding areas from slips, whose generation and displacement are hindered by trees and bushes. In this manner, about 85% of the zones subject to risk have been identified by scientists.

A considerable role in the processes underlying mountain rock displacements on slopes is also played by snow cover as it melts. Features of its location and melting are decoded in satellite images. Satellite images have been used by scientists from the Indian National Agency for seasonal and short-term forecasts of snow melting and floods in Himalayas. At reference (test) sites, high-resolution images were obtained from *Landsat* (TM), and, as a result, a digital classification was prepared that is now used as the basis for a snow-cover map (Reuben *et al.*, 1993).

So, it is impossible to obtain meaningful data about rock movements unless we use the many different sources of information on the region under study. These concern the number of terrain slopes, their ruggedness, drainage and water-retention of surface deposits, the existence of active tectonic disturbances, peculiarities of the slopes' relief, land use, and snow-cover development. It is also important to know to what extent the region is forested. Satellite information on the regions known to be susceptible to rock could be used as good geological and geographical material for further analysis, enabling us to combine the analysis of satellite data, for a particular region, with the general information on rock displacements obtained from Geological Information Systems (GISs). Such GISs, possessing data on rock displacements in a number of regions, including the Himalayas and Andes, have already been created.

In Colombia, satellite images of the Andes have been used for assessment of landslide hazards (Lerol *et al.*, 1992). Using images from *Landsat* and *SPOT*, landslide-prone areas and steep areas of slopes (satellite data were checked by means of on-land observations). Using digital methods to classify the images from *SPOT*, seven types of natural object, including four types of plant cover and three types of land-use (i.e. soils) were identified.

Satellite images were used to build a numerical relief model to determine quantitative data about the relief (slopes, slope exposure, drainage network). Then, using

GIS and information about the potentially dangerous factors that contribute to landslide activity (slope gradation, rock lithology, soil types, surface movements of mountain rocks, and drainage network) the corresponding critical conditions were assessed for all landslide-prone areas. These investigations have enabled us to produce in this way 1:25,000 up to 1:50,000 scale maps of the landslide hazard.

In a very similar way, GIS data were successfully used for assessment of the landslide hazard in the Himalayas (Frank, 1976). This assessment, like the one in the Andes was based on satellite survey data and the air photographs that supplement them. To forecast the likelihood of landslide activity such factors as rock lithology, land-use methods, and proximity to active tectonic structures were taken into consideration. The latter (i.e. active rupture-type tectonic disturbances, clearly decoded in satellite images) are especially significant, bccause surface oscillations due to geological movements at active fractures are associated with the movement of soil and rocks on slopes.

From information gained about 522 landslides in the region, a link was established between the factors involved in landslides, as presented above, and those involved in the risk of landslides (expressed as the ratio between the number of landslides within one category of terrain and the average number of landslides in diffcrent categories). The magnitude of the rated landslide risk factor of less than 0.67 is typical for regions of low risk of landslide generation, and magnitudes of 0.67–1.33 and more than 1.33 are representative of regions at moderate and high landslide risk, respectively.

As a whole, satellite images that study areas where landslides develop have yielded a lot of information and saved a lot of money as well. This is, for example, testified by the experience gained by applying satellite imagery to mapping mud flows in thc mountains of Central Asia (in the former USSR) (Challaev and Abdullaeva, 1989). Satellite images have enabled us to decrease land-based observations considerably, to reduce the amount of work needed to decode images, and at the same time outline more precise mud-flow routes.

The data about the economics of satellite data application are significant. The average cost of mapping mud flow-prone regions in the mountains of Central Asia by usual methods (air photographs and land-based observations) was approximately seven times higher than actual expenditure for work performed using satellite survey data (computations were carried out over an area of 55,000 km^2). Similarly, satellite data proved to be more economical (compared with traditional data) at forecasting mud flows in this region.

The application of satellite information monitor regions susceptible to landslides is the most efficient way of doing this within the framework of international projects at the present time. The large field of view of satellite images can be used to identify areas at risk in neighbouring countries as well as the country under study.

One such successful undertaking was the project that mapped mountain hazards in the Andes 'Geo-information for remote management of natural resources', which was adopted in 1988 as part of a UNESCO program. Sscertaining the risk of landslides was the focus of attention of this project (Rengers and Soeters, 1993), and a number of those countries in the Andes region that are prone to landslides were interested in its development (Bolivia, Peru, Ecuador, Colombia).

3.5 REFERENCES

Aberson, S. D. (2001) The ensemble of tropical cyclone track forecasting models in the North Atlantic basin (1976–2000). *Bull. Amer. Meteorol. Soc.* **82**(9), 1895–904.

Above-average (2001) Atlantic hurricane season continues recent upturn in activity. *Bull. Amer. Meteorol. Soc.* **82**(2), 329–31.

Abushenko, N. A., Altyntsev, D. A., Mazurov, A. A., and Minko, N. P. (2000) Estimation of areas of big forest fires from AVHRR/NOAA data. *Studying the Earth from Space* **1**, 87–93.

Adriatic Sea and the Western Mediterranean during the last Millenium (1996) *Climatic Change* **46**(1–2), 209–23.

Afanasyev, N. F. and Gololobov, Yu. N. (1994) Geo-information structures: forecast in geology and seismology from remote data. Airspace methods of geological and ecological studies. Thesis papers given at *Int. Conf. 30 May–4 June 1994*. Nauka, St. Petersburg, pp. 31–2.

Aforin, V. V., Molchanov, O. A., Kodama, T., Hayakawa, M., and Akentieva, O. A. (1999) Statistical study of ionospheric plasma response to seismic activity research for reliable result from satellite observation. In: M. Hayakawa (ed.) *Atmospheric and Ionospheric Electromagnetic Phenomena Associated with Earthquakes*. Terra, Tokyo, pp. 597–617.

Alcamo, J., Leemans, R., and Kreileman, F. (eds) (1998) *Global Change Scenarios of the 21st Century. Results from the IMAGE 2.1 Model*. Elsevier Science, Amsterdam, 296 pp.

Alcamo, J., Henricks, T., and Rösch, T. (2000) *World Water in* 2025, Kassel world water series Report No. 2. Center for Environmental Systems Research, Kassel, 48 pp.

Alcamo, J., Endejan, M., Kaspar, F., and Rösch, T. (2001) The GLASS model: A strategy for global modelimng of environmental security. *Environ. Sci. Policy* (in press).

Aller, D. (1991) Tropical cyclones in 1997. *Weather* **54**(5), 152–4.

Anon, (1991) World has 68 significant earthquakes during 1990. *Episodes* **14**(1), 79.

Anthes, R., Fellows, J., Hooke, W., and McPherson, R. (2001) AMS/UCAR Outreach to the Bush Administration and the 107th Congress. *Bull. Amer. Meteorol. Soc.* **82**(5), 987–993.

Antonopoulos, G. (1996) Modeling the geoelectric structure of Zante island from MT-measurements. Ph.D. thesis, University of Athens, Physics Department, Greece, pp. 1–287.

Arnell, N. W. (2000) Thresholds and response to climate change forcing: The water sector. *Climatic Change* **46**(3), 305–16.

Astafyeva, N. M., Pokrovskaya, I. V., and Sharkov, E. A. (1994) Large-scale properties of global tropical cyclogenesis. *Papers. Russian Acad. Sci.* **337**(7), 517–20.

Astafyeva, N. M., Pokrovskaya, I. V., and Sharkov, E. A. (1994) Hierarchical structure of global tropical cyclogenesis. *Studying the Earth from Space* **2**, 14–23.

Bartha, I., HorÃnyi, A., IhÃiz, I. (2000) The application of ALADIN model for storm warning purposes at Lake Balaton. *Idöiárás* **104**(4), 219–39.

Bates, N. R., Knap, A. H., and Michaelis, A. F. (1998) Contribution of hurricanes to local and global estimates of air-sea exchange of CO_2. *Nature* **395**(6697), 58–61.

Beal, R. C. (2000) Toward an international storm watch using wide swath SAR. *John Hopkins APL Technical Digest* **21**(1), 12–20.

Belgibayev, M. E. (1993) Impact of eolian processes on the dynamics of soil cover of Kazakhstan. Abstract of Doctoral Geographical dissertation, Moscow University, 61 pp.

Bengtsson, L. (2001) Hurricane threads. *Science* **293**(5529), 440–1.

Bengtsson, L., Botzet, M., and Esh, M. Will greenhouse gas-induced warming over the next 50 years lead to higher frequency and greater intensity of hurricanes? *Tellus* **48A**, 57–73.

Bluestein H. D. (1999) *Tornado Alley* (192 pp.). Oxford University Press, New York.

Board on Natural Disasters (1999) Natural hazards and policy: Mitigation emerges as major strategy for reducing losses caused by natural disasters. *Science* **284**(5422), 1943–4.

Bodechtel, J. and Gierloff-Emden, H. G. (1969) *Weltraum bilder der Erde* (176 pp.). Munich: Paul List Verlag.

Bonneville, A. and Gonze, P. (1992) Thermal survey of Mount Etna volcano from space. *Geophys. Res. Lett.* **19**(7), 725–8.

Bottorff, H. and Mayney, T. (1994) GPS, GTS and the Great Flood of 1993. *Satell. Commun.* **18**(1), 21–5.

Bousquet, J.-C. and Philip, H. (1990) Géologie et risques seismiques. Les enseignements du seisme de Spitac (Arménie, 7 dec. 1988). *Catastrophes et risques naturels*, 61–77.

Bove, M., Zierden, D., and O'Brien, J. (1998) Are Gulf landfalling hurricanes getting stronger? *Bull. Amer. Meteorol. Soc.* **79**, 1327–8.

Brebbia, C. A., Ratto, C. F., and Power, H. (eds) (1999) *Air Pollution* (Vol. VII, 1112 pp.). WIT Press, Southampton, UK.

Brebbia, C. A., Longhurst, J. W. S., and Power, H. (eds) (2000) *Air Pollution* (Vol. VIII, 900 pp.). WIT Press. Southampton, UK.

Brood, K. and Agrawala, S. (2000) The Ethiopia food crisis and limits of climate forecasts. *Science* **289**(5485), 1693–1694.

Bryson, R. A. (2000) Typhoon forecasting, 1944, or, The making of a Cynic. *Bull. Amer. Meteorol. Soc.* **81**(10), 2393–2398.

Buckle, B. W., Leslie, L. M., and Wang, Y. (2001) The Sydney hailstorm of April 14, 1999: Synoptic description and numerical simulation. *Meteorol. Atmos. Phys.* **76**(3–4), 167–182.

Burroughs, W. J. (1997) *Does the Weather Really Matter? The Social Implications of Climate Change* (230 pp.). Cambridge University Press.

Cao Shuhu (1991) Real-time and quasi-real time monitor system of remote sensing information on the flood risk and losses. *Bull. of Flood Damage Evaluation Information System and Wetland Use (Beijing)*, 38–46.

Carlson, T. N. (1979) Atmospheric turbidity in Saharan dust outbreaks as determined by analyses of satellite brightness data. *Monthly Weather Rev.* **107**(9), 322–335.

Carlson, T. N. and Prospero, J. (1972) The large-scale movement of Sahara air outbreaks over the Northern Equatorial Atlantic. *J. Appl. Meteorol.* **11**(2), 283–297.

Caviedes, C. (1991) Five hundred years of hurricanes in the Caribbean: Their relationship with global climatic variations. *Geo. J.* **23**, 301–310.

Challayev, L. V. and Abdullayeva, Z. G. (1989a) Assessment of the efficiency and use of satellite photo-information for mapping Earth flows. *Studying the Earth from Space* **5**, 31–35.

Challayev, L. V. and Abdullayeva, Z. G. (1989b) Earth flow phenomena in Issyk Kul depression and their assessment using satellite surveying. *Studying the Earth from Space* **3**, 84–90.

Chang, E. K. M. (2001) GCM and observation diagnoses of the seasonal and interannual variations of the Pacific storm track during the cool season. *J. Atmos. Sci.* **58**(13), 1784–1800.

Chang, J. C. L., Duan, Y., and Shay, L. K. (2001) Tropical cyclone intensity change from a simple ocean-atmosphere coupled model. *J. Atmos. Sci.* **58**(2), 154–172.

Changda Dai., Lingli Tang., and Cang Chen (1992) Flood damage estimation using TM image. *World Space Congress: 43rd Congress of the International Astronautical Federation (IAF) and 29th Plenary Meeting of the Committee for Space Research (COSPAR), Washington, DC, 28 August–5 September 1992* (Book Abstract A.3, Section 306). World Space Congress, Washington, DC.

Changnon, D. and Changnon, S. A. (1998) Evaluation of weather catastrophe data for use in climate change investigations. *Clim. Change* **38**(4), 435–445.

Changnon, S. A. and Changnon, D. (2001) Long-term fluctuations in thunderstorm activity in the USA. *Clim. Change* **50**(4), 489–503.

Changnon, S. A. and Easterling, D. R. U. S. (2000) Policies pertaining to wealther and climate extremes. *Science* **289**, 2053–2055.

Changnon, S. A. and Kunkel, K. E. (1999) Rapidly expanding uses of climate data and information in agriculture and water resources: Causes and characteristics of new applications. *Bull. Amer. Meteorol. Soc.* **80**(5), 821–830.

Changnon, S. F., Fosse, E. R., and Lecompte, E. L. (1999) Interactions between the atmospheric sciences and insurers in the USA. *Clim. Change.* **42**(1), 57–67.

Changon, S. A. (2001) Assessment of historical thunderstorm data for urban effects: The Chicago case. *Clim. Change* **49**(1–2), 161–169.

Charman, D. J. and Hendon, D. (2000) Long-term changes in soil water tables over the past 4500 years: Relationships with climate and north Atlantic atmospheric circulation and sea surface temperature. *Clim. Change* **47**(1–2), 45–59.

Chepurnykh, N. V. and Novoselov, A. L. (1996) *Economy and Ecology: Development, Catastrophes* (272 pp.). Nauka, Moscow.

Chmyrev, V. M., Sorokin, V. M., and Pokhotelov, O. A. (1999) Theory of small scale plasma density inhomogeneities and ULF/ELF magnetic field oscillations excited in the ionosphere prior to earthquakes. In: M. Hayakawa (ed.) *Atmospheric and Ionospheric Electromagnetic Phenomena Associated with earthquakes* (pp. 759–776). Terrapub, Tokyo.

Choudhury, A. M. (1980) Study of floods in Bangladesh and India with the help of meteorological satellites. *Contributions of Space Observations of Water Resource Management, Proc. Symp. 22nd Plenary Meeting of COSPAR, Bangalore, 1979* (pp. 227–230). Oxford University Press.

Chu, P.-S. and Clark, J. D. (1999) Decadal variations of tropical cyclone activity over the Central North Pacific. *Bull. Amer. Meteorol. Soc.* **80**(9), 1875–1881.

Clark, C. (1999) The great flood of 1768 at Bruton, Somerset. *Weather* **54**(4), 108–113.

Clark, C. (2000) Storms, floods and erosion: The consequences of the storm of 13 May 1998 at Hadspen, Somerset. *Weather* **55**(1), 17–25.

Colleau, A., Girault, F., and Scanvic, J. Y. (1991) *Mouvement de terrain aménagement et imagerie SPOT methodologie de gestion dans un système d'information geographique* (Principaux resultata scientifiques et techniques 1989, pp. 81–82). Bureau de Recherche Geologiques et Minières (BRGM), Paris.

Conaty, A. L. (2001) The structure and evolution of extratropical cyclones, fronts, jet streams, and the tropopause in the GEOS general circulation model. *Bull. Amer. Meteorol. Soc.* **82**(9), 1853–1868.

Crippen, R. E. and Blom, R. G. (1992) The 1992 Landers eathquake. *Earth Observations from Space* **43**(364).

Curtis, S., Adler, R., Huffman, G., Nelkin, E., and Bolvin, D. (2001) Evolution of tropical and extratropical precipitation anomalies during the 1997–1999 ENSO cycle. *Int. J. Climatol.* **21**(8), 961–972.

Dafait, J.-P. (1994) La France sous les yeux vue par SPOT. *Ciel et espace* **289**, 18–19.

Davies, P.(2000) *Inside the Hurricane: Face to Face with Nature's Deadliest Storm* (264 pp.). Henry Holt and Co.

de Blij, H. J., Glantz, M. H., Harris, S. L., Hughes, P., Lipkin, R., Rosenfeld, J., and Williams, R. S. (1997) *Restless Earth. Disasters of Nature* (288 pp.). National Geographical Society, Washington, DC.

Degg, M. (1992) Natural disasters: Recent trends and future prospects. *Geography* **77**(3), 198–209.

Degg, M. (1993) Earthquake hazard, vulnerability and response. *Geography* **78**(2), 165–169.

Diaz, H. F. and Pulwarty, R. S. (1997) *Hurricanes. Climate and Socioeconomic Impacts* (292 pp.). Springer-Verlag.

Dircke Weltraumbild-Atlas (1981) Georg. Westerman Verlag, Braunschweig, Germany, 176 pp.

Disaster Management Support Project (1998) Progress Report. GEOS NOAA/NESDIS, Silver Spring, MD, 183 pp.

Dorland, C., Tol, R. S. J., and Paluticof, J. (1999) Vulnerability of the Netherlands and Northwest Europe to storm damage under climate change. *Clim. Change* **43**(3), 513–535.

Dotto, L. C. (1999) *Storm Warming: Gambling with the Climate of Our Planet* (332 pp.). Doubleday, Canada.

Drabb, E. E. (1991) The world landside problem. *Episodes* **14**(1), 52–61.

Droogomlier, K. K., Smith, J. D., Binsinger, S., Daswell, C., Doyle, J., Duffy, C., Forfoula-Georgion, E., Graziano, T., James, L. D., Krajewski, V., La Mone, M., Lettenmatter, Mass, C., Pielke, R. Sv., Ray, P., Patledge, S., Schaake, J., and Zipser E. (2000) Hydrological aspects of weather prediction and flood warnings: Report of the Ninth Prospectus Team of the U.S. Weather Research Program. *Bull. Amer. Meteorol. Soc.* **81**(11), 2665–2680.

Druyan, L. M., Lonergan, P., and Eichler, T. A. (1999) GCM investigation of global warming impacts relevant to tropical cyclone genesis. *Int. J. Climatol.* **19**(6), 607–617.

Durman, C. F., Gregory, J. M., Hassell, D. C., Jones, R. G., and Murphy, J. M. (2001) A comparison of extreme European daily precipitation simulated by a global and a regional climate model for present and future climates. *Quart. J. Roy. Meteorol. Soc.* **127**(573), 1005–1016.

Dusmukhamedov, F. Ya. (1989) Lithospheric links before severe earthquakes in Middle Asia: Theory, methods and practice of geoindicative studies. Paper given at *Space Aeroindication '89, 3rd All-Union Meeting. Kiev, 16–18 May 1989* (pp. 42–43). Moscow.

Dust storm tracked across U.S, (1977) *Space World* **11**(167), 31–32.

Easterling, D. E., Diaz, H. G., Douglas, A. V., Hugg, W. D., Kunkel, K. E., Rogers, J. C., and Wilkinson J. F. (1999) Long-term observations for monitoring extremes in the Americas. *Clim. Change* **42**(1), 285–308.

Easterling, D. R., Meehl, G. A., Parmesan, C., Changnon, S. A., Karl, T. R., and Mearns, L. O. (2000) Climate extremes: Observations, modelling, and impacts. *Science* **289**, 2068–2074.

El Amir El Nur, M., Ellis N., Grimes, D. I. F. *et al.* (1933) The use of satellite derived rainfall estimates for operational flood warning and river management of the Blue Nile. *Proc. Int. Symp. Operational Remote Sensing, Enschede, The Netherlands*, Vol. 4, pp. 51–62.

Elsner, J. B. and Kara, A. B. (1999) Hurricanes of the North Atlantic. *Climate and Society* (504 pp.). Oxford University Press.

Emmanuel, K. A. (1999) Thermodynamic control of hurricane intensity. *Nature* **401**, 665–669.

Enomoto, Y. and Hashimoto. (1990) *Nature* **346**, 641–643.

Enomoto, Y., Shimamoto, T., Tsutsumi, A., and Hashimoto, H. (1994) Transient electric signals prior to rock fracturing: Potential use as an immediate earthquake precursor. In: M. Hayakawa and Y. Fujinawa (eds) *Electromagnetic Phenomena Related to Earthquake Prediction* (pp. 253–259). Terrapub, Tokyo.

Environment in the EU at the turn of the Century (1999) *EEA Newsletter* **19**, 1–3.

Etkin, D. and Brun, S. E. (1999) A note on Canada's hail climatology. *Int. J. Climatol.* **19**(12). 1357–1374.

Etkin, D., Brun, S. E., Shabbar, A., and Joe, P. (2001) Tornado climatology of Canada revisited: Tornado activity during different phase of ENSO. *Int. J. Climatol.* **21**(8), 915–938.

Evangelista, A. M., Reyes, P. I. D., Saito, G., and Imai, H. (1995) Application of satellite image analysis for the estimation of MT Pinatubo mudflow distribution. *Soil Sci. A Plant Nutr.* **41**(2), 367–370.

Finley, C. A., Cotton, W. R., and Pielke, R. A. Sr (2001) Numerical simulations of tornado-genesis in a high-precipitation supercell. Part 1: Storm evolution and transition into a bow echo. *J. Atmos. Sci.* **56**(13), 1597–1629.

Fischbach, D. B. and Nowick, A. S. (1958) Some transient electrical effects of plastic deformation in NaCl crystals. *J. Phys. Chem. Solids* **5**, 302–315.

Fitzpatrick, P. J. (1999) *Natural Disasters: Hurricanes* (286 pp.). ABC-CLIO.

Folland, C. K., Miller, C., Bader, D., Crowe, M., Jones, P., Plummer, N., Richman, M., Parker, D. E., Rogers, J., and Scholefield, P. (1999) Workshop on indices and indicators for climate extremes. Ashville, NC, 3-6 June 1997. Breakout Group C: Temperature indices for climate extremes. *Clim. Change.* **42**(1), 31–43.

Forbes, G. S. and Bluestein, H. B. (2001) Tornadoes, tornadic thunderstorms, and photogrammetry: A Review of the contributions by T. T. Fujita. *Bull. Amer. Meteorol. Soc.* **82**(1), 73–96.

Fovell, R. G. and Tan, P.-H. (2000) A simplified squall-line model revisied. *Quart. J. Roy. Meteorol. Soc. Part A* **126**(562), 173–188.

Franeau, B., Achacte, J. (1935) Détection du glissement de terrain de Saint-Etienne-de-Tinée par interférometrique SAR et modélisation. *C.R. Acad. Sci. France Ser. 2 Fasc A.* **320**(9), 809–816.

Frank, D. (1976) Debris avalanches at Mount Baker Volcano, Washington. *Geol. Surv. Profess. Pap.* **29**, 120–122.

Franke, C., Fraedrich, L., and Lunkeit, F. (2000) Low-frequency variability in a simplified atmospheric global circulation model: Storm-track induced 'spatial resonance'. *Quart. Roy. Meteorol. Soc.* **126**(569), 2691–2708.

Franzke, C., Freedrich, K., and Lunkeit, F. (2001) Teleconnections and low-frequency variability in idealized experiments with two storm tracks. *Quart. Roy. Meteorol. Soc.* **127**(574), 1321–1339.

Frater, H. (1998) *Natural Disasters. Cause, Course, Effect, Simulation* (CD-ROM). Springer, Berlin.

Freund, F., A. Gupta, S. J. Butow, and S. Tenn (1999) Molecular hydrogen and dormant charge carries in minerals and rocks. In: M. Hayakawa (ed.) *Atmospheric and Ionospheric Electromagnetic Phenomena Associated with Earthquakes* (pp. 859–871). TerraPub, Tokyo.

Friedman, K. S. and Li, X. (2000) Monitoring hurricanes over the ocean with wide swath SAR. *John Hopkins APL Technical Digest* **21**(1), 80–85.

Fujinawa, Y. and Takahashi, K. (1993) Anomalous VLF subsurface electric field changes preceding earthquakes. In: M. Hayakawa, and Y. Fujinawa (eds) *Electromagnetic Phenomena Related to Earthquake Prediction* (pp. 131–148). Terrapub, Tokyo.

Gardner, J. V. (1994) Landsat and the Midwest flood of 1993. *Proc. 2nd Them. Conf. Remote Sens. Mar. A. Coast Environ 'Needs, Solutions and Applications', New Orleans, LA* (pp. 1/80–1/84). Ann Arbor, MI.

Garg, J. K., Narayana, A., Arua, A. S. *et al.* (1996) Landside hazard zonation around Tehri dam using remote sensing and GIS techniques. *Proc. Int. Conf. on Disasters and Mitigation* (Vol. I, PAU-UI-A4-46). Anna University, Madras.

Gershenzon, N. I., Gokhberg, M. B. Karakin, A. V. Petvianshvili, N. V., and Rykunov, A. L. (1989) Modelling the connection between earthquake preparation processes and crustal electromagnetic emission. *Physics of the Earth and Planetary Interiors* **57**, 129–138.

Ghazi, A., Balabanis, P., Casale, R., and Yeroyanni, M. (eds) (1998) *Highlights of Results from Natural Hazards Research Project* (14 pp.). European Comission, Brussels.

Gokhberg, M. B., Morgounov, V. A. and Pokhotelov, O. A. (1995) *Earthquake Prediction, Seismo-Electromagnetic Phenomena* (pp. 1–193). Gordon and Breach.

Goldenberg, S. B., Landsea, C. W., Mestas-Nuez, A. M., and Grey, W. M. (2001) The recent increase in Atlantic hurricane activity: Causes and implications. *Science* **293**(5529), 474–479.

Gololobov, Yu. N., Yevlannikov, L. V., and Smirnova, I. D. (1994) Is it possible to forecast epicentre of an earthquake from remote data? Aerospace methods for geological and ecological studies. Thesis paper given at *Int. Conf. 30 May–4 June 1994* (pp. 164–165). St Petersburg.

Golovko, V. A. and Kozoderov, V. V. (2000) Radiation balance of the Earth: New applications to studying natural calamities from space. *Studying the Earth from Space* **1**, 26–41.

Gorbatikov, A.V., Molchanov, O.A. Hayakawa, M. Uyeda, S. Hattori, K. Nagao, T., and Nikolaev A.V. (1999) Acoustic emission response to earthquake process. *J. Geophys. Res.* (submitted).

Graham, N. E. and Diaz, H. F. (2001) Evidence for intensification of North Pacific winter cyclones since 1948. *Bull. Amer. Meteorol. Soc.* **82**(9), 1869–1894.

Grigoryev, Al. A. (1975) *Space Indication of the Earth Landscapes* (160 pp.). Leningrad University, Leningrad.

Grigoryev, Al. A. and Kondratyev, K. Ya. (1980) Atmospheric dust observed from space. Part I. Analysis of pictures. *WMO Bull.* **29**(4), 250–255.

Grigoryev, Al. A. and Kondratyev, K. Ya. (1981) Atmospheric dust observed from space. Part II. Quantitative assessments of dust content. *WMO Bull.* **30**(1), 3–9.

Grigoryev Al. A. and Kondratyev K. Ya. (1981) *Dust Storms on Earth and Mars* (64 pp.). Znaniye, Moscow.

Grigoryev, Al. A. and Kondratyev, K. Ya. (1993) Aerospace natural resources and ecological studies in China. *Studying of the Earth from Space* **2**,118–122.

Grigoryev, Al. A. and Kondratyev, K. Ya. (1993) Remote sensing of the environment and natural resources in India. *Studying of the Earth from Space* **3**, 118–122.

Grigoryev, Al. A. and Kondratyev, K. Ya. (1996) Satellite monitoring of natural and anthropogenic disasters. *Studying of the Earth from Space* **3**, 68–78.

Grigoryev, Al. A. and Kondratyev, K. Ya. (2000a) Natural and anthropogenic environmental catastrophes, classification and basic characteristics. *Studying the Earth from Space* **2**, 72–82.

Grigoryev, Al. A. and Kondratyev, K. Ya. (2000b) Natural and anthropogenic environmental catastrophes: Problems of risk. *Studying the Earth from Space* **3**, 5–12.

Grigoryev, Al. A. and Kondratyev, K. Ya. (2001) *Environmental Disasters* (687 pp.). St Petersburg University.

Grigoryev, Al. A. and Lipatov, V. B. (1974) *Dust Storms from Satellite Data* (31 pp.). Hydrometeoizdat, Leningrad.

Grigoryev, Al. A. and Lipatov, V. B. (1978) *Smoke Pollution of the Atmosphere from Satellite Observations* (36 pp.). Hydrometeoizdat, Leningrad.

Grigoryev, Al. A. and Lipatov, V. B. (1979) Dust storms in Priaralje from satellite surveying data. *Development and Transformation of the Environment* (pp. 94–103). LSPI, Leningrad.

Grigoryev, Al. A. and Lipatov, V. B. (1982) Dust storms over northern Prikaspian regions from satellite observations. *Remote Methods of Geological and Geographical Study of the Earth* (pp. 18–28). Geographical Society of the USSR, Leningrad.

Grigoryev, Al. A. and Zhogova, M. L. (1992) Heavy dust outbreaks in Priaralye. *Pap. Russian Acad. Sci.* **324**(3), 672–675.

Grigoryev, Al. A., Ivlev, L. S., and Lipatov, V. B. (1976) *Analysis of Meteor-4 Satellite TV Picture of Dust Storm in the Precaspian Region* (No. 16, pp. 29–34). COSPAR Space Research, Berlin.

Grigoryev, Al. A., Ivlev, L. S., and Lipatov, V. B. (1976) Analysis of TV images of dust storms taken from 'Meteor-4' satellite in the region of northern Predkavkazje. *Problems Atmos. Phys.* **14**, 75–86.

Groisman, P. Ya., Karl, T. R., Easterling, D. R., Knight, R. W., Jamanson, P. F., Hennessy, K. J., Suppich, R., Page, C. M., Wibig, J., Fortuniak, K., Razuvaev, V. N., Douglas, A., Forland, E., and Zhai, P.-M. (1999) Changes in the probability of heavy precipitation: Important indicators of climatic change. *Clim. Change* **42**(1), 243–283.

Gruza, G., Rankova, E., Razuvaev, V., and Bulygina, O. (1999) Indicators of climate change for the Russian Federation. *Clim. Change* **42**(1), 219–242.

Gufeld, I., Gusev, G., and Pokhotelov, O. (1994) Is the prediction of earthquake dates possible by the VLF radio wave monitoring method? In: M. Hayakawa, and Y. Fujinawa (eds) *Electromagnetic Phenomena Related to earthquake Prediction* (pp. 381–390). Terrapub, Tokyo.

Guirlet, M., Chipperfield, M. P., Pyle, J. A., Goutail, F., Pommereau, J. P., and Kyro, E. (2000) Modeled Arctic ozone depletion in winter 1997/1998 and comparison with previous winters. *J. Geophys. Res.* **105**(D17), 22185–22200.

Gupta, R. P. and Joshi, B. C. (1989) Landside risk assessment using remote sensing and other geodata sets. *7th Themat. Conf. Remote Sens. Explor. Geol., Calgary, 2–6 October 1989* (p. 135). Ann Arbor, MI.

Gupta, R. P., Saraf, A. K., Saxena, P., and Chandler, R. (1994) IRS detection of surface effects of the Uttarkahi earthquake of 20 October 1991, Himalayas. *Int. J. Remote Sens.* **15**(11), 2153–2156.

Hadjicontis, V. and Mavromatou, C. (1995a) Electric signals recorded during uniaxial compression of rock samples: Their possible correlation with preseismic electric signals. *Acta Geophys. Pol.* **XLIII**(1), 49–61.

Hadjicontis, V. and Mavromatou, C. (1995b) Laboratory investigation of the electric signals preceding earthquakes. In: J. Lighthill (ed.) *The Critical Review of VAN: Earthquake Prediction from Seismic Electric Signals* (pp. 105-117). World Scientific, Singapore.

Hadjicontis, V. and Mavromatou, C. (1996) Laboratory investigation of the electric signals preceding earthquakes. In: J. Lighthill (ed.) *The Critical Review of VAN: Earthquake Prediction from Seismic Electric Signals* (pp. 105–117). World Scientific, Singapore.

Hansen, J., Ruedy, R., Glascoe, J., and Sato, M. (1999) GIS analysis of surface temperature change. *J. Geophys. Res.* **104**(D24), 30997–31022.

Hayakawa, M. (1999) *Atmospheric and Ionospheric Electromagnetic Phenomena Associated with Earthquakes*. Terrapub, Tokyo.

Hayakawa, M. (2000a) Electromagnetic phenomena associated with earthquakes. *Bull. Univ. Electro-Communications* **13**(1), 1–6,

Hayakawa, M. (2000b) *Results of Earthquake Remote Sensing Frontier Research* (Part 2, preliminary report). NASDA.

Hayakawa, M., and F. Fujinawa, F. (eds) (1994) *Electromagnetic Phenomena Related to Earthquake Prediction* (p. 677). Terrapub, Tokyo.

Hayakawa, M., Kawate, R., Molchanov, O. A., and Yumoto, K. (1996a) Results of ultra-low-frequency magnetic field measurements during the Guam earthquake of 8 August 1993. *Geophys. Res. Lett.* **23**, 241–244

Hayakawa, M., Kawate, R., and Molchanov, O. A. (1996b) Ultra-low-frequency signatures of the Guam earthquake on 8 August 1993 and their implication. *J. Atmos. Electr.* **16**, 193–198

Hayakawa, M., Molchanov, O.A. Ondoh, T., and Kawai, E. (1996c) The precursory signature effect of the Kobe earthquake on subionospheric VLF signals. *J. Comm. Res. Lab.* **43**, 169–180.

Hayakawa, M., Ito, T., and Smirnova, N. (1999) Fractal analysis of ULF geomagnetic data associated with the Guam earthquake on August 8, 1993. *Geophys. Res. Lett.* **26**, 2797–2800.

Heamen-Rodrigo, D., Chorowicz, J., Deffontaines, B. *et al.* (1993) Cadre structural et risques geologiques Études à l'aide de l'imager spatiale: le region du Colca (Andes du sud Pérou). *Bull. Soc. Geol. France* **164**(6), 807–818.

Heino R., Brezdil, R., Furland, E., Tuomenvirta, H., Alexandersson, H., Beniston, M., Pfister, C., Rebetez, M., Rosenhagen, G., Rusner, S., and Wibig, J. (1999) Progress in the study of climate extremes in Northern and Central Europe. *Clim. Change.* **42**(1), 151–181.

Hellin, J., Haigh, M. J. (1999) Rainfall in Honduras during hurricane Mitch. *Weather* **54**(11), 350–359.

Henderson-Sellers, A., Zhang, H., Berz, G., Emanuel, K., Gray, W., Landsca, C., Holland, G., Lighthill, J., Shieh, S.-L., Webster, P., and McGuffie, K. (1998) Tropical cyclones and global climate change: A post-IPCC assessment. *Bull. Amer. Meteorol. Soc.* **79**(1), 19–38.

Hewitt, K. (1997) *Regions of Risk. A Geographical Introduction to Disasters* (389 pp.). Longman, Harlow, UK.

Hisdal, H., Stahl, K., Tallaksen, L. M., and Demuth, S. (2001) Have streamflow droughts in Europe become more severe or frequent? *Int. J. Climatol.* **21**(3), 317–334.

Hodell, D. A., Brenner, M., Curtis, J. H., and Guilderson, T. (2001) Solar forcing of drought frequency in the Maya Lowlands. *Science* **292**(5520), 1367–1369.

Holt, C. P. (2000) Characteristics of the Northampton rainstorm, 9th April 1998. *Weather* **55**(3), 78–84.

Holt, C. P. (in press) *Drought Severity in Wales* (Special publication). British Hydrological Society.

Holt, M. A., Hardaker, P. J., and McLelland, G. P. (2001) A lightning climatology for Europe and UK, 1996–1999. *Weather* **56**(9), 290–296.

Horton, E. B., Folland, C. K., and Parker, D. E. (2001) The changing incidence of extremes in worldwide and central England temperatures to the end of the twentieth century. *Clim. Change* **50**(3), 267–295.

Houghton, J. T., Meira Filho, L. G., Callander, B. A., Harris, N., Kattenberg, A., and Maskell, K. (eds) (1996) *Climatic Change 1995, The Science of Climate Change* (572 pp.). Cambridge University Press.

Hurricane Research and Forecasting. Policy Statement (2000) *Bull. Amer. Meteorol. Soc.* **81**(6), 1391–1346.

Huth, R., Kysely, J., and Pokorna, L. (2000) A GCM simulation of heat waves, dry spells and their relationships to circulation. *Clim. Change* **46**(1–2), 29–60.

Idso, S. B. (1976) Dust storms. *Scientific American* **235**(4), 108–114.

Ing, G. K. T. (1972) A dust storm over Central China, April 1969. *Weather* **27**(4), 136–145.

Ishakov, M. H., Loziyev, V. P., Saidov, M. S., and Urinov, B. O. (1990) Results of application of aerospace photographs in the study of Khusorsk earthquake consequences. *Studying of the Earth from Space* **5**, 59–65.

Jackson, J. D (1995) *Classical electrodynamics* (2nd edn). John Wiley and Sons, New York.

James, H. R. (1998) Atlantic hurricanes hardly happened in 1997. *Weather* **53**(12), 433–436.

Jighan Jiang and Shuhu Cao. (1990) *All Weather Flood Monitoring and Loss Estimation by Using Space Techniques* (Preprint, 10 pp.). Space Science Application Research Center, Acad. Cinica, Beijing.

Jin Zhang and Qi Tuan. (1990) Research and establishment of the nationwide natural environment information system. *Proc. 2nd Int. Workshop on Geographical Information Systems, Beijing* (pp. 198–208).

Jones, D. (1993a) Environmental hazards in the 1990's: Problems, paradigms and prospects. *Geography* **78**(2), 161–165.

Jones, D. (1993b) Landsliding as a hazard. *Geography* **78**(2), 185–190.

Jones, D., Horton, E. B., Folland, C. K., Hulme, M., Parker, D. E., Basnett, T. A. (1999) The use if indices to identify changes in climatic extremes. *Clim. Change* **42**(1), 131–149.

Kamogawa, M. and Ohtsuki, Y.-H. (1999) Plasmon model for origin of earthquake related electromagnetic wave noises. *Proc. Japan Acad. Ser. B* **75**, 186–189.

Karakin, A. V. and Lobkovsky, L. I. (1985) On derivation of equations for a three-component visco-deformable medium (crust and the asthenosphere). *Izv. Akad. Nauk SSSR, Fiz. Zemli (Solid Earth)* **12**, 3–13 (in Russian).

Karl, T. R. and Easterling, D. R. (1999) Climate extremes: Selected review and future research directions. *Clim. Change* **42**(1), 309–325.

Karl, T. R., Nicholls, N., and Ghazi, A. (1999a) CLIVAR/GCOS/WMO Workshop on indices and indicators for climate extremes (Workshop summary). *Clim. Change* **42**(1), 3–7.

Karl, T. R., Nicholls, N., and Ghazi, A. (ed.) (1999b) *Weather and Climate Extremes – Changes, Variations and a Perspective from the Insurance Industry* (349 pp.). Kluwer Academic, Dordrecht, The Netherlands.

Karol, I. L. (2000) Impact of the world transport aircraft flights on the ozonosphere and climate. *Meteorology and Hydrology* **7**, 17–22.

Kawate, R., Molchanov, O. A., and Hayakawa, M. (1998) Ultra-low-frequency fields during the Guam earthquake of 8 August, 1993 and their interpretation. *Phys. Earth Planet. Int.* **105**, 229–238.

Kedar, E. Y., Paterson, W., and Hsu Shin-yi. (1973) earthquake-risk mapping. *Photogramm. Eng.* **39**(8), 831–837.

Keeman, T., Rutlege, S., Carbone, R., Wilson, J., Takahashi, T., Moy, P., Tapper, N., Peleatt, M., Hacker, J., Sekelesky, S., Moncrieff, M., Saito, K., Holland, G., Crook, A., and Gage, K. (2000) The maritime continent thunderstorm experiment (MCTEX): Overview and some results. *Bull. Amer. Meteorol. Soc.* **81**(10), 2433–2456.

Kelley, M. D. (1989) *The Earth's Ionosphere* (280 pp.). Academic Press, New York.

Khovanski, V. N. (1991) Application of satellite information to forecast earthquakes and volcanic eruptions. Thesis paper given at *13th Interdepartm. Meeting on the Study of Current Earth Crust Motion at Tallinn Geodynam. Test Area, 13–18 May 1991* (133 pp.).

Kidder, S. Q., Goldberg, M. D., Zehr, R. M., De Maria, M., Purdom, J. F. W., Verden, C. S., Grody, N. C., and Kusselson, S. S. (2000) Satellite analysis of tropical cyclone using the Advanced Microwave Sounding Unit (AMSESs). *Bull. Amer. Meteorol. Soc.* **81**(6), 1241–1259.

Kondratyev, K. Ya. (1985) *Volcanoes and Climate, Scientific and Technical Review* (Meteorology and climatology series, 205 pp.). VINITI, Moscow.

Kondratyev, K. Ya. (1988) *Climatic Shocks: Natural and Anthropogenic* (296 pp.). Wiley, New York.

Kondratyev, K. Ya. (1992) Comprehensive monitoring of Pinatubo volcanic eruption. *Studying of the Earth from Space* **1**, 111–122.

Kondratyev, K. Ya. (1999) *Ecodynamics and geopolicy*. Vol. 1. *Global Problems* (1036 pp.). St Petersburg Scientific Centre, Russian Academy Science, St Petersburg.

Kondratyev, K. Ya. (2002) Global climate change: A reality, suggestions and illusions. *Studying the Earth from Space* **1**, 3–23.

Kondratyev, K. Ya. and Cracknell, A. P. (1999) *Observing Global Climate Change* (562 pp.). Taylor and Francis, London.

Kondratyev, K. Ya. and Galindo, I. (1997) *Volcanic Eruptions and Climate* (382 pp.). A. Deepak, Williamsburg, VA.

Kondratyev, K. Ya. and Grigoryev, Al. A. (2000) Natural and anthropogenic ecological disasters: meteorological calamities and catastrophes. *Studying the Earth from Space* **4**, 3–19.

Kondratyev, K. Ya. and Varotsos, C. A. (2000) *Atmospheric Ozone Variability: Implications for Climate Change, Human Health, and Ecosystems* (614 pp.). Springer/Praxis, Chichester, UK

Kondratyev, K. Ya., Barteneva, O. D., Vasilyev, O. B. *et al.* (1976) Aerosol in the region of ATEP and its radiation property. *MGO Transactions* **361**, 67–130.

Kondratyev, K. Ya., Grigoryev, Al. A., Rabinovich, Yu. I., and Shulgina, E. M. (1979) *Meteorological Sensing of the Underlying Surface from Space* (218 pp.). Hydrometeoizdat, Leningrad.

Kondratyev, K. Ya., Grigoryev, Al. A., Pokrovski, O. M., and Shalina, E. V. (1983) *Satellite Remote Sensing of Airborne Aerosol* (218 pp.). Hydrometeoizdat, Leningrad.

Kondratyev, K. Ya., Grigoryev, Al. A., and Zhvalev, V. F. (1987) The impact of desertification on the atmosphere. *Geofisica Int.* **26**(3), 341–358.

Kossin, J. P., Schubert, W. H., and Montgomery, M. T. (2000) Unstable interactions between a hurricane's primary eyewall and a secondary ring of enhanced vorticity. *J. Atmos. Sci.* **57**(24), 3893–3917.

Kumar, J. K. (1992) Studies on railway track disaster using IRS-1A imagery. *Natural Resource Management* (pp. 400–503). New Perspect, Bangalore, India.

Kundzewicz, Z. W. (1998) Floods of the 1990s: Business as usual?. *WMO Bull.* **47**(2), 155–160.

Kunkel, K. E., Pielke, R. A., and Chagnon, S. A. (1999) Temporal fluctuations in weather and climate extremes that cause economic and human health impacts: A review. *Bull. Amer. Meteorol. Soc.* **80**(6), 1077–1125.

Lan, K. M. and Weng, H. (2000) Remote forcing of summertime U. S. droughts and floods by the Asian monsoon. *GEWEX News* **10**(2), 5–6.

Landsea, Ch. W., Pielke, R. A., Jr., Mestas-Nucez, A. M., and Knaff, J. A. (1999) Atlantic basin hurricanes: Indices of climatic change. *Clim. Change* **42**(1), 89–129.

Lerol, E., Rouzeal, O., Scanyic, J.- Y., Weber, C. C., and Vargas, C. G. (1992) Remote sensing and GIS technology in landslide hazard mapping in the Colombian Andes. *Episodes* **15**(1), 32–35.

Lighthill, J. (ed.) (1996a) A brief look back at the review meeting's proceedings. *The Critical Review of VAN: Earthquake Prediction from Seismic Electric Signals* (pp. 349–356). World Scientific, Singapore.

Lighthill, J. (ed.) (1996b) A brief look forward to future research needs. *The Critical Review of VAN: Earthquake Prediction from Seismic Electric Signals* (pp. 373–376). World Scientific, Singapore.

Lighthill, J., Holland, G., Grey, W., Landsea, C., Craig, G., Evans, J., Kurihara, Y., and Guard, C. (1994) Global climate change and tropical cyclones. *Bull. Amer. Meteorol. Soc.* **75**(11), 2147–2157.

Lyons, W. A. (1980) Evidence of transport of hazy air masses from satellite imagery. *Aerosols: Anthropogenic and Natural Source and Transport* (pp. 418–433). New York.

Ma Zonjin and Gao Xianglin. (1991) *Application of Space Techniques to Earthquake Hazard Reduction in China* (Preprint, pp. 1–15). Institute of Geology, Beijing.

MacClenahan, P., McKenna, J., Cooper, J. A. G., and O'Kane, B. (2001) Identification of highest magnitude coastal storm events over western Ireland on the basis of wind speed and duration thresholds. *Int. J. Climatol.* **21**(7), 829–842.

McDonald, J. R. T. (2001) Theodore Fujita: His contribution to tornado knowledge through damage documentation and the Fujitscale. *Bull. Amer. Meteorol. Soc.* **82**(1), 63–72.

McLeod, N. H. (1974) The use of ERTS imagery and other space data for rehabilitation and development programs in West Africa. Preprint presented at the *COSPAR Seminar on Space Applications of Direct Interest to Developing Countries, San. José dos Campos* (11 pp.).

Maloney, F. D. and Hartmann, D. L. (2000) Modulation of hurricane activity in the Gulf of Mexico by the Madder-Julian oscillations. *Science* **287**(5460), 2002–2004.

Mann, M. E., Bradley, R. S., and Hughes, M. K. (1999) Northern hemisphere temperature during the past millenium: Inferences, uncertainties, and limitations. *Geophys. Res. Lett.* **26**, 759–762.

Marsh T. J. (2001) The 2000/01 floods in the UK–a brief overview. *Weather* **56**(10), 343–345.

Marsigli, C., Montani, A., Nerozzi, F., Paccagnella, T., Tibaldi, S., Molteni, F., and Buizza, R. (2001) A strategy for high-resolution ensemble prediction. II: Limited-area experiments in four Alpine flood events. *Quart. Roy. Meteorol. Soc.* **127**(576), 2095–2116.

Martin, J. W. (1975) Identification, tracking and sources of Saharan dust – an inquiry using SMS. *Gate Rep.* **14**, 217–24.

Mason, S. J., Goddard, L., Graham, N. E., Yulaeva, E., Sun, L., and Arlin, P. A. (1999) The IRI seasonal climate prediction system and the 1997/98 EI Niño event. *Bull. Amer. Meteorol. Soc.* **80**.

Meloche, S. D. (1999) An eyewitness account of Hurricane Hugo, 21–22 September 1989// Weather.–V. 54, N 4.–P. 102–106.

Miller, E. W. and Mille, R. M. (2000) *Natural Disasters: Floods* (292 pp.). ABC-CLIO, Oxford.

Mock, C. J. and Birkeland, K. W. (2000) Snow avalanche climatology of the Western USA mountain ranges. *Bull. Amer. Meteorol. Soc.* **81(10)**, 2367–2392.

Molchanov, O. A. (1999) Fracturing as underlying mechanisms of seismo-electric signals. In: M. Hayakawa (ed.) *Atmospheric and Ionospheric Electromagnetic Phenomena Associated with Earthquakes* (pp. 349–356). Terrapub, Tokyo.

Molchanov, O. A. and Hayakawa, M. (1995), Generation of ULF electromagnetic emissions by microfracturing. *Geophys. Res. Lett.* **22**, 3091–3094.

Molchanov, O. A. and Hayakawa, M. (1998a) Subionospheric VLF signal perturbations possibly related to earthquakes. *J. Geophys. Res.* **103**(A8), 17489–17504.

Molchanov, O. A. and Hayakawa, M. (1998b) On the generation mechanism of ULF seismogenic electromagnetic emission. *Phys. Earth Planet. Int.* **105**, 201–210.

Molchanov, O. A., Kopytenko, Yu. A., Voronov, P. M. Kopytenko, E. A., Matiashvili, T. G., Fraser-Smith, A. C., and Bernardy A. (1992) Results of ULF magnetic field measurements near the epicentres of the Spitak ($M_s = 6.9$) and Loma Prieta ($M_s = 7.1$) earthquakes: Comparative analysis. *Geophys. Res. Lett.* **19**, 1495–1498.

Molchanov, O. A., Hayakawa, M., Ondoh, T., and Kawai, E. (1998) Precursory effects in the subionospheric VLF signals for the Kobe earthquake. *Phys. Earth Planet. Int.* **105**, 239–248.

Monmonier, M. (1999) *Air Apparent: How Meteorologists Learned to Map, Predict, and Dramatize Weather* (309 pp.). University of Chicago Press, Chicago.

Mook, H. A., Pengcheng Dal, Dogan, F., and Hunt, R. D. (2000) One-dimensional nature of the magnetic fluctuations in Yba2Cu3O6.6. *Nature* **404**, 729–731.

Moon, W. M., Won, J. S., Li, B., and Slaney, R. V. (1990) Application of airborne C-SAR and SPOT image for geological investigation of the Nahanni earthquake area. *IGARSS '90: 10th Annual Int. Geoscience and Remote Sensing Symp. 'Remote Sens. Sci. Nineties'. Weather* **3**, 2105–2110.

Morozova, L. I. (1993) Atmospheric indication of earthquakes of Near East. *Studying of the Earth from Space* **6**, 81–83.

Moulain, C., Dulas, F., Breon, F. M., Andre, J. M. *et al.* (1994) Remote sensing of airborne desert dust mass over ocean using Meteosat and CZCS imagery. *Collog. Fond. Singer–Polignac 'Nouvelles Front télédétéct ocean' Monaco. Mem. Inst. Oceanogr.* **18**, 35–43.

Muchlbaerger, W. R. and Wilmarth, V. R. (1977) The Shuttle era: a challenge to the Earth. *American Scientist* **65**(N2), 153–158.

Nagarajon, R., Mukherjec, A., Roy, A. (1996) Landslide hazard assessment using remote sensing. *Procced. Int. Conf. On disasters and migration Madras: Anna Univ. Madras.–Vol. 1.–P. A4-78–A-4-84.*

Nakajima, I. W. a.o. (1975) Environmental quality pattern mapping from space data. *Proc. 11th Int. Symp. Space Technol., and Sci. Tokyo.–P. 927-928.*

Nakata, J. K., Wilshire, H. G., and Barnes G. G. (1976) Origin of Mohave Desert dust plumes photographed from space. *Geology* **4**(11), 644–648.

Nalivkin, D. V. (1969) *Hurricanes, Storms and Tornadoes. Geographical Features and Geological Activity* (488 pp.). Nauka, Leningrad.

National Environment Satellite, Data and Information Service. (1998) *Highlights of 1998* (53 pp.). NOAA/NESDIS, Suitland, MD.

Nechayev, Yu. V., Rogozhin, E. A., and Bodachkin, B. M. (1993) Peculiarities of Rachinsk earthquake manifestations (1991) in the field of tectonic scatterness (from satellite data). *Physics of the Earth* **3**, 64–69.

Nerushev, A.,F., Petrenko, B. Z., and Kramchanilova, E. K. (2001) Determination of parametres of tropical cyclones from SHF-radiometer SSM/I data. *Studying the Earth from Space* **2**, 61–68.

New Views of the Earth Scientific Achievements of ERS-1 (1995) European Space Agency, 162 pp.

Nicholls, N. and Murray, W. (1999) Workshop on indices and indicators for climate extremes, Ashville, NC, 3–6 June 1997 (Breakout Group B: Precipitation). *Clim. Change* **42**(1), 23–29.

Nicholls, N., Landsea, C., and Gill, J. (1998) Recent trends in Australian region tropical cyclone activity. *Meteorol. Atmos. Phys.* **65**(3–4), 197–206.

Nicholls, R. J., Hoozemans, F. M. J., and Marchand, M. (1999) Increasing flood risk and wetland losses due to global sea-level rise: regional and global analysis. *Global Environ. Change.* **9(1001), S69–S87.**

Nolan, D. S. and Farell, B. F. (1999) The structure and dynamics of tornado-like vortices. *J. Atmos. Sci.* **56**(16), 2908–2936.

Noorbergen, H. H. S. (1993) Multitemporal analysis of a braiding River. *Proc. Int. Symp. Oper. Remote Sens. Enschede* **3**, 161–168.

Nutter F. W. (1999) Global climate change: Why US insures care. *Clim. Change* **42**(1), 45–49.

Ohkura, H., Uchara, S., Yazaki, S., and Kumagai, T. (1992) Application of multiple satellite data to the study of natural disasters. *World Space Congr.: 43th Congr. Int. Astronaut. Fed. (IAF) and 29th Plen. Meet. Comm. Space Res. (COSPAR), Washington, DC, 28 August to 5 September 1992* (Book Abstr.). *Weather* **A8**, M.U.02.

Oike, K. and Yamada, T. (1993) Relationship between shallow earthquakes and electromagnetic noises in the LF and VLF ranges. In: M. Hayakawa, and Y. Fujinawa (eds) *Electromagnetic Phenomena Related to earthquake Prediction* (pp. 115–130). Terrapub, Tokyo.

Oppenheimer, C. (1996) Volcanism. *Geography* **81**(1), 65–81.

Osmanov, T. O. (1992) Kopet Dag Earth flows sources, their features and assessment from satellite images. *The Problem of Desert Exploration* **1**, 73–76.

Otterman, J. (1978) Point sources and steak of dust storms: a study of *Landsat* images and some inferences. *Israel Meteorol. Res. Pap.* **11**, 233–243.

Palecki, M. A., Changnon, S. A., and Kunkel, K. E. (2001) The nature and impacts of the July 1993 heat wave in the Midwestern USA: Learning from the lessons of 1995. *Bull. Amer. Meteorol. Soc.* **82**(7), 1353–1368.

Parfit, M. (1998) Living with natural hazards. *Nat. Geographic.* **194**(1), 2–39.

Paul, F. (1999) An inventory of tornadoes in France. *Weather* **54**(7), 217–219.

Penenko, V. V. and Isvetova, E. A. (2000) Analysis of the scales of anthropogenic impacts in the atmosphere. *Optics of the Atmosphere and Ocean* **13**(4), 392–396.

Perrie, W. (2001) The impact of climate change on marine storms. *1st Int. Conf. on Global Warming and the Next Ice Age, 19–24 August* (pp. 224–226). Dalhousie University.

Pewe, T. L. (ed.) (1981) *Desert Dust: Origin, Characteristics and Effect on Man* (Special Paper No. 180). Geological Society of America, Boulder, CO.

Pfaff, F., Broad, K., and Glantz, M. (1999) Who benefits from climate forecasts? *Nature* **397**(6721), 645–646.

Pichugin, A. P. (1985) Radar observations of river floods from Cosmos-1500 satellite. *Studying of the Earth from Space* **5**, 61–66.

Pielke, R. A., Jr (1997) *Hurricanes. Their Nature and Impact on Society* (298 pp.). Wiley, Chichester, UK.

Pielke, R. A., Jr (1999) Nine fallacies of floods. *Clim. Change* **42**(2), 413–438.

Pielke, R. A., Jr and Landsea, C. N. (1999) La Niña, El Niño and Atlantic hurricane damages in the USA. *Bull. Amer. Meteorol. Soc.* **80**(10), 2027–2033.

Plummer, N., Sallinger, M. J., Nicholls, N., Suppiah, M., Henessy, K. J., Leighton, R. M., Trewin, B., Page, C. M., and Lough, J. M. (1999) Changes in climate over the Australian region and New Zeland during the twentieth century. *Clim. Change* **42**(1), 183–202.

Pokrovskaya, I. V. and Sharkov, E. A. (1994) Poisson properties of global tropical cyclogenesis from satellite data. *Studying the Earth from Space* **2**, 24–33.

Pokrovskaya, I. V. and Sharkov, E. A. (1999a) *Catalogue of Tropical Cyclones and Tropical Disturbances of the World Ocean in 1983–1998* (160 pp.). Poligraph Service, Moscow.

Pokrovskaya, I. V. and Sharkov, E. A. (1999b) Structural features of tropical cyclogenesis of the Northern and Southern hemispheres as applied to the problems of satellite monitoring. *Studying the Earth from Space* **1**, 18–27.

Possekel, A. K. (1999) *Living with the Unexpected. Linking Disaster Recovery to Sustainable Development in Montserrat* (287 pp.). Springer, Berlin.

Prediction and Mitigation of Flash Floods. Policy Statement (2000) *Bull. Amer. Meteorol. Soc.* **81**(6), 1338–1340.

Prodi, F., Giorgio, F. *et al.* (1979) A case of transport and deposition of Saharan dust over the Italian Peninsula and Southern Europe. *J. Geophys. Res.* **89**(11).

Prospero, J. M. and Nees, R. T. (1977) Dust concentration in the atmosphere of the equatorial North Atlantic. Possible relations to the Sahelian drought. *Science* **196**, 1196–1198.

PSUCLIM Storm activity in ancient climates, Part 1 (1999) Sensitivity of severe storms to climate forcing factors on geologic timescales. *J. Geophys. Res.* **104**(D22), 27277–27294.

PSUCLIM Storm activity in ancient climates, Part 2 (1999) An analysis using climate simulation and sedimentary structures. *J. Geophys. Res.* **104**(D22), 27295–27320.

Putting NASA's Earth Science to Work Remote Sensing Applications (1998) Ratheon Systems, Upper Marlboro, MD (36 pp.).

Qian, W. and Yang, S. (2000) Onset of the regional monsoon over Southeast Asia. *Meteorol. Atmos. Physics* **75**(1–2), 29–38.

Rango, A. and Anderson, A. T. (1974) Flood hazard studies in the Mississippi River basin using remote sensing. *Water Resource Bull.* **10**(5), 1061–1081.

Rao, U. R. (1996) *Space Technology for Sustainable Development* (564 pp.). New Dehli: McDraw-Hill.

Rappaport, E. N. (1999) Atlantic hurricane season of 1997. *Mon. Weather Rev.* **127**(9), 2012–2076.

Reason, C. J. C. and Murray, R. J. (2001) Modelling for frequency variability in Southern Hemisphere extra-tropical cyclone characteristics and its sensitivity to sea-surface temperature. *Int. J. Climatol.* **21**(2), 249–267.

Reed, R. J., Kuo, Y.-H., Albright, M. D., Gao, K., Guo, Y.-R., and Huang, W. (2001) Analysis and modelling of a tropical cyclone in the Mediterranean sea. *Meteorol. Atmos. Phys.* **76**(3–4), 183–202.

Rengers, N., Soeters, R. (1993) Satellite remote sensing and GIS for landslide hazard zonation. *Int. Symp. Oper. Remote Sens. Enschede, 19–23 April 1993.: Proc. Vol. 3. Enschede* (pp. 133–134).

Reuben, P. P., Kumar, V. S., Raman, R. Ch. *et al.* (1993) Operational remote sensing for snow cover, run-off and avalanche applications in the Himalayas. *Proc. Int. Symp. Oper. Remote Sens. Enschede* **4**, 71–79.

Reynard, N. S., Prudhomme, C., and Crooks, S. M. (2001) The flood characteristics of large UK rivers: Potential effects of changing climate and land use. *Climatic Change* **48**(2–3), 343–359.

Rice, C. J. (1981) Satellite observations of the Mount St. Helens eruption of 18 May 1980. *Proc. Soc. Photo-Opt. Instrum. Eng.* **278**, 23–31.

Rodger, C. J. and Clilverd, M. A. (1999) Modeling of subionosphereric VLF signal perturbations associated with earthquakes. *Radio Science* **34**, 1177–1185.

Rodgers, E. B., Adler, R. F., and Pierce, H. F. (2000) Contribution of tropical cyclones to the North Pacific climatological rainfall as observed from satellites. *J. Appl. Meteorol.* **39**(10), 1658–1678.

Rosenfeld, J. (1999) *Eye of the Storm: Inside the World's Deadliest Hurricanes, Tornadoes, and Blizzards* (312 pp.). Perseus Books, Reading, MA.

Rothery, D. A. (1992) Volcano monitoring using short wave length infrared satellite image data. *World Space Congr.: 43th Cong. Int. Astronaut. Fed. (IAF) and 29th Plen. Meet. Comm. Space Res. (COSPAR), Washington, DC, 28–29 August–5 September 1992* (Book Abstr.). COSPAR, Washington, DC.

Royer, J.-F., Chauvin, F., Timbal, B., Araspin, P., and Grimal, D. (1998) A GCM study of the impact of greenhouse gas increase on the frequency of occurrence of tropical cyclones. *Clim. Change* **38**, 307–343.

Samarkoon, L., Ogawa, S., Ebisu, N. *et al.* (1993) Inference of landslide suspectible areas by Landsat Thematic Mapper data. *IAHS* **217**, 83–90.

Sampson, C. R. and Schrader, A. J. (2000) The automated tropical cyclone forecasting system (Version 3.2). PSUCLIM Storm activity in ancient climates, 1, Sensitivity of severe storms to climate forcing factors on geologic timescales. *Bull. Amer. Meteorol. Soc.* **81**(6), 1231–1240.

Sarlis, N. and Varotsos, P. (2000) Magnetic field near the outcrop of an almost horizontal conductive sheet. *J. Geodynamics* **33**, 413–426.

Sarlis, N., Lazaridou, M. Kapiris, P., and Varotsos, P. (1999) Numerical model of the selectivity effect and the DV/L criterion. *Geophys. Res. Lett.* **26**, 3245–3248.

Saunders, M. A. and Harris, A. R. (1997) Statistical evidence links exceptional 1995 Atlantic Hurricane Season to record Sea warming. *Geophys. Res. Lett.* **24**(10), 1255–1258.

Schade, L. R. (2000) Tropical cyclone intensity and sea surface temperature. *J. Atmos. Sci.* **57**(18), 3122–3130.

Schmidlin, T. W. and Kosarik, J. A. (1999) A record Ohio snowfall during 9–14 November 1996. *Bull. Amer. Meteorol. Soc.***80**(6), 1107–1116.

Schmith, T., Kaas, E., and Li, T.-S. (1998) Northeast Atlantic winter storminess 1875–1995 reanalyzed. *Climate Dynamics* **14**(7–8), 529–536.

Scholz, C. H. (2000a) Evidence for a strong San Andreas fault. *Geology* **28**, 163–166.

Scholz, C. H. (2000b) A fault on the 'weak' San Andreas Theory. *Nature* **406**, 234.

Schreider, S. Yu., Smith, D. I., and Jakeman, A. J. (2000) Climate change impacts on urban flooding. *Climatic Change* **47**(1–2), 91–115.

Scofield R., Vicente G., and Hodges M. (2000) *The Use of Water Vapor for Detecting Environments that Lead to Convectively Produced Heavy Precipitation and Flash Floods* (Techn. Rept. NESDIS-99, 64 pp.). NOAA, Washington, DC.

Semakin N. K. and Nazirov M. (1977) *Application of Satellite Photo-information in Teaching Physical Geography* (144 pp.). Prosveshcheniye, Moscow.

Sharkov, E. A. (2000) *Global Tropical Cyclogenesis* (370 pp.). Springer/Praxis, Chichester, UK.

Sharma, R. P., Ogale, S. B. Zhang, Z. H. Liu, J. R., Chu, W. K. Veal, B., Paulikas, A. Zheng, H., and Venkatesan, T. (2000) Phase transitions in the incoherent lattice fluctuations in $YBa_2Cu_3O_{7-d}$. *Nature* **404**, 736–740.

Shneidegger, A. B. (1981) *Physical Aspects of Natural Disasters* (232 pp.). 'Nedra', Moscow.

Shultz, S. S. (1978) *Earth from the Outer Space* (114 pp.). 'Nedra', Leningrad.

Singh, O. P., Ali Khan, T. M., and Rahmann, Md. S. (2000) Changes in the frequency of tropical cyclones over the North Indian Ocean. *Meteorol. Atmos. Physics* **75**(1–2), 11–20.

Singh, R. B. (1998) *Ecological Techniques and Approaches to Vulnerable Environment–Hydrosphere–Geosphere Interaction* (368 pp.). IBH Publ., New Delhi, Calcutta.

Skinner, B. D., Changnon, D., Richman, M. B., and Lamb, P. J. (1999) Damaging weather conditions in the USA. A selection of data quality and monitoring issues. *Clim. Change* **42**(1), 69–87.

Slifkin, L. (1996) A dislocation model for seismic electric signals. In: J. Lighthill (ed.) *The Critical Review of VAN: Earthquake Prediction from Seismic Electric Signals* (pp. 97–104). World Scientific, Singapore.

Smith, F. (1999) Atlantic and East Coast hurricanes 1900–98: A frequency and intensity study for the twenty-century. *Bull. Amer. Meteorol. Soc.* **80**(12), 2717–2728.

Smith, K. (1993) Riverine flood hazard. *Geography* **78**(2), 182–185.

Smith, K. (1997) *Environmental Hazards. Assessing Risk and Reducing Disasters* (389 pp.). Longman, London.

Smith, K. (2001) *Environmental Hazards: Assessing Risk and Reducing Disasters* (392 pp.). Routledge, Florence, KY.

Smith, P. L. (1999) Effects of imperfect storm reporting on the verification of weather warnings. *Bull. Amer. Meteorol. Soc.* **80**(6), 1099–1105.

Solid Earth Physics Institute, University of Athens Seismology and Dept. of Earth Sciences, University of Uppsala (2001) *Introduction to Seismology, Earthquake Prediction and Geophysical Exploration Methods* (pp. 303–319,). Sinthesis, Athens.

Solow, A. R. (1999) On testing for change in extreme events. *Clim. Change* **42**(1), 341–349.

Sorokin, V. M. and Chmyrev, V. M. (1999a) The physical model of electromagnetic and plasma response of the ionosphere on the pre-earthquake processes. In: M. Hayakawa (ed.) *Atmospheric and Ionospheric Electromagnetic Phenomena Associated with Earthquakes* (pp. 819–828). Terrapub, Tokyo.

Sorokin, V. M. and Chmyrev, V. M. (1999b) Modification of the ionosphere by seismic related electric field. In: M. Hayakawa (ed.) *Atmospheric and Ionospheric Electromagnetic Phenomena Associated with Earthquakes* (pp. 805–818). Terrapub, Tokyo.

Sugiyama, T. (1999) *Damage of Climate Change to Japan. An Estimate of Damage Costs and Its Implications* (Rept. No. EY99001, Vols I–IV, pp. 1–14). CRIEPI.

Tanaka, H., Kamogawa, M., and Ohtsuki, Y. H. (1999) The interactions between bulk plasmons and electromagnetic waves assisted by surface roughness. *Proc. Japan Acad. Ser. B* **75**, 190–194.

Thapa, K. B. (1993) Estimation of snowmelt run-off in Himalayan catchments incorporating remote sensing data. *IAHS* **218**, 69–74.

Thorpe, A. and White, A. (1998) UWERN Report No 1: Extra-tropical cyclonic storms. *Weather* **53**(6), 190–193.

Thouret, J.-C. (1990) Active volaniques explosive et calotte glaciare: Le cas des lahars du Nevado del Ruiz, Colombie (13 Nov. 1985) et l'évaluation des risques volcano-glaciares. *Catastrophes et Risques Naturels (Montpellier)*, 29–60.

Thouret, J.-C., Guillande, R., Huaman, D. *et al.* (1994) L'activité actuelle du Nevado Sabancaya (Sud Pérou): Reconnaissance géologique et satéllitaire, évaluation et cartographie des ménaces volcaniques. *Bull. Soc. Geol. Fr.* **165**(1), 49–63.

Tiling, R. I. (1989) Volcanic hazard and their mitigation progress and problem. *Reviews of Geophysics* **27**(3) (May), 257–269.

Tol, R. S. J. and Langen, A. (2000) A concise history of Dutch river floods. *Clim. Change* **46**(3), 357–369.

Tranquada, J.M., Sternlieb, B.J. Axe, J.D. Nakamura, Y. and Uchida, S. (1995) Evidence for stripe correlations of spins and holes in copper oxide superconductors. *Nature* **375**, 561–563.

Travis J. (2000) Hunting prehistoric hurricanes. *Sci. News* -**157**(21), 333–335.

Trenberth K. E. (1999a) Conceptual framework for changes of extremes of the hydrological cycle with climate change. *Clim. Change* **42**(1), 327–339.

Trenberth K. E. (1999b) The extreme weather events of 1997 and 1998. *Consequences* **5**(1), 1–15.

Trenberth, K. E., Owen, T. W. (1999) Workshop on indices and indicators for climate extremes, Ashville, NC, 3–6 June 1997, Breakout Group A: Storms. *Clim. Change* **42**(1), 9–21.

Trigo, I. F., Davies, T. D., and Bigg, G. R. (1999) Objective climatology of cyclones in the Mediterranean region. *J. Climate* **12**(6), 1685–1696.

Tronin, A. A. (1996) Satellite thermal survey – a new tool for the study of seismoactive regions. *J. Remote Sens.* **17**(8), 1439–1455.

Tronin, A.A., (1999) Satellite thermal survey application for earthquake prediction. In: M. Hayakawa (ed.) *Atmospheric and Ionospheric Electromagnetic Phenomena Associated with earthquakes* (pp. 1439–1455). Terrapub, Tokyo.

United Nations Environment Programme (1991) *Environmental Data Report* (3rd edn, 408 pp.). Basil Blackwell, London.

Uyeda, S. (1996) Introduction to the VAN method of earthquake prediction. In: J. Lighthill (ed.) *The Critical Review of VAN: Earthquake Prediction from Seismic Electric Signals* (pp. 3–28). World Scientific, Singapore.

Uyeda, S., Nagao, T. Orihara, Y. Yamaguchi, T., and Takahashi, I. (2000) Geoelectric potential changes: Possible precursors to earthquakes in Japan. *PNAS* **97**, 4561–4566.

Valla, F. (1990) Les avalanches. *Catastrophes et Risques Naturels (Montpellier)*, 88–94.

Vallve, M. B. and Martin-Vide, J. (1999) Secular climatic oscillations as indicated by catastrophic floods in the Spanish Mediterranean coastal area (14th–19th centuries). *Clim. Change* **38**(4), 473–491.

Van der Vink, G., Allen, R. M., Chapin, J., Fraley, W., Krantz, J., Lavigne, A. M., Le Cuyer, A., MacColl, E. K., Morgan, W. J., and Ries, B. (1998) Why the USA is becoming more vulnerable to natural disasters? *Earth Observations from Space* **79**(44), 533–537.

Vandemenlebrouck J., Thoret J.-C., and Dedieu J.-P. (1993) Reconnaissance par télédétection des produits éruptis et des lahars sur et autour de la calotte glaciare du Nevado del Ruiz, Colombie. *Bull. Soc. Geol. Fr.* **160**(6), 795–806.

Varotsos, P. (2002) *The Physics of Seismic Electric Signals* (monograph). Terrapub, Tokyo.

Varotsos, P. and Alexopoulos, K. (1984a) Physical properties of the variations of the electric field of the Earth preceding earthquakes, I. *Tectonophysics* **110**, 73–98.

Varotsos, P. and Alexopoulos, K. (1984b) Physical properties of the variations of the electric field of the Earth preceding earthquakes, Determination of epicentre and magnitude, II. *Tectonophysics* **110**, 99–125.

Varotsos, P. and Alexopoulos, K. (1986) Stimulated current emission in the earth and related geophysical aspects. In: S. Amelinckx, R. Gevers, and J. Nihoul (eds) *Thermodynamics of Point Defects and Their Relation with Bulk Properties* (pp. 136–142, 403–406, 410–412, 417–420). North Holland, Amsterdam.

Varotsos, P. and Alexopoulos, K. (1987) Physical properties of the variations of the electric field of the earth preceding earthquakes, III. *Tectonophysics* **136**, 335–339.

Varotsos, P. and Lazaridou, M. (1997) Latest aspects of earthquake prediction in Greece based on Seismic Electric Signals. *Tectonophysics* **188**, 321–347.

Varotsos, P., Alexopoulos, K. and Lazaridou, M. (1993) Latest aspects of earthquake prediction in Greece based on seismic electric signals, II. *Tectonophysics* **224**, 1–37.

Varotsos, P., Lazaridou, M., Eftaxias, K., Antonopoulos, G., Makris J., and Kopanas, J. (1996a) Short term earthquake prediction in Greece by Seismic Electric Signals. In: J. Lighthill (ed.) *The Critical Review of VAN: Earthquake Prediction from Seismic Electric Signals* (pp. 29–76). World Scientific, Singapore.

Varotsos, P., Eftaxias, K., Lazaridou, M., Nomicos, K., Bogris, N., Makris, J., Antonopoulos, G., and Kopanas, J. (1996b) Recent earthquake prediction results in Greece based on the observation of Seismic Electric Signals. *Acta Geophysica Polonica* **XLIV**(4), 301–327.

Varotsos, P., Sarlis, N., Lazaridou, M., and Kapiris, P. (1996c) A plausible model for the explanation of the selectivity effect of seismic electric signals. *Practika Athens Academy* **71**, 283–354.

Varotsos, P., Sarlis, N., Lazaridou, M., and Kapiris, P. (1998) Transmission of stress induced electric signals in dielectric media. *J. Appl. Phys.* **83**, 60–70.

Varotsos, P., Sarlis, N., and Lazaridou, M. (1999a) Interconnection of defect parameters and stress induced electric signals in ionic crystals. *Phys. Rev. B* **59**, 24–27.

Varotsos, P., Sarlis, N., Bogris, N., Makris, J., Kapiris, P., and Abdulla, A. (1999b) A comment on the DV/L-criterion for the identification of Seismic Electric Signals. In: M. Hayakawa (ed.) *Atmospheric and Ionospheric Electromagnetic Phenomena Associated with Earthquakes* (pp. 1–45). Terrapub, Tokyo.

Varotsos, P., Sarlis, N., Eftaxias, K., Lazaridou, M., Bogris, N., Makris, J., Abdulla A., and Kapiris, P. (1999c) Prediction of the 6.6 Grevena-Kozani earthquake of May 13, 1995. *Phys. Chem. Earth (A)* **24**, 115–121.

Varotsos, P., Eftaxias, K., Hadjicontis, V., Bogris, N., Skordas, E., Kapiris, P., and Lazaridou, M. (1999d) A note on the extent of the SESs sensitive area around Lamia (LAM), Greece. *Acta Geophysica Polonica* **XLVII**(4), 435–439.

Varotsos, P., Eftaxias, K., Hadjicontis, V., Bogris, N., Skordas, E., Kapiris, P., and Lazaridou, M. (1999e) A note on the extent of the SESs sensitive area around Lamia (LAM), Greece: Continuation I. *Acta Geophysica Polonica* **XLVII**(4), 441–442.

Varotsos, P., Eftaxias, K., Hadjicontis, V., Bogris, N., Skordas, E., Kapiris, P., and Lazaridou, M. (1999f) A note on the extent of the SESs sensitive area around Lamia (LAM), Greece: Continuation II. *Acta Geophysica Polonica* **XLVII**(4), 443–444.

Varotsos, P., Sarlis, N. and Skordas, E. (2000b) Transmission of stress induced electric signals in dielectric media, Part III. *Acta Geophysica Polonica* **XLVIII**(3), 263–297.

Varotsos, P., Sarlis, N., and Lazaridou, M. (2000a) Transmission of stress induced electric signals in dielectric media, Part II. *Acta Geophysica Polonica* **XLVIII**(2), 141–177.

Varotsos, P., Sarlis, N., and Skordas, E. (2001a) Spatio-temporal complexity aspects on the interrelation between seismic electric signals and seismicity. *Practica of Athens Academy* **57**, 11–22.

Varotsos, P., Sarlis, N., and Skordas, E. (2001b) Magnetic field variations associated with the SES before the 6.6 Grevena-Kozani earthquake. *Proc. Japan Acad.* **77B**, 93–97.

Varotsos, P., Sarlis, N., and Skordas, E. (2002) Long-range correlations in the electric signals that precede rupture. *Phys. Rev. E*, to be published 1 July 2002.

Vasiloff, S. V. (2001) Improving tornado warnings with the Federal Aviation Administration's terminal Doppler weather radar. *Bull. Amer. Meteorol. Soc.* **82**(5), 861–874.

Vitart F., Anderson, J. L., Sirutis, J., and Tuleya, R. E. (2001) Sensitivity of tropical storms simulated by a general circulation model to changes in cumulus parametrization. *Quart. J. Roy. Meteorol. Soc.* **127**(571), 25–52.

Wagner, D. (1999) Assessment of the probability of extreme weather events and their potential effects in large conurbation. *Atmos. Environ.* **33**(24–25), 4151–4155.

Walsh, K. and Pittock, A. B. (1999) Potential changes in tropical storms, hurricanes, and extreme rainfall events as a result of climate change. *Clim. Change* **39**(2–3), 199–213.

Warner, T. T., Brandes, E. A., Sun, J., Yates, D. N., and Mueller, C. K. (2000) Prediction of a flash flood in complex terrain. Part 1: A comparison of rainfall estimates from radar, and very short range rainfall simulations from a dynamic model and an automated algoritmic system. *J. Appl. Meteorol.* **39**(6), 797–814.

WASA Group (1998) Changing waves and storms in the northeast Atlantic? *Bull. Amer. Meteorol. Soc.* **79**, 741–760.

Weaver J. F., Grunfest, E., and Levy, G. M. (2000) Two floods in Fort Collins, Colorado: Learning from a natural disaster. *Bull. Amer. Meteorol. Soc.* **81**(5), 2359–2366.

West, J. J., Small, M. J., and Dawlatadi, H. (2001) Storms, investor decisions, and the economic impacts of sea level rise. *Climatic. Change.* **48**(2–3), 317–342.

Williams, A. A. J., Karoly, D. J., and Tapper, N. (2001) The sensitivity of Australian fire danger to climate change. *Climatic Change.* **49**(1–2), 171–191.

Willoughby, H. E. (1999) Hurricane heat engines. *Nature* **401**, 649–650.

Wilson, R. M. (1997) Comment on 'Downward trends in the frequency of intense Atlantic hurricanes during the past 5 decades' by C. W. Landsea *et al. Geophys. Res. Lett.* **24**(17), 2203–2206.

Wobber, F. J. (1972) The use of orbital photography for Earth resources satellite mission planning. *Photogrammetria.* **28**, 35–59.

Wolff, E. A. (2000) A good millennium?. *Weather* **55**(1), 2–7.

Yates, D. N., Warner, T. T., and Laevesley, G. H. (2000) Prediction of a flashflood in complex terrain. Part II: A comparison of flood discharge simulations using rainfall input from radar, a dynamic model, and an automated algorithmic system. *J. Appl. Meteorol.* **39**(6), 815–825.

Yohe, G. (2000) Assessing the role of adaptation in evaluating vulnerability to climate change. *Climatic Change* **46**(3), 371–390.

Zaanen, J. (1999) Self-organized one dimensionality. *Science* **286**, 251–252.

Zaanen, J. (2000) Stripes defeat the Fermi liquid. *Nature* **404**, 714–715.

Zebrowski, E., Jr. (1997) *Perils of a Restless Planet. Scientific Perspectives of Natural Disasters* (320 pp.). Cambridge University Press.

Zhai, P., Sun, A., Ren, F., Lin, X., Gao, B., and Zhang, Q. (1999) Changes of climate extremes in China. *Clim. Change* **42**(1), 203–218.

Zhu, H., Smith, R. K., and Ulrich, W. (2001) A minimal three-dimensional tropical cyclone model. *J. Atmos. Sci.* **58**(14), 1924–1944.

Zschau, J. and Küppers A. N. (eds) (2001) *Early Warning Systems for Natural Disaster Reduction* (800 pp.). Springer Verlag.

4

Some anthropogenic disasters and their consequences

4.1 LAND SURFACE POLLUTION

4.1.1 Land surface pollution

Compared with other dangerous phenomena, land surface pollution is one of the most important. Most often it progresses slowly and hence for a long time it proceeds unnoticed. However, at some time concentration of noxious pollutants in soil becomes higher than maximum permissible standards and normal ecological conditions are violated. In such cases dangerous ecological situations arise in one or another region of the Earth. It manifests itself in different ways, for example in poisoning living organisms, mass-scale diseases, and destruction of plant cover.

Pollution of the earth surface and soil occur as a result of waste discharge into rivers, lakes, upon the soil surface (in allocated places), and accidents.

Waste is really a problem today (Grigoryev and Kondratyev, 1994; UNEP, 1991). Judging from selected and rather scant data in some countries (there are no such data for the former USSR) the USA is the main source of solid waste in the world. Japan is the second. The distribution (in 1985) of waste (in 103 t/year) in the USA and Japan by waste type is as follows: municipal (178,000 and 41,530), industrial (628,000 and 312,000), agricultural (1,400,000 and 90,544), mining (1,300,000 and 26,017), wastewater dry residue (8,400 and 2,003). Waste in the paper industry prevailed among municipal waste in the USA in 1985 (34.7%).

Hazardous waste (highly noxious and toxic, related to metal working, chemical industries, mining and processing of mineral raw materials) is very important. 'Record-holders' in production of this waste in 1980 were the USA ($265{,}000 \times 10^3$ t/year) and India ($36{,}000 \times 10^3$ t/year).

The biggest amounts of liquid radioactive waste are discharged in the USA, Canada, and in Europe – in France (there are no available data for the former USSR). The period from 1981 and to 1984 inclusive shows a rise in amounts of such waste, while in 1985 a sharp decline in their growth took place.

Every year about 200,000 m^3 of low level radioactive waste and about 10,000 m^3 of high level radioactive waste are released by nuclear power production. The annual total amount is increasing in 1990 it reached 80,000 t as compared to 30,000 t at the beginning of the 1980s (Keating, 1994).

Estimates of waste in different sectors of industrial production obtained for a number of countries of the world as of 1990 (O'Riordan,1995) are of considerable interest. Comparable data are available only for 18 countries: Austria (1990), Canada (1985), Czechoslovakia (1987), Finland (1987), France (1990), FGR (1991), Greece (1990), Hungary (1989), Ireland (1991), Japan (1985), Luxembourg (1990), Holland (1990), New Zealand (1990), Norway (1988), Poland (1990), Portugal (1989), Spain (1990), and the USA (1990).

Together, these countries discharge almost all the (up to ten million tons) and solvents organic substances. The USA occupies first place (70 million t). Other countries showing high discharge levels in this type of waste include FRG (454,489 t) and Canada (202 thou t).

The amount of waste in the above mentioned 18 countries due to rubber and raw rubber manufacture is also estimated as tens of millions of tons (although, approximately half the amount for solvents for organic substances). In the USA these discharges are about 41 million tons, and in Japan are about 2894 thousand tons.

Waste from other sectors of industrial production (related to production and processing of oil, metals and plastic materials, manufacture of paints and so on) make up about 35 million tons. However, it should be taken into account that ecological problems are affected more by toxic properties (with regards to living organisms) than by actual waste amounts. The above mentioned countries are responsible for only several tens of tons of dangerous biological waste (including 13,216 in the USA and 10,300 t in Hungary), but they may be the cause of not only diseases but death of many living organisms.

Among different types of earth surface pollution that are associated with agriculture, of special note is soil pollution with chemical herbicides and pesticides. They are applied in order to control weeds and various pests, to increase agricultural yield. In the beginning of the sixties it was found that application of chemical herbicides and pesticides had significant consequences. Noxious chemical substances eventually poison people, animals and plants. Chemicals washed off into the sea found their way into the organisms and inhabitants of the Antarctic – penguins. *Silent Spring* by Carson (1963) contained information on the consequences of herbicide and pesticide application. Later on a revised edition of the book was published (Carson, 1991) in many countries of the world, including the USSR, in which their application practice in agriculture was revised and restricted.

In the former USSR sizable damage to nature was caused by unreasonable soil application of pesticides in the republics of Middle Asia and Moldavia (Grigoryev, 1991). In Moldavia for example, as reported by chemists of Moscow Society of Naturalists, 21–22 kg of herbicide and pesticide per hectare were applied annually during the period 1981–1985. This began to adversely affect the health of people. In the regions of maximum chemicalization, sickness rate among children ncreased to 3.5 times higher than in other places. Thus, high yield produc-

Table 4.1. Emissions and deposition of sulphur in European countries, 1988 (UNEP, 1991).

	Deposition			Source of depositions (%)		
Country	Emission (10^3 t/year)	Total (10^3 t/year)	Per unit area (kg/hectare per year)	National	Foreign	Unknown
Albania	25	28.5	10.4	13	67	19
Austria	57	176.4	21.3	7	85	8
Belgium	208	86.7	26.4	47	48	5
Bulgaria	515	213.0	19.3	57	35	8
Czechoslovakia	1400	658.0	52.5	53	44	3
Denmark	121	60.9	14.4	22	69	9
Finland	151	202.2	6.6	23	60	17
France	608	514.4	9.3	39	47	14
Germany (GDR)	2629	786.4	74.7	81	17	2
Germany (FRG)	650	538.5	22.0	36	58	6
Greece	250	122.5	9.4	37	45	18
Hungary	609	280.3	30.4	54	41	4
Iceland	3	9.8	1.0	5	46	49
Ireland	74	35.5	5.2	43	42	15
Italy	1205	473.7	16.1	60	29	11
Luxembourg	6	3.9	15.1	26	67	8
Netherlands	139	86.3	25.4	24	71	6
Norway	33	151.2	4.9	5	73	21
Poland	2090	1241.7	40.8	54	43	4
Portugal	102	45.0	4.9	46	41	13
Romania	100	351.9	15.3	9	80	11
Spain	1570	478.5	9.6	81	12	8
Sweden	107	259.7	6.3	12	73	16
Switzerland	37	70.4	17.7	13	75	12
Turkey	199	220.7	4.5	25	42	33
United Kingdom	1832	541.2	22.4	88	8	3
USSR	5062	3570.8	9.8	62	24	14
Yugoslavia	800	525.0	20.6	47	44	9

tion by application of pesticides has caused damage to health of people on a regional scale.

Earth surface pollution occurs also as a result of sulphur deposition from the atmosphere (Table 4.1) (Figure 4.1). In Europe a network of 102 stations is used for trans-boundary pollution transport monitoring of sulphur.

Sulphur is one of the main 'airborne' pollutants of soil. Its amount (dry and wet deposition) in Europe (1988) changed from 10 kg/hectare in Scandinavia to 50 kg/hectare in East European countries.

The greatest amounts of sulphur deposited per unit area in 1988 were in Germany, as well as in Czechoslovakia and Poland. The study of sulphur dioxide concentrations in precipitation for 6–8 years (1978–1988, 1981–1986) has revealed

(a)

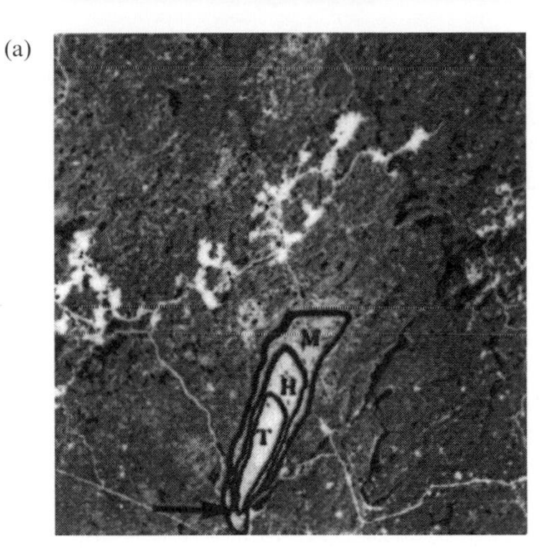

(b)

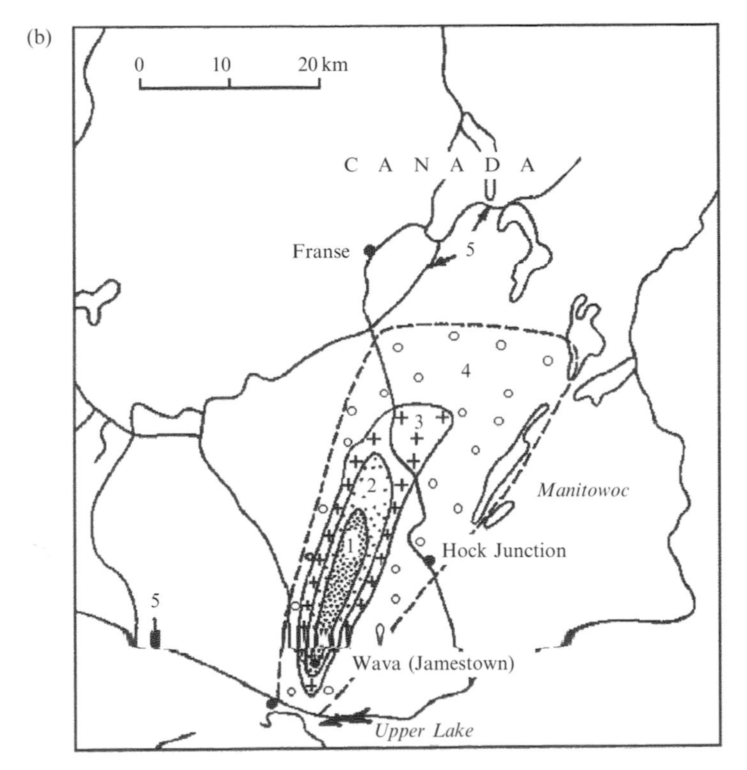

Figure 4.1. Plant cover poisoning as a result of toxic pollutant deposition (sulphur dioxide) from smoke emissions by metallurgical combined works in the town of Wava (Jamestown near the Upper Lake, Canada) (Grigoryev, 1985). (a) Satellite picture obtained on 8 September 1973 from *Landsat-1*; (b) Picture decoding diagram: 1–4 represent zones of forest damage of very high, high, medium and low degree, respectively (zone 4 is not seen in the satellite image and was identified by means of on-land observations; 5 – roads (Grigoryev, 1985).

Table 4.2. Damage caused to forests in the countries of Europe as a result of atmospheric pollutant fall out: 1988–1989. From observation data on pilot sites (UNEP, 1991).

Country	Coniferous affected trees (%) (more than 25% of needles affected)		Deciduous affected trees (%) (more than 25% affected)	
	1989	Changes between 1988–1989	1989	Changes between 1988–1989
Austria	4.1	+0.1	6.7	+1.2
Bulgaria	32.9	+25.3	16.2	+7.4
Czechoslovakia	32.0	+5.0	37.0	+7.9
Denmark	24.0	+3.0	30.0	+16.0
Finland	18.7	+1.7	12.6	+4.7
France	7.2	−1.9	4.8	−0.5
Germany (GDR)	17.5	+2.0	12.9	+3.9
Germany (FRG)	13.2	−0.8	20.4	+3.9
Greece	6.7	−1.0	18.4	−10.1
Hungary	13.3	+3.9	12.5	+5.5
Ireland	13.2	+8.4		
Luxembourg	9.5	−1.6	13.9	−11.6
Netherlands	17.7	+3.2	13.1	−12.3
Norway	14.8	−6.0		
Poland	34.5	+10.3	17.7	+10.6
Portugal	9.8	+8.1	8.6	+7.8
Spain	3.5	−3.8	3.2	−3.6
Sweden	12.9	+0.6		
Switzerland	14.0	−1.0	6.0	−1.0
United Kingdom	34.0	+7.0	21.0	+1.0

increases in the eastern part of the Gulf of Finland and in the southern part of Sweden.

Summarized world data on accumulation of soil pollution due to fall out of atmospheric aerosols is very fragmentary. For example, in Europe (1988) in the territory of four states – Denmark, Holland, Germany and France about 330,000 ha of land (4% of the whole area) were polluted to one extent or another, with 185 ha polluted considerably.

Forests are a good indicator for Earth surface pollution caused by industrial emissions (Table 4.2). In 1989 among the countries whose coniferous forests became most degraded were Poland (34.4% of forests), Bulgaria (32.9%), Czechoslovakia (32%), Yugoslavia (34%), and Denmark (24%). The less degraded were those in Spain (3.5%), Norway (14.6%), and Finland (18.7%).

Particularly rapid growth of the share of damaged coniferous trees for the period 1968–1989 were reported in Bulgaria, Poland, and that of deciduous species – in Denmark, Holland, Poland and Greece.

The tendency of forest damage (mainly due to atmospheric pollution) to increase has been maintained even after 5 years. This is evidenced in the map of European forests, showing extent of damage caused to arboreal (both coniferous and deciduous) species (up to 1994) (World Resources, 1996–1997). In this map within the boundaries of 29 countries of Europe (including Ukraine, Belorussia, Lithuania, Latvia, Estonia, and Moldavia but without Russia) on average about 26.4% of trees were subject to damage (either their leaves or needles). The most seriously damaged were forests in Poland (41–60% affected), and also Ukraine, Belorussia, Latvia, Norway, Denmark, Bulgaria, and Albania (26–40% affected). The healthiest were forests in France and Portugal (0–10% affected).

In Russia large forest tracks affected by atmospheric pollution are located around cities – industrial centers where there are metallurgical, petrochemical and similar industrial enterprises, an example of which is the 'Pechenganickel' combined plants (Figure 4.2). Emissions to the atmosphere from nickel processing plants around Norilsk ruined 350,000 ha of forest, and partial damage was caused to forests in the surrounding area of 140,000 ha. Damage caused to forests was the result of the fact that local plants emit 2.3 million tons of sulphur to the atmosphere annually, which is 5 times more than Swedish emissions (Kravtsova, 1981).

However, data on acid rain impact on European and North American forests and their considerable degradation are open to question (Sedio, 1998). Forest degradation is becoming associated with other factors.

Certain types of soil (forest) poisoning are associated with fall out of radioactive precipitation. The largest spot of pollution of this kind (discontinued type, though) is located in Ukraine, Belorussia and Russia. It is associated with the accident at the Chernobyl nuclear power plant in 1986. Its affected area is estimated to be 4 million hectares (World res., 1996).

Every year about 5.2 million people including 4 million children die as a result of diseases caused by soil pollution with industrial waste (Keating, 1994). Chemical compounds hazardous to health, entering soil and then agricultural products (in the course of waste discharge into water) are not removed from the food chain. For example, in developing countries less than 10 % of municipal waste undergoes cleaning.

Pesticides are among the most important environmental pollutants. Soil application of pesticides serves to kill various invaders of crops, and thereby to increase crop yield. There is evidence that accumulation of pesticides in the soil and therefore living organisms, including human beings (O'Riordan, 1995) causes degradation and diseases. In China in 1993, 10,000 farmers died because of pesticides poisoning. According to some estimates 3–25 million people in the world have suffered from the effects of pesticides. Therefore, several countries cut down their application in agricultural practices (for example, in Sweden pesticide use has been reduced by 50% since 1991).

Hazardous waste poisoning of the environment, and soil in particular is especially bad for ecosystems and public health. The highest amounts of these substances are emitted (at the end of the eighties) in the USA – about 270 million t/year. In OECD – member countries (Europe) they make up about 20 million t/year, and in the countries of East Europe – about 13 million t/year (Keating, 1994).

Waste disposal is one of the most important problems related to improvement of

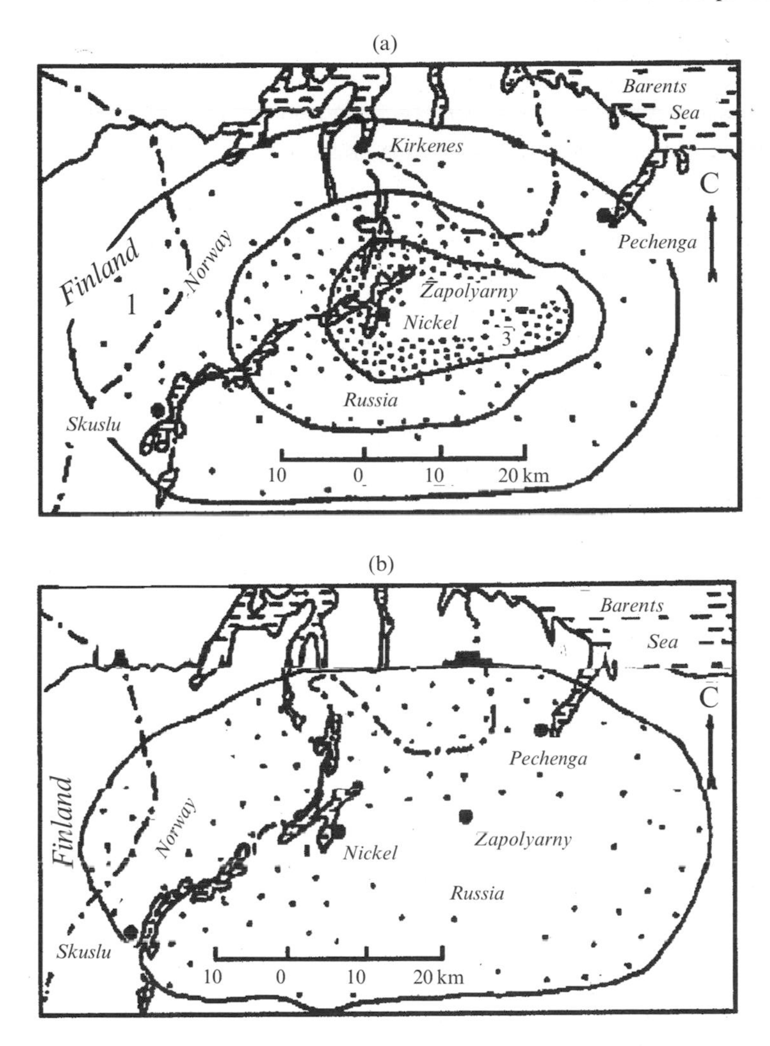

Figure 4.2. Trans-boundary impact on landscapes in the northern Finland and north-eastern Norway caused by industrial emissions from 'Pechenganickel' ore mining and processing enterprise located in the town of Zapolyarny, and settlement of Nickel in Cola Peninsula (Alekseyev, 1995). (a) zones of different extent of damage to Scotch pine stands: 1 – zone of low damage, 2 – moderate, 3 – heavy; (b) zone of nickel concentration in bilberries biomass in excess of maximum permissible concentrations for health.

ecological conditions. Municipal waste in many countries of Western Europe, Canada, and the USA is removed and stored in specially allotted sites. In this way these countries get rid of 70%, and England – up to 90% of municipal waste. Occasionally, municipal waste is processed in special installations (plants).

Hazardous production refuse in the USA (70%) and in the countries of European Community (50%) is also stored in specially allotted sites, with the remainder being processed. In developing countries processing of waste is difficult due to a shortage of equipment. To stabilize and reduce environmental pollution as a whole and soil pollution in particular, economic sanctions are being suggested, with the introduction of pollution taxes (O'Riordan, 1995).

4.1.2 Satellite observations of land surface pollution

Satellite images make information available about the areas of atmospheric pollution fall out including areas of plant cover poisoned by various pollutants entering the environment from both land and from the air.

One important indicator for soil and ground pollution in one or another region are spots of snow cover that are clearly tracked in satellite images (Grigoryev, 1985; Kondratyev *et al.*, 1983). On-land identification of snow pollution areas is very difficult. At the same time snow cover is a perfect accumulator of pollutants. Aerosols of industrial origin are the main pollutants for snow. However, atmospheric deposition may be of natural origin resulting from dust rising in the course of dust storm development. It may be transported large distances and deposited on snow, as was observed from outer space during the winter storm in Ukraine (Prokacheva *et al.*, 1988; Semakin and Nazirov, 1977).

Deposition of industrial aerosols on snow is clearly shown in satellite observations. In contrast to natural pollution, snow pollution of this kind is distinctly associated with anthropogenic sources (industrial enterprises). Anthropogenically induced snow pollution was revealed in satellite image data around single industrial plants, cities and along some highways. It was also discovered that areas of atmospheric deposition on snow exceeds by many times the area of emission sources (Grigoryev, 1985; Kondratyev *et al.*, 1983). Land surface pollution due to individual sources, for example plants, quite often merge and form vast single spots on snow. Satellite images were used to determine the area of atmospheric dust deposition on snow cover in the North-Moravian area (Czech Republic) and in Upper Silesia (Poland). The sources of pollution were emissions by the plants in the Ostrava–Karvina basin (Dworak, 1992).

Snow cover is not only an accumulator of pollutants but is also a good indicator for pollution. Clean snow is characterized by high reflectivity, its albedo reaches 0.7–0.9. Snow pollution gives rise to a decrease in albedo 0.2–0.3 (Kupriyanov, 1977). This accounts for the distinct images of polluted snow areas in the form of dark spots on satellite pictures of snow-covered territory. Image blackening depends directly on the extent of pollution of the territory.

In winter, snow near railways is polluted. This pollution is caused by deposition of atmospheric pollutants from trains. For example, soot particles emitted by steam locomotives, and particulate matter (especially coal) transported by trains in goods trucks. Depositing on both sides of the railway, particulate matter forms a surface pollution band up to several hundreds of meters wide. Particularly wide are pollution bands on snow cover along the railroads of mass carriage of coal (e.g. in Siberia and Kazakhstan). Such linear anomalies snow cover due to pollution are even visible in less detailed pictures. They are often of hundreds of kilometers long. For example,

Figure 4.3. Zones where there is heavy urban impact on snow cover. Produced from image data obtained by meteorological satellite (Prokacheva *et al.*, 1988). 1 – urban areas and urban agglomerations areas in different regions with firm snow cover; 2 – cities in the regions with unstable and ephemeral snow cover.

atmospheric pollutant deposition bands along the railroad between Omsk and Petropavlovsk.

Snow pollution changes surface albedo and thereby surface character – shades or colors on satellite images. Observational data show even small levels of snow pollution (of about $10\,g/m^2$) reduce albedo by 10–15% (Kupriyanov, 1977).

Satellite pictures help to clearly identify areas of atmospheric pollutant soil deposition (this mainly refers to winter pictures of land surface landscape covered with snow). There are no other means to obtain regional scale information about spatial distribution of atmospheric pollutants deposited around cities.

Zones of land surface pollution that occur near urbanized regions are evident even in the low resolution pictures depicting snow-covered landscapes, in particular around the cities of Petropavlovsk, Norilsk, Omsk (Prokacheva *et al.*, 1988).

Usually snow pollution around cities is distinguished in visible spectral range pictures as dark spots of irregular form whose sizes are considerably bigger than the area of the city itself (Figure 4.3).

The larger an urban center, the more highly developed are its industries and the larger the areas of pollution and pollution intensity. However, the above mentioned direct dependence may be broken because of the influence of other factors (including types of industry, application of smoke filtering, local meteorological conditions). It should be noted that changes in coloration – shade (color) of snow cover images around urbanized regions is an indicator for anthropogenic development of the landscape, not merely a surface pollution indicator. Although surface pollution is the main cause for changes in coloration of snow-covered surfaces, other factors also play a role in this respect. For example, the extent of development of different structures like buildings and roads (the higher their specific weight the darker the image in the visible spectral range).

From satellite pictures it is usually possible to single out zones showing surface pollution and others at different stages of snow cover melting. In one of the zones (the central one), we see the city itself and its immediate outskirts. This area (1.5–2 times the area of the city itself) is the zone showing the highest intensity of land surface pollution; snow melting is shown darker in the images. The central zone where the pollution source is located shows up as a core with larger zones surrounding it. As a rule, unlike the first zone, the zones surrounding it are lighter and do not contrast so well, which is due to the weakening pollution impact on the landscape – decreasing aerosol fall-out and slower melting of snow.

Configuration of land surface pollution's geographical location around a city is determined by many factors like specific features of the city and population density in the areas surrounding the city. The main factor of land surface pollution distribution is exerted by the wind direction (wind rose) during the period of pollution transport (Figure 4.4).

The length of 'on-land' pollution train depends on the prevailing wind direction in the area. For instance, cities of Cherepovets, Vologda are characterized by land surface pollution trains extended westwards, which is caused by prevailing easterlies (Kupriyanov, 1977). Trains from the cities located in tundra zone, such as Dudinka, Vorkuta, and Norilsk are extended southward resulting from northerlies.

One further important indicator for land surface pollution identifiable using satellite images is plant cover. Using images from the *Meteor-Priroda* satellite and also on-land research data, peculiar features of forest damage in the zone of impact of emissions from Norilsk integrated mining and iron and steelworks have been revealed (Kharuk *et al.*, 1995). Images in three spectral intervals were used – 0.5–0.6, 0.6–0.7, 0.7–0.8 μm with a resolution of 50 m.

The zone of damaged stands (larch forests prevail, but spruce, birch, and willow are also present) extends to 200 km. From the images several types of stands are singled out – dead, heavily damaged, damaged and conventionally healthy. At a distance of 8–100 km from Norilsk mining and iron and steel integrated works almost all the stands are dead.

The major source of forest stands degradation is SO_2 emissions (about 2 million t/year). Nitrogen oxides emission are insignificant (about 20 thousand t/year) and do not affect the stand considerably.

Satellite surveying data made it possible to detect increases in forest degradation rate at the end of the 1970s and beginning of the 1980s. This was connected with the start-up of a new plant and with the increase of the distance of pollutant transport as a result of high chimneys at the plant (to 250 m).

Degradation of forests occur mainly south-eastward from Norilsk integrated works. This is determined both by prevailing winds and obstacles (in the form of

Figure 4.4. Diagram of snow cover pollution range resulting from deposition of atmospheric emissions from Karaganda's territorial industrial complex. The pollution zone is many times larger than the area of urbanized territory. The diagram is reproduced from *Cosmos* satellite image data. The wind rose is shown by the dotted line (Grigoryev and Kondratyev, 2001).

elevations) to pollution transport. Damage caused to stands is less when forests grow on fertile soils and are outside the zone of direct impact of emissions (in river valleys, behind elevations).

Satellite images are useful to track features of land surface pollution in frontier regions being of interest to neighbouring countries, in particular Finland, Norway and Russia.

Multi-spectral images from *Landsat-TM*, *MSS* and *Spot-Pan* satellites were used to assess plant cover damage in the frontier regions of Norway and Russia on the Cola Peninsula resulting from industrial emissions – primarily SO_2 emissions from the city of Nickel (polar region) (Roberts *et al.*, 1994). To reveal anomalies in plant cover, combinations of images obtained in different spectral regions were used.

The images showed zones of deposition of lethal and sub-lethal (for plant cover) atmospheric pollutants. The dynamics of change in plant cover damaged by pollutants was tracked in this territory (13,250 km^2) for 15 years. It was discovered that an area of lichen cover (2,500 km^2) reduced to one sixth of its former size in this 15 year period. The area of plant cover most heavily damaged by pollution increased 5 times in size.

There are some other important indicators for land surface pollution as well. However, application of a complex of indicators is optimal. Let's consider this taking pollution due to areas of extraction (and processing) of mineral resources as an example.

Large quarries are also potential sources of atmospheric dust pollution, especially during the course of blasting. Dust over a quarry migrates and falls out onto the surface of the earth in the zone adjacent to the quarry. This may result in pollution of soil and water bodies, and sometimes damage to plants and agricultural lands. Satellite images combined with on-land data are useful for studying the poorly known environmental impact of quarrying. In Russia, for example, the environmental impact of one of the largest regions in the world where, the big quarries are concentrated has been estimated. These quarries are located within open-cut mining fields of the Kursk magnetic anomaly (Mironova and Mikhailova, 1982).

One of the largest areas showing man's impact on the landscape in Russia, associated with mining, is located at Khibini in the Cola Peninsula. In satellite images the 'traces' of such impacts are tracked. The images show areas of mountain slopes, quarries, piles of mountain rock (all devoid of vegetation) that often surround industrial zones (Kravtsova, 1981). 'Tongues' of material from such piles sometimes give rise to anthropogenic mud flows. The images also show settling basins of concentrating mills where pulp is discharged to.

The locations of settling basins are not always chosen wisely. For example, a designated part of Imandra Lake (one of the biggest lakes of the Cola Peninsula) was made into a settling basin. The images showed that the pulp material was not restricted to the settling basin, 12-km-long tongues of polluted water were extending out of the settling basin into the lake.

In Khibini over quarries and 'hills' of waste rock, dust is entering the atmosphere and then being deposited around mineral resource extraction areas. In this manner zones of polluted landscapes are clearly seen in winter satellite images. In this region it is not only quarries that are the sources of dust that pollutes the snow around industrial zones. Industrial enterprises, and primarily mining and concentrating mills (Kravtsova, 1981) contribute a lot to this pollution.

Similar phenomena are observed around the Vorkuta coal basin mines. Coal dust deposition on snow speeds up its melting by 20–25 days.

Vast zones of pollution fixed by snow cover are formed in the regions of mineral resources processing. From surveying data of *Landsat (MSS)* and *Meteor–Priroda* satellites a snow cover pollution halo has been revealed to be associated with deposition of pollutants transported from Norilsk integrated mining and iron and steelworks (Kharuk *et al.*, 1995). It was found out that this halo corresponds to the zone where heavy metals (Cu, Ni, Co) are concentrated in the soil. Soil concentration of metals at the distance of 40 km from the integrated works exceeds the background level by 10–1000 times. Active accumulation of metals is fixed in the upper soil horizon (depth of 10 cm).

Destruction of plant cover in the areas of location of many large quarries is also clearly identifiable by means of satellite images. In particular, when taking pictures in several spectral intervals it is possible to reveal and map openings filled with water which is usually polluted. In coal quarry openings in the Appalachians (USA) waters are acid. Their acidification results in poisoning and even destruction of plant cover in the areas adjacent to the openings. Already dead or floating vegetation is hardly visible under examination of multi-spectral pictures with the naked eye. However, using computer processing techniques areas of vegetative (green) and dying or dead plants around openings are distinctly identified thanks to the difference in their spectral characteristics (Chase, 1973; Dworak, 1991). This data are important to plan activities on reclamation of lands disturbed by mining.

Remote sensing data on oil production areas can be used to monitor soil pollution due to spilled oil and vegetation contamination. Satellite observations were used to study anthropogenic dynamics of landscapes in certain areas of oil- and gas-production in West Siberia (Kozin, 1982).

Pollution of dry land ecosystems (landscapes) had attained regional and even sub-continental extents by the end of the 20th century. This is evidenced, for example by data on damaged and even dead forests that was caused (may be partly) by acid rain. In Europe this zone embraces many countries of West and central Europe, especially in the band from England to Lithuania. Such a large-scale land surface pollution requires new assessment methods. Satellite observation data have proved to be especially important in this respect.

The above-mentioned area of damaged European forests is undoubtedly associated with vast clouds of smoke pollutants periodically observed in this region from outer space. One such cloud extended over Western Europe from Southern England to Czechoslovakia to a distance of 700 km and covered an area of 400,000 km^2 (Grigoryev, 1985).

Analysis of satellite images in Eastern Europe, Eastern Siberia and Kazakhstan has revealed quite a number of zones of snow cover pollution with industrial aerosols, including two of the most considerable ones. One of them is about 180,000 km^2 and is located in Southern Ukraine – primarily Donbass (Figure 4.5), the other one is about 130,000 km^2 in area, located at the center of European Russia, near to Moscow's urban agglomeration (Prokacheva *et al.*, 1988). Results strongly suggest that satellite observation data is very helpful for the evaluation of large-scale land surface pollution.

4.2 FOREST AND OTHER FIRES

4.2.1 Forest fires in the 20th century

Among catastrophes, fires both natural and anthropogenic, are not among the most terrible disasters (as compared, say with earthquakes) in view of their small number of victims. The major consequence of fire is damage to forests and their inhabitants.

Fire can pose a real threat (not remote one as in the case of damage caused to ecosystems by fires) to settlements and even cities. However, in these circumstances fire causes damage primarily to constructions, human victims as a rule are not high in numbers.

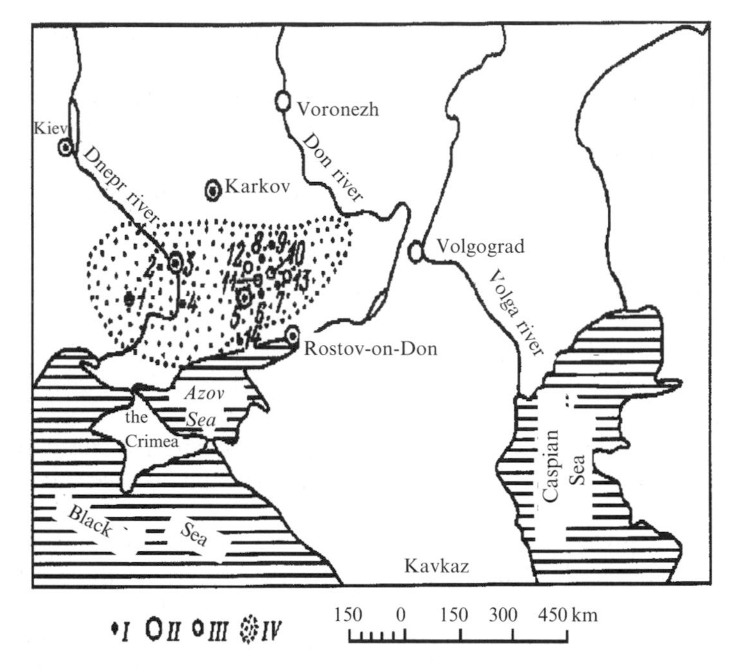

Figure 4.5. Vast area of snow pollution that was formed as a result of merging of smoke pollutant deposition from industrial centres of a South European highly urbanized area. *Cosmos* satellite image (Prokacheva *et al.*, 1988). I – cities with maximum pollution and elevated mortality of population since 1988; II – large urban agglomerations (million inhabitants); III – large urban agglomerations (half a million inhabitants); IV – areas of polluted snow. Cities: 1 – Krivoi Rog, 2– Dneprodzerzhinsk, 3 – Dnepropetrovsk, 4 – Zaporozhye, 5 – Donetsk, 6 – Makeyevka, 7 – Kommunarsk, 8 – Lisichansk, 9 – Severodonetsk, 10 – Stakhanov, 11 – Gorlovka, 12 – Kramatorsk, 13 – Voroshilovgrad (Lugansk), 14 – Mariupol.

4.2.1.1 Causes and factors of natural and anthropogenic fires

Fires are caused by many factors. Among them are meteorological conditions, such as, high temperature, strong wind, prolonged drought, and low moisture of soil and ground. Forest cover features like type of stand (as combustible material), stand density and so on, are essential for fire development.

Fires occur in the natural way, through people's fault as a result of carelessness in handling fire, through the fault of a man being a deliberate incendiary. It should be noted that the number of fires of natural origin (as a result of lightning, peat inflammation) is progressively decreasing with time (of course, relatively) since more and more often fires occur through the fault of man. This is illustrated by the situation in the countries of Mediterranean. Here the number of natural fires does not exceed 0.5% of the total number of fires. 25–35% of fires occur because of carelessness in treating fire, and 1–5% of all fires occur as a result of ill-intentioned arsons. The rest of the fires (42–68%) in the Mediterranean are of unknown (unidentified) origin (Bertrand, 1990).

In the vast flat territories of Australia, less densely populated compared with Mediterranean countries, the number of natural fires is certainly higher (10%), but still not too high (Smith, 1997). Anthropogenic factors in fire occurrence is well documented – California as an example. The number of fires here is permanently increasing, especially when masses of citizens leave the city at weekends (Grigoryev and Kondratyev, 1997; Grigoryev and Lipatov, 1977).

There are two types of anthropogenic fire worth singling out. One of them is fires occurring as a consequence of intended burning of forests or savanna (steppe) plants for further agricultural use. These fires can soon get out of control. Ranges of fires of this type are tied-up with landscapes of tropical, subtropical and equatorial forests and savannas.

Burning of vegetation as a rule is carried out in limited areas. It is difficult to block the way of fire, and under strong winds it envelopes neighboring territories.

Tropical and sub-tropical forests, jungles of the Amazon basin, forest tracts in the Hindostan Peninsula, Central Africa and other regions of the earth have long been affected by human beings. This was connected with the expansion of agricultural lands. Certain areas of forest were burnt to release land for agricultural works.

Burning has been practiced for centuries and even for thousands years on a planetary scale (not only in individual regions, like Hindostan), but it did not have an effect globally on forests. In recent years burning of forests has increased considerably in conjunction with population growth in developing countries and increasing demand for foodstuffs.

The second variety of forest fire is associated with carelessness of people in the course of other types of activities (hunting, recreation, and etc.). Fires of this sub-group occur everywhere except for Arctic regions, but they are more frequent in the moderate climate belts. The number of fires of this variety are permanently growing due to the growth in numbers of hunters, fishermen, tourists, geologists etc.

Fires of this kind are typical for Siberia and the Far East. In the 1980s, according to data from the Russian Ministry of Agriculture they totalled 85–90% of all fires that occurred in these regions (Grigoryev, 1991).

An important role in fire development can be played by the specific economic conditions in a big region. An example of this is the Mediterranean. Fire development here is favoured by both natural conditions (frequent dry summer months, strong winds, high flammability and combustibility of forests), and lack of proper forest ecosystem management in conditions of high population density. Forest fires ravaged areas around 20 cities in the Mediterranean (the natural conditions in all these countries are more or less similar). Data on frequency of fires and the areas of fire sites in EEC-member countries for 1985 are presented in Table 4.3 (Bertrand, 1990).

From the considered data it follows that the highest number of inflammations is observed in Italy, but as to the area of fire sites Portugal occupies the first place. Total area of fire sites in 1985 here exceeded 1% of the territory of the country.

Table 4.3. Fires in some countries of the Mediterranean in 1985 (Bertrand, 1990).

Country	Number of fires	Area of all fires (ha)	Area of only forest fires (ha)	Area of the country (km^2)
Italy	16,903	160,935	62,512	301,200
France	5,596	59,850	35,050	551,500
Greece	727	71,865	37,511	133,000
Spain	9,770	355,998	147,235	507,000
Portugal	5,459	13,570	81,475	92,000
Total	38,455	784,218	363,786	

Forests injured by fire occupied almost 1% of the territory of the country. Among the enumerated countries the less suffered from fires was France. The area hit by fire did not exceed 0.1% of the territory of the country (although the number of fires was approximately the same as in Portugal).

4.2.1.2 Forest fires range

Forest fires occur in rather different landscapes – in forests of moderate, sub-tropical and tropical belts, in steppes, savannas, semi-deserts, forest-tundra and tundra. Landscapes of the Mediterranean, South-West of the USA and Australia are especially favorable for fire development. Australia has both the greatest number of fires and the largest scales. Forest combustion is encouraged here by the presence of eucalyptus that release oils during combustion.

Large steppe and forest fires in Australia and forest fires in Siberia and Canada may envelope territories of more than 100,000 ha. Fire scale criteria in different countries are different. In Siberia large fires are considered to exceeding 200 ha. In European Russia a fire is considered large if its area exceeds 25 ha. In Western Europe fires of areas of more than 100 ha are recognized as catastrophic, whilst in the USA the criterion are greater than 120 ha and duration of more than 24 hours (Valendik, 1990). There are countries and regions where fires occur very frequently. These are Australia, the USA, Canada, France, and Italy. In France on Corsica and in the Var département each year fires envelop about 5,000 ha (Smith, 1997). An example of a large fire in the USA occurred in the summer in 1987 in the west of the USA – mainly in California and Oregon States – where due to 4500 lightning discharges about 2000 fires occurred. Fires in Alaska in 1969 enveloped territory of 160,000 ha, and in 1970 in California – 200,000 ha (Valendik, 1990).

Scales of fires globally provide the following data: in the 1980s on the planet more than 800,000 forest fires were reported. The majority of the fires were due to man's activities – carelessness in dealing with fire. But the number of fires of natural origin (due to lightning) made up $<10\%$. Every year in the countries of European Mediterranean fires envelop 500–1,000,000 ha of land, half of which are forest tracts.

The scale of Russian fires are incomparable with Western European ones,

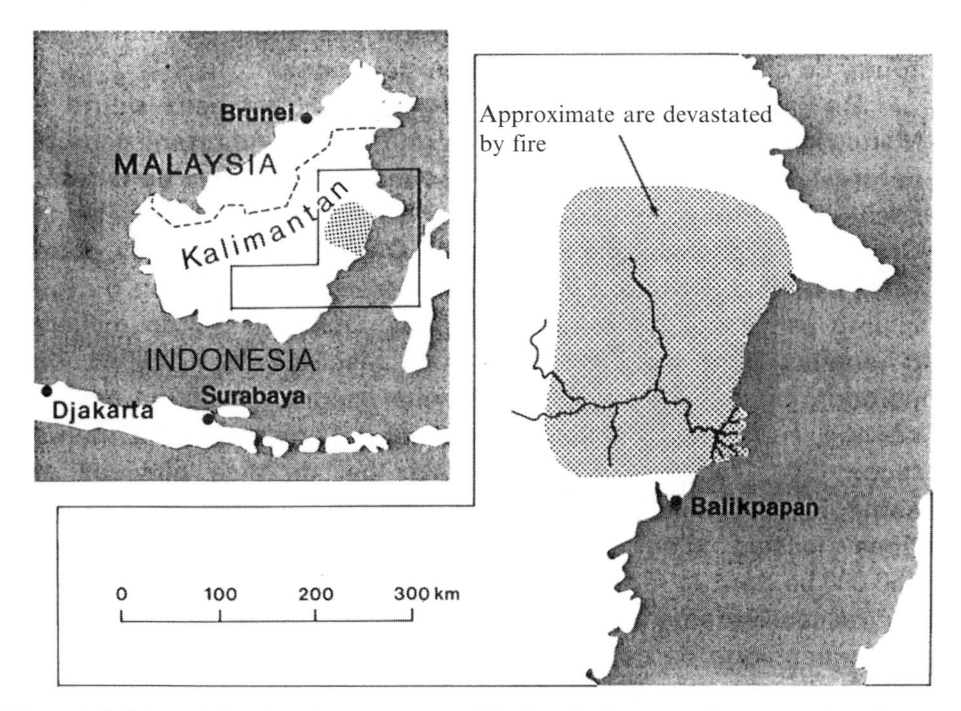

Figure 4.6. Map of the gigantic range covered by fires in forests and swamps of Kalimantan Island (Borneo) in Indonesia. In 1982–1983 an area the size of Switzerland was affected (Ramade, 1987).

including those in the Mediterranean. The Khatanga region of Krasnoyarsk territory alone in 1986 had more than 100 forest fires. The area of some forest tracts engulfed by fire exceeded 1,000 ha. In the Irkutsk area in 1985, 208,000 ha of forests were burnt.

Animals and birds perished in the fire, and enormous damage was caused to the fur trade, and, of course, to the forests themselves. The territory of burnt forests in 1986 in Siberia reached millions of hectares. In India the area destroyed during the period 1980 to 1985 as a result of 18,000 fires totalled more than 50 million ha (Rao, 1996).

Let's describe the two largest fires of the 20th century. One of them occurred in 1982–1983 in Indonesia in Kalimantan Island (Borneo) in the provinces of East Kalimantan and North Borneo (Figure 4.6). The fire hit tropical forests and swamps of areas of 3.5 and 1.0 million ha, respectively (Mallingreau *et al.*, 1985). Forests and swamps were burning for several months. Total damage from the fire was estimated at 5 billion dollars.

In these virgin, moist tropical forests (800,000 ha), swamp (550,000 ha), forests subject to felling (1.2 million ha) and forest tracts of swidden farming (750,000 ha), not only rare plant species, but thousands of animals including rare species such as orang-utans, leopards and bears perished in the fire. It is assumed that such large devastating fires may occur here once every 50–100 years.

The fires that originated in 1982–1983 in Kalimantan are assumed to be associated with superposition of a number of factors. One of them is passive, prolonged drought from July to November 1982 probably caused by recurrent El Niño near the shores of South America.

Long periods of no precipitation, drying of foliage and surface plant cover, decline of ground water table, dry winds in vast territories of the northern and eastern Kalimantan prepared favorable conditions and propagation of fire.

Another active factor of fire propagation was carelessness of people who were engaged with routine agricultural activities and, in particular with burning forest areas for agricultural lands.

Fire development was assisted by other human activity, namely selective felling. In cut out forest corridors plant remains and dry debris layers had transformed into an excellent combustible material. Through these forest corridors fire caused by people's carelessness advanced.

The second largest fire in the 20th century occurred in 1987 in China. An area of about 1 million ha – 650,000 ha of which were occupied with forest – was enveloped with fire. In the fire 191 people were killed, and 56,000 people were evacuated from the zone dangerous for life. 12,000 houses were destroyed (Peart, 1988). Drought was the main factor that promoted the fire. It was believed that the origin of the fire was carelessness of man's activity rather than lightning strikes.

We know how big the material damage can be from both considerable (usually multi-source) and small fires. The losses reach hundreds of millions of dollars. For instance, 3 million people were killed, 80,000 hectares of forest were burnt and 1,171 buildings were ruined, and total losses caused by the 21 fires that broke out in autumn 1993 in California reached 1 billion dollars (Smith, 1997).

Damage caused by fire does not consist only of the loss of wood and devastation of settlements. Significant damage is connected with disturbance, degradation and destruction of forest and other ecosystems. This is especially true for forest ecosystems. As forest cover is destroyed, other components of forest ecosystems are effected, such as, shrub vegetation cover, biocenosis, and soil cover. In the places of burnt out forests even the hydrological regime is disturbed. Even under normal soil moistening conditions in flat areas bogging processes are developed. In hilly and mountainous areas formation of gullies and soil losses are quite possible. Similar disturbances of ecosystems occur in steppes, savannas, and tundra. In the latter case these processes are especially prolonged because of the fragility of tundra ecosystems. Quite often they threaten settlements and even ruin them. In some countries (Spain and the USA) fires periodically encroach close to cities, causes damage to their environs, in particular, to those where there are forest 'spots' and waste grounds covered with shrubs.

Special programs have been prepared to decreasing the scales of fire calamities in a number of fire hazardous regions of the world (USA, France, Canada, Russia, and Australia). They include fire hazardous situation assessment, including zoning a country, fire development observations and finally, fire suppression measures.

4.2.2 Satellite information about forest fires

Satellite images may be used for a number of purposes, such as, timely detection of fire focuses; for revealing fire spreading; for planning fire control measures; for determining sizes and conditions of old and new burns; for determining and adjusting the boundaries of major regions of permanent or regular burning out of plants; for searching for cloudiness that can be used to provoke precipitation over fires (Kondratyev *et al.*, 1983; Bertrand, 1990).

4.2.2.1 Satellite information about scales and location of fires

Satellite observation data can be effectively used to discover fires. They are especially important for revealing fires in areas of reduced access and thinly populated regions (e.g. taiga or forest-tundra in Siberia). Daylight fires are recognized in visible spectral ranges by smoke plumes. Entire zones of mass fires – forest in particular may be easily outlined. This was the case with forest and swamp fires in the summer and the autumn of 1972 that emerged during a dry summer in a densely populated center of European Russia. From meteorological satellite images the entire zone of forest fire focused on the left bank of the Oka River. The area of this zone was marked by tens of smoke plumes reaching 200,000 km^2 (Kondratyev *et al.*, 1983).

Smoke clouds being indicators for fires may have different sizes and configurations, depending on the nature of fires, their intensity and atmospheric circulation. In the course of development of fires in the center of European Russia, using satellite images (27 August 1972) single smoke plumes are seen to stretch to distances of 75–400 km, their widths being of 15–25 km. During the periods of the big fires that developed in August 1971 over the Central Siberian Plateau, satellite images showed (*ESSA-8* AES) sizes of smoke plumes of the 15 biggest fires reached from 30 to 200–250 km in width. Simultaneously more detailed images (*Meteor-12*) showed up to 40 fires were fixed with plume sizes from 15 to 300 km (Grigoryev and Lipatov, 1977).

Satellite observation is the only way to carry out the continuous tracking of entire zones of large-scale fires in Siberia and the Far East, which quite often cover thousands of square kilometres and have durations of many weeks. Information of this kind is obtained by means of low-resolution pictures, including those taken by geo-stationary satellites.

The biggest fires in North America were caused mainly by lightning strikes during 1980–1982 in West Canada. Judging from *NOAA-7* satellite images, smoke from the fire propagated some 3,500 km, and in some cases the smoke cloud extended in the direction of the Atlantic Ocean (over Quebec and Newfoundland or over New York and Newfoundland) as far as 5,000 km (Chung, 1984). The area of the smoke cloud on 27 and 28 August 1981 measured 1.2 million km^2. Its size exceeded the size of the dust and smoke cloud ejected on 21–22 May 1980 during the course of the Mount St Helens volcanic eruption. The length of some smoke plumes reached 100–200 km.

Fire smoke was tracked using the images taken by this satellite in the wave interval of 0.58–0.68 μm. At the same time, for general assessment of fire hazard conditions, pseudo-color images were used, obtained by mixing images in the following three spectral ranges: 0.58–0.68; 0.72–1.1 and 10.5–11.5 μm.

Images taken from the *NOAA-9* satellite (AVHHR apparatus) made it possible to estimate the ranges of forest and swamp fires that enveloped the Amazon basin in 1987 (Setzer and Pereira, 1991). During the course of the fire's development, smoke dispersed over the continent and adjacent South Atlantic over an area of several millions of square kilometers. During the course of these fires development it was to possible to estimate emissions into the atmosphere (due to burning forests and swamps). It was discovered that the emissions contained 94 million tonnes of CO, 1,700 million tonnes of CO_2, 0.1 million tonnes of CH_3CL, 1 million tonnes of NO_x, and 10 million tonnes of CH_4.

Pictures taken by the *NOAA-7* satellite made it possible to assess from smoke plumes the scales of the fires that developed in 1982–1983 on Kalimantan Island in Indonesia (Mallingreau *et al.*, 1985). Smokes from single fire focuses merged into clouds whose area reached 350,000 km^2. Smoke plumes extended sometimes to the Malay Peninsula (to Singapore) a distance of more than 1,500 km. According to satellite data, a thin haze layer extended even further to South China.

In the course of development of large-scale fires in 1972 in the central part of European Russia, satellite images also registered a huge smoke cloud. In meteorological satellite images it was seen as a curved zigzag light band having dull texture and being less bright than background cloud fields. In the area of fires southeastward of Moscow the cloud had a stripy structure. The widest section of the smoke cloud visible in the pictures was 400 km wide, and the narrowest one was 150–200 km. The total length of the smoke cloud was 5,600 km (Figure 4.7).

Joint application of satellite images of different types obtained from *DMSP* and *NOAA* satellites made it possible to reveal fire range pattern in 1987 in Africa. Images taken by *DMSP* satellite made it possible to judge upon fire development over entire continent, while infrared images obtained by *NOAA* satellites made it possible to estimate the spatial range of fire in the savannas of South Africa (Cahoon, 1994).

Changes in atmospheric smoke content due to fires make it possible to judge upon the dynamics of the latter. For example, in the period 1973 to 1988, inclusive sharp increase of biomass combustion, and thereby forest and swamp fires range in Amazon basin was detected by images obtained by *Skylab-3* and the *Space Shuttle* in 1988 (Heffert and Lulla, 1990). It was found out that the area of smog formation over inflammation zones increased from 300,000 km^2 in 1973 to 3,000,000 km^2 in 1985 and 1988.

4.2.2.2 Satellite information about meteorological conditions during fires

Satellite images make it possible to have different information about atmospheric conditions during fires. These data, in particular concerning wind direction and

(a)

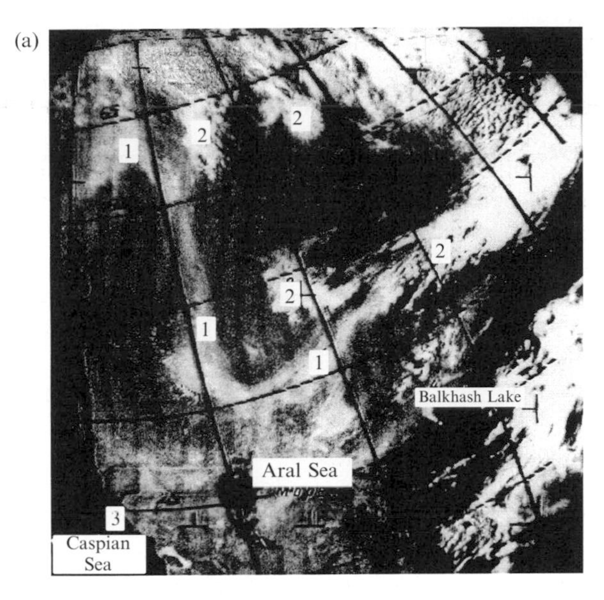

(b)

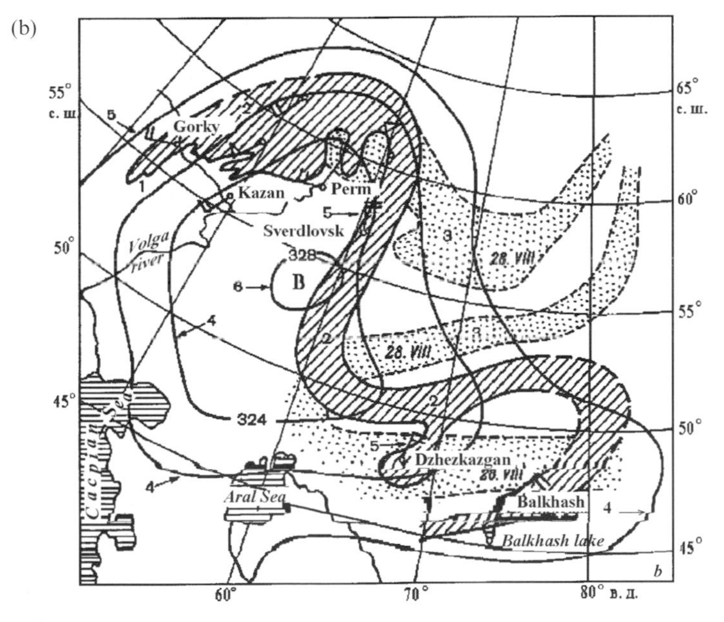

Figure 4.7. (a) Smoke plume of sub-continental scale (more than 5,000 km) caused by forest and swamp fires in the center of the European part of the former USSR. The picture was taken from a height of 1,400 km by the *ESSA-8* satellite on 27 August 1972; 1 – smoke plume, 2 – clouds, 3 – Caspian Sea, 4 – Balkhash Lake (Kondratyev *et al.*, 1983). (b) Diagram of smoke plume migration prepared using pictures taken on 27 and 28 August 1972; 1 – zone of fire focuses; 2 and 3 – smoke plumes, 4 – isohypses (0 in decameters in AT map – absolute isobaric topography 700 mb), 5 – wind direction and speed, 6 – high pressure area (Grigoryev and Lipatov, 1978).

cloud development (especially of thick rain clouds) is important for assessment of fire hazard situations.

Direct symptom – indicator for forest (and any other) fire is a smoke plume identified in satellite images. Smoke plumes are recognized even in low resolution pictures and are very distinct in medium resolution ones. To differentiate them from clouds one may use images obtained in visible (0.5–0.7 μm) and near infrared (0.8–1.1 μm) spectral ranges. Smoke plumes due to fires differ from clouds in that the typical form of a smoke cloud is usually extended windward.

Analysis of the the form of smoke plumes due to fires makes it possible to judge the character of micro- and meso-scale changes in wind and conditions of the lower troposphere. Satellite images of the fires which occurred in 1971 in the area of the Central Siberian Plateau may serve as an example. Images obtained from meteorological satellites were analysed together with synoptic data and radar sounding data (Kondratyev *et al.*, 1983).

In the images obtained in the visible spectral range on 8 August 1971 smoke plumes were seen as parallel bands expanding as the distance from the source increased. Then at some distance different for different sources, smoke plumes became blurred, and over the central part of the Eastern Siberian Plateau they were like a haze with slightly twisting contours. As synoptic analysis showed, at that time the region under consideration was in the south-west periphery of an anticyclone and atmospheric stratification was stable. Wind speed at the level of smoke propagation (1.5 km) measured 15 m/s.

On 10 August the synoptic situation changed a little. Portions of more humid air arrived in this region from the south, and cloudiness increased. Here and there thunderstorms occurred. Only westward, in the direction of the Yenisei mountain-ridge did strong night-time cooling occur. The wind regime also changed. Wind speed did not grow as considerably with height. At the height of 1.5 km wind speed measured 7 m/s.

Vertical temperature lapse rate reached dry adiabatic values. Wind and temperature regimes of this kind determined specific features of smoke propagation. Smoke plumes had the appearance of wavy expanding bands with the spots of increased brightness which was due to 'dragging' of the smoke plume in convective clouds. These parts of the plume were seen as brighter spots in the images.

As analysis of satellite pictures has shown, smoke transported by atmospheric circulation can travel large distances. This makes it possible to study atmospheric circulation in the area of smoke propagation. For example, from the series of satellite images from meteorological satellite *ESSA-8* changes in atmospheric circulation in July–August 1972 during the largest fires in European Russia have been traced (Kondratyev *et al.*, 1983).

Aerological sensing in the area of these fires during the period of their maximum development from 26 to 28 August 1972 were used to estimate vertical sizes of the smoke layer and particulate matter transport rate. Smoke cloud configuration on 27 August matched the circulation features at the level of 700 hPa. However, the upper boundary of the smoke layer was at the height of 3 km.

From the data on the vertical atmospheric cut crossing the smoke plume along

56°N it was seen that at the level of 2.8 km in the area 37°–50°E there was an isothermal layer of about 300 m depth. Wind speed in the lower troposphere under this layer reached maximum values (up to 20 m/s). Thus, satellite images of smoke clouds due to fires make it possible not only assess the sizes of fires, but also to study the character of circulation on lower atmospheric layers – which is also important for assessment of fire development. In this case it was discovered that there was an intense air flow in the lower troposphere in the periphery of the anticyclone. Similar flow in the lower troposphere was discovered from meteorological satellite images in the area of Central Siberian Plateau in summer of 1971 (Kondratyev *et al.*, 1983).

Satellite images, even the most small-scale ones obtained from meteorological satellites (in particular, from satellites of the *Meteor* series) may be useful for assessment of a number of other parameters of meteorological criteria that determine fire hazard conditions. They may be used for example to determine soil humidity characteristics. Satellite images are also important for revealing thunderstorm centers. Finally, satellite images may be used for identification of so called cloudiness development with precipitation, important for artificially produced precipitation in the areas of fire development (Valendik, 1990; Kondratyev *et al.*, 1983; Kondratyev *et al.*, 1972).

4.2.2.3 Satellite information about fire focuses

To specify boundaries of fire focuses it is efficient to use infrared images. Optimum spectral intervals are 3.5–4.2 and 4.4–5.0 μm where maximum radiation contrast between fire focus and background is observed (Kondratyev *et al.*, 1972). Infrared images are also useful to track fire focuses at night, for underground fires and fire focuses in swamps.

Development of forest and swamp fires in 1983 in Indonesia was successfully tracked using infrared images from the *NOAA-7* satellite. Smoke-screening did not prevent tracking them in this spectral area. These images also revealed focuses of underground fires (in swamps), including those that remained after rainfall (Mallingreau *et al.*, 1985).

Images from the *NOAA* satellite obtained in the 3.1–3.7 μm spectral range have been used since 1994 to identify forest fires in the Irkutsk area, Siberia (Zherebtsov *et al.*, 1995). In the infrared image of the Bratsk and Nizhneilimsk regions on 16 May 1994, 12 focuses of forest fires of areas from 0.01 to 200 ha were identified. All of them were confirmed by aircraft data. Continuous satellite observations of forest fires during these 16 days showed that, on average, about 70% of fires were identified correctly. Only a few areas of former fires from 1993, some sand quarries and fires that had already been extinguished resulting in unvegetated soil were mistaken for fires. Minimum areas of inflammation that could be identified in satellite images was 0.2–0.5 ha.

Currently for forest, as well as swamp and steppe fires, satellite monitoring best suited to the task is by the *NOAA* satellite surveying data (AVHRR radiometer) that characterizes their own and reflected radiation in five spectral ranges 0.58–0.68, 0.725–1.1, 3.55–3.93, 10.3–11.3 and 11.5–12.5 μm (Kennedy *et al.*, 1994). Although, relevant images are of rather low spatial resolution (about 1 km) they allow us to estimate (in infrared range) temperature of fire focuses with the accuracy of about 1°C. Simultaneously satellite surveying is effective over a vast territory (the width of the band surveyed along the satellite route is 2500 km).

Spectral interval 3.1–3.7 μm is the most suitable for revealing fire focuses. Maximum infrared radiation due to fire focuses falls on this interval. In this interval fire focuses are detected, which are smaller areally than the spatial resolution of AVHRR radiometer.

Data for two other infrared intervals, 10.3–11.3 and 11.5–12.5 μm are useful for estimation of soil humidity. Such information about humidity and temperature is important for analysis of fire hazard situation development.

Data from the *NOAA-9* satellite (AVHRR radiometer) was used to assess the ranges and focuses of forest and swamp fires in 1987 in the Amazon basin (Mallingreau *et al.*, 1985). From 46 satellite images 350,000 individual forest fires were identified. Their total area made up almost 20,000 ha. Images obtained on different days (from March to May inclusive) made it possible to judge whether forest fire focuses were increasing or decreasing in numbers.

Judging from satellite images, smoke emissions over the Amazon basin in May decreased sharply. This was explained by attenuation of fire intensity due to showers during a 2-week period at the beginning of May. Unexpected expansion of forest fire focuses at the end of May (detected in thermal infrared images) was associated with the fire burning activities of farmers.

At the final stage of fire development it was noticed that fires in the swamps were burning longer than elsewhere.

Currently localization of the focuses of burning forest (with temperatures of 500–1000 K°) from the *NOAA* satellite images for medium and large fires (areal extent greater than 40 ha) are performed with the accuracy up to 10–12% (Chuvieco and Martin, 1994). It is important to emphasize, that from *NOAA* data it is possible to monitor forest fires with a temporal interval of 12 hours.

In China where the number of inflammations annually reaches about 4000, images from *NOAA* meteorological satellites have been successfully used for forest fire monitoring starting from 1985. In particular, they were used to obtain information about disastrous forest fires in May–June 1987 in the mountainous area of Ksinan.

Satellite images covering territories of different sizes, and having different resolutions (from *Landsat*, *NOAA* and *SPOT* satellites) made it possible not only to trace the dynamics of zones of fire ranges but also to reveal new sources of fire (Xu Guanhua *et al.*, 1991).

Forest and other fires are also identified by their radio-thermal radiation. Over fire focuses, microwave radiation increases – a result of increasing soil and plant cover temperature. Microwave radiation is characterized by its capacity to penetrate through smoke (as compared to infrared radiation). This made it possible to judge zones of fire under conditions of high intensity smoke cover (Zherebtsov *et al.*, 1995). However, fire surveying in the microwave range has not yet found a wide utility in practice because of its low spatial resolution.

4.2.2.4 *Satellite information about burned-out forests and forest disturbance after fires*

Diagnostics of the state of forests and freshly burnt out places after fires is best using large-scale spectral zonal aerial photographs (of the scale 1 : 10,000 – 1 : 15,000) (Furyaev, 1993; Furyaev *et al.*, 1997). At the same time less detailed but wider field of view satellite images are useful for prompt assessments of freshly burnt out places, and estimation of their areal size.

To reveal burnt out places in Russia, scanner satellite images (0.5–0.7 and 0.8–1.1 μm) are used. It is convenient to use them to obtain information about burnt out places in, difficult to access, unprotected regions of the northern taiga forests and deer's pastures (in tundra and forest-tundra) in north-east Russia. Scanner satellite images detect only large burnt out places (200 ha and more) where forests are completely dead (Breido *et al.*, 1995; Furyaev, 1993; Furyaev *et al.*, 1997).

Small-scale satellite images of this kind (scale 1 : 2,000,000) proved to be suitable for assessment of material damage caused by fires. To assist in this task it is necessary to have special maps containing data on species composition of a forest and on arboreal resources.

Similar phenomena – destruction of large forest tracts as a result of the carelessness of man in the course of burning out vegetation are revealed from satellite pictures in the tropical zone. Catastrophic situations have arisen in Africa. Huge forest tracts have been burnt out southward from the Sahel band in the area from Tanzania to the Ivory Coast. The area of some of burnt out places on the coast of the Gulf of Guinea as can be calculated to reach several hundreds of square kilometers. Older burnt out places are distinguished in the images by their lighter coloring – fresh burnt out places are seen as dark spots.

Surveying from the *NOAA* satellite has changed our ideas considerably on the scales of fires in Brazilian Amazon basin. In 1987, from FAO data the area of burnt out forests there was estimated as 54,280 ha. Satellite surveying of the same territory has revealed fresh burnt out places of an area of 20 million ha (200,000 km^2) – some 400 times greater than FAO estimation (Chuveico and Martin, 1994). Such a considerable difference in the estimates of the areas of burnt out places highlights the imperfection of traditional prompt control of fire consequences.

In Russia small-scale satellite images obtained by satellites of *Cosmos* series were used to study forest dynamics after fires. For example, for the Ob and the Yenisei inter-stream area a map of forests was made using satellite and aerial images as well as on-land data. This map shows forests damaged by fires to varying extents (slightly, moderately and highly disturbed by fires). In doing this, landscape approach to differentiation of forest tracts as used, pyrological features of natural complexes were also taken into consideration (especially frequency of occurrence of fires) (Furyaev, 1987).

Selection of initial satellite images depends on the scale of maps of forests disturbed by fires (Furyev *et al.*, 1997). To prepare small-scale review maps multi-zonal satellite MSC-4 images with resolution up to 100 m are used. To prepare mid-scale maps MSU-E scanner photographs were used (spectral intervals 0.5–0.6 and 0.6–0.7 μm with resolution up to 45 m). And finally, to prepare maps of larger scales it is recommended to use SPhC–1000 spectral zonal images with resolution up to 10 m.

To ensure efficient decoding of forest fire satellite images different processing methods are used. For example, for mapping (on the scale 1 : 250000) forest fire consequences in Provence (France) from Landsat satellite images, the techniques of coding into 16 levels, chromatic coding, plotting histograms, interactive mode alteration of image contrast range and sizes were used (Housson, 1994).

4.2.2.5 Forecasting fires using of satellite information

Different methods are being developed in different countries to use satellite information for determination of fire risk – primarily forest fire risk. For example, in Elba Island (Italy) images from the Thematic Mapper data (*Landsat* satellite) here used. Control over development of fire hazard conditions was exercised by means of analysis of spectral reflectivity of earth cover (a special spectral index has been established). Simultaneously, development of fire hazard conditions was monitored by a traditional method that takes into account changes in plant and soil cover and position in the area. Assessment of forest fire risk using satellite information proved to be as efficient as the traditional one from on-land data, but also proved more reliable.

It is possible to assess the possibility of fire development or even to forecast them from satellite data on the state of vegetation characterized by the normalized vegetation index (NDVI). Analysis of *NOAA-7* data on NDVI during large-scale fire development in Indonesia, and also before their origination and after their decay revealed decreases in the index from 0.5 in April 1982 to 0.4 by the autumn of 1982, when drought developed (Mallingreau *et al.*, 1985). After the fire, the value of NDVI decreased to 0.3. Only in the rainy season 1983–1984 when the forest began to regenerate actively did the vegetation index increase to 0.35–0.4.

Infrared satellite images obtained from *NOAA* data were used to track drought development as the most important factor for fires in the summer of 1989 in France (Graux, 1990). This drought was the culmination of the dry years of 1985–1987, and in some parts of the country to 1988. Drought was caused by a sharp reduction of precipitation. This led to crop failure – more than 50% of harvest was lost. The drought also prepared the ground for fire development.

Images obtained daily from the *NOAA* satellite made it possible to watch deterioration of plant cover in a number of regions of France using changes in vegetation index. It was possible to estimate soil moisture storage using a well known correlation between plant cover state and soil moisture. However, as French experts believe, resolution of *NOAA* data (about 1 km) is at the limit of feasibility for obtaining data for small, and for inhomogeneous (as regards landscape structure) regions.

In Russia, works have been performed on application of satellite survey materials for mapping forecasted forest disturbance due to fires. Maps of this kind are prepared on the basis of landscape using materials of aerial and satellite surveys and modeling space and time dynamics of forest communities (Furyaev *et al.*, 1997). Application of satellite information ensures prompt creation of forecast maps for large territories. Elaboration of this kind was performed for Angara–Yenisei region in the territory of the Central Siberian Plateau which was rather prone to fire development.

4.2.2.6 Satellite information about global fire development mechanisms

On the basis of forest fire and other natural and anthropogenic fire surveys from US manned spacecraft during the period from 1965 to 1992 new information was obtained about their spacial and temporal features on a global scale (Andrea, 1993). For example, more often fires emerge over the perimeter of the Amazon basin and in West Africa, mainly in the deforested areas than in forest regions of these continents.

This is presumably associated with anthropogenic development of these regions. Large fires occurred particularly often in Central America, Mexico, in the pampas of South Brazil, Paraguay, and Argentina, and also in a number of mountainous countries like Bolivia, Peru, and Ecuador. In Africa fires appeared to be especially typical in the sub-Saharan belt.

Maximum concentration of fires judging from satellite images was detected in the savannas of South Africa, particularly in Zaire, Angola, Zambia, Tanzania, and in Madagascar. In Australia the majority of fires was observed in the provinces of the Northern Territories and Queensland.

In the extra-tropical zones of the Northern Hemisphere fires were rare. This is partly due to insufficient information for Siberia and the Far East of Russia because of the specificity of American satellite orbits. In the USA main fire focuses are located in forests and farming lands of the Pacific Coast of the country, Gulf of Mexico, and Atlantic Coast in the south and swamps of Florida.

In Europe large fires were recorded in Mediterranean area (in the maquis – evergreen hard-leaf brushwood in the islands of Corsica and Rhodes) and in Central European Russia. Fire clusters were recorded in Nepal, in the south of Western Siberia (Figure 4.8) and in Eastern Siberia.

It is interesting to look at the distribution of fires over different regions of the planet. The majority of fires (among 1,030 detected by surveying) took place in Africa (42.6%) and South America (26.5%). Less fires were observed in Europe (0.8%), North America (4.2%), Central America and Mexico (6.8%). It seems that the value of 12.5% for Asia is underestimated.

Satellite surveying made it possible to obtain some other information about fire development with time. In the Southern Hemisphere in this period the majority of fires occurred in September and October. This was reflected in simultaneously observed high concentrations of atmospheric ozone over the South Atlantic in the same months of the year.

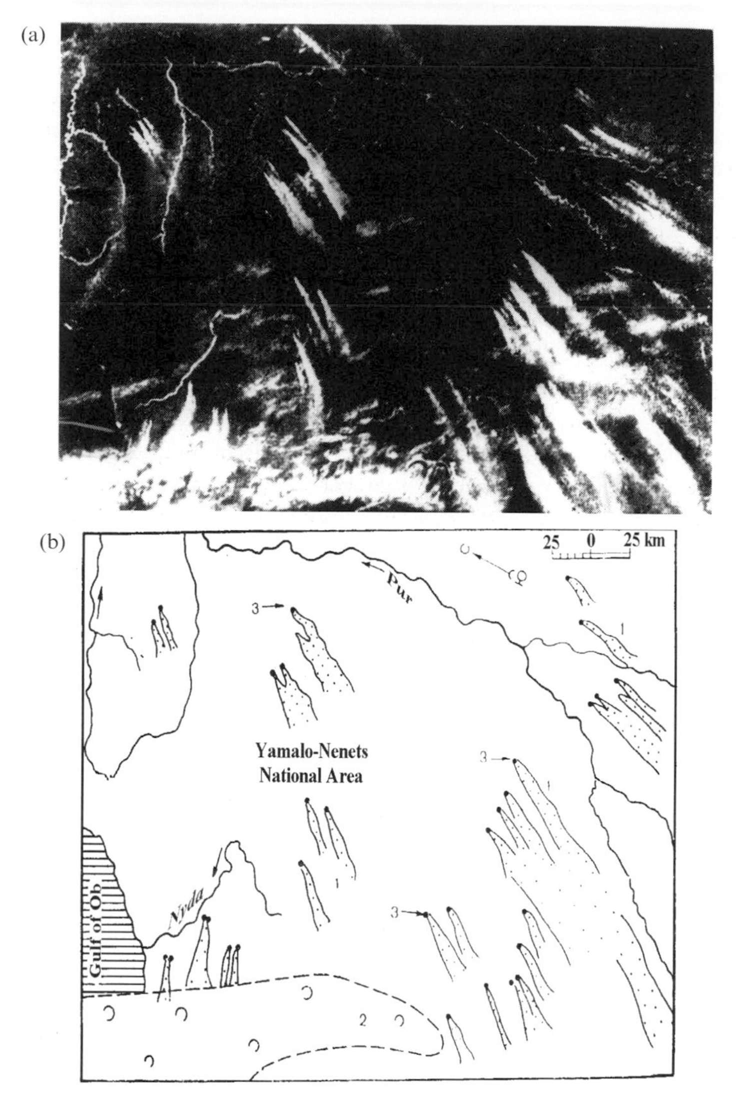

Figure 4.8. Smoke plumes in the south of the Western Siberian Lowland resulting from mass-scale burst's of forest and swamp fires during thunderstorm (Grigoryev and Kondratyev, 2001). (a) image from the *Meteor* satellite obtained on 5 August 1977, (b) picture location; 1 – smoke 'tongues', 2 – clouds, 3 – fire focuses.

Detailed analysis of satellite surveying data disproves the idea that fire emergence is associated mainly with fire hazard seasons. It was found that many large fires in both hemispheres occur all year round.

In the Southern Hemisphere, high-latitude fires occur more frequently in summer and autumn, while in tropical areas fires are associated with dry seasons.

In the tropical regions of Brazil the majority of fires occur in August–October. This is characterized by the driest weather. During this season in tropical regions of the country drying of soil and dehydration of plants take place which makes it easier to cut forests. Information of this kind is supported by satellite surveying data for the region (Richter *et al.*, 1986).

Fires in the pampas of Argentina and in the south-east of Brazil are connected with other times of the year (October to February). Similar situations are observed in Africa. Here fires 'migrate', in particular from northern and central Zaire southward following dry season development. In March–June they break out in northern Zaire. In June–July fires range across southern Zaire, Angola, Zambia, and Zimbabwe (this being the end of dry season in the continent).

In the Northern Hemisphere fires are clearly tied-up to high latitudes, and in the end and beginning of year to tropical latitudes.

In view of its incompleteness, information presented above does not make it possible to draw a reliable conclusion on tendencies of the amount of fires to either increase or to decrease in one or another region for the period under consideration. Most clearly the tendency of fires to increase is seen in the Amazon basin, this is supported by satellite data for the period from 1973 to 1985 (Heffert and Lulla, 1990).

So, satellite observation data are valuable tools to assess fire propagation, to pinpoint their location, to analyse smoke development, to identify places that have been totally destroyed by fire, and even to predict fires.

4.3 TECHNOLOGICAL ACCIDENTS AND DISASTERS

4.3.1 Technological accidents and environmental destabilization

Among anthropogenically induced ecological disasters there are disasters that are provoked by technological impacts (in contrast to, for instance, desertification, deforestation, biodiversity decline and so on), including technological accidents. There is a wide scope of ecological disasters caused by accidents at industrial plants. As a rule, these are usually rapidly-developed or sudden incidents caused for example by transport facilities accidents, breech of dams and pipelines, as well as explosions and fires. Figure 4.9 illustrates a fire in St Petersburg. Accidents may involve not only victims and cause material damage (due to fire, blast waves etc.) but can lead to mechanical disturbance of terrain, and can also lead to pollution that sometimes causes considerable damage to ecosystems. Among the most widely spread sources of technological ecological disasters resulting in water pollution belongs to tanker accidents giving rise to oil and oil product spillage. From world statistical data (Lloyd's) for 26 years (from 1973 to 1989) there have been tanker accidents accompanied by considerable oil spillage. On average, there were 7 accidents each year (Alayev and Monina, 1990) (Figure 4.10).

The largest accident took place near Alaskan shores in 1989 with the tanker Exxon Valdez which ran against hidden rocks because of failure of the automatic

Figure 4.9. Ancient engraving depicting a fire in St Petersburg in 1737 (Grigoryev and Kondratyev, 2001).

Figure 4.10. Global distribution of ocean pollution (with oil) resulting from large tanker accidents (more than 10 thousands barrels) between 1974–1989 (Grigoryev and Kondratyev, 2001).

control and also incorrect procedures of the operator of the ship. Oil spread over an area of 2,600 square miles and polluted the coastline to the distance of 1,000 miles from the shore. It killed many thousands of sea animals and birds. Fishermen who suffered were paid compensation of 200 million dollars. Several billion dollars were spent to clean the coast and save its ecosystems.

Disastrous ecological incidents at sea occur quite often as a result of accidents at oil wells. The largest of such accidents took place in the area of Santa Barbara in California in 1969 and at Ecofisk oil field in the North Sea in 1977.

Land accidents often occur at settling basins for tailings and discharges. Breaks of protection banks of settling basins are a serious ecological hazard. One of the most significant accidents of this kind occurred in September 1983 in the former USSR near the town of Dragobych in Ukraine (Grigoryev, 1991). It took place at a settling basin for residuals of potash fertilizers that originated from the Stebnick plant. As a result of a settling basin dam break, 4.5 million m^3 of brine were discharged into the Dnestr River. A similar accident happened in Hungary in February 2000. Toxic brine effluent caused an ecological disaster, polluted the river to a distance of 500 km up to the dam of Novodnestrovsk storage basin. One million tons of salt – 10–12 cm thick layer extending over the floor of the 50 km storage basin – was deposited. Ecosystems of a large part of the Dnestr River were poisoned, 920 t of market fish and 1,300 t of fry were killed. Dozens of cities located in the basin of Dnestr River, experienced water shortages (both potable water and water to be consumed for industrial purposes). Not only cities in areas of Ukraine experienced water 'dearth', but also a number of cities and settlements in mid-Moldavia.

Large-scale pollution is caused by accidents at oil pipelines. In August 1994 an accident of this kind occurred as a result of a pipeline break in the north-east of European Russia in the Komi Republic not far from the town of Usinsk. Oil leakage was estimated to be between 14,000–60,000 tons (data from the Ministry of the Environment of Komi Republic and from estimates of Hydrometservice of Russia, respectively). Not only small rivers – tributaries to Pekhra-river, but lakes, swamps and pastures in the area of 60 km^2 were polluted with oil. It is important that tundra ecosystems are especially delicate and vulnerable. Damage caused to the ecosystems was estimated as 17 billion dollars.

The second half of the 20th century is characterized by one more type of specific accidents associated with emissions of radioactive substances and radioactive environmental pollution. Accidents of this kind took place in the ocean (nuclear submarines), in outer space (satellites) and on land (at nuclear power plants and military plants). The most significant accidents of this kind took place in the former USSR – in 1957 near the town of Kyshtym (in the Urals not far from the city of Chelyabinsk) (Figure 4.11) at the military industrial plant and in 1986 at Chernobyl nuclear power plant in Ukraine (Figure 4.12).

The ecological consequences of the accident at the Chernobyl Nuclear Power Plant were disasterous (see Chapter 1). The radioactive cloud enveloped considerable parts of Europe – from Scandinavia to France and Turkey (Figure 4.13). Radioactive pollution of landscapes in Belorussia and Ukraine, as well as in several areas

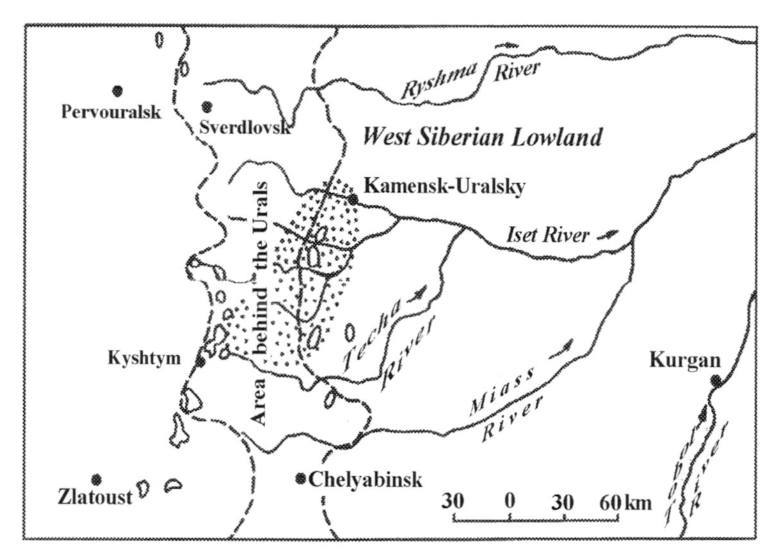

Figure 4.11. Map of traces of radioactive pollution due to an accident that took place at a military industrial plant near the town of Kyshtym in the Urals (Grigoryev, 1991).

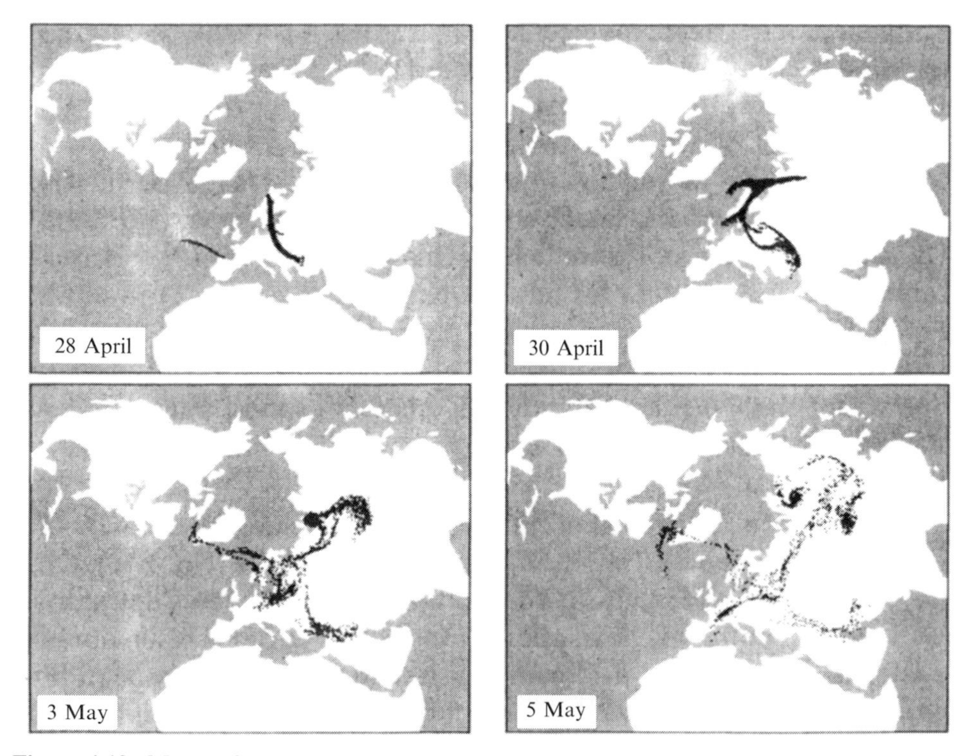

Figure 4.12. Maps of the global distribution of the radioactive cloud resulting from an accident on 26 April 1986 at the Chernobyl Nuclear Power Plant (Ukraine) monitored to heights of up to 1,200 m (Ramade, 1987).

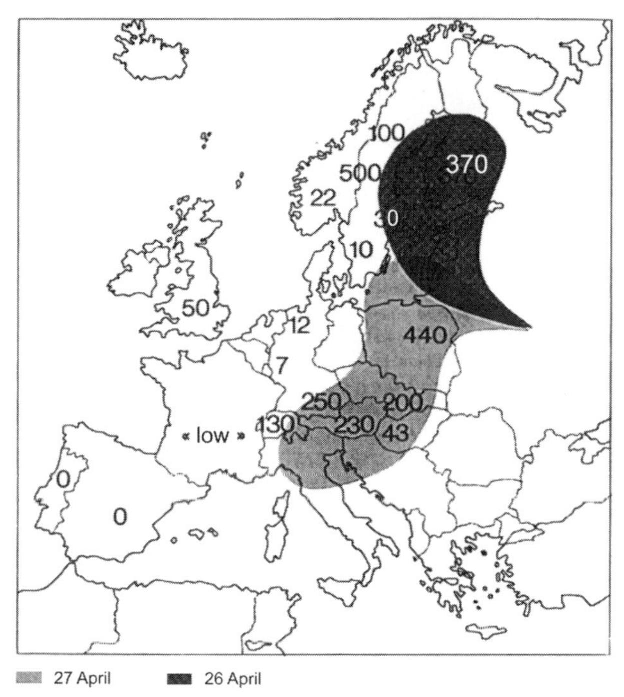

Figure 4.13. Map of radioactive pollution distribution in Europe on 26 and 27 April 1986 resulting from the accident on 26 April 1986 at Chernobyl nuclear power plant in Ukraine (only maximum doses mR/h) (Ramade, 1987).

of Russia (primarily the Bryansk area) (Figure 4.14) were particularly hazardous to people's health as well as damaging to the ecosystem. In Belorussia several hundreds of settlements found themselves within the 'spots' of radioactive pollution.

In 1991 from unofficial data, about 7,000 people died and 50,000 out of 600,000 accident liquidators fell ill as a result of radioactive exposure (Scherbak, 1991). According to information from the public association 'The Hope of the World', the exposure standard of 30 REM was exceeded for hundreds of thousands of children (Gritsman, 1993). In Belorussia alone (from approximate data of its Academy of Sciences) 500,000 people suffered from irradiation. Let's note that assessments of ecological consequences of nuclear accident at this power plant (including those who fell ill due to irradiation, disturbance of ecosystems etc.) are rather contradictory and due to different reasons incorrect (they may be either overestimated or underestimated for the sake of 'national' interests).

Many technological accidents that lead to ecological disasters occur in cities and industrial centres that have large populations and vast amounts of industrial components housing toxic materials (Grigoryev, 1991). Emergency situations have occurred, for example, with rolling stock. For instance, an accident happened at the railways in Missisaugi in Canada in 1979. In this city, contrary to all safety requirements, railway trucks whose tanks were filled with toxic gases were left in

Figure 4.14. Zones of intense radioactive pollution of the Earth surface with caesium-137 after the accident at the Chernobyl Nuclear Power Plant in the territory of Belorussia, Ukraine and Russia (Grigoryev, 1991).

sidings. Some of these tanks exploded. As a result, toxic pollutants were released into the atmosphere and rapidly covered the whole city. The danger of mass-scale poisoning arose. On short notice 230,000 inhabitants were evacuated from the city, an affected area of $120\,km^2$ was cordoned off, and an ecological catastrophe was narrowly prevented.

The biggest accident at a chemical plant took place in 1984 in Indian city of Bhopal. Leakage of toxic gases occurred at the plant of the Union Carbide corporation. Emission of tens of thousands of tons of pesticides suddenly leaked into the atmosphere. A toxic cloud enveloped the sleeping city. About 2,500 people died immediately, and 70,000 people were poisoned.

The ecological effects of accidents in a city can extend far beyond the city limits. This may be illustrated by the accident at the chemical company Sandos in 1986 in the city of Basel (Switzerland). Here a big fire occurred. As a result, up to 30 t of toxic substances were discharged to the Rhine, including insecticides, herbicides, solvents, and about 70 kg of mercury. Ambient air was also poisoned due to an explosion of barrels containing sulphur dioxide, nitrogen oxide and sulphur hydrocarbon compounds. A toxic cloud emerged over Basel.

The main consequence of the accident was pollution of the Rhine up to its mouth along many hundreds kilometers. River ecosystems were disturbed, dozens of tons of dead eels, trout and other species were cast on bank. The catastrophe that took place in Switzerland assumed international significance, since it had affected ecological conditions and vital functions of the population in the Rhine areas in other countries, such as, Germany, France, and Holland. In Germany and Holland fresh water supply from the Rhine was interrupted for some time (because of difficulties in removing the pollutants). With regards to the ecological consequences this catastrophe was considered as one of the most significant in Europe.

Ecological disasters practically always accompany another specific type of people's activity – military operations. Wars are an integral feature of civilizations, they lead millions of people to death, destruction and impoverishment. Total losses among civil and military people during the First World War numbered 55 million. During the war in Indo-China about 7 million people died. These were mainly peaceful people (Militarism, 1985).

There were also considerable economic losses. Especially large-scale losses took place during the Second World War in the territory of the former USSR. Here 1,710 cities and settlements, and more than 70,000 villages were totally ruined. Deliberate destruction of industrial enterprises and cities has resulted in mass-scale pollution of water and soil, and large-scale fire development. Both in the past and at the present time military operations are always accompanied by fire. Great fires ensued, for example, when Hamburg was bombed in July 1943. Separate fire spots merged. Dozens of square kilometres of the city were engulfed in fire. A flame vortex originated in the city, and its smoke reached a height of 8–12 km (Budyko *et al.*, 1986).

Large fires during the course of the Second World War occurred when Dresden, Kassel and Darmstadt were bombed. In Dresden they lasted for about a week. Fire ruined 75% of buildings over an area of 12 km^2. Equally devastating were fires in Hiroshima and Nagasaki occurring after the nuclear explosions. In Hiroshima 13 km^2 and in Nagasaki 7 km^2 of urban buildings were burnt out in the fires (Ginzburg, 1988).

Especially large-scale impacts of fires on nature were observed in the course of prolonged wars in Vietnam. Fire from bomb explosions was used specially for burning out forest tracts (in order to destroy jungle areas to suppress partisan detachments, American army applied special 'scorched earth' tactics).

Considerable ecological disturbances also give rise to various mechanical impacts on ecosystems. After the First World War and especially after the Second World War the lands of the majority of European countries were scarred by bomb craters. This distortion of natural relief still makes land use difficult. The density of bomb craters is particularly high in a number of regions of Vietnam. In Southern Vietnam there are more than 10 million craters left by bomb explosions.

Chemical weapons cause particular damage to the environment and people. Chemical weapons were widely used in Indo-China (Westing, 1976; *Ecocide in* ..., 1985). In Southern Vietnam about 72 million litres of herbicides and defoliants were dispersed. As a result, jungles and other forests were damaged over a vast area. Some 12% of all forests in the Southern Vietnam are poisoned (Figure 4.15). Mangrove

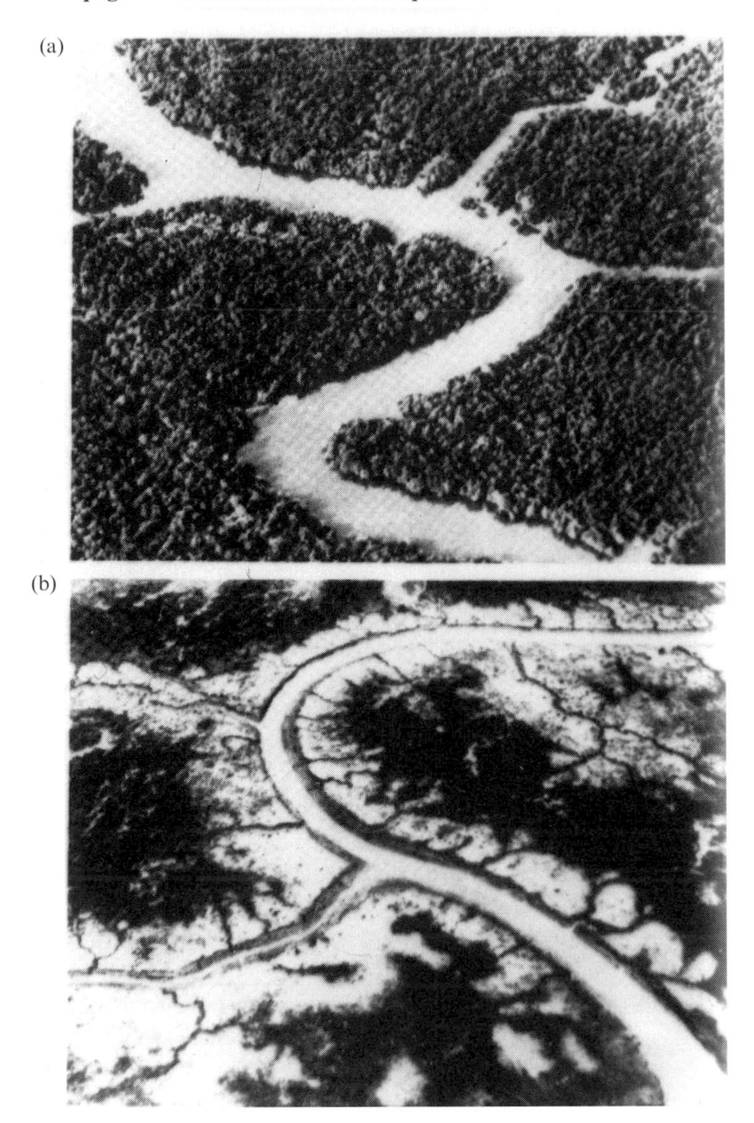

Figure 4.15. Tropical forest area in Vietnam (a) before and (b) after having been poisoned with special substances dropped from American aircraft (Grigoryev, 1985).

forests were severely damaged. From 1965 to 1970, 41% of mangrove brushwood were poisoned. According to experts, it will take several tens of years to restore forests in Vietnam, Laos and Kampuchea.

Specific ecological disasters occur as a result of nuclear-testing. It will suffice to mention, for example, pollution of the water area in the Arctic Ocean near Novaya Zemlya, as well as on one of the Marquesas Islands in the Pacific Ocean, in semi-deserts and deserts in eastern Kazakhstan.

Some technological, ecological disasters occur as a result of 'delayed action', associated with major constructions that slowly bring about changes in the environment. Ecological situations of this kind sometimes occur as a result of construction of large canals (for example, Kara Kum Canal in the Turkmen Republic). Catastrophes of this kind develop as a result of engineering construction within the natural landscape. Ecological consequences of these catastrophes are associated with enormous material damage (e.g. fisheries on the Volga River affected by dams of hydro-electric power plants that divided river ecosystem into separate water basins).

A critical ecological situation arose in the Neva Inlet in the middle of the 1980s. At that time it was misinterpreted as being associated with the construction of a dam in the eastern part of the Gulf of Finland (started in 1979). Construction of the dam was undertaken to protect St Petersburg from floods that from time to time cause considerable material damage to the city. As is well known, surge-type floods, caused by storm-force winds of the deep cyclones that pass over the Baltic, took on an especially disastrous character in 1777, 1824, and 1924. Water rose by 3.21 m, 4.21 m, and 3.80 m, respectively, and a part of the city was inundated. Particularly dangerous rises in water level in the Neva rivermouth (between 211 and 299 cm) have already taken place 66 times. The last time was on 30 November 1999 when water rose 262 cm at a time when the dam was under construction (Klevanny, 2000).

The body of the detention dam (total length of 25.4 km including 22.2 km over the water area) is designed to have two structures (110 and 210 m wide) to let ships pass and six structures to let water pass through (each of 10–12 openings being 24 m). Preliminary examination data has shown that cross-sections of all water-pass openings in the dam (1.5 times larger than the existing cross-section of the entire Neva mouth) ensures a completely hydrodynamic situation in the gulf and will not aggravate ecological conditions (Usanov, 1989).

In the middle of the 1980s, the ecological situation in the Neva river inlet was aggravated (Env. protection ..., 1997). One of the causes of this sudden deterioration in the ecological situation was flagrant violation of the procedure of closing the northern and southern dam sites. For some time the northern dam site was closed completely. As a result, its conveyance capacity was only 18% instead of its designed 60%, and the conveyance capacity for the southern dam site increased from 40% to 82%. Because of violated construction schedule, water quality deteriorated sharply, beaches were polluted, water alga blooms occurred, water exchange became delayed and so on (Lesogorov, 2000; Protection ..., 1998).

The main cause of such conditions at the Neva River inlet was not the dam at all and not even the above-mentioned violations of the construction schedule. It was heavy pollution (which has been observed for several decades) of the Neva inlet and eastern part of the Gulf of Finland due to St.Petersburg sewage water. For several years this exceeded MPC (maximum permissible concentrations) of different chemical and biological ingredients. This was emphasized by a commission of experts of the Presidium of Leningrad Scientific Center of USSR Academy of Sciences for 1989. The coli-index was 100–1000 times higher than the average, microbe pollution in the bottom layer was tens of times higher and in bottom grounds it was hundreds to thousands of times higher than in the surface water layer (Resolutions of ... , 1989). Simultaneous excess of MPC for such pollutants as copper, mercury, oil products and so on were also recorded.

The commission noted that practically 100% of the fish population were affected with toxicosis. Similar symptoms were observed in the eastern part of the Gulf of Finland behind the unfinished dam. It was clear that the cause of this unfavorable situation in the Neva Inlet was not caused by construction of the dam but instead by the low efficiency of the wastewater treatment works. It is clear that the dam has aggravated the already sensitive ecological situation (Resolutions of ... , 1989).

Politicization of this problem has been seen as a victory of those who were against the dam construction. The latter was slowed down sharply (*Concepts of* ... , 1997b).

In recent ten years (in the 1990s) ecological conditions in the Neva Inlet stabilized and even improved (*Concepts of* ... , 1997a; *Protection* ... , 1998; *Ecological Situation* ... , 1998). However, eutrophication of the eastern part of the Gulf and Neva Inlet has remained. The ecological conditions remain critical (*Protection* ... , 1998). Under conditions of evident industrial recession, the volume of sewage water entering the Neva inlet has decreased. From 1980 to 1996 it decreased from 1,963 to 1,664 million m^3, respectively (*Protection* ... , 1998). However, discharge of unpurified sewage is still the main cause of unsatisfactory conditions in the Neva Inlet. International commissions (1980 and 1994–95) with participation of experts from the UK and the Netherlands (states that are used to dealing with dam erection) proved once again the ecological safety of the dam being erected. According to assessment of the project and field studies, the dam has not changed either the hydrodynamics or the ecological conditions in the eastern part of the gulf and inlet in any significant way (Klevanny, 2000).

The disastrous conditions of the unfinished dam is still a threat to St Petersburg. But it is not the dam that is the source of threat, it is the possible inundation of the city. Data show the following probability of water rise in the Neva mouth: up to 3.5 m once in 100 years, to 4.75 m once in 1,000 years, and 5.4 m once in 10,000 years (*Concepts* ... , 1997b). It is believed that a particularly severe flood is likely to occur this century. If the dam is not finished, major parts of the city will be inundated.

Aggravation of the ecological situation at the Neva rivermouth is probable if, at a time of restored and increased industrial production, the required efficiency of the wastewater treatment works is not reached. So, it follows that it is necessary to improve monitoring of water area conditions not only for the Neva inlet but for the whole system Ladoga–Neva–Gulf of Finland. It is necessary to do so while the dam is under construction, and in future.

Among technological ecological disasters, especially large-scale are those provoked by natural hazards (e.g. seismic shocks, floods). A catastrophe of this kind occurred in spring 1990 in the city of Ufa (Russian Federation). Drinking water in the southern part of the city was polluted with phenol, and led to mass-scale illness. The phenol was discharged from the Chimprom Chemical Plant into flood waters that suddenly raised the water level in the river.

Concern about the risk of a catastrophe at a nuclear power plant in Armenia, due to the high seismic potential of the region and, more importantly, the devastating consequences of the earthquake in 1988, led to it being shut down temporarily. Currently, it has temporarily recommenced. Similarly, construction of a number of nuclear power plants was put on hold in different areas of the former USSR.

Correlation of many technological catastrophes with natural phenomena calls for special safety measures when choosing future construction sites to reduce a minimum the influence of various dangerous natural phenomena (earth shocks, landslides, floods, etc.). Circumstances of this kind must be taken into consideration when forecasting technological disasters as must the frequency of dangerous phenomena by seasons and by years.

It should be emphasized once again that all technological disasters are primarily connected with natural factors. Only some of these factors are obvious (floods, landslides, and so on). In other cases, technological disasters are usually associated with human error (safety engineering, automatic equipment failure, and so on). However, it seems likely that even human errors are to some extent of natural origin, as evidenced by results of statistically processed data on accidents (Epov, 1994). These data were obtained for large territories that are environmentally rather inhomogeneous and located within different countries (Russia, France, the USA). They strongly suggest the cyclical occurrence of accidents and, therefore, that their causes are so far unknown.

4.3.2 Satellite observations of technological environmental disasters

4.3.2.1 Ecological disasters caused by accidents and fires

Observations from outer space are useful for showing the onset of the development of certain rapidly progressing technological catastrophes (e.g. assessment of major fires at industrial plants, and transport in cities and their environs; this is particularly true for the situation in Los Angeles in 1992).

Analysis of satellite data on the major fire in Los Angeles (at the end of April 1992) has shown the undeniable suitability of their application to monitor such situations (Dousset *et al.*, 1993). Thermal infrared images of Los Angeles under a wavelength of 3.7 μm obtained at night on 30 April 1992 and during the day on 2 August 1985 were compared. During the fire a vast thermal anomaly, of area 85 km^2, was detected in the first of the images in the southern part of the city. Its mean temperature reached 48.6°C, while the temperature in the rest of the city was 13°C. Fire focuses were identified by changes in brightness temperature in channels 10.8 and 12 μm (from *NOAA* satellite) and from a synthesized image obtained by *SPOT* satellite survey on 15 June 1985.

Of course, fires and explosions at industrial sites, including nuclear power plants, may not always be noted from satellites in the initial phase of their development because of delay (maybe several hours) in receiving satellite information. However, information of this kind, through delayed (e.g. about the explosion at the Chernobyl

Nuclear Power Plant or about the fire in Los Angeles) proved to be useful for timely assessment of the ecological situation.

Satellites can trace several specific features of oil pollution such as its range in the Gulf of Mexico caused by an accident at a large oil well – *Ixtok-1*. It happened as a result of an explosion in June 1979 near the Yucatán Peninsula not far from the Mexican shore. Oil was continuously discharged into the Gulf for almost 10 months (to 24 March 1980) till the spouting well was shut off.

Airplane observations have shown that surveys in the ultraviolet and blue part of spectrum are the most suitable for oil and oil spillage identification, and, to a smaller extent, those in the near-infrared wavelength range. Oil spots may also be discovered by temperature contrast with the background during thermal infrared spectral range surveying.

To determine the depth of spilled oil and the boundaries of its extent, special microwave radiometric equipment is used, especially in combination with sensors operating in the infrared range. Radar sounding of water areas makes it possible to watch the areas of spilled oil at any time of day and in any weather. The polarimetric method is also promising because it can detect the contrast between clean water and oil film.

The Gulf of Mexico, during and after the accident at the *Ixtok-1* well, became a testing ground for a long time of satellite and airplane oil pollution monitoring. Oil spills in the Gulf were discovered from five satellites operating at that time (i.e. *GOES*, *Nimbus-7*, *Tiros-N*, *Landsat-2*, and *Landsat-3*: Johnson and Munday, 1983). The visible and infrared ranges of sensors mounted on these satellites were sometimes restricted by the cloudiness that often develops over this area.

The low spatial resolution of images also interfered, to some extent, with reliable identification of oil pollution in the course of meteorological satellite survey (from all five satellites). Nevertheless, many oil slicks fixed as light-grey spots in the images taken by these satellites were confirmed by more detailed aerial surveying data.

It was discovered that, in the visible range (0.55–0.90 μm), oil pollution is seen distinctly. In thermal infrared images (10.5–11.5 μm), it is almost invisible, probably because of low spatial resolution (Cracknell *et al.*, 1983). High-resolution images obtained from *Landsat* are very useful. Oil slicks were clearly seen (both as light and dark spots). In pseudocoloured images (MSS), it is seen even better than in more detailed images (RBV). These images were used to track the migration of oil spills over the Gulf water area for several thousands of square kilometres. They proved that the oil pollution that appeared in August 1979 near the coast of Texas was not Texan in origin, but was transported as a result of oil migration from the *Ixtok-1* well. Aerial observations cannot obtain this information because of low coverage, as compared with satellite data. *Landsat* not only provides information on the direction of coastal currents but makes it more precise. This information has been used in oil migration forecast models.

Satellite observations are useful for control of accidental oil spillage on land. Pseudocoloured images from *SPOT* satellites' obtained by mixing images in different visible and near-infrared spectral ranges, detected dark areas of spilled oil in the tundra during the accident at the oil pipeline in the Komi Republic in Russia in 1994.

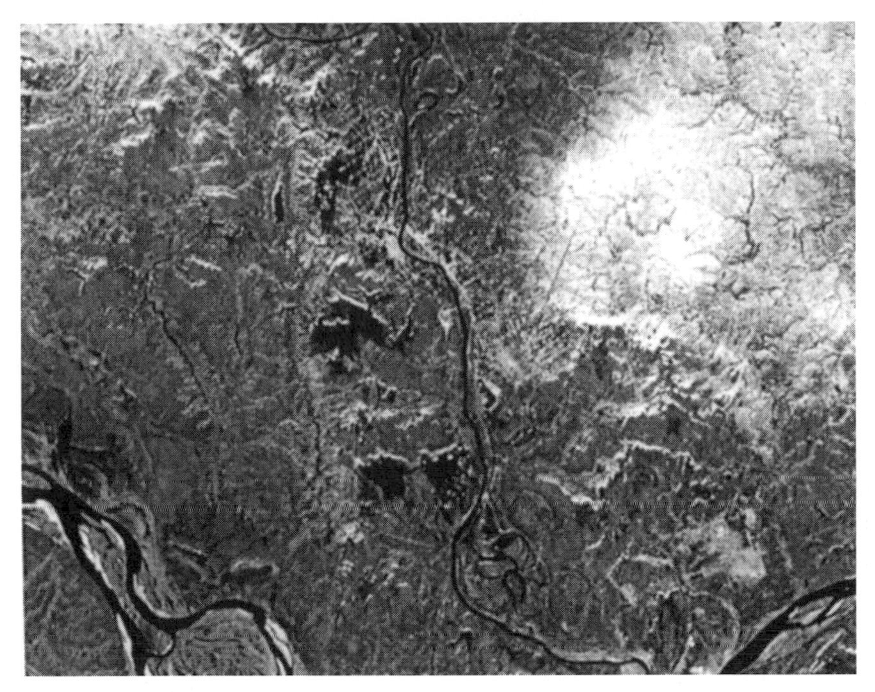

Figure 4.16. Satellite image of oil spillage in tundra in the north-west of Russia (Komi Republic) obtained on September 1994 from *SPOT* satellite. Oil spillage (black spots), as a result of an accident to the oil pipeline, is clearly seen in the background. The snowy landscape is drained by the Pechora, Kolva, and Usa Rivers (left, centre, and right, respectively, in the image) (Grigoryev and Kondratyev, 2001).

Areas of oil were clearly seen in the image of the tundra, partly covered by snowfall in the course of the survey. The high resolution of images (10 m) obtained from this satellite not only determines the area of oil spillage but its possible (albeit slow) propagation over the terrain and risk of entering rivers (Figure 4.16).

Satellite observations are used to assess the ecological consequences of nuclear accidents. The nuclear catastrophe at Chernobyl Nuclear Power Plant was detected, at the moment the nuclear reactor exploded (26 April 1986), by the *DMSP* satellite by its bright flash. This information proved to be extremely important because meteorological conditions forecast diffusion of radioactive emissions primarily over Europe (Ramade, 1987).

A very hot spot was discovered in the area of Chernobyl by infrared thermal survey by the *NOAA-9* satellite at the end of April 1986. Results from *Landsat-5* satellite data confirmed this. So, the possibility of monitoring power plant emergencies from satellites in the wavelength of 3.7 was proved.

Thermal infrared images of this territory obtained by the *Landsat-5* (TM) satellite also proved that, 3 days after the explosion (i.e. on 29 April 1986), the situation remained dangerous (Richter *et al.*, 1986). This was seen in the high temperature of the underlying surface of the power plant (1,000–1,300°K).

On the basis of satellite information, studies into radioactive pollution assessment for different natural media have been carried out. For example, images taken by the *Kosmos* satellite on 27 July 1989 in the Chernobyl area were used to assess the radioactive pollution of vegetation cover. Using analysis data of the images, obtained in three visible spectral ranges, it was shown that all plant communities decrease their capacity to absorb luminous fluxas pollution increases. In this way, the correlation between Cs-137 content and spectral characteristics of plants was revealed (Lyalko *et al.*, 2000).

Investigations in Byelorussia have shown that areas of radioactive pollution correlate with zones of a particular geological structure that, in turn, are seen in satellite images (Gridin, 1993). Radioactive pollution of terrain sharply increases in areas with shallow bedding of dense solid rocks. In these areas, according to the hypothesis of the experts from Byelorussia, fallout of nuclides is favoured by a growth in gravitational forces.

An increase in radioactive pollution of the terrain is recorded in places where fractures (faults) in dense mountain rocks are located and where gravitational force intensification occurs. Fractures are clearly seen in satellite images as rectilinear contours of landscape elements (ridges, river reaches, valleys, and so on). In the Pripyat Depression, not far from Chernobyl, the contrary is the case: the impact of gravitational forces on the earth surface is minimal and almost zero radioactive pollution of terrain is recorded.

Experts from Byelorussia (Gubin, 1993) also noted that the high-permeability zone of the Earth's crust, in particular, is in contact with the boundary of the Mikashevsk and Polesye oil bearing blocks (identified in satellite images) and correlate with radioactive precipitation distribution. In such places, the level of radioactive pollution with Cs-137 totals 5 Ku/km^2.

4.3.2.2 Slow ecological disasters caused by technological activity

One such catastrophe took place in Kara Bogaz Gol in the 1980s after construction of a dam (dike) separating the gulf from the Caspian Sea waters that feed it. The dam was built in 1980 to decrease the influx of sea water from the Caspian, which it was hoped would stem the catastrophic decline in its level.

Satellite observations made it possible to track how Kara Bogaz Gol had developed both before and after construction of the dam. As estimated by A.I. Dzen-Litovsky in conjunction with *Tiros-5* surveying data of the beginning of the 1960s, the water area of the gulf decreased to 10,000 km^2 (Terziyev *et al.*, 1981). It was not a casual phenomenon, since a strait connects Kara Bogaz Gol with the Caspian, and therefore a decline in the sea level does affect the gulf. It was responsible for the decrease in its water area. This process started at the beginning of the 1930s and intensified in succeeding years.

Traces of such a decrease in water area were seen in satellite photographs taken in the 1970s from the manned *Soyuz-9*. Even at that time the coast receded eastward, especially in the northern part of the gulf by several tens of kilometres. The water area was 1.5 times smaller than that shown on maps, and over the 10 years that preceded cosmonaut flight it further reduced in area by 7,000 km^2.

It should be noted that a decrease in water amount in the gulf leads to an increase in the salt content, the gulf is already salty, and, importantly, to alteration of the composition of salt fall-out in this huge natural chemical laboratory. Salts are extracted from the gulf by an enterprise located near the town of Bekdash. Changes in brine composition caused concern to both experts and economic executives.

In the dam project provision was made for stabilization of the Caspian level as a result of stopping the flow of water arriving at the gulf. It was expected that the gulf would drain a little. However, the 100-m long dam was built as a blind dam without any water conveyance openings for controlling water arrival. Contrary to the forecasts, the gulf began to dry up rapidly since it turned to a big salt lake. From analysis of satellite survey data, during the evaporation season of 1980 the water area decreased by approximately 3–5% (Fedin, 1982) (Figure 4.17a, b, c).

Before construction of the dam, 5–7 km^3 of water arrived at Kara Bogaz Gol annually. A considerable part of it evaporated quickly. The Gulf is shallow. In 1980 its maximum depth was 3.5 m. Soon after construction of the dam its depth did not exceed 1.5 m. The Gulf began to dry out rapidly. According to satellite images, in 1983 the surface area of water in the Gulf decreased from 20,000 km^2 to 6,000 km^2.

Soon after construction of the dam profound changes in salination took place in the gulf (growing density of surface brines and salt deposition processes). These changes jeopardized both the operation of a large salt processing works and the normal way of life in the town of Bekdash, with a population of about 10,000. Although, in 1984, water conveyance devices were installed in the dam, the Gulf, as evidenced by satellite observations, continued to dry out. By 1982 it had dried out almost completely.

More frequent salt storms and vortexes are the consequences of drying out of the gulf, as is the formation of a huge saltmarsh at its lowest lying point. Such a vortex, raised from the bottom of the former gulf, was discovered for the first time on 20 September 1964 by a *Nimbus-1* image. After construction of the blind dam salt storms have increased. Several salt storms, several hundred kilometres long, have been observed over the Caspian Sea, in particular during their propagation towards the city of Baku (Fedin, 1987).

Releases of sulphates from the bottom of the Gulf were especially dangerous for pastures surrounding the Kara Bogaz Gol depression. These toxic salts were responsible for degradation of plants, diseases, and loss of cattle.

Satellite observations of Kara Bogaz Gol in an area that is difficult to access and organize continuous macro-scale field studies were the main source of information about the ecological changes caused by building a dam without giving it adequate thought. Satellite survey data combined with land-based and aerial observations provided very real evidence of a large-scale ecological disaster here. The intensity of salt storms (similar to those that formed on the shores of the Aral Sea) in the dried out bottom of the Gulf may be responsible for it. In an attempt to restore the Gulf's ecosystem and to maintain full-scale Caspian water supply in this area, openings allowing water to pass were built in the dam.

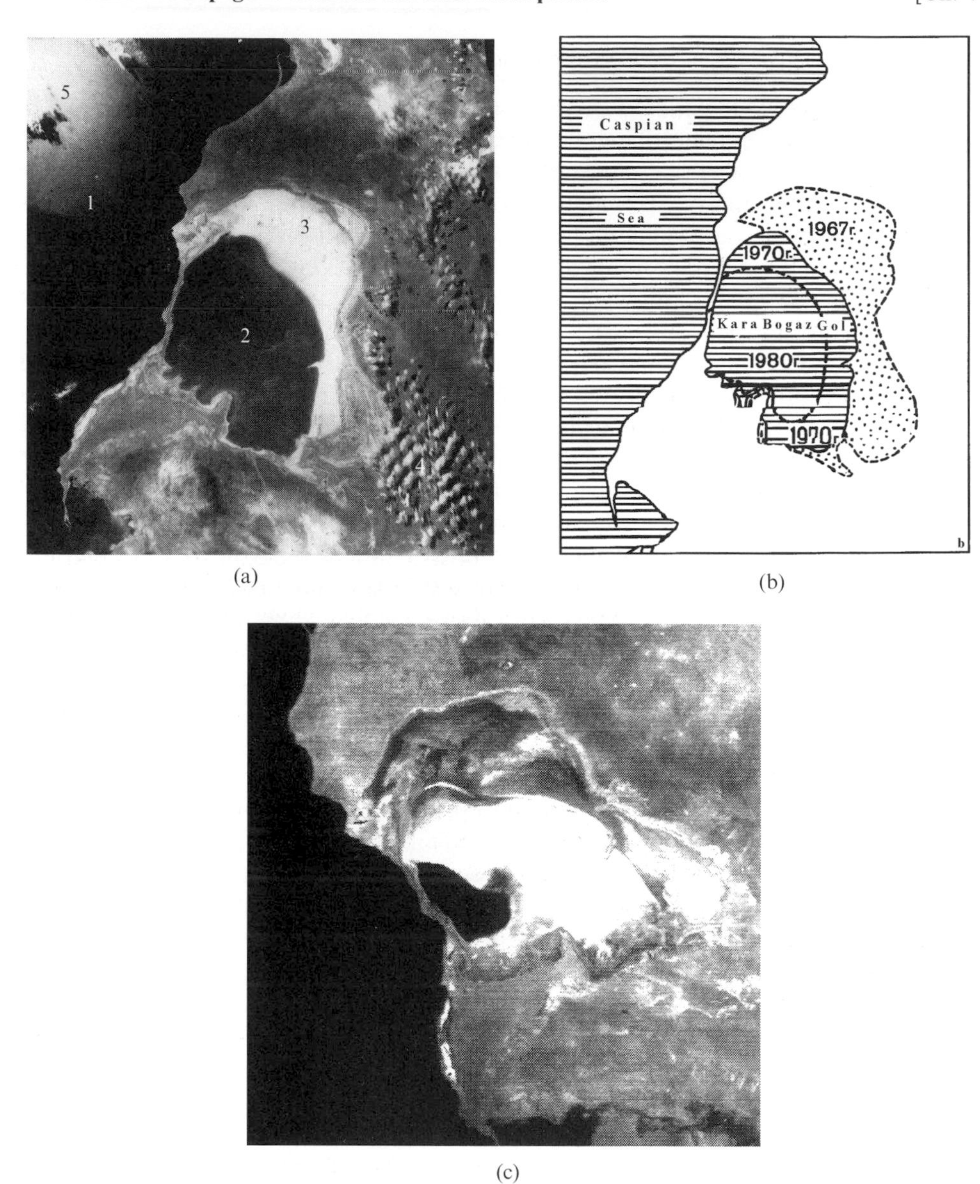

Figure 4.17. (a) Kara Bogaz Gol aridification as a result of anthropogenic activity, from comparison of satellite images and map. Picture taken from the manned *Soyuz-9* (summer 1970) (Grigoryev, 1991): 1 – Caspian Sea, 2 – Kara Bogaz Gol Gulf, 3 – dried out bottom of the Gulf (sheet of salt), 4 – clouds, 5 – patch of light on water surface. (b) Diagrammatic representation of the photograph. Water area contours are shown as of 1980 (from satellite survey data) and 1967 (Grigoryev, 1991). (c) Kara Bogaz Gol almost completely aridified after construction of a blind dam in the channel connecting it to the Caspian Sea. Photograph taken on 12 October 1987 in the spectral interval 0.7–1.1 μm from *Meteor-30* (Grigoryev and Kondratyev, 2001).

4.3.2.3 Ecological disasters caused by military operations

Satellite images enable us to assess the sources and transport of atmospheric pollution caused by military operations in Indochina and in the Persian Gulf in 1983 and 1991, respectively. As a result of the tactics of burning down jungles by US pilots during the war in Vietnam, forests and the smoke plumes that stretched from them for tens and even hundreds of kilometres were tracked from outer space.

A smoke cloud, as a result of fire in these jungles, could be seen in the images taken by *Landsat* on 2 January 1973 (Grigoryev, 1985). It stretched over the mouth of the Mekong River in the form of a band of inhomogeneous brightness and was more than 100-km long and about 20-km wide. Sometimes, smoke from the bombardment of settlements, industrial plants, and oil storage tanks added to smoke of this kind. In the images taken on 2 January 1973, we can see a black smoke plume from burning oil. This oil spilled out of defective tanks near Saigon after bombardment.

Even greater was the scale of smoke pollution during military operations in the Persian Gulf. Some areas of the military conflict between Iran and Iraq in 1983 were enveloped with smoke from burning oil storage tanks. A smoke plume of this kind was observed by Soviet cosmonauts when they were flying over this region on board *Salyut-7*.

The propagation of fires due to military operations was closely tracked from outer space (there were no other ways to do it) during military operations in 1990–1991 in the period of the Iraq–Kuwait war. At that time about 500 oil wells were set on fire by the Iraqis. Vast smoke clouds, as evidenced by meteorological satellite survey, were detected in January 1991 over these smoke sources (i.e. fires near Basra in Kuwait and in the oilfields). Sometimes, their area covered 800 km^2.

At the end of May 1991 smoke clouds were observed from outer space moving in various directions from the sites of this military conflict – northward (in southern areas of the former USSR), westward (in the Mediterranean), south-westward (in Ethiopia), and eastward (over Hindustan). They had reached distances of 1,000, 2,000, 2,000, and 3,000 km from the wells that were on fire, respectively. Their propagation over Hindustan resulted (as estimated) in cooling the monsoon system by 1°K (and in places up to 9°K) (Rao and Chandrasekhar, 1991).

Satellite observation has also proved to be useful for tracking the development of such ecological disasters as the pollution of seawater areas by oil, again as a consequence of military action. Observations were especially actively carried out over the Persian Gulf during military operations in the 1980s and 1990s. At that time vast oil spillages were a permanent feature in the Gulf. They originated from tankers, oil storehouses, oil refineries, and oil wells that had been bombed. During the Iraq–Kuwait war in 1990–1991 oil spillages occurred as a result of the deliberate discharge of huge amounts of oil from pipelines by the Iraqis. Iraq intended to impede US and Allied warships in the armed conflict with Iraq in the Gulf. Oil pollution was fatal for many sea dwellers, in particular, marine terrapins and dugongs. A further consequence of water pollution was its effect on desalination works.

During the military action between Iraq and Iran in the Persian Gulf, Iran initiated the so-called tanker war. High-speed Iranian craft attacked tankers that belonged not only to Iraq but practically to all other countries maintaining economic relations with Iraq and its allies as well. As a result of these attacks, many tankers were destroyed and their oil discharged into the sea.

A special survey of the Gulf was carried out by the *Shuttle* (from 24 March to 2 April 1991) at a height of 300 km. The colour images of the water areas observed allowed us to track the propagation of water polluted by oil (Ackleson *et al.*, 1992).

Surveys from *NOAA* satellites were also very useful. Oil pollution was tracked by the satellites both at day and at night. Oil films detected in thermal infrared images (AVHRR equipment) at night were warmer than films of freshwater, whereas they were cooler at day time. The propagation of oil film could therefore be measured by satellite from the dynamics of temperature contrast. The area of oil pollution determined in this way on 29 January 1991 covered 144 km^2 (Cross, 1992). Note that it was only possible to estimate oil pollution in the Gulf in the course of military operations by satellite data.

Satellite images have been used to track the migration of oil spills in the gulf. For example, one near the Kuwait coast stretched for about 70 km in January and by February had increased to 360 km. In the course of the same satellite survey oil spills of smaller sizes were also detected (Rao and Chandrasekhar, 1991).

Earth-clearing tactics used by the American army during the war in Indochina led to the destruction of large tracts of jungle. Clearing was done mechanically (by means of huge bulldozers) or by spraying forests with herbicides from airplanes. Tracts of dead or partly damaged forest (altogether about 2.5 million hectares of jungle were damaged in South Vietnam as a result of such military action) were detected by satellite images thanks to the contrast characteristics of healthy and affected forests (Grigoryev, 1985).

Satellite observations have been used to detect (from flashes) nuclear weapon tests, for example, near the lake of Lobnor (in China) on land and near the African shore in the ocean. This information is important to identify and track (together with other methods) radioactive dust clouds. One such cloud, as a result of the nuclear explosion near the lake of Lobnor in October 1980 traversed the ocean and hung over the USA and Canada.

4.3.2.4 Urban ecological disasters (man-made)

Satellite observation data are useful in assessing the ecological situation as a result of accidents, explosions, and fires in urban settings. In some cases, they can pinpoint the precise moment of an accident (e.g. the explosion at Chernobyl nuclear power plant). However, urban survey by most satellites is not a continuous process, but takes place every few hours. Because of this their effective operational use is restricted to estimating the aftermath of accidents.

It is considerably more effective to use space observation to monitor slowly developing ecological catastrophes or critical situations that are man-made in origin that occasionally crop up in cities (Figure 4.18). One such situation

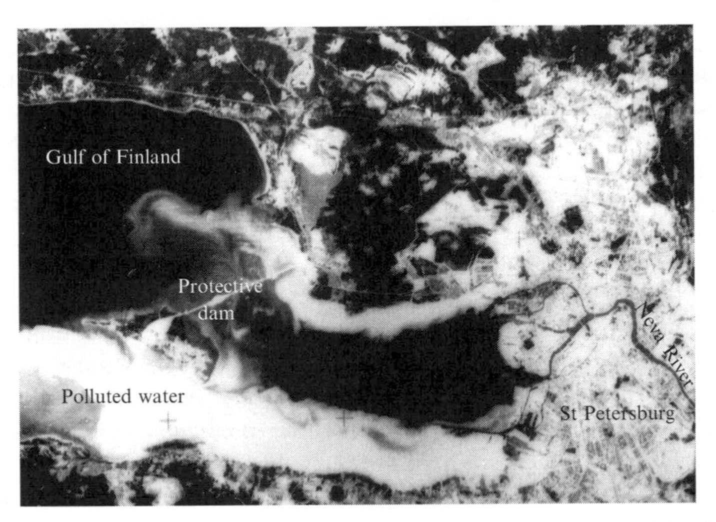

Figure 4.18. Pollution of water area along the coast of the Gulf of Finland caused by industrial sewage and water pollution in the course of sand extraction. Satellite photograph taken at a height of 300 km (Grigoryev and Kondratyev, 2001).

occurred in the 1980s in the eastern part of the Gulf of Finland near St Petersburg in rather unusual conditions: the construction of a protective dam and continuous discharge of untreated sewage. The latter was the result of failing to construct wastewater treatment facilities either on time or to a sufficiently high standard.

Satellite observations in the Neva inlet and eastern part of the Gulf of Finland adjacent to the dam (the critical area) gave us information about the pollution (Viktorov *et al.*, 1989; Kondratyev *et al.*, 1989). These photographs from space enabled us to assess the upwelling phenomena that cause vertical movements of water masses (Viktorov *et al.*, 1989). From the photographs, we see that upwelling waters in the Neva inlet are 3–10 km distant from the shore, and those in the Gulf are 30–40 km from the shore. The data were used to assess upwelling intensity from the difference in temperatures of the open part of the Gulf and the upwelling zone, the difference reached 6–8°K.

Multi-spectral satellite images can detect differences in water mass motion through indicators such as phytoplankton, suspended matter, and ice formations. Satellite observations showed the development of vortex-shaped structures in the Gulf of Finland, including one adjacent to the Baltic Sea and another near the dam (Viktorov *et al.*, 1989).

Vortex- and mushroom-shaped structures in satellite images are indicators of suspended mineral matter (Kondratyev *et al.*, 1989). Areas with cleaner water are clearly seen in the images. Satellite survey in the Neva inlet revealed comparatively small (0.6–3.2-km long), muddy, mushroom-shaped structures. Their formation is explained by the flow of river water slowing down due to viscosity and changes in flow rates near dams (Viktorov *et al.*, 1989).

Some satellite images clearly testified to the fact that the Neva inlet water area was more heavily polluted than the shallow eastern part of the Gulf of Finland, westward of the dam (Kondratyev *et al.*, 1989). Contradictory information about Neva inlet pollution, obtained by comparison of individual satellite images, can be explained. Satellite images reflect the instant situation, while thermodynamic features and the wind situation in the inlet and in the Gulf are characterized by greater temporal variability.

The unquestionable value of satellite observations of the water system of Ladoga–Neva–Gulf of Finland, primarily useful for tracking meso- and macro-scale features of the space–time variability of ecological conditions, is optimal when using a series of satellite images. In this case, satellite observations are a very significant constituent of the ongoing monitoring of the critical ecological situation in this region. Without it, assessment and management of the environment in light of the complex interaction between its natural (Ladoga–Neva–Gulf of Finland) and man-made (primarily the dam, wastewater treatment facilities) components would be impossible.

Satellite observations help us to detect the occurrence of smog over industrial areas which can sometimes be a serious health hazard for the population. It is only possible to judge the likelihood of a health hazard by means of land-based data. Satellite observations are useful for operational assessment of the development of any such ecological disaster, including determination of the scale of the phenomenon and its dynamics.

For example, smog over Los Angeles has many times given rise to critical situations for the health of its inhabitants. Satellite images showed that smog enveloped an area covering 3,000–13,000 km^2, and could be observed far beyond the boundaries of the city (*Manual* ... , 1983). Synchronous, differentiated survey of several intervals of the visible spectral range shows that the maximum brightness of smoke pollution falls in the range 0.4–0.5 μm. This is because atmospheric dissipation is at its maximum in this spectral interval.

Surveying in the near-infrared spectral range lets us see the spindrift-type clouds of the smoke, which have low transparency when surveyed using other spectral intervals. Moreover, at times of high smoke content, surveying in the near-infrared spectral range enables the topographic coupling of smoke with water bodies to be more clearly seen in such images (Grigoryev, 1991).

Large-scale regional smoke pollution caused by the intermingling of smoke from many cities and industrial areas was first seen on 5 August 1970 in TV images from *ESSA-8* over West Europe (Grigoryev, 1985; 1991). It occurred in moist maritime air at a low height (about 900 hPa) at a time of gentle, near-surface winds. The smoke cloud emerged at the western edge of an anticyclone under a layer of temperature inversion. It stretched a distance of more than 700 km and was not less than 200 km wide. The area exceeded 400,000 km^2 (Figure 4.19).

As the smoke cloud developed, visibility decreased sharply. Land-based, meteorological data that were obtained by weather stations located in the pollution diffusion area, showed visibility varying from 1.5 to 6 km. Prior to the advent of the smoke cloud, visibility in the north of Germany ranged from 25 to

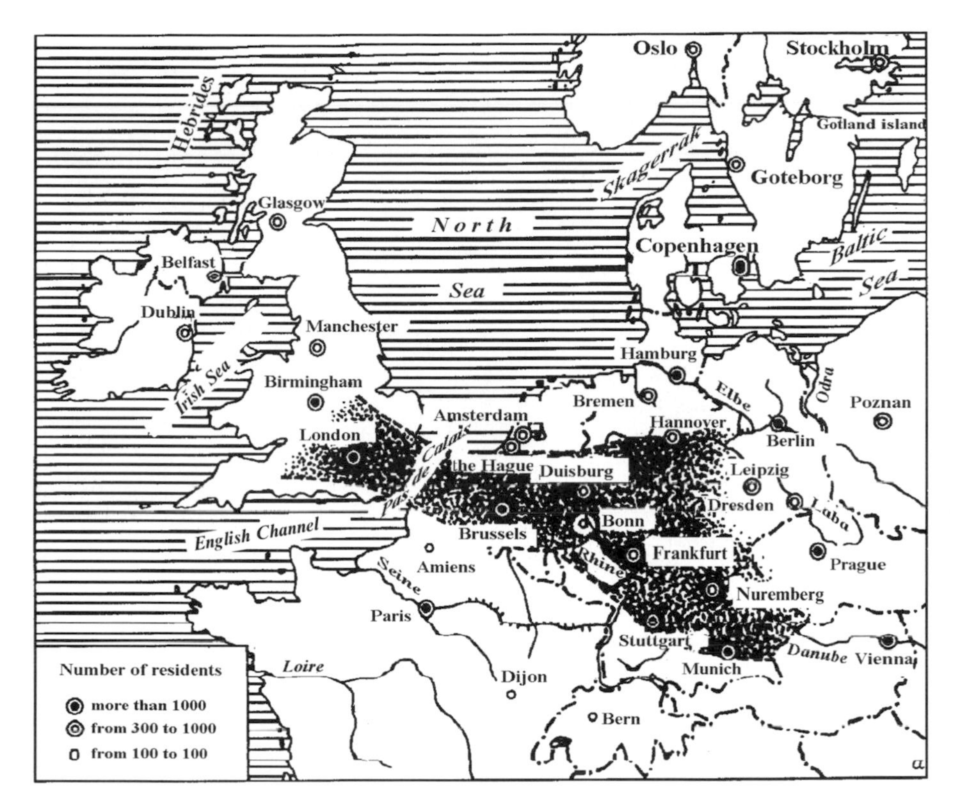

Figure 4.19. Large-scale distribution of smog over West Europe. Industrial areas of several countries were the source of this pollution. The diagram was produced using data obtained by analysis of imagery from *ESSA-8* on 5 August 1976 (Grigoryev, 1985).

40 km. All the meteorological stations detected this. Separate, land-based data are generally only linked by means of satellite observations.

Similar, unprecedented (as far as their scale is concerned) outbreaks of smog, caused by the simultaneous impact of many cities on the atmosphere, were observed from satellite survey data for the industrial region in the east of North America (Figure 4.20). Smog propagated at its greatest extent a distance of 1.5 million kilometres and lasted 13 days (Kondratyev *et al.*, 1983; Manual ... , 1983).

Changes in the brightness characteristics of smoke pollution images correlate with the features of their structure and composition. For example, in satellite images of the huge clouds of smog over the USA on 26 and 28 August 1976, areas of elevated brightness, first in the area of Chesapeake Bay and second in the area of Pennsylvania and New York states, were clearly seen. Land-based observations confirmed the areas of elevated brightness corresponded to zones of lower visibility, higher sulphates, and ozone concentration. In particular, in the second case, the area of elevated brightness in the images corresponded to the actual areas where sulphate concentration was at its highest, reaching 30–50 mg/m^3. Outside this zone, although

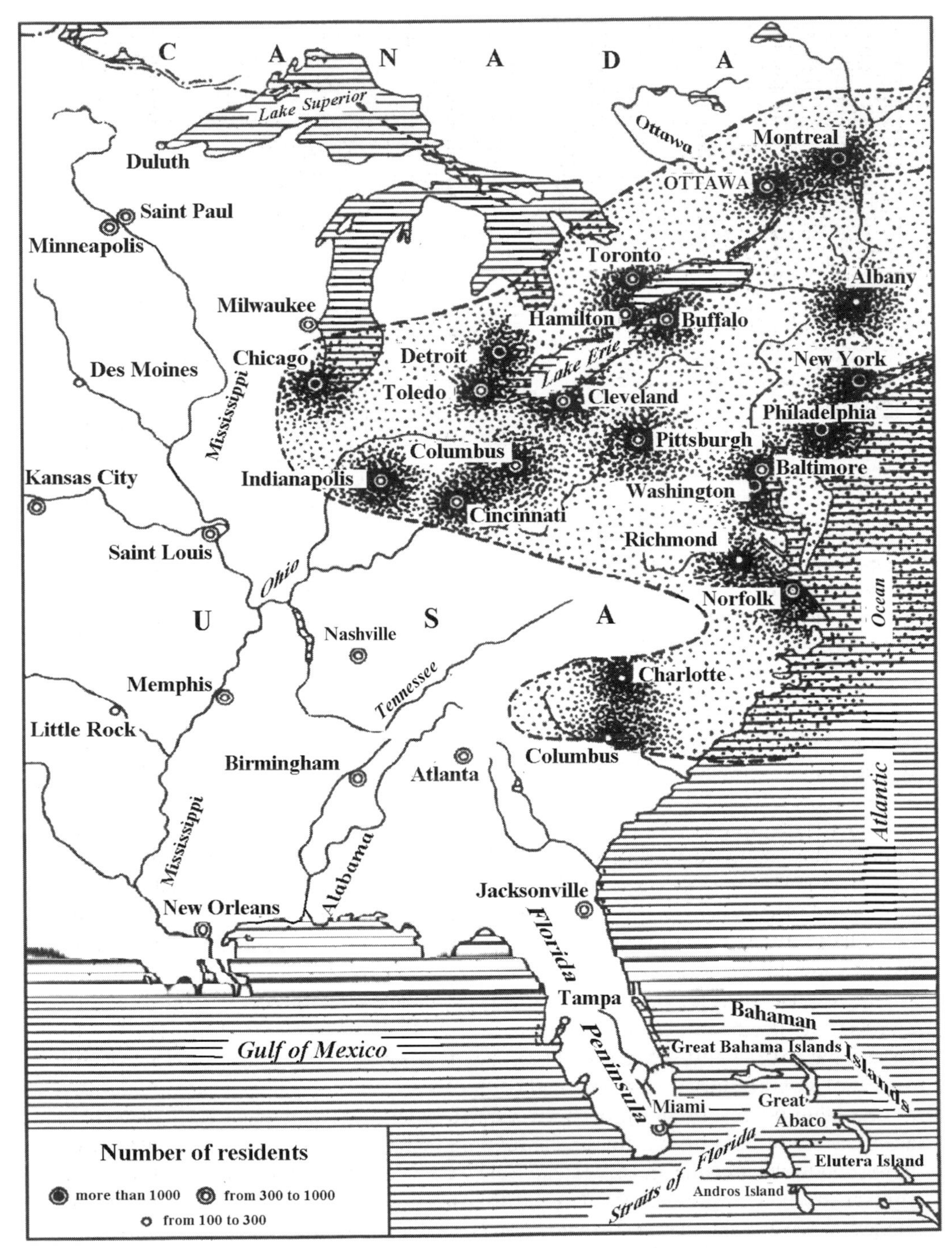

Figure 4.20. Satellite image decoding diagram. Propagation of smog, resembling clouds, is confirmed by land-based observation data (Grigoryev, 1982).

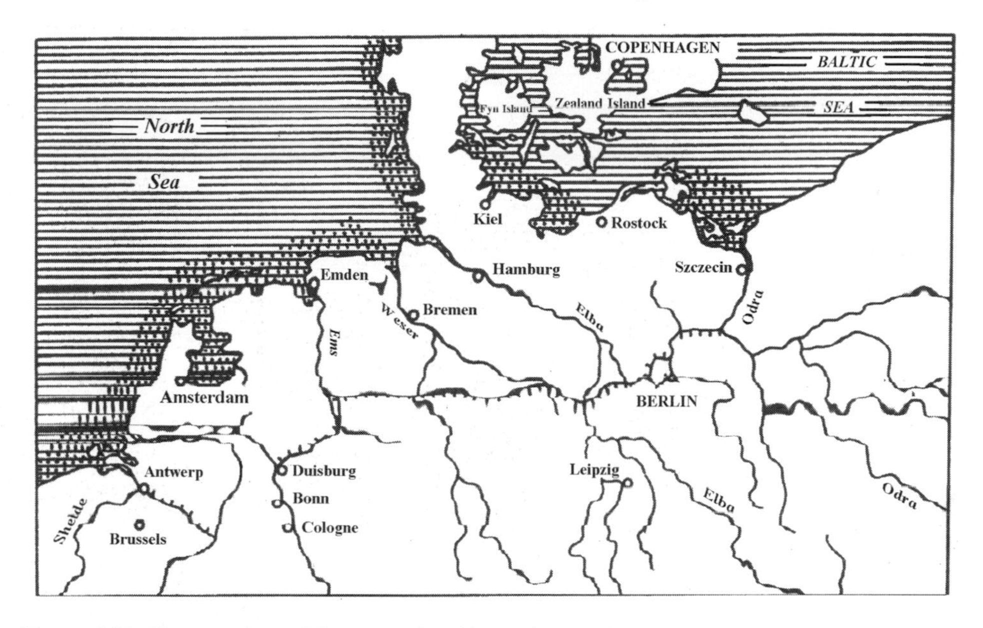

Figure 4.21. Propagation of the anomaly of heated water in coastal regions (near the shores of the North and Baltic Seas). The heat originates from sewage waters from cities located along the coast and on rivers. The diagram was prepared from analytical data of thermal infrared images obtained at the spectral interval 3.4–4.1 μm in June from *Nimbus-3* (Grigoryev, 1982).

sill high but still less, it made up in average 22.7 mcg/m^3, and beyond smog boundary it made up 5.7 mcg/m^3 (74,75,76)/Lyons *et al.*, 1978; Lyons, 1980; Lyons and Husar, 1976).

Such atmospheric pollution events are caused by industrial activity in cities under certain meteorological conditions. Strictly speaking, they are not catastrophes and certainly not on the scale of the infamous London smog in 1952 that affected and even killed many citizens. However, under certain circumstances the chance of calamities caused by pollution does arise, highlighting the importance of satellite observation data to track situations leading to such circumstances.

Cities also pollute water basins and large-scale pollution of this kind that can give rise to dangerous ecological situations that could lead to catastrophes are increasingly frequently observed. Once again, such pollution can be tracked by means of satellite survey data (Figure 4.21).

Satellite observations can also be used to study natural phenomena, including floods, volcanic eruptions, landslides, and so on. For example, satellite images have been used to detect fractures at the surface of the earth directly below which tectonic stresses are reaching criticality. One of the first maps of this kind, prepared from images taken by the manned *Apollo-7* spacecraft, shows areas in the Los Angeles locality that fall within the active seismic zone. Such maps enable us to evaluate those zones most likely to suffer earthquakes (Campbell, 1976) (Figure 4.22a, b).

(a)

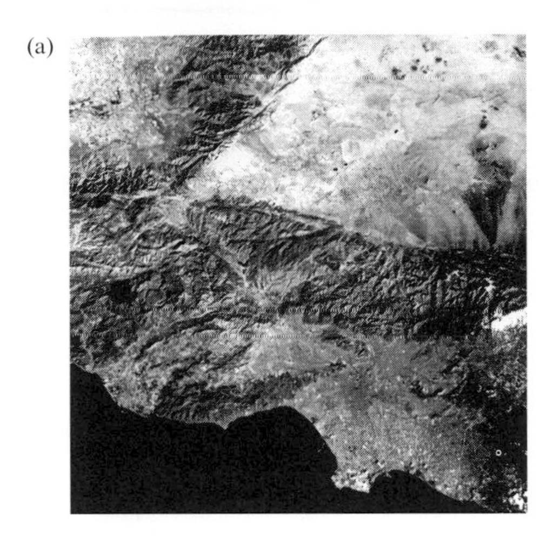

(b)

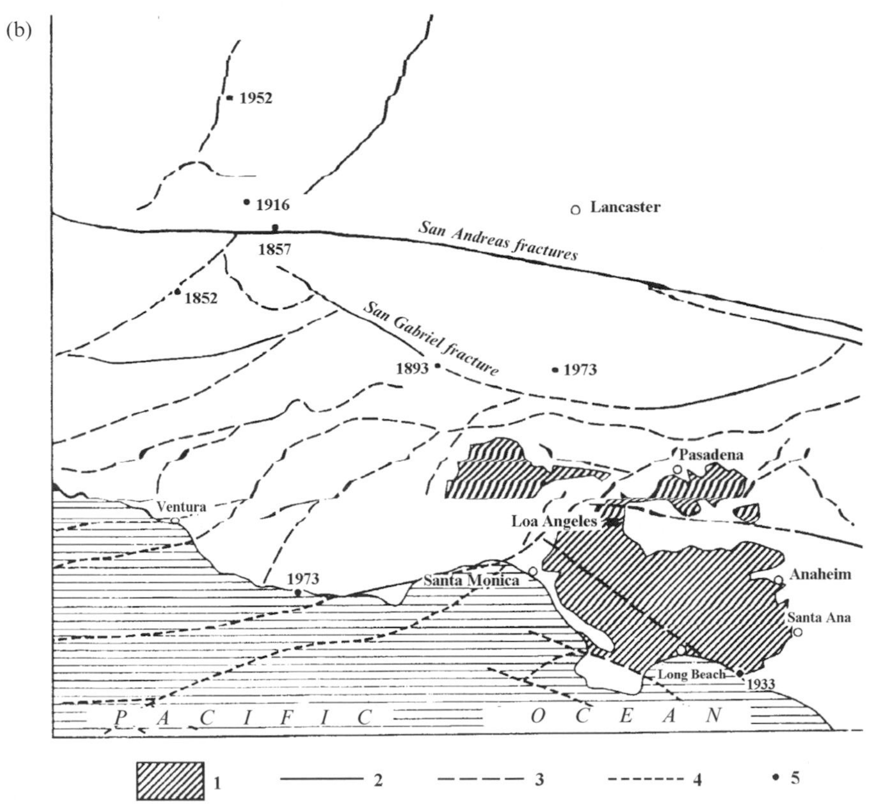

Figure 4.22. (a) *Landsat-1* image of urban areas around Los Angeles that are located in the tectonic seismic zone of the Cordilleras (Campbell, 1976). (b) Diagram of the urban territory and fracture system prepared from a satellite image. 1 – urban areas; 2–3 – fractures clearly see (2) and poorly seen (3) in the image; 4 – fractures at the seabed not seen in the image but detected from other data; 5 – earthquake epicentres (Campbell, 1976).

Satellite images are also used to choose sites where nuclear power plants may be built (i.e. sites that do not run a risk of natural calamities).

4.4 BIBLIOGRAPHY

Ackleson, S. G., Pitts, D., and Smith, R. S. (1992) Astronautical observation of the Persian (Arabian) Gulf during STS-45. *Geocarto Int.* **7**(4), 59–69.

Alayev, E. B. and Monina, Yu. I. (1990) Accident rate of tanker shipping and World Ocean pollution. *Proc. Acad. Sci. USSR. Ser. Geogr.* **6**, 39–46

Alekseyev, A. S. (1995) Modeling of anthropogenic impacts on forest ecosystems of urban territories. Author's abstract of his doctoral dissertation, St Petersburg State University, 32 pp.

Andrea, M. O. (1993) Global distribution of fires seen from space. *Earth Observed from Space* **74**(12),1129–44.

Barosh, B. J. (1980) Landsat data use in evaluating nuclear reactor sites. *Case Stud. Appl. Adv. Data Collect. Manag.* **4**, 383–90.

Bertrand, M. (1990) Incendies de forêt en zone Mediterranéenne une catastrophe peu naturelle. *Catastrophe et risques naturels*, 217–35.

Breido, M. D., Popik, A. G., Rakov, D. B., and Starostenko, D. A. (1995) Consequences of large forest fires recorded from satellite scanner patterns. *Studing the Earth from Space* **1**, 115–26.

Budyko, M. I., Golitsyn, G. S and Izrael, Yu. A. (1986) *Global Climatic Disasters* (166 pp.). Hydrometeoizdat, Leningrad.

Cahoon, D. R. Jr, Levin, J. S., Cofer, W. R. III, and Stocks, B. J. (1994) The extent of burning in African savannas. *Adv. Space Res.* **14**(11), 447–54.

Campbell, K. K. (1976) Active faults in the Los Angeles–Ventura area of South California. *ERTS-1. A New Window on Our Planet.* (US Geol. Survey Prof. Paper 929). Washington, DC.

Carson, R. L. (1991) *Silent Spring* (231 pp.). Cambridge University Press.

Chase, P. E. (1973) ERTS-1 investigation of ecological effects of strip-mining in Eastern Ohio. *Symposium on Significant Results Obtained by ERTS-1* (Vol. 1, pp. 561–8). University of Michigan, Washington, DC.

Chung, Y. S. and Le, H. V.(1984) Detection of forest-fire smoke plumes by satellite imagery. *Atmospheric Environm.* **18**(10), 2143–51.

Chuvieco, E. and Martin, P.(1994) Global fire mapping and fire danger estimation using AVHRR imagery. *Photogramm. Eng. Remote Sensing.* **60**(5), 563–70.

Concepts of St Petersburg Development (1997a) (Vol. 2, 439 pp.). Nauka, St Petersburg.

Concepts of St Petersburg Development (1997b) (Vol. 3, 392 pp.). Nauka, St Petersburg.

Cracknell, A. P., Muirhead, K., Calison, R. D., and Estas, J. E. (1983) Satellite remote sensing environmental monitoring and the offshore oil gas industries. *Proc. EAR Sel/ESA Symp. Remote Sensing Application, Environmental Studies (Brussels, 26–28 April 1983)* (pp. 163–71), European Space Agency, Paris.

Cross, A. M. (1992) Monitoring marine oil pollution using AVHRR data: Observation of the coast of Kuwait and Saudi Arabia during 1991. *Int. J. Remote Sensing* **13**(4), 781–6.

Dousset, B., Flament, P. and Bernstein, R. (1992) Los Angeles fires seen from space. *Earth Observations from Space* **74**(3), 37–8.

Dworak, T. Z. (1992). Okreslenie wplywu przemysly Morawskiej Ostravy na rejon Ciezynski na podstanwie interpretacji obrazyw satelitarnych. *Zcsz. Nayk AGN in S. Staszica. Sozol Isozotechn.* **32**, 19–23.

Ecocide in the Policy of American Imperialism (1985) (translated from English). Progress, Moscow.

Ecological Situation in St Petersburg and Leningrad Area in 1997 (1998) (290 pp.). Committee on Environmental Protection in St Petersburg and Leningrad Area, St Petersburg.

Epov, A. B. (1994) *Accidents, Catastrophes and Natural Calamities in Russia.* Finizdat, Moscow.

Esenov, R. (1988) Lesson. *Nature and Man* **6**, 39.

Fedin, V. P. (1982) Investigation results and the current problem of Kara Bogaz Gol. *Problem of Development of Deserts* **3**, 13–18.

Furyaev, V. V. (1987) Application of aerospace pictures to the study and evaluation of forest fire consequences. *Studying Forests by Means of Aerospace Methods* (pp. 85–98.). Nauka, Novosibirsk.

Furyaev, V. V. (1993) Application of satellite scanner patterns for revealing large burnt out forests and damage due to forest fires. *Studying the Earth from Space* **4**, 83–93.

Furyaev, V. V., Kireyev, D. M. and Zlobina, L. P. (1997) *Revealing, Mapping and Forecasting Disturbances of Taiga Forests of Siberia Due to Fires* (Review information, 24 pp.). VNII Forest Resources, Moscow.

Ginzburg, A. S. (1988) *Planet Earth in the After-nuclear Epoch* (102 pp.). Progress, Moscow.

Graux, E. (1990) La secheresse de 1989 en Haute-Garonne et ses conséquences sur les sociétés paysannes. *Catastrophes et risques naturels*, 205–14.

Gridin, V. I. (1993) Chernobyl from outer space. *Green World* **18**, 6

Grigoryev, Al. A. (1982) *Cities and the Environment* (Satellite studies, 120 pp.). Mysl, Moscow.

Grigoryev, Al. A. (1985) *Anthropogenic Impacts on the Environment from Satellite Observations* (240 pp.). Nauka, Leningrad.

Grigoryev, Al. A. (1991) *Ecological lessons taught by historical past and present time* (252 pp.). Nauka, Leningrad.

Grigoryev, Al. A. and Kondratyev, K. Ya. (1994) Present-day state of global environment and natural resources (1991–1992) (UNEP report review). *Izv. of the Russian Geogr. Society* **126**(3), 12–26.

Grigoryev, Al. A. and Kondratyev, K. Ya. (1997) Satellite monitoring of natural and anthropogenic disasters. *Studying the Earth from Space* **3**, 68–78.

Grigoryev, Al. A. and Kondratyev, K. Ya (2001) *Environmental Disasters* (687 pp.). Scientific Centre, Russian Academy of Sciences, St Petersburg (in Russian).

Grigoryev, Al. A. and Lipatov, V. B. (1977) Remote sensing of forest fires/smoke from space. *Remote Sensing of Earth Resources* **5**, 369–83.

Grigoryev, Al. A. and Lipatov, V. B. (1978) *Smoke Pollution of the Atmosphere Observed from Outer Ppace* (36 pp.). Hydrometeoizdat, Leningrad.

Gritsman, Yu. (1993) To reveal and proclaim crying contradictions. *Green World* **18**, 5.

Gubin, V. N. (1993) Aerolandscape indication of geodynamic morphogenetic processes (Byelorussia taken as an example). Author's abstract of his doctoral geography dissertation, Minsk University, 48 pp.

Gupta, V. and Rich, D. (1996) Locating the detonation point of China's first nuclear explosive tests on 16 October 1964. *Int. Journal Remote Sensing* **17**(10), 1969–74.

Haydn, R., Volk, P. (1987) Erkennung von Umweltproblemen in Luft- und Satellitenbild. *Geogr. Rundschau* **6**(June), 316–27.

Heffert, R. and Lulla, K. P.(1990) Mapping continental-scale biomass burning and smoke polls over Amazon basin as observed from the Space Shattle. *Photogr. Eng. Remote Sensing* **56**(10), 1367–73.

Housson, A. (1994) Example d'utilisation de la télédétection en France: Le cartographe de feux de forêt. *Geomètre* **127**(3), 52–4, 62.

Jonson, R. W. and Munday, J. C. (1983) *The Marine Environment* (Manual of remote sensing, Vol. 2, pp. 1371–496. American Society of Photogrammetry, Washington, DC

Keating, M. (1994) *The Earth Summit Agenda for Change* (A plain language version of Agenda 21 and the other Rio agreements, 70 pp.). Centre for our Common Future, Geneva.

Kennedy, P. J., Belward, A. S., and Gregoire, J.-M. (1994) An improved approach to fire monitoring in West Africa using AVHHR data. *Int. J. Remote Sensing* **15**(11), 2235–55.

Kharuk, V. I., Vitenberg, K., Tsibulsky, G. M., and Serov, V. S. (1995) Analysis of technogenic degradation of near-tundra forests from satellite surveying data. *Studying the Earth from Space* **4**, 91–97.

Klevanny, K. A. (2000) Floods in St Petersburg in conditions of uncompleted flood protection works. *Protection of St Petersburg from Floods: Construction, Economy, Ecology*. Hydromerteoizdat, St Petersburg.

Kondratyev, K. Ya., Dyachenko, L. N., Binenko, V. I., and Chernenko, A. P. (1972) Detection of small fires and mapping of large forest fires by infrared imagery. *Proc. 8th Int. Symp. of Remote Sensing of the Environment* (Vol. 2, pp. 1297–310). University of Michigan, Ann Arbor, MI.

Kondratyev, K. Ya., Grigoryev, Al. A., Pokrovsky, O. M., and Shalina, E. V. (1983) *Satellite Remote Sensing of Atmospheric Aerosol* (216 pp.). Hydrometeoizdat, Leningrad.

Kondratyev, K. Ya., Filatov, N. N. and Zaitsev, L. V. (1989) Assessment of water exchange and zones of water body pollution from satellite surveying data. *Papers of Academy Science* **304**(4), 829–32.

Kosin, V. V. (1982) Anthropogenic landscape mapping in oilfield regions on the basis of satellite surveying data. *First All-Union Conf. Biosphere and Climate from Satellite Research Data* (pp. 240–3) Elem, Baku.

Kravtsova, V. I. (1981) decoding of the consequences of technogenic impact on nature from satellite images (Khibini taken as an example). *Aero-satellite Methods in Geographic Studies of Siberia and the Far East* (pp. 57–64). Institute of Geography of Siberia and Far East, Irkutsk, Russia

Kupriyanov, V. V. (1977) *Hydrological Aspects of Urbanization* (144 pp.). Hydrometeoizdat, Leningrad.

Lesogorov, V. B. (2000) Construction of flood protection works for St Petersburg. *Materials of the 4th Congress of St Petersburg Union of Societies of Scientists and Engineers, 24 August to 16 September 1999, St Petersburg*. Nauka, St Petersburg.

Lyalko, V. I., Khodorovsky, A. Ya., and Sakhatsky, A. I. (1994) Application of multi-zonal satellite images for monitoring of plant pollution with radionuclides (in connection with the accident at Chernobyl nuclear power plant. *Chernobyl '94: The 4th Int. Scientific and Technical Conf.* (Results of 8-year works on liquidation of the consequences of the accident at Chernobyl nuclear power plant). Theses, Green Cape University, 61 pp.

Lyons, W. A. (1980) Evidence of transport of hazy air masses from satellite imagery. *Aerosols: Anthropogenic and Natural Source and Transport* (pp. 418–33). McGraw-Hill, New York.

Lyons, W. A. and Husar, R. B. (1976) SMS/Goes visible images detect a synchronal scale air pollution episode. *Monthly Weather Rev.* **104**(12), 1623–41.

Lyons, W. A., Dooley, J. C., and Withby, K. T. (1978) Satellite detection of long-range pollution transport and sulfate aerosol hazes. *Atmos. Environ.* **12**, 625–40.

Mallingreau, J. P., Stephens, G., and Fellows, L. (1985) Remote sensing of forest fires: Kalimantan and North Borneo in 1982–1983. *Ambio* **14**(6), 314–21.

Manual of Remote Sensing (1983) (Vol. 2). American Society of Photogrammetry, VA.

Melentyev, V. V. and Pozdnyakov, D. V. (1996) Aerospace methods and means of environmental monitoring. *Ecodynamics and Ecological Monitoring of St Petersburg Region in the Context of Global Changes* (pp. 388–431). Nauka, St Petersburg.

Militarism. Figures and Facts (1985) (288 pp.). Nauka, Moscow.

Mironova, A. V. and Mikhailov, Z. N. (1982) Application of aersace data for the study of anthropogenic landscapes. Presentations and the program of *The All-Union Scientific Symp. 'Methods for Investigations of Anthropogenic Landscapes, Voronezh 1982* (p. 19)Geographical Society of the USSR, Leningrad.

O'Riordan, T. (ed.) (1995) *Environmental Science for Environmental Management* (pp. 221–42). Longman Group, New York.

Peart, M. K. (1988) Forest fires in China. *Geograph. J.* **73**(Pt 2, No. 319), 152–4.

Prokacheva, V. G., Chmutova, N. P., Abakumenko, V. P., and Usachev, V. F. (1988) *Zones of Snow Cover Pollution around Cities in the Terriory of the USSR* (125 pp.). Hydrometeoizdat, Leningrad.

Protection of the Environment, Nature Use and Provision of Environmental Safety in St Petersburg in 1997 (1988) (Vol. 1, 309 pp.; Vol. 2, 520 pp.). Chemistry and Scientific Research Institute, St Petersburg State University.

Ramade, F. (1987) *Les Catastrophes Ecologiques* (318 pp.). McGraw-Hill, Paris.

Randerson, D. (1976) Overview of regional scale numerals models. *Bull. Amer. Meteorol. Soc.* **57**(7), 797–804.

Rao, U. R. (1996) *Space Technology for Sustainable Development* (564 pp.). McGraw-Hill, New Delhi.

Rao, U. R., Chandrasekhar M. G. *et al.* (1991) Environmental impacts of the Persian Gulf oil spill and oil fire smoke. *Current Science* **60**(8), 486–92.

Richter R., Lehmann F., Haydn R., and Volk P.(1986) Analysis of *Landsat* TM images of Chernobyl. *Int. J. Remote Sensing* **7**(12), 1259–67.

Roberts, D., Karpuz, M., Tommervik, H. *et al.* (1994) Spectral detection of mining-induced environmental stress in the Nikel-Zapolyarny region of western Kola Peninsula, NW Russia and adjacent areas of Norway. *Proc. 10th Them. Conf. of Geological Remote Sensing, San Antonio, TX, 19–22 May 1994* (Vol. 2, pp. 109–20). Environmental Research Institute of Michigan, Ann Arbor, MI.

Rumyantsev, V. A. and Rodionov, V. Z. (eds) (1989) *Resolution of the Committee of Experts on Ecological Conditions of Neva Inlet and Eastern Part of Gulf of Finland* (Preprint, 136 pp.). USSR Academy of Science, Leningrad.

Sedio, R. A. (1998) Forests. Conflict signals. In: E. Balley (ed.) *The State of the Planet* (pp. 177–209). Free Press, New York

Semakin, N. K. and Nazirov, M. (1977) *Application of Satellite Photo-information in Teaching Physical Geography* (144 pp.). Prosvescheniye, Moscow.

Setzer, A. V. and Pereira, M. C. (1991) Amazonian biomass burnings in 1987 and estimate of their tropospheric emission. *Ambio* **20**(1), 19–21.

Shcherbak, K. Yu. (1991) Chernobyl as a new phenomenon in the history of civilization. *Oecumena* **3**, 24–30.

Smith, K. (1997) *Environmental Hazards. Assessing Risk and Reducing Disasters* (389 pp.). Longman, London.

Tersiyev, F. S. and Goptarev, N. P. (1981) Kara Bogaz Gol Bay and the Caspian Sea problem. *Meteorology and Hydrology* **2**, 19–27.

UNEP (1991) *Environmental Data Report* (3rd edn, 408 pp.). Basil Blackwell, London.

Usanov, B. P. (1989) *Dialogue of City and the Sea* (32 pp.). Znaniye, Leningrad.

Valendik, E. N. (1990) *Big Forest Fire Control* (193 pp.). Nauka, Novosibirsk.

Victorov, S. V., Bychkova, I. A., Lobanov, V. Yu., and Suchova, O. A. (1989) The problem of satellite ecological monitoring of the Gulf of Finland. *The Study of the Sea from Space* (Express information No. 4-89, pp. 13–26). Scientific Council on Space Research for National Economy, USSR Academy of Science, Leningrad.

Westing, A. H. (1976) *Ecological Consequences of the Second Indochina War* (199 pp.). Almquist and Wiksell [a SIPRI book], Stockholm.

Wobber, F. J., Russel, O. R. and Deely, D. J. (1975) Multispectral aerial and orbital techniques for management of coal mines areas. *Photogrammetria* **31**(4), 40–56.

World Resources 1996–1997. A Guide to the Global Environment (1996) (365 pp.). Oxford University Press.

Xu Guanhua, Sun Siheng and Zhang Jugui (1991) *The Afforestation in China and Its Monitoring by Remote Sensing Data* (Preprint, pp. 1–6). National Laboratory of Resources and Environmental System and Chinese Academy of Forestry, Beijing.

Zherebtsov, G. A., Kokurov, V. D., Koshelev, V. V., and Minko, N. P. (1995) Application of AVHRR *NOAA* satellite data for forest fires detection. *Studying the Earth from Space* **5**, 74–7.

Znamensky, V. A. (1999) *Ecological Safety of Water System of St Petersburg* (119 pp.). Nauka, St Petersburg.

5

Global and regional 'slow' disasters

In accordance with its very definition, a disaster is perceived as a sudden phenomenon (e.g. an earthquake, a volcanic eruption, etc.). However, disasters are actually the result of a long-time development of the relevant processes. The 20th century is characteristic of the appearance of human-induced potential disasters of 'slow' (or 'delayed') action. An example is that of so-called 'global warming' – a surface air temperature rise as a result of increased concentrations of greenhouse gases. Another example is total ozone depletion, due to the emission into the atmosphere of ozone-destroying substances (above all, freons). We discuss these problems in more detail in the following pages.

5.1 STRATOSPHERIC AND TROPOSPHERIC OZONE CHANGES

5.1.1 Introduction

It is generally acknowledged that the ozone layer in the stratosphere protects life on Earth from the Sun's destructive ultravlolet (UV) radiation. This layer, which consists of about 90% ozone, is located in the stratosphere some 10–50 km above the Earth's surface. Due to continuing human-induced destruction of this ozone layer, UV radiation on the earth's surface has increased and, in the long run, may become ruinous for both humans and the biosphere. P. J. Crutzen (1970) was among the first to reveal that nitrogen oxides emitted to the atmosphere as a result of using agricultural fertilizers, together with nitrogen oxides emitted by subsonic and supersonic aircraft, can seriously damage the ozone layer. Later, M. Molina and F. Rowland (1974) revealed the ozone destructive role of chlorofluorocarbons (CFCs); their conclusions were later confirmed by subsequent studies and Rowland and Molina (as well as Crutzen) were to win public recognition, culminating in being awarded the Nobel Prize in 1995.

Tropospheric ozone, which constitutes about 10% of the total ozone content (TOZ), is an important greenhouse gas. Changes in stratospheric and tropospheric ozone have affected climate and ecosystems in different ways. An important feature of the processes involving tropospheric ozone is the formation of photochemical smog; the specific impacts of stratospheric and tropospheric ozone form a kind of basis to qualify stratospheric ozone as 'good' and tropospheric ozone as 'bad' (Low, 1998).

Obviously, ozone is an important atmospheric component which deserves special attention. In recognition, certain international documents have been adopted: these include the Montreal Protocol (1987), concerning substances destructive to the ozone layer; the Vienna Convention (1985), on the protection of the ozone layer; and subsequent amendments and appendices to the Protocol (Koehler and Hajost, 1990; Sarma, 1998). In accordance with the Helsinki Declaration (2 May 1989) on the protection of the ozone layer, individual states were supposed to reduce the production and consumption of CFCs (listed in the Montreal Protocol) as soon as possible, but not later than the year 2000.

The London Amendment to the Montreal Protocol (Kondratyev and Varotsos, 2000a) contains recommendations about the total elimination of emissions to the atmosphere of ozone destructive substances, taking account of assessments made in 1989. The discovery of the hole in the ozone layer above the Antarctic in 1985, and conclusions drawn on the basis of assessments made in 1991, have been conducive to the adoption of the Copenhagen Amendment (1992). This Amendment broadened the list of ozone depleting substances (ODS) by adding hydrofluorocarbons (HCFCs) and methylbromide. The decade after the adoption of the Montreal Protocol (1987), was marked by acceptance of other amendments and adjustments (London 1990, Copenhagen 1992, Vienna 1995, Montreal 1997, Beijing 2000) stipulating the elimination of emissions of the following categories of substances (at the moment, the Protocol envisages the control of 95 chemical compounds):

- Chlorofluorocarbons: CFCs, halogens, hydrochlorocarbons; halogenated bromofluorocarbons (HBFCs), other halogenated chlorofluorocarbons (HCFCs) tetrachlorocarbon, 1,1,1-trichloroethane; methylchloroform.
- Hydrochlorofluorocarbons and hydrobromofluorocarbons: HCFCs, HBFCs and methylbromide.
- Control measures for chemical substances envisaged (for industrially developed states): elimination of halogens by 1994; CFCs, tetrachlorocarbon, methylchloroform and HBFCs by 1996; methylbromide by 2005; HCFCs by 2030.
- For the developing world, relevant dates were determined as: elimination of HBFCs by 1996; halogens, CFCs, tetrachlorocarbon by 2010; methylchloroform and methylbromide by 2015; HCFCs by 2040.

It should be mentioned here that the situation concerning methylbromide (CH_3Br bromomethane, MeBr) is not quite clear even today. Yates *et al.* (1998) have demonstrated, for instance, that a recently developed new technology is capable of almost completely eliminating MeBr emissions from soils, thus making unnecessary the problem of its prohibition.

In 1994, UNEP published a report which clearly revealed that the increase of concentrations of CFC-11, CFC-12, halon 1301 and halon 1211 was slowing down (UNEP, 1994). And, according to WMO/UNEP (1998), the total ODS amount in the lower atmosphere reached its peak in 1994, after which it began a gradual decrease; the peak of total concentration of combined stratospheric bromide and chlorine in 1992 was only about 60 ppt/year (1.6%), whereas in 1980 it constituted 110 ppt/year (2.9). The amount of tetrachlorocarbon is also diminishing. On the other hand, as might be anticipated, the amount of some substitutes of HCFC is increasing.

As a result of the above, the increase of halogen carbons in the stratosphere began slowing down, this being the first example of positive human impact on the atmosphere. According to UARS satellite observations (Kondratyev, 1999a; Kondratyev and Varotsos, 2000a), the global averaged (for the 70°N–70°S belt) trends of chlorine and fluorine at the height of 55 km decreased during the 1991–7 period from 124 to 116 ppt/year and from 92 to 82 ppt/year.

Bojkov (1998) emphasized that the long period of increasing total chlorine abundance in the troposphere – primarily from CFC, carbon tetrachloride and methyl chloroform – has ended: chlorine from major CFC is still increasing slightly and most halons continue to increase substantially (e.g. halon-1211, 5% per year up to the present time). The abundance of HCFC and HFC is increasing as a result of their CFC substitutes. Stratospheric halogens loading lags behind tropospheric loading by up to six years: the maximum concentrations of chlorine and bromine are expected to peak at –3.7 ppb around the year 2000 and keep close to this over the next one or two decades. Even with a complete and immediate global elimination of all ODS, emissions would result in stratospheric halogen loading returning to pre-1980 values by the year 2033. Some major scientific findings in WMO/UNEP (1999) can be summarized as follows:

- The rate of decline in stratospheric ozone at mid-latitudes has slowed; hence, the projections of ozone loss made in the 1994 UNEP Assessment are larger than has actually occurred. In the northern polar latitudes, in six out of the last nine boreal winter/spring seasons, ozone has declined in some months by 25–30% below the 1960s average. Bojkov *et al.* (1998) have shown that during the 1990s over the 35–50° mid-latitude belts the ozone deficiency in the Southern Hemisphere was less than over the Northern Hemisphere by about 39%. The observed TOZ losses from 1979 to 1997 were about 5.4% and 2.8% for the northern mid-latitudes in winter/spring and in summer/autumn respectively; and 5% in the southern mid-latitudes all the year round. There are no statistical TOZ trends in the equatorial region (20°S, 20°N).
- The springtime Antarctic ozone hole continues unabated. Although the extent of ozone depletion was strongest in 1998, in recent years it has remained generally similar to that in the early 1990s with monthly TOZ values in September and October continuing to be 40–55% below the pre-ozone hole values (with up to 70% decrease for periods of about one week).
- The late winter/spring ozone values in the Arctic were unusually low in six out of nine years (before 1998). These six years, were characterized by unusually cold

and protracted stratospheric winters with the minimum Arctic circumpolar vortex (CPV) temperatures near the threshold for large chlorine activation. Ozone has declined during some spring months by 25–35% below the pre-1976 average. Elevated stratospheric halogen abundance over the next decade or so would imply the Arctic's continuing vulnerability to further large ozone losses.

- Over northern mid-latitudes the downward TOZ trend is largest between 40 km and 15 km (>7% per decade) and is smallest near 30 km (2% per decade). The bulk of TOZ decline was between the tropopause and 25 km.
- Stratospheric ozone losses may have caused part of the observed cooling of the lower stratosphere in the polar and upper mid-latitudes and global average negative radiative forcing of the climate system. Much of the observed downward trend in lower stratospheric temperatures (about 0.6°C per decade from 1979 to 1994) is attributed to the ozone loss in the stratosphere. The stratospheric ozone losses since 1980 may offset about 30% of the positive greenhouse gas effect over the same time period.
- The understanding of the reaction between increasing surface UVB radiation and decreasing column ozone has been further strengthened by ground-based observations and newly developed satellite methods show promise for establishing global tends in UV radiation.
- Based on past emissions of ODS and the projection of the maximum allowances under the Montreal Protocol into the future, the maximum ozone depletion is estimated to happen within the current decade or the next two decades, but its identification and the evidence for the recovery of the ozone layer are likely to remain unknown for some time. It was pointed out in WMO/UNEP (1999) that 'a full recovery of the Earth's protective ozone shield could occur by the middle of the next century, but it would require that the Protocol is fully implemented'.
- The increase of ozone in the troposphere since pre-industrial times is estimated to have contributed 10–20% of the warming due to the increase in long-live greenhouse gases during the same period.

Sarma (1998) emphasized the role of the Montreal Protocol, as well as subsequent Amendments and Adjustments, in avoiding negative environmental consequences such as at least 50% ozone depletion at mid-latitudes in the Northern Hemisphere and 70% depletion at mid-latitudes in the Southern Hemisphere. That is, about 10 times larger than in 1998 and at least a doubling of surface UVB radiation at mid-latitudes in the Northern Hemisphere and a quadrupling at mid-latitudes in the Southern Hemisphere compared with an unperturbed atmosphere. Sarma's figures compare with increases of about 5% and 8% respectively in the Northern and Southern Hemispheres since 1980.

The implications of ozone depletion could be horrendous with nearly 19 million additional cases of non-melanoma skin cancer by the year 2060 – and some 3 million more cases by the year 2030. The number of eye cataracts could increase by about 130 million cases by the year 2060 – about 50% of this increase being experienced in developing countries. There is, too, a number of unquantifiable effect, such as damage to the immune system, adverse impact on animals, lower productivity of crops, damage to aquatic ecosystems and so on. It must be pointed out, however, that a number of scientists (Ellsaesser, 1992, 1994; Maduro and Schauerhammer, 1992) have warned against over-emphasizing the dangers, of ozone depletion.

The global scale of the ozone problem has become an essential part of international study programmes. These include the World Climate Research Programme (WCRP) and its SPARC (Stratospheric Processes and Their Role in Climate) Project (see Chanin and Geller, 1998); the International Geosphere-Biosphere Programme (IGBP) which includes the IGA (International Global Atmospheric Chemistry) Core Project; and many other ozone-related projects (see Kondratyev, 1998a).

Marked progress has been achieved in the development of satellite remote sensing aimed at obtaining information on global-scale spatio-temporal ozone and ODS variability and trends (Fortuin and Kelder, 1999; Herman *et al.*, 1999). A number of comprehensive observational programmes have been carried out in the high latitudes of the Northern Hemisphere and in the Antarctic to study the stratosphere and ozone dynamics. As a result, not only has the well-known 'ozone hole' in the Antarctic been discovered, but also ozone 'mini-holes' and substantial depletion of ozone over vast territories of Siberia (Geophysical, 1986; Tuck *et al.*, 1989; Pyle and Harris, 1991; Harris and Hudson, 1998). Additional reliable assessments of the atmospheric effects of aircraft emissions on both stratospheric and tropospheric ozone have been obtained (see Kondratyev and Varotsos, 2000b). Important progress has been made in the use of interactive modelling of general atmospheric circulation and ozone dynamics (Hólm *et al.*, 1999; Takigawa *et al.*, 1999). A new analysis of ozonesonde data for the troposphere and the lower stratosphere has been carried out (Logan, 1999a, 1999b) as well as further studies about the nature of the Antarctic ozone hole (MacKenzie *et al.*, 1999; Rex *et al.*, 1998). New studies have been conducted to assess the influence of ozone depletion on surface ultraviolet radiation (UVR) and the subsequent impact on human health and the biosphere (Zerefos *et al.*, 1998; Ziemke *et al.*, 1998b).

Problems associated with tropospheric ozone change have recently attracted much more attention than before, especially in the field of numerical modelling (observations still remain inadequate). Important progress has been made in interactive simulation of troposphere dynamics and ozone change (Høv, 1997; Høv *et al.*, 1999; Isaksen, 1998). The impact of both stratospheric and tropospheric ozone on climate has been studied (Ozone depletion, 1989; Kondratyev and Varotsos, 2000b). Special attention has been paid to heterogeneous chemical reactions on the surfaces of aerosol particles in the stratosphere (especially in the case of polar stratospheric clouds) and troposphere (Kondratyev, 1998b, 1999b; Kulmälä and Wagner, 1996; Seinfeld and Pandis, 1998).

As has been pointed out in WMO/UNEP (1999):

- Enhancing anthropogenic emissions of ODS (nitrogen oxides, carbon monoxide and hydrocarbons) lead to large-scale production of ozone, which, through long-range transport, influences the ozone concentration in large regions of the troposphere in both hemispheres. Such changes are characterized by strong spatio-temporal variability.

- The increase of ozone in the troposphere since pre-industrial times is estimated to have augmented the average radiative forcing (RF) by $0.35 \pm 0.15\,W/m^2$, which is thought to have contributed 10–20% of the warming owing to the increase of long-lived GHG during the same period.

Undoubtedly, marked progress has been achieved in studying atmospheric ozone and anticipating the consequences of its changes; however, quite a number of related problems remain unsolved thus far. Many aspects of ozone research have been discussed; two of especial interest are, firstly, the UNEP (1998) paper 'Scientific Assessment of Ozone Depletion: 1994', and, secondly, one of a series of NATO ASI monographs 'Atmospheric Ozone Dynamics' (see Varotsos, 1997) and 'Atmospheric Ozone' (Bojkov and Visconti, 1998).

5.1.2 Stratospheric ozone

5.1.2.1 Observations and interpretations of observation data

Although TOZ observations started about 150 years ago, with the discovery of the existence of the ozone layer in the stratosphere, the global archive of TOZ data has been accumulated mainly during the past few decades by means of satellite observations. Various remote sensing techniques have been used for this purpose utilizing satellite measurements of backscattered diffuse solar UV radiation, attenuation of UV radiation by the atmosphere under conditions of 'occultation geometry', and thermal emission in the 9.6 μm ozone absorption band (Kondratyev, 1989; UNEP, 1994; WMO/UNEP, 1999; EC, 2001).

The most important results from observations are detection of the mean global TOZ decrease and the appearance of the 'ozone hole' in the Antarctic, as well as a short-term but sharp TOZ decrease in the Northern Hemisphere high latitudes.

Over the past few years, substantial data on ozone depletion in the Arctic have been obtained. For example, Hansen and Chipperfield studied ozone depletion at the edge of the Arctic stratospheric polar vortex in the winter of 1996/97, when the Arctic stratospheric polar vortex was extremely long lived; the ozone lidar at the Arctic Lidar Observatory for Middle Atmosphere Research (ALOMAR) operated from mid-December 1996 until mid-May 1997, when the vortex was centered at the pole, and its edge was almost permanently located over Northern Scandinavia. Ozone depletion of up to 40% was observed at the 475 and 550 K levels. Maximum depletion occurred around 5 May at levels up to 550 K and around 20 April at 675 K. Model analysis performed by Hansen and Chipperfield (1999) showed that while much of the early spring ozone depletion was due to halogen chemistry, associated with chlorine activation on polar stratospheric clouds, the ongoing depletion in late April and early May was due to 'summertime' $NO_x(NO + NO_2$ chemistry. The unusual persistence of the vortex, with the isolation of high-latitude air masses until early May, allowed this depletion to occur.

Eichmann *et al.* (1999) discovered, through analysis of ERS-2 satellite data for the Arctic spring periods of 1997 and 1998, substantial stratospheric ozone depletion. Extensive regions of low ozone total column were observed, and it has been shown that the major decrease is dominating in the lower and middle stratosphere inside the polar vortex. In the spring of 1998 an ozone mini-hole was observed (ozone profiles under mini-hole conditions being derived for the first time).

In order to explore the mechanisms of Arctic ozone depletion and relevant chemical-radiative interaction, MacKenzie and Harwood (2000) have undertaken numerical simulation of three Northern Hemisphere winter/spring seasons with contrasting thermal and dynamic characteristics using a mechanistic model of the middle atmosphere with an interactive chemistry scheme. The calculations indicate that cooling of the lower stratosphere induced by the enhanced ozone loss ranges from virtually zero in the warmest winter to $\sim$5 K in the coldest one. An analysis of the role of polar stratospheric clouds has shown that chemical-radiative interaction via polar stratospheric clouds (PSC) processes has had only a small impact on ozone distribution in recent Arctic winters, but it has the potential to become more important should an unprecedented cold winter occur while atmospheric halogen concentrations remain artificially high.

Guirlet *et al.* (2000) have used a three-dimensional stratospheric model to simulate Arctic ozone depletion in three winters with quite different dynamic conditions, from 1995/96 to 1997/98. Despite the very different meteorological conditions for these three winters, by mid-February the model-calculated mean vortex was similar in each year, at around 20% at the 480 K level. But, by late March, relevant total ozone losses were 53 Dobson Units (DU) in 1998, 71 DU (1996), and 80 DU (1997).

Rinsland *et al.* (1999) discussed observations made inside the November 1994 Antarctic stratospheric vortex and inside the April 1993 remnant Arctic stratospheric vortex using the Atmospheric Trace Molecule Spectroscopy (ATMOS) Fourier transform spectrometer. By comparing vortex and extravortex observations of NO_y (total reactive nitrogen) obtained at the same N_2O volume mixing ratios, the Arctic vortex denitrification of 5 ± 2 ppbv at 470 K ($\sim$18 km) has been inferred (these results are robust for a wide range of winter conditions).

Since stratospheric ozone is an intensive absorber of solar radiation, and tropospheric ozone is a greenhouse gas (Kondratyev, 1998; Low, 1998; Wang and Isaksen, 1995), there is no doubt that changes in the content of ozone in the atmosphere must affect the climate.

The reports published by UNEP (1994) and WMO/UNEP (1999), along with other publications, categorize the following as the most important aspects of the problem:

- From 1979, the rate of mean annual TOZ decline in mid-latitudes of both hemispheres was within 2.6–3.7% over a period of 10 years. Analysing ozone decline in the northern polar and mid-latitudes during winter and spring, Bojkov *et al.* (1998) concluded that the ozone mass deficiency (defined from the pre-1976 base average and area extent with negative deviations greater than $\sim 2\sigma$ and

Table 5.1. Total ozone trends (% per decade since January 1979) calculated as averages of individual station trends in the indicated regions and groups of months (ground-based observations). All values are negative and statistically significant at 2σ.

Region	DJF	MAM	JJA	SON	Year
Arctic	7.9	7.7	2.5	3.6	5.7
35–60°N	4.1	5.7	2.9	1.6	3.7
35–36°S	2.9	2.2	3.4	2.0	2.6
Antarctic	6.3	2.4	6.5	20.0	8.9

$\sim 3\sigma$) integrated for the first 105 days of each year, has increased dramatically from ~2,800 Mt in the winter/spring periods of 1993 and 1995. Bojkov (1999) po-inted out that total ozone levels in the latitudes 60°N, 60°S were at their lowest in 1993 because of the large increase in stratospheric aerosol caused by the eruption of Mount Pinatubo in 1991 (Kondratyev and Galindo, 1997; Kondratyev, 1999a). Since 1992–93, TOZ values over this part of the globe have been variable around a fairly constant level. Table 5.1 illustrates the observed TOZ trends (Bojkov, 1999).

- The amplitude of the annual cycle of ozone at mid- to high latitudes has decreased by 15% in the last two decades, because larger declines of maximum ozone values have occurred during winter and spring.
- Ozone depletion increases with latitude, particularly in the Southern Hemisphere. Little or no downward trends are observed in the tropics (20°N, 20°S). Analysis of global TOZ data through early 1998 showed substantial decreases in ozone in all seasons at mid-latitudes (35–60°) of both hemispheres. There were downward trends in the Northern Hemisphere of about 6% per decade between 1979 and 1998, observed in winter and spring, and about 1.5–3% per decade were observed in summer and fall. In the Southern Hemisphere the difference between seasons was smaller and the mid-latitude trends averaged within 2.0–3.4% per decade. Satellite and ozonesonde data show that much of the downward trend in ozone levels occurs basically below 25 km.

According to WMO/UNEP (1999), TOZ decreased significantly at mid-latitudes (25–60°) between 1979 and 1991 with estimated linear downward trends of 4.0%, 1.8% and 3.8% per decade, respectively, for northem mid-latitudes in winter/spring, northem mid-latitudes in summer/fall and southern mid-latitudes year round. However, since 1991 the linear trend observed during the 1980s has not continued, but rather TOZ was almost constant at all mid-latitudes in both hemispheres after recovery from the 1991 Mount Pinatubo eruption. Present-day understanding of how changes in halogen and aerosol loading affect ozone explains why linear extrapolation of the pre-1991 ozone trend to the present is not suitable.

- A complex three-dimensional spatial structure of ozone distribution was recently discovered taking the form of lamination (Varotsos, 1997) and formation of low-ozone air pockets during the northern winter in the middle stratosphere outside the polar vortex.

- Nair *et al.* (1998) pointed out that rapid ozone loss localized in pockets is due to the isolation of air at high latitudes (and high solar zenith angles). Thus, low-ozone levels are due to a decrease in the odd oxygen production rates and not an increase in the loss rate by reaction with halogen species, as in the classical ozone hole. For an explanation of low-ozone pockets see Morris *et al.* (1998): they developed a simulation model that takes account of photochemical processes and stratospheric dynamics.
- Russel and Lastvicka (1998) concluded, on the basis of ozonesonde data, that the number of thin ozone layers per one ozone vertical profile decreased from 1970 to 1993 from about 1.9 to 1.0 (with depletion of the ozone content in such layers of about 50%).
- The lowest TOZ values were obtained in 1992–93 at the 'ozone hole' in the Antarctic and the lowest TOZ values in the whole observational period were obtained over the densely populated regions of the Northem Hemisphere. The conclusion that anthropogenic chlorine and bromine compounds, coupled with surface chemistry on natural polar stratospheric particles, are the cause of polar ozone depletion, has been further strengthened. The links to halogen chemistry have been established with regard to ozone losses detected in the Arctic winter stratosphere.
- According to WMO/UNEP (1999), the large ozone losses during the spring in the Southem Hemisphere continued unabated with approximately the same magnitude and areal extent in the early 1990s. In Antarctica, the total monthly ozone in September and October has continued to be 40–55% below the pre-ozone hole values of approximately 320 m-atm cm (DU) with up to a 70% decrease for the period of a week or so. This depletion occurs primarily over the 12–20 km altitude range, most of the ozone in this layer disappearing during early October.
- Observations made after the eruption of Mount Pinatubo in 1991 resulted in the discovery that such a major volcanic eruption has a substantial impact on stratospheric ozone (Kondratyev and Varotsos, 2000a; Timmreck *et al.*, 1999).
- Impacts of subsonic and supersonic aircraft on upper tropospheric and lower stratospheric ozone have been confirmed (Vay *et al.*, 1998; Weisenstein *et al.*, 1998; Penner *et al.*, 1999; Kondratyev and Varotsos, 2000b).

Earlier assumptions about the substantial role of methyl bromide as an ozone destroying substance were now proved, the basic sources being crop fumigation, biomass burning and car exhausts. The most recent data, published by WMO/ UNEP (1999) indicate, however, that the contribution of methyl bromide has proved to be lower than that estimated earlier (UNEP, 1994), although appreciable uncertainties remain as the science of atmosphere methyl briomide is complex and not yet well understood.

- Planned countermeasures to control emissions of ozone depleting compounds will not greatly change the maximum level of TOZ decline forecast to be reached during the next 10 years.
- According to model simulations, much of the observed downward trend in lower stratospheric temperatures (about 0.6° per decade over 1979–94) is attributed to ozone loss in the lower stratosphere (WMO/UNEP, 1999). The decrease in stratospheric ozone since 1980 may have offset about 30% of greenhouse forcing, which favoured global climate warming. However, the increase in tropospheric ozone in the same period has intensified the atmospheric greenhouse effect in the Northern Hemisphere approximately by ~20%. The problem of ozone trends with opposite signs in the stratosphere and in the troposphere deserves special attention with respect to TOZ variation impact on UV radiation.
- From some published data (Michaels *et al.*, 1994), the TOZ level decline in 1992–93 was not followed by any increase in UVB radiation. The necessity of a thorough analysis of data on the trends of TOZ, UVB and related biological consequences was expressed by Ellsaesser (1992, 1994). However, Feister and Grewe (1995) observed higher UV radiation from low ozone values at northern mid-latitudes in 1992 and 1993 and pointed out that this could lead to adverse biological effects. Numerous further studies have confirmed this conclusion.
- WMO/UNEP (1999) concluded that the understanding of the relationship between increasing surface UVB radiation and decreasing TOZ has been further strengthened by ground-based observations and that newly developed satellite methods show promise for establishing global trends in UV radiation. The satellite estimates for 1979–92 indicate that the largest UV increases occur during spring at high latitudes in both hemispheres.

Logan (1999b) presented an analysis of ozonesonde data with the aim of using them to evaluate models of transport and chemistry. These data show that between 10–30% of the ozone column is located between 100 hPa and the tropopause at mid- and high-latitudes and this region drives much of the seasonal TOZ variation. The sonde data quantify the build-up of ozone in the lowermost stratosphere of the Northern Hemisphere in winter and its loss in late spring and summer. The amount of ozone between the tropopause and 100 hPa in the Northern Hemisphere decreases from 175 to 75 Tg from March to September, with the maximum rate of decrease in May to June, about half of the decrease being caused by the increase in the height of the tropopause. There is a lag of about one month, when ozone starts accumulating in the lowermost stratosphere from September near 40°N to October or November near 80°N, and a similar lag when ozone starts decreasing from February or March near 40°N to March or April near 80°N.

Despite caution against overemphasis of the ozone impact on the environment and ecosystems, this problem undoubtedly deserves very serious attention. Prather *et al.* (1996) posed a number of important questions in this context:

(1) What would have happened if CFC use had followed a free-market growth into all sectors and countries before the observation of stratospheric ozone depletion?

(2) What if the ozone hole had been discovered in 1985 (as indeed it was) without the prior suspicion about chlorine from CFCs?
(3) When should we expect to see the recovery of the ozone layer as provisions of the Montreal Protocol begin to show their effect on the global atmosphere?
(4) How much stratospheric chlorine and associated ozone depletion could the global atmosphere have been committed to in a worst-case scenario?

Answering these questions, Prather *et al.* (1996) emphasized that:

(1) Whether we had ignored the scientific evidence or not, ozone depletion would be dramatically worse than that we are experiencing today.
(2) If CFC had followed free-market growth until 2002, the Antarctic ozone hole would be a permanent fixture throughout the twenty-first century, instead of disappearing by 2050 as predicted in the Copenhagen 1992 scenario.
(3) Ozone depletion is expected to reverse and recover measurably in the first decades of the current century, unambiguous detection of the recovery will take a decade or more.

When discussing ozone problems, it is appropriate to quote the Policy Statement on Atmospheric Ozone accepted by the American Meteorological Society Council, 28 January 1996 (see Hales, 1996):

(1) The AMS recognizes ozone as an atmospheric constituent that has a number of important beneficial as well as detrimental effects on the atmosphere, on the surface ecosystems and on humankind. These include effects on both ultraviolet and longwave radiation, resulting effects on atmospheric wind systems and direct impacts on plants and animals. Moreover, ozone is recognized as the dominant progenitor of much of the trace-gas chemistry that occurs in both the troposphere and the stratosphere. The AMS also notes that extensive couplings exist among these chemical, radiative and dynamical components of ozone's behavior and add substantial uncertainty to many of the currently available assessments of ozone's impact.
(2) Despite many uncertainties, ample evidence exists to substantiate that atmospheric ozone has been affected in important and even critical ways by human activity. In the case of the stratosphere, the existence of the Antarctic ozone hole is unquestionable and our evidence that it results from human-produced halocarbons is overwhelming. Ozone depletion in the mid-latitude stratosphere is less dramatic; however, the general agreement of the large numbers of available column and profile measurements make it reasonable to speculate that such depletion is indeed occurring and that it is largely human-induced. Anthropogenic activities also significantly influence tropospheric ozone. Humankind is directly responsible for the excessive ozone levels that often occur near the surface in and down-wind from populated areas. These effects are also felt throughout significant regions of the free troposphere. The effects in both the stratosphere and troposphere are sufficiently profound to mandate substantial concern, on both a local and a global basis.

of chemically non-active gases is represented by dichlorofluoromethane (CCl_2F_2, F-12), trichlorofluoromethane (CCl_3F, F-11), carbon tetrachloride (CCl_4) and fully halogenated ethanes ($C_2Cl_3F_3$, F-113; $C_2Cl_2F_4$, F-114; C_2ClF_5, F-115). It may be supposed that about 80–85% of chlorine in the Antarctic atmosphere is human-induced. The most essential natural gases are methylchloride and chloroform.

During a period of 20-year-long observations, the maximum annual mean concentration of chlorine in the Antarctic atmosphere reached approximately 3 ppb in 1993, after which it began gradually decreasing. The concentration is defined as a sum of concentrations (mixing ratios) of individual gases multiplied by the number of chlorine atoms in each molecule of the gas. The variation in time of the concentrations of chlorine-containing gases in the Antarctic is similar to that observed throughout the world, but with a one to two year shift because the gas sources are mainly located at the middle latitudes of the Northern Hemisphere. Those compounds used presently and serving as substitutes for CFC, can be placed in the category of chemically active substances.

Considerable differences in the circulation and air exchange between the atmospheres of the northern and southern polar zones explain the absence of an 'ozone hole' in the Arctic. The Arctic circumpolar stratosphere is not as isolated from the lower latitudes as is that of the Antarctic stratosphere; therefore, air masses from the south are warmer and richer in ozone and reach the Arctic stratosphere throughout winter.

Unlike the Antarctic winter stratosphere there is not enough time for an isolated zone of cold air to be formed near the North Pole, where, with the participation of polar stratospheric clouds, the chlorine radicals, carried there by freons, destroy the ozone layer with such vigour that an 'ozone-hole' is formed. The Arctic Ocean, although covered with ice, is much warmer and it heats the atmosphere mainly through water openings and polynyas. Thus, for several months – the time necessary for substantial photochemical destruction of the ozone layer – the Arctic Ocean prevents the formation of an isolated air mass in the stratosphere near the Pole.

However, during the colder half-year in the Arctic, there also exist regions where TOZ is lower or higher than its average level. Over north-eastern Siberia and Chukotka, as well as over the adjacent seas of the Arctic Ocean, there is a region of a warm stratosphere and greater TOZ than, on the average, in the rest of the Arctic stratosphere. The regions of a colder stratosphere and lower TOZ are mainly situated over the north of Europe, the Barents and Norwegian Seas. These regions usually appear in October and exist until April–May, being horizontally displaced but remaining within the mentioned Arctic sectors.

The symptoms of continuous decrease of TOZ at the middle latitudes and in the tropics of the Northern Hemisphere were also observed. A considerable TOZ drop in Europe and North America began in November 1982 and reached 5–6% by the 1990s. The drop in TOZ also encompassed almost the entire area of Asia. Ozone destruction in the lower stratosphere was caused by aerosol particles formed out of the gas emitted to the stratosphere during a powerful eruption of the El Chichon volcano in southern Mexico. The mechanism of such a destruction of ozone is not clear, but it differs from the processes on the particles of polar stratospheric clouds.

At present, it may be assumed that there is a certain correlation between quasi-biannual oscillations (QBO) of atmospheric circulation and TOZ. The QBO cyclicity is most pronounced in the change of zonal (along latitudinal circles) wind direction in the equatorial stratosphere. If the air flow is directed from the east to the west (the eastern phase), then the inter-latitude air exchange is more intensive than in the years with the western QBO phase. Therefore, in such conditions, larger amounts of ozone and heat are transferred to the Antarctic atmosphere from the middle latitudes of the Southern Hemisphere in winter and spring. In this case, the 'ozone hole' is less intense and is fast filled with ozone. That is why all periods of TOZ drops fall on the western QBO phase, during which the inter-latitude exchange in the stratosphere is weakened.

5.1.2.3 Satellite observations

The discussion on the respective measurement data obtained by TOMS (Total Ozone Mapping Spectrometer) and SAGE (Stratospheric Aerosol and Gas Experiment) instruments is summarized in Kaye (1998), McCormick *et al.* (1992) and McCormick (1998). Until recently, four TOMS instruments were in space and provided reliable data on TOZ and also on the tropospheric ozone content (TrOZ); however, the latter problem is not so easy to solve.

The launching of TOMS instruments made it possible to obtain data for a period from late 1978 to early 1998. One of the applications of the data obtained was the study of TOZ long-term trends and inter-annual variability in the mid-latitudes (with particlular emphasis on the contribution of dynamics-conditioned factors). The results obtained have shown that variations in lower stratospheric temperature and (or) tropopause height could have greatly contributed to the decrease of the ozone content level at the mid-latitudes detected with the help of TOMS instruments. It has been demonstrated that short-term meteorological events could significantly affect the zonal mean distribution of TOZ, especially in February, when winter-to-winter variations can be large enough.

To test the retrieval algorithms, data have been used that were obtained from instruments within the SOLSE (Satellite Ozone Limb Sounding Experiment) and LORE (Limb Ozone Retrieval Experiment) programmes. The instruments were carried into space on the piloted orbital space Shuttle launched by the USA in the autumn of 1997. Measurements of the concentrations of some components (BrO, ClO and SO_2) from the UV radiation have been performed, using the GOME (Global Ozone Monitoring Experiment) instrument on board the *ERS-2* satellite.

In the course of continued SAGE instrument observations, three new ozone-measuring instruments for SAGE-III have been developed (McCormick, 1998) and have been launched during the period 1999–2002. Using 'occultation' techniques, the SAGE-III data will allow the retrieval of profiles of ozone, aerosols, water vapour, temperature, nitrogen dioxide, and cloud characteristics in the middle atmosphere, with a height resolution of about 1 km. In addition, observations made during lunar eclipses will make it possible to obtain vertical profiles of nitrogen and chlorine dioxide.

For validation of SAGE-III data, the SOLVE (SAGE-III Ozone Loss and Validation Experiment) programme is now running, which includes Arctic expeditions to study those processes that influence stratospheric ozone in the zone from polar to middle latitudes. A network of ozone sondes will be very important in the solution of problems encompassed by the programme. Guirlet *et al.* (1998) described the THESEO (Third European Stratospheric Experiment on Ozone) programme conducted in the years 1991–92 as a continuation of the SESAME (Second European Stratospheric Arctic and Mid-Latitude Experiment) programme. The purpose of the programme was to achieve a better understanding of the processes responsible for stratospheric ozone trends in the mid-latitudes; the basis of these studies was investigating the interactions between chemical and dynamical processes. The aim was to study ozone formation both within and outside of the circumpolar vortex. The observations have been conducted within a wide range of latitudes from the tropics to the Arctic; ozonesondes, lidars, small balloons, and aircraft have been employed.

As noted by Harris and Hudson (1998), the main uncertainty concerning assessment of the impact of CFC emissions on the ozone layer is relevant to an evaluation of the trends of ozone content in the 15–20 km layer.

Analysis of SAGE satellite observation data (WMO/UNEP, 1994), has revealed a trend of a $-20 \pm 8\%$ drop for 10 years at the mid-latitudes of the Northern Hemisphere, whereas it follows from the ozone sonde data that the trend reached only $-7 \pm 3\%$ for 10 years.

In 1996, within the framework of joint efforts made by investigators working for the SPARC programme (ozone impact on climate) and the International Ozone Commission, a revision of the above-mentioned assessments was undertaken. To estimate the long-term trends of ozone abundance, observation data used were obtained with the help of four methods: SAGE (I and II); SBUV/2; the Dobson conversion (U/D) method; and direct ozonesonde observations.

The data of SAGE satellite 'occultation' soundings embrace the period from February 1979 to July 1996 with a three-year break (starting in November 1981). The results of satellite observations of UV radiation backscattering (SBUV–SBUV/2), relevant to the period from 1978 to the present, and of ground-based observations (U/D) based on the Dobson conversion method for the period from 1957 up to the present time, show that restoration of ozone abundance in the middle and upper stratospheres is possible. The data of ozonesonde measurements cover the period from the early 1960s until now and are relevant to the lower stratosphere and troposphere.

In view of the revision of the data array on ozone discussed above (for comparison purposes, more short-term *UARS* satellite data have also been used), measurement and retrieval methods were subjected to detailed analysis. In processing SAGE data, for example, a correction for aerosols has been introduced, which is important at heights below 20 km, and a 'pumping' correction for ozonesondes; the SAGE-II data, obtained some 18–30 months after the Mount Pinatubo volcano eruption and relevant to levels below 10, have been excluded.

Analysis of observation data has revealed that a possible sensitivity drift for all types of instruments does not exceed 5% during 10 years (within 2σ). The error of global-averaged SAGE-I and SAGE-II data on trends in the 20–40 km layer (at the level of 95% statistical significance) is 0.2% per year, and that of SAGE-I data on trends in the lower stratosphere about 0.25% per year. Analysis of all data on trends revealed the presence of statistically significant trends of ozone amount drops at all heights in the 12–50 km layer, with two pronounced maximums of the drop near 15 km and 40 km.

With regard to the upper stratosphere (that is, 30–50 km), it had been earlier predicted that it must be the layer of ozone content decrease (chemical processes are governed there by gas-to-particle conversion reactions). According to fresh data, the decrease constitutes from –6% to –8% during 10 years at heights of 40–45 km, with considerable annual variation having a maximum in winter, but without any noticeable differences between the hemispheres. In the lower stratosphere (below 30 km), where SAGE (20–30 km) and ozonesondes (below 27 km) mainly contribute to observation data, the trend at 20 km varies (for different observation sites) from –3% to –11% during 10 years. The annual variation (which is absent above 20 km) shows up mostly in the 10–20 km layer, but it is subject to geographic variability (according to the data of European stations, the maximum of the drop occurs in winter/spring; according to the data of Canadian stations, in spring/summer).

Data relevant to tropospheric ozone are insufficient in volume and therefore are difficult to summarize. From the data of Canadian stations, the trends are negative or close to zero during the period 1970–96, whereas it follows from observations at three European stations that a significant positive trend is characterisitc for the same time period. However, data from two other European stations indicate that during the period 1980–96, the trend was weaker according to the data of SAGE-I and SAGE-II than that of TOMS. Satellite and ozonesonde observations are in agreement in determining annual ozone content variation, which mainly takes place at 10–20 km at the middle latitudes, with a distinct maximum in the Northern Hemisphere in winter/spring. In the Southern Hemisphere, the annual variation is much less pronounced.

The processing of the results of TOZ remote sensing from observation data, obtained with the help of TOMS and instruments for measuring UV solar radiation backscattering (SBUV), provided information on TOZ global dynamics for two decades. Although the use of simultaneous data from TOMS and SAGE made it possible to retrieve the values of total ozone content in the troposphere, direct retrieval of TrOZ became possible after the GOME instruments for global ozone monitoring were launched on the ERS-2 satellite in April 1995.

The GOME spectrometer for the UV and visible spectral regions allows the measurement (with a spectral resolution of 0.2–0.4 nm) of outgoing (backscattered) shortwave radiation in four spectral intervals of the 237–790 nm range, two of which,

1A (237–307 nm) and 2B (312–405 nm), contain information for retrieving the vertical profile of ozone concentration (VPOC), and hence TrOZ.

Munro *et al.* (1998) discussed the results of VPOC retrieval for a number of points on the globe from the data on reflectivity spectrum for a part (325.5–334.5 nm) of the 2B band, making use of the GOMETRAN algorithm for radiation transfer *a priori* in formation on VPOC (as an initial approximation for subsequent iterations). Comparison of the results of retrieval with the ozonesonde data has shown that, in the absence of clouds (or above the lower layer cloudiness), the GOME data provide a reliable enough retrieval of VPOC both in the stratosphere and troposphere. The presence of upper-layer clouds makes the problem more difficult to resolve. The values (found from retrieved VPOC) of tropospheric ozone concentration reveal its strong spatial-temporal variability. For instance, a TrOZ increase up to 200% was detected in Africa in areas of biomass burning.

Analysis of long-term observations of tropospheric ozone concentrations in polar regions (both Arctic and Antarctic) show the presence of a regular minimum in spring, which is, apparently, of natural origin. Simultaneous measurements of bromine concentrations revealed their increased levels before and after the 'events of decreased tropospheric ozone content'. Measurements of the BrO mixing ratio yielded values up to 30 ppt. It has become clear by now that the traces of TrOZ are mainly due to catalytic reaction cycles with the participation of BrO, with less significant contribution of reactions with the participation of ClO and IO. This conclusion can be confirmed by numerical modelling results, which demonstrate that drops in TrOZ can be explained if the presence of a sufficient amount of halogens is assumed.

What is not clear as yet is the source of bromine compounds during polar spring, and these may include organic compounds of the $CHBr_3$ type and sea salt aerosols. In this connection, Richter *et al.* (1998) analyzed the results of satellite observations of the backscattering of the outgoing shortwave radiation (OSR) in spring and summer of 1997 in the latitude belt (40°, 90°N) with a double monochromator GOME (global monitoring of concentrations of ozone and other trace gases) installed on the ERS-2 satellite. The measurements were aimed at retrieving the content of BrO.

The above-mentioned instrumentation provides spectral measurements of OSR near the nadir, in the wavelength interval 240–793 nm with a resolution 0.2 nm (wavelengths less than 400 nm) or 0.4 nm (wavelengths more than 400 nm). The spatial resolution (at a solar zenith angle less than 85°) is 40×320 sq km. To obtain global information at the equator, two days are required and at the latitude 67°, one day. The retrieval of the total concentration of BrO is based on using the method of differential absorption (DOAS); the BrO content in the troposphere was calculated by subtracting values measured in the longitude range that is representative because it characterizes background conditions.

Analysis of observation data enabled arriving at a conclusion that there is an increased concentration of BrO over Hudson Bay and the Canadian Arctic during the period from February to early March. It was not noticed before and may be accounted for by the influence of a powerful local source of bromine (it looks like biogenic bromine incomes through cracks in the ice cover). From March till May, there were smaller in amplitude and shorter-term drops in BrO concentrations in the troposphere along the coastline of the Arctic Ocean and over polar ice. These drops are in agreement with the results of ground-based observations of positive anomalies in BrO concentrations at several Arctic stations.

Because the ozone concentration in the lower stratosphere depends on the distribution of the radiative flux divergence (RFD), Williams and Toumi (1999) analyzed this kind of dependence using the results of calculations by means of a two-dimensional global model of chemical processes-radiation transfer-dynamics of an atmospheric layer at 0–95 km, with a height resolution of 3.5 km and horizontal latitude resolution of 9.5°. The module of chemical processes includes O_x, NO_x, HO_x, ClO_x, BrO_x and CHO_x.

The results obtained indicate that, in the presence of upper-layer cloudiness in the upper troposphere in the lower stratosphere there appears to be a negative RFD (radiative cooling), which causes the enhancement of meridional atmospheric circulation, which, in its turn, leads to a decrease in TOZ in the tropics. Here it is important that there is no linear relationship between the cloud-induced temperature changes due to RFD variations and the radiative flux divergence itself owing to cloudiness.

The mechanism of the impact of clouds on RFD in the stratosphere is such that, on the one hand, cirrus clouds hamper the advance of the earth's surface thermal radiation to the stratosphere, which originates longwave cooling. On the other hand, the reflection from clouds of incoming shortwave radiation is followed by ozone-absorbed radiation growth.

Williams and Toumi (1999) considered the possible relationship between the upper-layer cloud amount growth and the TOZ drop observed in some regions at the time of strong El Niño episodes, and analyzed the possible impact on total ozone content of the global trend of the upper-layer cloud amount. The results discussed reflect the importance of taking account of cloudiness and radiative properties of clouds when the effect of the upper-layer cloudiness on ozone formation is investigated.

The reliability of data on the ozone content in the upper stratosphere is a fundamental condition necessary for a reliable analysis of the impact of human-induced chlorine compounds upon the stratospheric ozone. The chemical composition of the upper stratosphere is determined by chemical reactions which are faster than relevant dynamic processes. The characteristic photochemical lifetime of ozone at heights above 40 km is less than 24 hours (with the sun being near its zenith), whereas the characteristic lifetime of dynamic processes varies from several weeks to several years. Therefore, dynamic processes affect the distribution of only long-living components, such as CH_4, N_2O, H_2O, etc.

Although, on the face of it, such circumstances facilitate numerical modelling of the O_3 concentration field in the upper stratosphere, marked discrepancies have been found between the results of model computations (started in the mid-1980s) and observations: thus, ozone concentrations at heights 40–50 km have been underestimated by 40–60%. The discrepancy was due to the 'disbalance' between the processes of ozone formation and destruction (it should be noted, however, that the latest numerical modelling results are in better agreement with observational data).

The situation has been defined as 'the problem of ozone deficit'. On the other hand, approximate 'box' assessments, using the data of HALOE instruments for remote 'occultation' sounding of the stratosphere, installed on the *UARS* satellite (temperature, the concentration of gases-ozone sources, and key radicals responsible for ozone destruction), not only revealed ozone deficit but also discovered a certain surplus of ozone (however, within the errors of observation and model input parameters). In this connection, Groos *et al.* (1999) made new 'box' calculations, using more precise HALOE data ('Version 8') corresponding to a higher rate of ozone depletion than that obtained before; in their case there was no significant deficit of ozone. The most important source of errors of 'box' assessments is the uncertainty of reaction rates for the HO_x cycles. Data on the O_2 photolysis, the temperature and concentration of ozone in the upper stratosphere should also be made more precise.

Gas emissions into the atmosphere at great heights, caused by satellites and supersonic transport aircraft, are the only kind of direct human-induced emissions; among such emissions are those produced by solid-propellant missiles which form an important component. Emissions of nitrogen oxide by the engines of supersonic transport aircraft will be of practical significance when flights of aircraft-rockets of the Sanger and NASP type may start (the presence of nitrogen oxide is due to the formation of a shock wave and its heated boundary layer).

Stuhler and Frohn (1998) analyzed possible consequences of the impacts of SST upon the upper atmosphere by calculating a large amount of kinetic energy transformed into the energy of translational and rotational degrees of freedom of molecules after the shock wave. At large Mach numbers, the temperature immediately after the shock wave front may reach many thousands of °C, which stimulates chemical reactions of nitrogen oxide formation.

With supersonic aircraft, the formation of nitrogen oxide due to engine operation exceeds the level of NO formation after the shock wave front and the boundary layer. As the returning aircraft enters the atmosphere at speeds corresponding to high Mach numbers, 'cut-off' shock waves are formed. Estimates relevant to the piloted orbital space Shuttle show that the NO concentration behind the shock wave front, averaged over the cross section $10^6\,m^2$, exceeds the background concentration of NO by a factor up to 800. During every descent of the Shuttle, about 5,000 kg of NO is formed at heights from between 5 km and 60 km. Therefore, Shuttle descent episodes are connected with local changes in the atmospheric chemical composition and, in particular, with ozone removal in accordance with reaction: $NO + O_3 \rightarrow NO_2 + O_2$. In conditions of the contemporary intensity of SST launches, the resulting excess of the background concentration of NO (on a global scale) due to natural sources is significant.

5.1.2.4 Biological implications of total ozone depletion

The biological consequences of the decreasing TOZ trend are connected mainly with the enhancement of biologically active UV solar radiation at the earth's surface. It is proper in this context to quote the basic conclusions published in the UNEP Scientific Assessment (UNEP, 1994):

- Large increases in ultraviolet radiation have been observed in association with the ozone hole at high southern latitudes. The measured UV enhancements agree well with model calculations.
- Clear sky UV measurements at mid-latitude locations in the Southern Hemisphere are significantly larger than at a correponding site in the Northern Hemisphere, in agreement with expected differences due to TOZ and sun-earth separation.
- Local increases in UVB were measured in 1992–93 at mid- and high-latitudes in the Northern Hemisphere. The spectral signatures of the enhancement clearly implicate the anomalously low ozone observed in those years, rather than variability of cloud cover or tropospheric pollution. Such correlations add confidence to the ability to link ozone changes to UVB changes over relatively long time scales.
- Increases in clear-sky UV over the period 1979 to 1993 due to observed ozone changes are calculated to be greatest at short wavelengths and at high latitudes. Poleward of 45°, the increases are greatest in the Southern Hemisphere.
- Uncertainties in calibration, influence of tropospheric pollution, and difficulties of interpreting the data from broadband instruments continue to preclude the unequivocal identification of long-term UV trends.
- Scattering of UV radiation by stratospheric aerosols from the Mount Pinatubo eruption did not alter total surface-UV levels appreciably.

Recent information on the biological implications of total ozone depletion has been discussed at the European Conference on Atmospheric UV Radiation in Helsinki (Abstracts, 1998) as well as in the monograph by Martens (1998) and in numerous journal publications.

As far as impacts of variable UV radiation on humans and ecosystems are concerned (Kondratyev and Varotsos, 2000a), the most important aspects of the problem include human health (one of the most dangerous consequences is skin cancer) and the health of terrestrial and marine ecosystems.

5.1.2.4.1 Terrestrial ecosystems

Not many observation data of surface UV solar radiation have been discussed in Kondratyev and Varotsos (2000); WMO/UNEP (1999). The development of techniques for restoring surface UV radiation from satellite observations has been a new step in this context. Thus, for example, Udelhofen *et al.* (1999) having processed TOMS Australian data for the period 1979–92, discovered a statistically significant increase in erythema UV radiation of the order of 10% over 10 years in the summer months in the tropics, which was caused by the joint effect of TOZ decline and cloud amount decrease. The UV radiation trend in the mid-latitudes was absent, but the relationship with quasi-biannual fluctuations was detected.

In view of fragmentary observations of UV radiation flux at the earth's surface level, Sabziparvar *et al.* (1999) calculated the global distribution of radiation fluxes (daily doses) in UVB (280–320 nm) and UVA (320–400 nm) wavelength ranges, and on this basis (with consideration for biological spectra of action) the effective radiation producing erythema, cataract, and keratitis. Calculations of spectral UV radiation fluxes with clear sky and under the overcast conditions (at a resolution equal to 5 nm) have been made with the use of four-flux parameterization of radiation transfer for the given fields of ozone content, cloud amount, surface pressure, underlying surface albedo, temperature and very schematic aerosol characteristics (a model of continental aerosol has been given with the optical thickness of 0.1 at the wavelength of 400 nm). The calculations were based on the model of 17-layer or 19-layer atmosphere (depending on the height above sea level), and the horizontal resolution is 5° latitude × 5° longitude.

The comparisons of calculated and measured values of erythema dose with clear sky have shown quite satisfactory agreement. The analysis of calculation results has demonstrated that the most important factor determining UV radiation fluxes is the sun's height (for this reason the locations of the maxima of UV radiation daily doses and the minima of total ozone content do not coincide). Atmospheric pressure drop plays an important role at high altitudes above sea level (in mountains). The presence of cloud amount causes the decrease of UV radiation fluxes (as compared to clear sky conditions) in the range from several per cent in arid and semi-arid regions to 45% in mid-latitude regions where frequent cyclonic situations occur. The results obtained have been used for estimating the effect of changes in the stratospheric and tropospheric ozone content, as well as aerosol, upon UV radiation doses.

Plumb and Ryan (1998) have made a calculation to analyze the effect on biologically active ultraviolet radiation coming to the earth's surface which is exerted by changes in the TOZ due to the impact of aircraft emissions of ODS on the ozone layer. To this end, there have been proved scenarios of ODS and their effects on the ozone layer with the use of a two-dimensional model of atmospheric dynamics including chemical reactions – developed at the Commonwealth Scientific and Industrial Research Organization in Australia (CSIRO) – as well as schemes of UV radiation transfer in the atmosphere.

The analysis of numerical modelling results has shown that similar TOZ values corresponding to different vertical profiles of ozone concentration are not equivalent from the viewpoint of biologically active UV radiation (BUVR) inflow. Such a situation is related to the fact that while the effect of freon ODS shows up mainly in the lower and middle stratosphere, the impact of aircraft emissions is limited mostly to the upper troposphere. The ratio of BUVR flux changes to TOZ changes proves to be larger in the case of aircraft emissions due to the influence of multiple scattering in the troposphere. For the same reason the effect of the assumed fleet of supersonic aircraft on BUVR should be weaker than the effect of subsonic aviation. Although the presence of aerosol in the troposphere causes in all cases a decrease of BUVR, it exerts a minimal effect on the ratio of BUVR changes to TOZ changes, since aerosol is located below the levels at which aircraft emissions or freons affect ozone.

A limited number of studies have been made so far concerning the effect of UVB radiation (wavelengths 230–320 nm) on different agricultural crops (mainly in conditions of special chambers and hotbeds, and only in some cases in field conditions). In this case, individual plants were primarily studied and not populations, communities, or ecosystems, though it is of no doubt that, generally speaking, long-term studies of the response of entire ecosystems to the increased level of UVB radiation (Kondratyev and Varotsos, 2000a; WMO/UNEP, 1999) will be required. It is impossible to estimate reliably the relationships and feedbacks between separate types of ecosystems, if we have data only for some types. Therefore, studying separate types should be considered only as an initial stage of further, more complete, studies. One of the unexpected and surprising results of the consequences examined is a very strong variability in the sensitivity of different kinds (different samples of plants within the given species) to UVB radiation changes. Such characteristics as vegetation growth and biomass distribution were used as quantitative sensitivity criteria. As a rule, the sensitivity of plants to UVB radiation changes proves to be weaker in field conditions than in chambers and hotbeds.

Among the particular manifestations of plant sensitivity is the decrease of biomass due to weakening of photosynthesis processes at the growth of UVB radiation intensity. In some cases the decrease of plant biomass occurs as a result of reducing the extension of leaf cover, but in other cases morphological changes occur which show up as the growth of the number of branches, the change of thickness and increase of the number of leaves, and the shortening of trunk length. The response to the increase of UVB radiation can weaken due to different chromophores interacting between each other (e.g. nucleinic acid, phytohormones and pigments). In some cases, the functioning of specific receptors of UVB radiation has been discoverd which absorb light primarily at wavelengths of about 290 nm.

Very few studies of trees have shown, for instance, that UVB radiation increase depresses pine (*Pinus tacda*) development, the accumulation of negative consequences being observed during three years.

Observations made in the USA in 1950 showed the presence of vegetation damage done by atmospheric pollution, in particular chlorosis of Ponderosa pines in the San Bernardino mountains, due to the effect of photochemical oxidants (mainly ozone). Until now, the damage done to vegetation by surface ozone has been revealed both in the USA and in Europe. In this connection Salardino and Carroll (1998) carried out investigations on visible damage caused by ozone to needles of Ponderosa and Jeffrey pines growing on the west slopes of the Sierra Nevada mountains in California. Observations were made at 11 stations in national parks and forests on the mountain slopes for 1992–94; to estimate the exposure to ozone effect, data were used on the hourly ozone concentration in summer (1 June–15 October).

Analysis of the results obtained have revealed a clear relationship between vegetation damages (characterized by chlorosis and flocculent surface) and exposure to ozone; in this case, the correlation has proved to be highest when using the complex index of damages and with consideration for the exposure duration (for the whole day).

The conditions existing on the Antarctic continent open the possibilities for studying climate effect on plants (see Day *et al.*, 1997). An appreciable total ozone content decline observed here in spring and early summer produces an essential increase in biological active solar ultraviolet radiation in the UVB region (wavelengths 280–320 nm). Besides, over the last 45 years, surface air temperature in summer has risen by more than 1°C. In November–March 1995–97 field observation experiments near station Palmer were carried out to study the impact of artificially varied levels of UVB and UVA radiation (320–400 nm) on two types of vascular plants existing in Antarctica: *Deschampsia antarctica* (Antarctic hair grass) and *Colobanthus quitensis* (Antarctic 'pearl' vegetation). The temperature was also changed and (sometimes) nutrients or water were introduced into the soil. The rate of photosysnthesis in leaves and that of phytomass growth were an indicator of the impacts.

The results show evidence that the decrease (using light filters) of UV radiation level did not cause appreciable changes in photosynthetic productivity (Pn). The value of Pn increased in conditions of artificial warming. Warm sunny days (at the temperature in a vegetation layer above 20°C) were, however, an important exclusion when in all cases the Pn value turned out to be negligibly small (high temperature caused depression of Pn related with a strong growth of plant respiration).

Special experiments aimed at studying the possibility of plant acclimatization have yielded negative results. It has turned out, however, that phytomass growth is more intensive at higher (20°C) than at lower (12°C) temperatures. This means that Pn cannot be a direct indicator of the rate of plant growth and, on the whole, vital activity. The functions of sexual reproductivity changed greatly under warming conditions. The concluding result of the studies is the inference that in conditions of continuing regional climate warming in the vegetation period, the values of Pn and carbon assimilation by plants should decrease, though a more intensive phytomass growth can occur. Such tendencies, however, would be restrained by the effect of UVB radiation increase.

Since in future one can expect not only the increase of UVB radiation but also the growth of CO_2 concentration and climate warming, the joint effect of these factors becomes actual. According to data available, the increase of CO_2 concentration affects mainly the C3 species, while C4 species are much less sensitive in terms of their photosynthetic response. The experiments with combined effect of the above factors have found in the case of C3 type plants such as rice and wheat that with CO_2 concentration increase the biomass growth does not occur if plants respond to CO_2 at the increased UVB radiation level. When exposed to solar radiation, the temperature rise in the case of maize (C4) crops, or CO_2 concentration growth in the case of sunflower (C3) could compensate for the negative effect caused by UVB radiation increase. If, however, an increase in UVB radiation is combined with CO_2 concentration growth over salt marshes, then the effect of the interaction between these two factors is not practically observed. Thus, the response of plants to the joint influence of CO_2 and UVB depends essentially on the plant species.

The effect of UVB radiation increase on the chemical composition of plants shows up mostly in the growth of concentration of pigments (first of all flavonoids) absorbing UVB radiation, the function of which is protection of plants from destructive effect due to UVB radiation. Thus, for example, the accumulation of such pigments in the epidermis of rye protects leaves from photosynthesis depression. The growth of UVB radiation intensity can also result in changes of concentrations of such minor components as alkaloids and coumarins, which can affect the quality of crops production, the rate of organic chemistry decomposition, and the intensity of minor gas constituents emission into the atmosphere.

A separate aspect of the problem is UVB effects on the depressed plants from the ecosystem. It has been established, for instance, that plants developing in conditions of moisture deficit respond to the increase of UVB radiation more weakly than those well supplied with moisture.The sensitivity of plants growing in conditions of phosphorus deficit (the effect of other nutrients deficit remains to be studied) to UVB radiation is weaker. Taking into consideration the interaction between soil contamination effect and UVB radiation increase has proved to be important. The following problems to be solved, related to the effects of an increased UVB radiation level, should be considered as a top priority:

- Estimations of the effect on the primary productivity of different agricultural and non-agricultural ecosystems.
- Estimations of the effect on the morphological changes in the architecture of vegetation cover and biomass distribution, as well as the corresponding consequences.
- Estimations of the direct and indirect effect on the users of ecosystem production and the possible influence of feedbacks on the ecosystems.
- Analysis of the genetic basis of resistance to the effects of UVB radiation changes.

5.1.2.4.2 Water ecosystems

Water ecosystems cover a broad spectrum of inhabitance – from oceans to large and small ponds, rivers, estuaries, and wetlands – the condition of which varies within the range from undisturbed ecosystems to practically complete anthropogenic transformation. With a great variety of water systems, there is an involvement of cycles of energy and matter with the decisive role of the variability of solar radiation income (as energy source) and its spectral distribution. The central link of photosynthesis processes occurring due to the sun's energy is primary production, the level of which is inseparably related to the dynamics of food chains and biochemical cycles of different compounds determining a wide range of phenomena – from water environment chemistry to regional and global conditions of weather and climate. The changes of spectral regime caused by UVB radiation increase mean the appearance of substantial disturbances (first of all related with photoprocesses in water ecosystems) in the environmental conditions which have existed during a long period of water ecosystems evolution.

By now many factors have been established which are indicative of the destructive effect of UVB radiation on water organisms. The effect of UVB radiation increase on primary productivity in Antarctica was, in particular, revealed. This effect influences not only phytoplankton resistance, but also causes disturbances in the polychromatic balance of light field, which affects the interaction of the processes at molecular, biochemical, physiological, and trophic levels, as well as being likely to result in the changes of natural communities evolution. However, the consequences of UV effects remain unclear for regional and global water primary production and the related processes in ecosystems. In this connection, it is very important to obtain perspective in the near future on the data of both satellite and ground-based observations, which would enable us to have available information on the global distribution of UVB radiation over the oceanic surface and to estimate its penetration into the depths. In this context the problem is also actual to obtain data on photosynthetically active radiation (PAR). The problem can be solved by analysing the effect of UVB radiation increase on the communities and ecosystems (the community is determined as a part of the ecosystem which is an ensemble of organisms exisitng within an ecologically definite region), though it is clear that the response of water ecosystems can vary over a wide range – from non-existent to catastrophic. This determines the necessity of:

- Obtaining more complete and reliable data on UVB radiation increase on local, regional, and global scales.
- Finding out about water ecosystems potentially more sensitive to UVB effects.
- Studying the relationship between the response of ecosystems to the effects and their subsequent restoration.
- Quantitative estimates of effects showing up on different spatial-temporal scales.
- Revealing the processes in ecosystems which is subject to UVB radiation increase effect.
- Studying possible global positive feedbacks, taking into account the changes of cloud cover, temperature and the processes of biological adaptation which can be produced by global changes at the level of UVB radiation coming to the earth's surface.

5.1.2.4.3 Biogeochemical cycles

The cyclic variations of solar radiation showing up as daily and annual variations as well as long-term variability (the 11-year cyclicity is its most prominent manifestation) due to solar activity exert a considerable effect on biogeochemical cycles determined by the complex interaction of biological, chemical, and physical processes of matter and energy transformation. A well-studied example of the manifestation of biogeochemical cycle recurrence can be the annual variation of carbon dioxide concentration in the atmosphere caused by the cyclicity of global biomass appearing under the effect of annual variation of solar radiation income and, respectively, vegetation processes. An important contribution is also made, in this connection, by anthropogenic emissions of carbon dioxide which, however, despite their appreciable value (about 6 Gtons of C per year) are by an order of value smaller than the CO_2 exchange between the atmosphere and the land biosphere (approximately 60 Gt of C/year), as well as the close-in value CO_2 exchange between the atmosphere and the ocean. The latter means that even due to a small variation in biospheric fluxes, substantial changes in the atmospheric CO_2 concentration can occur.

When analysing the average global trend of CO_2 concentration, our attention is engaged by the fact that in spite of slowing down the growth rate of anthropogenic CO_2 emissions, the ratio of the observed growth rate of CO_2 concentration to the growth rate of emissions due to combustion of carbon fossil fuels has increased considerably. This can be interpreted as a consequence of UV radiation increase at the Earth's surface level for the last 15 years which caused the depression of photosynthetic activity and, therefore, the primary production decrease and weakening of carbon dioxide assimilation from the atmosphere (however, the effect of other factors is not excluded either: for instance, changes in land use in the tropics).

The increase of UVB solar radiation can lead not only to depression of photosynthetic activity, but also to intensification of dead organic substance decomposition. Over long time intervals, the variations of the productivity of ecosystems can affect the dynamics of nutrients cycles. The cycle rate of some compounds' decomposition processes (lignin, humus substances) can, in particular, increase. Observation evidence, for example, shows that UVB radiation increase favours the decomposition of water humus substances. The close interrelationship of biogeochemical cycles determines the inevitability of accompanying changes in the cycles of nitrogen, oxygen, suphur, halogens, and metals playing an important role in cell metabolism. Three basic manifestations of the effect of UVB radiation growth on the cycles of the above elements are possible:

- Direct influence on biological processes can arise due to intensification of the disintegration of DNA and other biological molecules (first of all chlorophyll), which result in changes of photosynthetic activity and land vegetation respiration, as well as such processes, directly or indirectly related with sea phytoplankton, as microbe respiration, nitrification, sulphur cycle, and redox reactions with participation of metals. These effects can be related with such events as global climate warming and CO_2 concentration growth, and PAR changes.
- UVB radiation increase affects strongly the geochemical processes in the sea and on the land. Among these are: producing 'greenhouse' and chemically active gases CO_2, CO, NO, COS (carbonisulphide), NMHCs (non-methane hydrocarbons); nitrogen and phosphorus mineralization; transformation of organic substance to biologically accessible organic compounds with smaller molecular weight; deeper penetration of UV radiation into the depth of sea water; the change of redox-condition of the oceanic upper layer *et al.*

- Chemical processes produced by UVB radiation growth in the atmosphere show up as changes: in oxidizing capacity of the troposphere; in formation (more intensive) of such greenhouse gases as methane and ozone; aerosol concentration and cloud condensation nuclei; concentration of ozone-destroying gases (including different halogenated compounds). Phenomena of this kind should strongly interact with the emissions of chemically active gases resulting from biomass burning (organic substances) and other anthropogenically-induced processes.

All the above changes can be accompanied by the appearance of feedbacks leading to an increase or decrease in the formation of greenhouse gases and aerosol in the atmosphere, which affect the dynamics of global climate. Below are briefly formulated the most important considerations on the perspectives for further studies; of course, the basic unsolved problem is the poorly known effects of UVB radiation on the entire ecosystem. That is why it is very important to study the UVB radiation effect on:

(1) Land ecosystems, including:
  (a) The primary productivity of individual agricultural and non-agricultural ecosystems.
  (b) The morphology (architecture) and spatial structure of vegetation biomass distribution.
  (c) Population structure and its stability.
  (d) Influence on chemical composition and corresponding studies at the ecosystem level.
  (e) The influence on long-term vegetation evolution with consideration for genetic dynamics.

(2) Water systems, including:
  (a) Variability of UVB radiation, as well as the ratios UVB:UVA:PAR at different depths, and respective effects on ecosystems.
  (b) Spatial-temporal sensitivity of ecosystems (primary productivity, substance and energy balances, adaptability) and UVB radiation changes.
  (c) Revealing the role of possible feedbacks (caused, for example, by changes in cloud amount and temperature) which can affect the process of increased UVB radiatio impact on ecosystems.

(3) Biogeochemical cycles and global changes (see Table 5.2), including:
  (a) Carbon and nitrogen cycles in land ecosystems.
  (b) Carbon and nitrogen cycles in water ecosystems.
  (c) Sulphur cycle in water ecosystems.
  (d) Cycles of biologically important metals and phosphorus in the water environment.
  (e) Formation and degradation of organogenes.
  (f) Disturbances brought into atmospheric biogeochemical cycles, which is of interest in terms of the impact on climate (CO_2, CH_4, N_2O, SO_2, DMS, O_3),

Table 5.2. Direct and indirect effects of UVB radiation on biogeochemical cycles.

Carbon Land ecosystems	1. Photosynthesis, plant species shifts. 2. Emissions of nonmethane hydrocarbon compounds. 3. Bedding composition, organic substance decomposition.
Carbon Water ecosystems	1. Degradation and cycle of organic substance. 2. Spectral properties of water environment. 3. Synthesis of organic carbon and assimilation of CO_2 by phytoplankton and macrophites.
Carbon Atmosphere	1. Exchange of carbon-containing gases. 2. Rate of chemical transformation, particularly during radical formation.
Nitrogen Land ecosystems	1. Nitrogen fixation. 2. Nitrification, N_2O and NO emissions. 3. Composition and disintegration of bedding. 4. Emissions of inorganic N as a result of degradation of organic substance and subsequent effect on soil fertility, CH_4/CO sinks.
Nitrogen Water ecosystems	1. Nitrification, N_2O and NO_2 emissions. 2. Formation of inorganic N as a result of degradation of organic substance and subsequent effect on nitrients accessibility. 3. Formation of radicals from NO_3 and NO_2.
Nitrogen Atmosphere	1. Exchange of nitrogen-containing gases (N_2O, NO, NH_3) between atmosphere and other natural environments. 2. Chemical transformation rate, particularly during radical formation.
Sulphur Land ecosystems	1. Assimilation and emissions of COS by vegetation.
Sulphur Water systems	1. Assimilation and emissions of precursors of DMS, COS, and S by water organisms. 2. Formation and consumption of DMS. 3. Photooxidation of DMS.
Sulphur Atmosphere	1. Exchange of sulphur-containing gases (COS, DMS, SO_2) between atmosphere and other natural environments. 2. Chemical transformation rate, particularly during radical formation.
Metallophosphoric Land ecosystems	1. Emissions of inorganic N as a result of degradation of organic substance compounds and subsequent effect on soil fertility.

(continued)

Table 5.2—*(continued)*

Metallophosphor Water ecosystems compounds	1. Presence of metal traces (e.g., Fe, Cu, Mn). 2. OH formation through reaction with participation of Fe.
Metallophosphoric Atmosphere compounds	1. Chemical transformation rate, particularly during radical formation.
Oxygen Land ecosystems	1. Photosynthesis. 2. Bedding decomposition.
Oxygen Water ecosystems	1. Photosynthesis and oxydation of POB. 2. Formation and desintegration of H_2O_2. 3. Formation of HO_2 and OH.
Oxygen Atmosphere	1. Transformation of O_2 to O_3.
Halogens Land ecosystems	1. Formation of organohalogens.
Halogens Water ecosystems	1. Formation of nutrients and consumption of organohalogens
Halogens Atmosphere	1. Desintegration of methylhalides, chlorofluorocarbon compounds and their substitutes with subsequent ozone destruction.

the tropospheric air quality (NO_x, SO_2, CH_4, CO, O_3, NMHCs), and stratospheric chemistry (COS, CH_4, metal halides).

(g) Development and use of techniques for remote sensing to estimate the effects of UVB radiation increase on ecosystems at global scales, with due regard for other related processes.

(h) Development and use of techniques for simulation numerical modelling of the natural environment dynamics (first of all global biogeochemical cycles) in conditions of anthropogenic impacts with complex account of such interactively affecting factors as the decline of the total stratospheric ozone content and the growth of tropospheric ozone, UVB radiation increase, rise of the concentration of CO_2 and other greenhouse gases, intensification of the greenhouse effect of the atmosphere, climate warming *et al.*

5.1.3 Tropospheric ozone

Increasing pollution of the troposphere has stimulated studies of anthropogenic impacts on tropospheric chemical composition (Kondratyev and Varotsos, 2001a, 2001b). In this context, tropospheric ozone occupies a special place in view of its importance for human health and ecosystem functioning as well as its importance as a greenhouse gas. Pszenny and Brasseur (1997) pointed out that surface ozone is a human respiratory irritant. Relatively small ozone amounts can cause chest pain, coughing, nausea, throat irritation and congestion in healthy people. It may also worsen bronchitis, heart disease, emphysema, and asthma. Surface ozone is also phytotoxic. It can produce foliar injuries, reductions in crop yield and biomass production, and shifts in competitive advantages of vegetation species in mixed populations. Photolysis of tropospheric ozone by UV radiation in the presence of water vapour, the primary source of hydroxyl radicals (OH), is very important. Hydroxyl radicals are responsible for the removal of many trace gases (such as CH_4, HFCs, HCFCs) through oxidation.

A number of fundamental publications (UNEP, 1994; Høv, 1997; Isaksen, 1998; WMO/UNEP, 1999) contain basic information on various features of tropospheric ozone dynamics and relevant problems. Basic achievements up to 1994 are shown in UNEP (1994):

- Recent measurements of the NO_y/O_3 ratio have basically confirmed earlier estimates of the flux of ozone from the stratosphere to be in the range of 240–280 Tg (O_3) yr^{-1}, which is in reasonable agreement with results from general circulation models.
- The observed correlation between ozone and alkyl nitrates suggests a natural concentration of 20–30 ppb in the upper planetary boundary layer (at 1 km altitude), which agrees well with the estimate from the few reliable historic data.
- Measurements of the gross ozone-production rate yielded values as high as several tens of ppb per hour in the polluted troposphere over populated regions, in good agreement with theoretical predictions. Likewise, the efficiency of NO_x in ozone formation in moderately polluted air masses was found to be in reasonable agreement with theory.
- Direct measurements of hydroxyl and peroxy radicals have become available. While they do not survey with the aim to establish a global climatology of OH, they do provide a test of our understanding of the fast photochemistry. Today, theoretical predictions of OH concentrations (from measured trace-gas concentrations and photolysis rates) tend to be higher than the measurements by up to a factor of two.
- Measurements of peroxy radical concentrations in the remote free troposphere are in reasonable agreement with theory; however, significant misunderstanding exists with regard to the partitioning of odd nitrogen and the budget of formaldehyde. Measurements have shown that the export of ozone produced from anthropogenic precursors over North America is a significant source of ozone in the tropics during the dry season. These findings show the influence of human activities on the global tropospheric ozone balance.
- Photochemical net ozone destruction in the remote atmosphere has been identified in several experiments. It is likely to occur over large parts of the troposphere with rates up to several ppb per day. Consequently, an increase in UVB radiation (e.g. from stratospheric ozone loss) is expected to decrease tropospheric ozone in the remote atmosphere, but in some cases it will increase production of ozone in and transport from the more polluted regions. The integrated effect on hydroxyl concentrations and climate is uncertain.

It was emphasized in UNEP (1994) that uncertainties in the global tropospheric ozone budget, particularly in the free troposphere, are mainly associated with uncertainties in the global distribution of ozone itself and its photochemical precursors, especially CO and NO_x. The role of heterogeneous processes including multiphase chemistry in the troposphere is not well characterized, and the catalytic efficiency of NO_x in catalyzing ozone formation in the free troposphere has not been confirmed by measurements.

A new phenomenon, discovered during the last decade in the Arctic, is episodes of tropospheric ozone depletion (ODE) in the inversional layer of about a few hundred metres thick. A very severe drop of ozone concentration (up to zero level) was observed in a few hours. Rockmann *et al.* (1999) have pointed out that there is a high correlation of ODE with filtered bromine concentration.

At Alert in Canada, Foster *et al.* (2001) have undertaken observations of Br_2, BrCl and Cl_2 to find an adequacy of the hypothesis that bromine atoms play a central role in the depletion of surface-level ozone in the Arctic at polar sunrise. The measurements indicate that, in addition to Br_2 at mixing ratios up to ~25 ppt, BrCl was found at levels as high as ~35 ppt, but molecular chlorine was not observed, implying that BrCl is the dominant source of chlorine atoms during polar sunrise (this is consistent with recent modelling studies).

Recent results and problems of tropospheric ozone research have been discussed by Høv (1997), in connection with the completion of the first phase of the Tropospheric Ozone Research (TOR) Project which is a part of the EUROTRAC Programme, as well as by Pszenny and Brasseur (1997) in the context of the development of tropospheric-ozone studies within the IGAC (International Global Atmospheric Chemistry) Core Project of the IGBP (Bates *et al.*, 1998b; Kondratyev, 1998d).

Pszenny and Brasseur (1997) pointed out that the most important tropospheric ozone precursors are nitrogen oxides ($NO_x = NO + NO_2$), methane and other 'non-methane' hydrocarbons (NMHCs) and carbon monoxide. All of these precursors are products of fossil fuel and biomass burning, but each also has significant sources from the biosphere. The only ozone sink is depositions to surfaces (vegetation, soil, oceans) chemical destruction *in situ*, and some export back to the stratosphere in the tropics. It is believed that the increase of ozone observed in near-surface air, which may also be occurring in the free troposphere, is the result of increasing NO_x emission.

Cox (1999) developed a one-dimensional box model to describe the ozone budget and NO_x chemistry in the marine boundary layer. Results using small prescribed NO_x concentrations gave compensation points, where ozone loss by photolysis and physical removal is balanced by its production, via NO_x chemistry, of ~15 ppt at the two sites considered (Mace Head, Ireland and Cape Grim, Tasmania), in line with conclusions from observational data.

According to WMO/UNEP (1999), trends in tropospheric ozone since 1970 in the Northern Hemisphere show large regional differences, with increases in Europe and Japan, decreases in Canada and only small changes in the USA. The trend in Europe since the mid-1980s has reduced to virtually zero (at two recording stations). In the Southern Hemisphere, small increases have been observed in surface ozone.

A number of field-observational programmes carried out during the 1990s obtained new important information on ozone dynamics. One of the programmes was the Pacific Exploratory Mission – West A (PEM – West A) which is a major component of IGAC's East Asia/North Pacific Regional Experiment (APARE) Activity (Pszenny and Brasseur, 1997). The principal aim of PEM – West A has been investigations of the tropospheric ozone distribution over the North Pacific and of the growing emissions of ozone precursors from eastern Asia, which will be a major contributor to expected ozone increase on a hemispheric scale in the coming decades. Observed ozone trends were best described in terms of two geographical domains: the western North Pacific rim and the western tropical North Pacific. For both regions, photochemical destruction decreased more rapidly with altitude than did photochemical formation. The ozone tendency was typically found to be negative below 6 km and positive from 6–8 km. On the basis of lidar sounding data from a DC-8 aircraft, Newell *et al.* (1997) studied the influence of Asian continental pollution on the western Pacific in September–October 1991 and February–March 1994.

Schultz *et al.* (1999) studied the budget of ozone and nitrogen oxides ($NO_x = NO \pm NO_2$) in the tropical South Pacific troposphere by photochemical point modelling of aircraft observations at 0–12 km altitude from PEM-Tropics A flights in September–October 1996. It was found, in particular, that chemical production of ozone equals one half of chemical loss in the tropospheric column over the tropical South Pacific. The net loss is 1.8×10^{11} molecules $cm^{-2} s^{-1}$. The missing source of ozone is matched by westerly transport of continental pollution into the region. Application of a global three-dimensional model corroborates the results from the point model and reveals the importance of biomass burning emissions in South America and Africa for the ozone budget over the tropical South Pacific (biomass burning increases average ozone concentrations by 7–8 ppbv throughout the troposphere). The NO_x responsible fo ozone production within the South Pacific troposphere below 4 km can largely be explained by decomposition of peroxyacetylnitrate (PAN) transported into the region with biomass-burning pollution at higher altitudes.

Fenn *et al.* (1999) discussed *in situ* and laser remote measurements of gases and aerosols, made with airborne instrumentation to establish a baseline chemical signature of the atmosphere above the South Pacific Ocean during PEM-Tropics A observations. Between 8 and 52°S, biomass-burning plumes containing elevated levels of O_3 over 100 ppbv, were frequently encountered by the aircraft at altitudes ranging from 2–9 km. Air with elevated O_3 was also remotely observed up to the tropopause, and these air masses were seen to have no enhanced aerosol loading. Frequently, these same air masses had some enhanced potential vorticity (PV) associated with them, but not enough to explain the observed O_3 levels. A relationship between O_3 and PV was developed from the cases of clearly defined O_3 of stratospheric origin, and this relationship was used to estimate the stratospheric contribution to the air masses containing elevated O_3 in the troposphere.

The aim of NARE's intensive summer investigations in 1993 was to determine how major sources of gases in the surrounding industrial regions affect ozone

and Atmospheric Chemistry near the Equator – Atlantic (TRACE-A); field studies of this project were also concluded in the 1992 BIBEX IGAC's Southern Atlantic Region Experiment (STARE) campaign.

The results of SAFARI confirmed that it is justified to consider biomass burning as a significant contributor to the overall increase in GHG that has occurred over the last 150 years, accounting for some 10–25% of current emissions. A major finding of TRACE-A was that widespread biomass burning in both the USA and Southern Africa is the dominant source of precursor gases necessary for the formation of the huge amounts of ozone observed over the southern Atlantic Ocean. The generation of ozone is usually enhanced in the upper troposphere, where relatively high concentrations of NO_x prevail.

It has been found that ozone in the tropical troposphere plays a key role in determining the global oxidizing power of the atmosphere. Most of the oxidation of long-lived gases by hydroxyl radicals takes place in the tropics, where intense sunlight and high humidity promote the formation of OH from photolysis of ozone.

The release into the atmosphere of biogenic gases produced by the African rainforest and savannah, as well as from biomass burning, and their role in the formation of tropospheric ozone and aerosol is the focus of the IGAC Experiment for Regional sources and Sinks of Oxidants (EXPRESSO). Under IGAC, the Global Tropospheric Ozone Project (GTOP) is now being formulated so that all forthcoming data can be used intelligently to provide an understanding of the global tropospheric ozone cycle (Pszenny and Brasseur, 1997). A major aim of IGAC is to understand the role of biological processes in producing and consuming atmospheric trace gases.

An important contribution to studying tropospheric ozone dynamics was made by the TOR sub-project of EUROTRAC (Borrell and Borell, 1999). As pointed out by Høv (1997), the main achievement of TOR was the development of suitable instrumentation for high-quality measurements of ozone precursors (NO_x, NO_y, VOC), intermediates (carbonyl compounds, RO_2) and photolysis rates as well as the implementation of quality assurance procedures for these measurements, the establishment of a large high-quality network and accumulation of a significant database. The high-quality datasets provide detailed mechanistic information on the chemical physical processes that control the budget of ozone and its precursors in the polluted boundary layer and in the free troposphere. The vertical soundings were analyzed for distribution and seasonal variation in ozone in the free troposphere over Europe and provide an insight into processes – such as the exchange between the boundary layer and the free troposphere and between the troposphere and the stratosphere.

Examples of the results obtained include the concept of O_x the use of alkyl nitrates to measure peroxy radicals, the role of individual hydrocarbons in ozone formation, and the use of the H_2O/O_3 correlation for the ozone budget in the upper troposphere.

A hierarchy of models was developed and/or applied in TOR for data analysis. Three-dimensional transport models with coupled chemistry modules were developed in TOR and have been used to study important transport processes – such as convections and for budget studies.

Of course, after the completion of the first phase of TOR many problems still remain unsolved because of the complexity of the processes that govern tropospheric ozone dynamics. A different aspect of the problem is the unseparable influence of transport and chemistry on the concentrations observed at a given location. It is also important that the net chemical balance of ozone depends in non-linear fashion on the concentrations of NO_x, VOCs, H_2O and O_3 itself, and on the UV-radiation flux, and thus it is closely coupled with the atmospheric lifecycles of other trace gases. At continental surface sites, dry deposition plays an important role in the net balance of ozone, in addition to advection and vertical exchange. It is obvious that future success may be achieved only on the basis of combined analysis of sufficiently complete observational data and results of simulation modelling with the use of fully coupled models of chemistry and transport.

In this context, very important progress was recently made in simulations of tropospheric ozone distribution using a chemistry-climate model. An illustration of this success is the analysis of tropospheric ozone over the Indian Ocean carried out by De Laat *et al.* (1999) using the European Centre Hamburg (ECHAM) chemistry-general circulation model. Such an approach allowed interactive dynamics of the atmospheric circulation and ozone to be studied. It was shown, in particular, that large-scale upper tropospheric ozone minima were caused by convective transport of ozone-depleted boundary-layer air in the intertropical convergence zone (ITCZ). Similarly, an upper tropospheric ozone minimum was caused by cyclone Marlene, south of the ITCZ. The mid-tropospheric ozone maxima were caused by transport of polluted African air.

As was pointed out earlier, satellite remote sensing data are an important source of information on tropospheric-ozone distribution (Borrell and Borrell 1999; WMO/UNEP, 1999). There are many techniques capable of retrieving tropospheric ozone from TOMS data (Kaye, 1998). One of them is the use of TOMS-SBUV residuals (in which tropospheric ozone is taken as the difference between TOMS total ozone and integrated stratospheric ozone from one of the Solar Backscatter Ultraviolet (SBUV) or the difference in ozone columns over mountains and nearby sea-level areas (as have been carried out over the Andes Mountains and the nearby eastern Pacific). Another residual technique suggested by Chandra *et al.* (1996) is based on the difference between TOMS total ozone and stratospheric ozone determined by combining the Microwave Limb Sounder (MLS) and HALOE instruments aboard the UARS satellite. Another, and similar, method known as the Convective Cloud Differential (CCD) technique, derives tropospheric ozone from the average difference between TOZ over clear regions and over regions of high clouds (in the latter case reliable cloud-top information becomes very important). Initial studies used TOMS data and TOMS/SAGE residuals only; recently the work has been based on a residual approach using TOMS and SBUV data (in this case determination of the tropopause height is very important).

The Tropospheric Emission Spectrometer (TES) planned for EOS CHEM spacecraft (Kaye and Rind, 1998) is a new method of obtaining tropospheric ozone data.

TES is a Fourier Transform Spectrometer measuring in the 650–3,000 cm^{-1} wavelength region with the use of both nadir and limb-viewing geometry, spectral resolution being 0.025 cm^{-1} in the limb mode and 0.1 cm^{-1} in the nadir mode. The primary focus of the TAS instrument is measurements of ozone and the precursors of ozone in the troposphere, although it has a capability for detecting a broad range of species because of its high resolution.

Munro *et al.* (1998) discussed a retrieval of vertical profiles of ozone in the troposphere from the data of the European Space Agency's GOME on board the *ERS-2* satellite launched in April 1995. GOME is an ultraviolet/visible spectrometer that measures solar radiation backscattered from the earth's atmosphere in four contiguous wavelength bands between 137–790 nm at moderate resolution (0.2–0.4 nm). An important advantage of GOME in comparison with the SBUV instrument is the availability of data for the ozone Huggins bands (310–340). With knowledge of the temperature profile, the temperature-dependent spectral structures in the Huggins bands yield additional ozone-profile information below the ozone peak (in the troposphere and lower stratosphere). The comparison between retrieval results obtained and ozonesonde data indicates that, in cloud-free scenarios and above low clouds, ozone profiles extending down through the troposphere can be retrieved from GOME measurements. Munro *et al.* (1998) demonstrated GOME's ability to detect tropospheric ozone produced by biomass burning over Africa.

Future tropospheric ozone faces a number of still unsolved problems, including:

(1) Accumulation of adequate global information on observed variability of tropospheric ozone for sufficiently long time periods (the lack of data is especially severe in the tropics and sub-tropics); observations of atmospheric-ozone percursors are equally important.
(2) Investigations on the chemistry and photochemistry of tropospheric ozone under cloudy conditions.

As emphasized by Crutzen (1998), the role of clouds as transporters of chemical consituents (such as reactive hydrocarbons, CO and NO and their oxidation products) from the boundary layer to the mid- and upper-troposphere (and possibly into the lower stratosphere) needs to be better understood and quantified, so that clouds can be parameterized for inclusion in large-scale photochemical models of the atmosphere. Similarly, the production of NO by lightning and its vertical redistribution by convective storms needs also to be much better quantified.

The interaction of chemical constituents emanating from the boundary layer with liquid and solid hydrometeors in the clouds, are of special importance. The influence of clouds on the photochemically active UV radiation field is a potentially important research topic. An issue of key significance is interactions between gases and atmospheric aerosols. Since the continental biosphere is a large source of hydrocarbons, quantification of relevant sources in terms of physical (e.g. temperature, humidity, light levels) and biogeochemical (solid physical and chemical properties, land use) parameters are urgently needed for inclusion in atmospheric models. The formation of ozone, carbon monoxide, partially oxidized gaseous hydrocarbons and organic aerosol may be better quantified and parameterized for inclusion in a chemical transport model, if the hydrocarbon oxidation mechanisms in the atmopshere were better understood.

Monod and Carlier (1999) have conducted a box-model daytime study of the multiphase photochemistry (both gas and aqueous phases) of Cl organic compounds within a non-precipitating cloud on a local scale. According to their results, when ozone accumulates in clear-sky conditions, as soon as a cloud is formed, the tropospheric ozone changes drastically: the net production decreases by a factor of two or more and, depending on NO_x, concentrations and pH values, can actually lead to a net chemical destruction. Monod and Carlier (1999) pointed out that both indirect and direct impacts, caused by the presence of the liquid phase, explain this. The indirect impact is the result of the higher solubility of HO_2 and RO_2 radicals than NO and NO_2 resulting in lower gas-phase effciency of the NO_x cycles in producing ozone. The direct impact is caused by a very fast reaction of ozone towards O_2 radicals within the liquid phase. The relative importance of these two impacts has been determined for different NO_x concentrations and pH values. When NO_x = 1 ppbv, the direct impact is of negligible importance at pH = 3 (for a total decrease in ozone production by a factor of two), but it accounts for 6% at pH = 4.17 (for a total decrease in ozone production by a factor of 6.6), 16% at pH = 5.2 (for a total net chemical destruction of ozone), and 28% at pH = 6 (for a total net chemical destruction of ozone). At this NO_x level and at pH = 5.2 and pH = 6, despite its small contribution, the effect of the direct impact is the net chemical destruction of ozone. When NO_x = 0.1 ppbv, both direct and indirect impacts contribute to the net chemical destruction at all pH values. However, the direct impact now contributes to a larger extent, accounting for 15% at pH = 3, and up to 77% at pH = 6. The direct impact involves the $O_3 \pm O_2$ – the reaction which leads to the production of aqueous-phase OH radicals. The latter species in turn react with dissolved organic compounds to produce additional HO_2 and O_2 radicals; therefore, the aqueous-phase sink of ozone is auto-catalytic in the presence of soluble organic compounds.

Warnings about potential surprises in the photochemistry of tropospheric ozone were made by Ravishankara *et al.* (1998): they discussed the role of hydroxyl radical OH as a natural atmospheric detergent responsible for cleansing the atmosphere of pollutants by means of their oxidation. They also pointed out that, in general, the reactions of electronically excited species are of negligible importance in the lower atmosphere, but the case of $O(^1D)$ is a notable exception. ts role is pivotal. Even though most of the $O(^1D)$ is deactivated to the ground state, $O(^3P)$ – a small fraction that survives to react with H_2O and CH_4 – turns out to be the major source of OH. Knowledge of how $O(^1D)$ is formed in the atmosphere is therefore critical in understanding the reaction of OH. Recent surprising findings are beginning to reveal the importance of the longer wavelength 'tail' in the chemistry of $O(^1D)$ formation. The longer wavelengths are important because stratospheric ozone screens most of the shortwave ultaviolet from the lower atmosphere.

Bojkov (1999) emphasized the significant uncertainties that remain in the budget of tropospheric ozone.

5.1.4 Conclusions

Investigations of changes in total ozone (especially in high latitudes) highlight the TOZ variability problem as a very important aim for the study of global-scale environmental dynamics, having great significance for humans and ecosystems. The recognition of such an environmental danger led to the signing of the Montreal Protocol and further Amendments to the Protocol to avoid the dangerous environmental consequences of ozone depletion. There are many lessons to be learned from the Montreal Protocol that can be applied to solving other global environmental issues (WMO/UNEP, 1999).

The primary lesson is an application of the 'precautionary principle': that is, taking necessary actions in time to prevent damage, rather than waiting for proof of the damage, by which time the damage could be great and irreversible. Another important lesson of the Protocol is how to react to issues when there is no scientific certainty (in 1982, when the Protocol was signed, many questions remained open). The solution has been to take successive steps to phase-out ozone-depleting substances and getting the scientific community to advise governments periodically on the further steps needed to protect the ozone layer and recommend alternative technologies. Over the last 10 years of the 20th century government have changed the Protocol four times in accordance with relevant scientific advice. Another lesson of the Protocol is that of promoting universal participation, including developing countries in the Protocol by recognizing 'common and differential responsibility'. Such a position has resulted in almost every country committing itself to protection of the ozone layer. One more important lesson is the integration of science, economics and technology both in developing control measures and implementing them.

Sarma (1998) emphasized that one measure of success of the Montreal Protocol and its subsequent Amendments and Adjustments was the forecast of 'the world that was avoided' by the Protocol:

- The abundance of ozone-depleting substances in 2050, the approximate time when the ozone layer is now projected to recover to pre-1980 levels, would be at least 17 ppb of equivalent effective chlorine (this is based on the conservative assumption of a 3% annual growth in ozone-depleting gases), which is about five times larger than today's value.
- Ozone depletion would be at least 50% at mid-latitudes in the Northern Hemisphere and 70% at mid-latitudes in the Southern Hemisphere, about 10 times larger than today.
- Surface UVB radiation would at least double at mid-latitudes in the Northern Hemisphere and quadruple at mid-latitudes in the Southern Hemisphere compared with an unperturbed atmosphere. This compares with current increases of 5% and 8% in the Northern and Southern Hemispheres, respectively.

As far as the Arctic ozone is considered, Ferguson and Wardle (1998) have pointed out that over the next few decades its total column composition will depend on climatic change. Not only anthropogenic impact due to ODS but various natural phenomena will lead to ozone changes, including weather systems, the quasi-biannual oscillation (a periodic reversal of the direction of stratospheric winds over the equator), El Niños, slight variations in solar radiation associated with the sunspot cycle, and volcanic eruptions.

A thematic issue of JGR has been devoted to the POLARIS mission (Photochemistry of Ozone Loss in Arctic Region In Summer) conducted in 1997 with special emphasis on photochemical investigation and detailed comparison of radical and reservoir measurements with models. On the other hand, Müller *et al.* (2001) used N_2O as a long-lived tracer to identify chemical ozone depletion in the Arctic vortex in the presence of ozone variations caused by dynamical effects. Observations conducted during the 1991/92 winter and early spring indicated (in consistence with the dynamical development of the polar vortex and with observed chlorine activation), that the major fraction of ozone decline had occurred before February 1992. The reduced ozone levels persisted over the lifetime of the polar vortex until late March 1992.

Shindell *et al.* (1998) have investigated polar ozone losses and delayed eventual recovery owing to increasing greenhouse-gas concentrations. Temperature and wind changes induced by increasing GHG concentrations alter planetary-wave propagation, reducing the frequency of sudden stratospheric warmings in the Northern Hemisphere. This results in a more stable Arctic polar vortex, with significantly colder temperatures in the lower stratosphere and concomitantly increased ozone depletion. Increased concentrations of GHGs might therefore be at least partly responsible for very large Arctic ozone losses observed in recent winters. Due to model calculations made by Shindell *et al.* (1998), the Arctic ozone losses will reach a maximum in the decade 2010 to 2019, roughly a decade after the maximum in stratospheric chlorine abundance.

Of serious importance may be the development of supersonic transport aircraft that will fly in the lower stratosphere. A fleet of 500 to 1,000 of these aircraft would release large quantities of nitrogen oxides, water vapour, and sulphates, all of which have the potential to increase ozone depletion. Because these substances contribute to PSC formation as well, their impact on the Arctic ozone levels could be particularly harmful (Kondratyev and Varotsos, 2000).

Special attention has been paid to problems of polar ozone changes during the Eight Session of the SPARC Scientific Steering Group and SPARC 2000 General Assembly (Chanin, 2001). An important subject was the internal variations in the stratosphere/troposphere coupled system – and the most pertinent issue was that of Arctic oscillation. The importance of the modulation due to large internal dynamical variations was repeatedly pointed out. Model calculations made by Tabazadeh *et al.* (2000) show that an important factor in Arctic ozone change may be widespread severe denitrification, which can enhance future Arctic ozone loss by up to 30%.

One of the main causes of the current focus on tropospheric ozone is its role as a greenhouse gas and toxic component. While retrieving the sources of change in radiative forcing and their influence on climate, Hansen *et al.* (1997) emphasized the need for information about the exact nature of change in the vertical distribution of ozone (especially to see if increases in tropospheric ozone may have countered decreases in stratospheric ozone) and the nature of aerosol particles (especially single-scattering albedo) as well as their height. Derwent *et al.* (2001) have emphasized the importance of considering indirect climatic impact due to tropospheric ozone, which dictates the necessity to take into account ozone precursors such as radiatively active trace gases. In this context further development of numerical climate modelling, taking into account various minor gaseous components, is urgently needed. Brasseur *et al.* (1998c) described simulations of ozone and related chemical tracers obtained by the Model for Ozone and Related Chemical Tracers (MOZART) developed at the National Center for Atmospheric Research (NCAR) in the USA. One of the results was the simulation of changes in surface ozone since pre-industrial time, which shows changes in ozone amounts of some 30–40 ppb in July over the USA and much of Europe along with those of 20–30 ppb over most of Eurasia. The average radiative forcing coming from these increases should be approximately 0.45 W/m^2 (higher in the Northern Hemisphere, lower in the Southern Hemisphere). Predictions for the future (2050) show that the largest changes of surface ozone are likely to come in the tropics.

Jacob *et al.* (1995) described the Chemistry, Aerosols and Climate Tropospheric Unified Simulation (CACTUS) effort at Harvard University. The present targets of this simulation include the climatic effects of changes in emissions of precursors on tropospheric ozone formation, the feedback of climate change on tropospheric ozone, the climatic effects of changes in emissions of sulphate precursors, and the response of sulphate aerosols to climate change.

Scenario calculations for the year 2025 made by Lelieveld and Dentener (2000) indicate that man-made emissions at low northern latitudes, in particular, in southern and eastern Asia, will become a very strong tropospheric O_3 source in the next decades.

5.2 GLOBAL CLIMATE CHANGE AND CARBON CYCLE

5.2.1 Introduction

Recent decades have witnessed the world community's attaching unprecedented importance to climate change problems, particularly those issues highlighted by the mass media. This undoubtedly has stimulated both scientific and applied developments through which further progress has achieved an understanding of the causes of present-day climate change and the regular features in paleoclimate. However, we still need a better vision of climate change in the future. The reliability of various scenarios and predictions of future change remain doubtful. Various speculative exaggerations and apocalyptic prognoses of human climatic impacts have, unfortunately attracted too much attention. As a result, the climate change problem conceptualized as human-üinduced global warming became an acute geopolitical issue (see, among many others: Kondratyev, 1992a, 1998a, 1998b, 2001a, 2001b; Bohmer-Christiansen, 1999, 2000; Hansen *et al.* 2000; Houghton, 2000, 2001). Paradoxically, presidents and prime ministers in various countries (e.g. in the USA) enter into the discussion on whether the Kyoto Protocol is to be treated as a scientifically sound document (From the Candidates, 2000).

This situation is getting even more complicated, particularly in the absence of sufficiently clear and uniform terminology. Without focusing on the very complex situation regarding definition (which warrants its own discussion), we are reminded that the Kyoto Protocol defined 'climate change' as human-induced climate change. Anthropogenic impact on climate is beyond question, but the contribution made by anthropogenic factors to global climate formation still need to be convincingly assessed, and this is the main problem.

In this situation, international documents analyzing modern views of the world's climate unnecessarily apply the term 'consensus'. Obviously, science has progressed along the path that compares various views and discusses, rather than votes on, specific issues. In the case of climate change, this progress is dictated not only by a lack of precise definitions in the terms being used but also by unclear and vague conceptual assessments for various aspects of climate problems.

Unfortunately, the notion of 'sustainable development' itself still needs an adequate definition; for example, the term is especially pertinent in its Russian version. In this connection we are reminded that despite the undeniable success of UNCED, the Second UN Conference on Environment and Development held in 1992 at Rio de Janeiro and the Special Session of the UN General Assembly, 'Rio + 5', which was held five years later (1997) in New York, these two forums consisted only in drawing the attention of governments and the public to global change and sustainable development problems. Unfortunately, both forums were ill-prepared, as most clearly evidenced by their failure to elaborate the 'Earth Charter intended for formulating and substantiating the priorities; instead, a very amorphous and declarative 'Rio Declaration' was adopted.

All of this is becoming of particular importance in the context of the World Summit on Sustainable Development which was held in the Republic of South Africa from 26 August to 4 September 2002. There is good reason to believe that this 'Rio + 10' conference will focus specifically on climate problems. Today, the focus of discussion should be concentrated on three global environmental problems:

(1) 'Global warming' – anthropogenically induced climate change.
(2) The fate of the ozone layer.
(3) The closed nature of global biochemical cycles and the concept of biotic regulation of the environment.

The primary importance of the third problem and the secondary significance of the two first-named problems have been convincingly substantiated in the scientific literature (Gorshkov, 1990; Gorshkov *et al.*, 2000; Kondratyev, 1990, 1992b, 1997a, 1998c, 1999a, 2000b, 2001a). A bitter paradox consists, however, in that despite all this the UNCED documents show inadequate understanding of the crucial importance, from a conceptual viewpoint, of the sequence of events: that is, socioeconomic development under conditions of the population growth → anthropogenic impact on the biosphere → environmental implications of these impacts (climate, ozone, etc.).

Such misunderstanding advanced to the foreground the phrase 'global warming' and this in turn resulted in the adoption of the International Framework Convention on Climate Change (FCCC). This framework document is rather inadequate and misleading in its unjust treatment of developing countries; its focus – without good reason – is on the anthropogenic origin of observed global climate change and recommended reductions in greenhouse gases emissions, above all of carbon dioxide.

In December 1997, 160 countries participated in the Third Conference of the FCCC signatory-states, which was held in Kyoto, Japan. The possibility of implementing the required 5%, on average, CO_2 emission reduction by 2008–12 relative to the level of the year 1990 was the subject of prolonged and heated debate. The latter and the, to date, lack of notable progress in CO_2 emission reduction clearly illustrate the absurd character of the Kyoto decisions. Global emissions are tending to increase – not only in developing but also in advanced industrial countries, including the USA – and this trend will be preserved in the future. Naturally, the attitude of developing countries is dictated by the highest priority for them – namely, raising people's living standard via industrial development rather than curtailment aimed at CO_2 emission reduction. However, this latter requirement was specifically the FCCC signing condition laid down by the USA and other 'golden billion' countries. The history of the FCCC is an example which illustrates how a primarily bureaucratic institution can devour hundreds of million dollars each year – according to GEF (Global Environmental Facility) data as of 30 July 1998, $US 1.9 bn was assigned for 267 GEF projects (Project, 1998). It should be remembered that there were some 10,000 participants at the Kyoto Conference alone. The Sixth Conference of the FCCC signatory-countries held in The Hague on November 13–24, 2000 and the Bonn Workshop (July 2001) also had large audiences and entailed large expenditures. The sums absorbed by these bureaucratic meetings could have been invested in the development of science to the greater benefit of humankind (Singer, 1997, 1998).

One might think that the situation stems from the poor development of scientific principles as regards global change problems. This conclusion is justified only in part since, as early as 1990, the key aspects of global environmental dynamics were discussed; for example, monographs by Kondratyev (1990, 1998c) and Gorshkov (1990) and Gorshkov *et al.* (2000) advanced and substantiated the basic concept of biotic regulation of the environment. Kondratyev (1990, 1992b, 1998c, 2001a) demonstrated the fruitlessness of solely focusing on the 'greenhouse' aspect of global warming, drawing attention to the requirement in the study of the climatic system 'atmosphere–ocean–land–ice cover–biosphere' to take into account the whole complex of feedbacks among its interactive components. The problem of global observation of the system was seriously analyzed in a number of publications (see Goody *et al.*, 1998; Kondratyev and Varotsos, 2000a; Reconciling, 2000; Goody, 2001; Goody *et al.*, 2001; Kondratyev and Cracknell, 2001), special attention being given to atmospheric ozone variability.

In the problem of human-induced global climate change, the main uncertainties stem from the following:

(1) Observation data are incomplete and insuffuciently reliable not only for quantitative assessment of the anthropogenic component of 'global warming' but even for its scientifically substantiated identification.
(2) Uncertainties involved in accounting for the climatic-forming role of atmospheric aerosol and in introducing the so-called 'flux correction' in numerical climate modelling, can amount to tens or even 100 W/m^2 (Kondratyev, 1998d, 1999a) being much more significant than a potential $\sim$4 W/m^2 enhancement of the atmospheric greenhouse effect due to the expected doubling of the atmospheric CO_2 concentration.
(3) The results of numerical modelling of climate changes using relevant models differ and cannot be adequately verified; nor can they be used for reliable assessment of 'global warming', because the results of such calculations depend on adjustment to observation data.
(4) Even full accomplishment of the Kyoto Protocol recommendations can ensure only a negligible decrease in the annual global average near-surface air temperature (SAT) of no greater than several hundredth of one degree (Wigley, 1999).

In this situation, the recommended greenhouse gas emission reductions make no sense, although their implementation can have far-reaching adverse socioeconomic consequences.

Recently, much attention has been given to analyzing the uncertainties (incompleteness) of numerical climate modelling. The major uncertainties, evidently, result from an inadequate account of interactive processes in the 'aerosol-clouds-radiation' system (see Kondratyev, 1992b, 1998c, 2002a). The fact that the most complex aspect of climate numerical modelling is that of taking into account the interactive biosphere dynamics is beyond question. This can be illustrated by two specific examples, which, although only to a small extent, certainly reflect the complexity of the problem discussed.

For explaining the decrease in the daily variation amplitude of the surface air temperature (DTR) by 3–5 K (which was observed during the period 1951–93) owing to a more rapid rise of the minimal, compared to maximal, temperature, it was suggested that various factors such as changes in the amount of clouds, water vapour and tropospheric aerosol content, as well as turbulence and soil moisture, be taken into account. Positive trends in the three first-named factors could be responsible for a global radiation reduction during the daylight hours and enhancement of atmospheric longwave radiation (LWD) at night. At the same time, changes in turbulent mixing intensity and soil moisture could be responsible for variations of heat and moisture exchange between the surface and the atmosphere, more substantial during the day than at night.

A fairly pronounced interactivity of climate-forming processes and inadequate parameterization of them in climate models significantly complicate the task of assessing contributions from various mechanisms that explain the DTR decrease. In this connection, Collatz *et al.* (2000) undertook numerical modeling on the response of the daily temperature variation of the vegetation-covered land surface to the changes in the external forcing and biophysical state of the vegetation cover. To this end, Collatz *et al.* used the SiB2 approximate land biosphere model at given meteorological conditions and under various scenarios so as to simulate the likely impact of the interactive dynamics of vegetation cover on the DTR.

The numerical modelling results showed that with increasing LWD the temperature of the air above the vegetation cover T_m tends to increase at night, thus decreasing the DTR. At the same time, changes in T_m or a rise in $T_m \pm$ LWD (this is specifically the case for global warming conditions) favour increases both in the minimal and maximal temperatures making these factors of little importance for DTR. This response is due mainly to the influence of daily variations in aerodynamic stability and radiation balance.

Many climate numerical modelling experiments utilize global atmosphere circulation models (GCM) coupled with land surface processes models (LSM). The results of these numerical experiments are essentially dependent on the specific features of interaction between the GCM and LSM models simulating the radiation, momentum, and energy exchanges between surface and atmosphere. In striving to cover the diversity of terrestrial ecosystems, LSM models were significantly complicated by introducing submodels, which take into account photosynthesis, vegetation cover dynamics, and biogeochemical cycles thus making the models radically more realistic.

In 1999, Kim *et al.* (2001) conducted various numerical modelling sensitivity experiments based on the Simple Biosphere Model SiB (an LSM version). The experiments were carried out on a rice paddy test in Thailand (17°03′N, 99°42′E) during the period 1–6 September, using SiB2 models and a modified SiB2-Paddy model coupled with a mesometeorological model GAME-Tropics. (GAME is the monsoon experiment in Asia conducted in the framework of the GEWEX global field experiment on energy and water cycles.) The Thailand findings indicated the importance of not only the sensitivity of SiB to many of the morphological parameters of the vegetation cover but also of the sensitivity of transpiration of high vegetation cover to parameters characterizing vegetation cover resistance. With the modified SiB model, Kim *et al.* took into account the biogeochemical processes governing water vapour, energy, and carbon dioxide exchange between the surface and the atmosphere. The results of numerical modelling of the processes were compared with the meteorological data derived in the rainy season over the period.

This comparison showed a good agreement between, on the one hand, the results simulated with the two SiB models and, on the other, the observed daily variation of the radiation balance and the latent heat flux. The only exception was the latent heat flux simulated by the SiB2 model. The SiB2-Paddy simulations show a satisfactory agreement with the observed fluxes of latent heat and heat flux in soils, as well as of the carbon assimilation rate, but SiB2 simulations entail major systematic errors. After certain adjustment of the paprameters, the SiB-Paddy model provides fairly reliable values of the soil, water, and vegetation cover temperatures. The simulations of the radiation balance, as well as of the energy and water balances, latent heat fluxes, and carbon dioxide assimilation rate yielded fairly adequate results and offers certain promise of an adequate account of the biosphere as an interactive component of the climatic system. De Rosnay *et al.*

(2000) assessed the reliability of the parameterization schemes for the processes occurring on the land surface, which are utilized in general atmosphere circulation models (GCMs) from the viewpoint of agreement between the observed and the calculated annual average energy and water fluxes as depended on the degree of detail in accounting the vertical structure of soil. The simulations evidence a fairly strong dependence of the fluxes on the vertical resolution. The 11-year scheme of parameterization of the heat and water transfer in soil proves fairly adequate for a 1 mm thick upper layer. The possibilities of realizing a scheme with this thin upper layer are unclear. However, if one takes into account the fact that the horizontal resolution of GCM is of the order of hundreds of kilometres, finding a solution to this kind of problem requires further effort.

A major component of the problem of climate numerical modelling is a complex of issues related to the chemistry of atmosphere. It is a well-known fact, for example, that the concentration field of such a greenhouse gas as tropospheric ozone (TO) under various conditions (city, region, and globe) is strongly affected by various short-lived minor gaseous components (MGCs) – ozone precursors – such as nitrogen oxides ($NO_x \times NO \pm NO_2$) and methane (CH_4), as well as many organic compounds, hydrogen, and carbon monoxide (CO). Each MGC has its specific natural (biospheric) and anthropogenic sources.

Since TO is a greenhouse gas, the MGC emitted can affect indirectly the formation of an atmospheric greenhouse effect by in turn affecting the TO concentration field. Also, MGC-TO precursors affect the hydroxyl concentration field and, thereby, the oxidizing capacity of the troposphere. The hydroxyl concentration distribution in the troposphere, in turn, governs the lifetime and, thereby, the global-scale concentration of methane.

All this is responsible for a complex interactivity of the processes responsible for both direct and indirect impacts on the formation of atmosphere greenhouse effect. Derwent *et al.* (2001) described the global three-dimensional Lagrangian STOCHEM model which simulates chemical processes taking into account the MGC transport. This model was used for simulating interconnected TO and methane concentration fields in case of emissions of short-lived precursors of tropospheric ozone such as CH_4, CO, NO_x, and hydrogen. In this case the radiative forcing (RF) of NO_x emissions varies with the emission location, be it near the surface or in the upper troposphere, or in the northern or southern hemisphere. For each short-lived MGC-TO precursor the global warming potential (GWP) is calculated using the data for the reaction between methane and tropospheric ozone under 100-year forcing. The introduction of GWP means that RF due to emission of 1 Tg of an MGC was estimated (for a 100-year period) as equivalent (in RF) of carbon dioxide emission. The combined impact of methane and TO led to the GWP of 23.3.

The simulations showed that indirect RF due to the methane and TO content is significant for all the MGC-TO precursors of interest. In the case of methane, the RF is determined primarily by emissions of methane itself, while in the case of TO by those of all the MGC-precursors, especially nitrogen oxides. The tropospheric ozone-induced indirect RF may be so large that MGC-TO precursors will need to be ranked among the MGC essential for assessing likely climate changes and identifying preventive measures.

In the USA, despite the anti-Kyoto statements made by President George Bush, many American newspapers published January 2001 articles with dramatic headings informing readers about the terrible warming predictions made by scientists. Articles in such media as the *Washington Post* and the *International Herald Tribune* warned of the threat of a global catastrophe in this new century posed by accelerating climate shift, as well as about how the earth's warming sent a new danger signal, etc. All this stemmed from new climate change scenarios for the twenty-first century, which predict that the changes will be more significant than expected earlier. For example, IPCC-2001, the Third Assessment Report of the Intergovernmental Panel on Climate Change, states that by the year 2100 the annual global average surface air temperature (SAT) can increase by 5.8°C relative to the present time against only 3.5°C according to the estimate of five years' standing (IPCC, 1996).

Kerr (2001b) correctly noted that of even greater importance is the widening range of the likely SAT rise compared to earlier estimates. For many experts in climate numerical modelling, however, this was a surprising observation as this field is still in its infancy; numerical modelling has to rely on a limited amount of observation data (even for SAT the dfata series is about 100 years long, being not globally complete).

Although the majority of experts tend to attribute observed global warming to the GHG concentration growth, the range of possible climate change assessments has expanded rather than conracted in certain respects.

The main uncertainties in climate change assessment involve the following:

(1) Global warming detection from observation data.
(2) Global warming attribution to anthropogenic factors.
(3) Future climate change prediction.

Kerr (2001b) believes that, owing to the new findings in IPCC-2001, the range of uncertainty with respect to the above two first-named aspects of the problem has contracted, but future climate predictions have become still less clear. IPCC-2001 estimated the observed global warming at $0.6° \pm 0.2°C$ (at the statistical significance level of 95%); the warming during the recent 50 years being probably (by 66–99%) due to GHG concentration growth. One of the main factors responsible for uncertainties in climate numerical modelling is, as has been mentioned, inadequate account of the climate-forming role of atmospheric aerosols and clouds.

Unfortunately, IPCC-2001 did not adequately assess the role of uncertainties in climate numerical modelling and this attracted criticism by many experts (among them: Singer, 1997, 1998, 1999; Woodcock, 1999, 2000; Han *et al.*, 2000; Soon *et al.*, 2000; Kondratyev and Demirchian, 2001; Schrope, 2001; Wojick, 2001). In the context of global change, assessments of present-day and likely future global climate change are undoubtedly of primary importance. Although the 'global warming' concept is still dominant (as evidenced by IPCC-2001), new assessments should be treated as due only to mechanical development of the earlier speculations with a far-from-scientific motivation, as convincingly demonstrated by Boehmer-Christiansen (1999, 2000).

The contradictions inherent in climate assessments can be illustrated by the radically opposing opinions expressed by the two candidates contesting the US Presidential election campaign of 2000 (From the Candidates, 2000). The Democrat, Al Gore, had long been known as an ardent advocate of the 'global warming' concept and the Kyoto Protocol, while his Republican opponent, George Bush, can be characterized as having rejected an environmental policy corresponding to the Kyoto Protocol. Bush argued that such a policy would result in a radical rise in the prices of oil, oil products intended for district heating needs, natural gas, and electricity, and that an agreement would expose the USA economy to a strongly increased load without proper protection against undesirable climate changes. For Bush, the Kyoto Protocol is an inefficient and inadequate document which unjustly treats the USA, as it excludes from implementation of the Protocol recommendations 80% of the world, including such major centres of population as China and India. He assigns the primary importance to development of new environmentally friendly technologies and to the use of market mechanisms, including regulation-free electricity and natural gas markets, taxation, and 'emission trading'. President Bush, as he now is, believes that natural gas and nuclear energy will play an important role in weakening the USA's perilous reliance on petroleum from abroad and will also provide the USA with energy resources in the twenty-first century. While sharing this radical criticism toward the Kyoto Protocol, we note that President Bush's opinion on market mechanisms is specific to the USA and, in the case of 'emission trading', disputable.

The unsoundness of the Protocol is even more clearly evident from the failure of the Sixth Conference of the FCCC signatory-countries held in The Hague on November 13–24, 2000. Speaking at the Conference, which was attended by 7,000 representatives of 182 governments, 323 inter-governmental and non-governmental organizations, and 443 mass media titles, Congressman J. Barton (Republican, Texas) said that if Bush won (which came true), his advice to the new president would be that the USA reject the Kyoto Protocol and start negotiations aimed at setting the USA's economy free from unsound environmental restrictions. Barton added that what they saw at The Hague was an extremely useless – or, at best, making-things-up – exercise. Therefore, in Barton's opinion, nothing from that week's discussion deserved a support by positive voting.

Interestingly, in this connection, K. Topfer, the UNEP Executive Director, rejected the suggestion to treat nuclear energy as an important energy prospect, while the delegates from the USA and Japan expressed their readiness to support the funding of nuclear energy projects in developing countries with a view to reducing carbon dioxide atmospheric emissions.

An important feature of The Hague discussions was a certain confrontation between the USA and member countries of the EC. The Europeans rejected the suggestion made by the Americans of making use of various possibilities, including restoration of forests as carbon sinks in the carbon balance and required that the USA follow the general recommendations on carbon dioxide emission reduction (this view changed later on to begin 'forest games').

One cannot qualify differently than absurd the recommendations of the second part of the Conference, which was held in July 2001 in Bonn. These recommendations introduced the notion of 'certified emission reduction' (CER); that is, the recommended real emission reduction targets were replaced by equivalent intensification of carbon sinks such as forests. This 'innovation' permits Japan, Russia, and Canada to accumulate CER owing to their forests. The absurd character of this recommendation is evident, above all, from the fact that the global carbon cycle problem is still far from being solved (Kondratyev and Demirchian, 2000), which makes absolutely non-realistic any reliable assessment of the role of CER as a global climate-affecting factor. As to the intensively debated subject of the three 'flexibility mechanisms': namely, joint implementation, emission trading, and clean development mechanism (of technologies), these can be regarded as rhetorical only.

The Bonn recommendations received extremely contradictory opinions. On the one hand, D. Kennedy, the Editor of *Science* was distressed (Kennedy, 2001) about the 'going it alone' policy of President Bush and the USA and qualified the Bonn agreements as 'breathing new life' into hopes for progress in solving the problem of global climate change. In the opinion of Sherman (2001), the Kyoto Protocol, in spite of its clear drawbacks, should be regarded as an important milestone in the history of climate protection. Sherman argues that the Protocol will, probably, exceed any other international agreement in its effect on the lives of all humans on this planet in the twenty-first century, and believes that the success achieved in Bonn is a direct consequence of the dialogue and mutual understanding, as well as of the sense of reconcilement and compromise adopted by delegates.

On the other hand, the opinion of a well-known Indian scientist, Dr A. Agrawal, Chairman of the Centre for Science and Environment of India, was absolutely different. To begin with, Agrawal said that the Kyoto Emperor was not wearing any clothes. Next, he characterized the Kyoto Protocol as a tiring and a fairly meaningless agreement on the problem of climate change which was agreed by the world within those two weeks. Agrawal pointed out that there are huge scientific uncertainties in assessments of greenhouse gas emission reduction efficiency and that made of the Kyoto compromise nothing more than a big shameful invention. This is not, however, surprising, in Agrawal's opinion, because climate negotiations concern economics rather than envionmental problems and each nation state does its best to protect its own rights regarding environmental pollution. In the same week as the euphoric declarations that the agreement achieved in Kyoto could save the world. the EC decided to prolong for another 10 years the implementation of its subsidies programme for coal (the most polluting of carbon fuels). Dr Agrawal concluded his statement by saying that the Kyoto compromise would cost the whole world and everyone in it more than new clothes for the Emperor.

The main point about the global change problem is that while global climate warming in the 20th century (especially, in its last quarter) was beyond question, the factors responsible and quantitative estimates of their contributions to global climate change remain the subject of heated debate. This point is even more pertinent to climate forecasts, taking into account anthropogenic impacts. In this connection, it is symptomatic that the authors of IPCC-2001 abandoned the 'climate change' term defined in the FCCC as due to anthropogenic factors only and agreed upon an adequate definition which takes into account both natural and anthropogenic factors responsible for climate change. This correction to the 'climate change' term makes senseless the Kyoto processes aimed at emission reduction, for the constant changes to the climate are due to many factors. To supplement this, such change also contradicts the traditional definition of climate as the phenomenon characterized by a 30-year average value of its parameters.

We now turn to the question of how the IPCC-2001 has affected views on the causes of climate change.

5.2.2 Observational data

The variance of the present-day climate and climate change assessments result primarily from the inadequate completeness and quality of available global observation data. Climate, of course, can be characterized by numerous parameters such as surface and atmospheric air temperature and humidity, precipitation (liquid and solid), amount, lower and upper boundary altitude, microphysical and optical characteristics of clouds, radiation, balance, and its components, microphysical and optical parameters of the atmospheric aerosol, chemical constituents of the atmosphere, and many others. At the same time, the empirical analysis of climatic data is usually confined to the appeal to SAT observation data, because only in this case the observational series cover a time period of 100–150 years. However, even these series are far from being homogeneous, which is particularly relevant to the global data file which is the main information source for 'global warrning' concept substantiation. It should also be mentioned that the long-term variations of annual average SAT relies mainly on imperfect observation data for the sea surface temperature.

In climate observation diagnostics, the focus must be on analyzing climate variability, where not just averages but higher order moments are taken into account. Unfortunately, this approach has not been attempted so far. The same is true of internal correlation for the observational series. McKirtrick (2002) analyzed SAT long-term variations and showed that, by subtracting the contribution to the temperature change within several recent decades by internal correlation (that is, climate system inertia), one reduces the temperature change virtually to zero. This points only to the fact that natural factors can affect the temperature, but the main problem still lies in singling out its anthropogenic component. The importance of this problem is dictated by the fact that, specifically, the global average SAT rise over the last 20–30 years serves the main argument in favour of the dominating anthropogenic contribution to climate change.

5.2.2.1 Surface air temperature

One of the SAT observational series for the period since 1860 showed that the global annual SAT average value increased by $0.60 \pm 0.2°C$. This exceeds, by $\sim 0.1°C$, the analogous value from IPCC-2001 (see Kondratyev, 2002c) which explains this increase by a higher SAT level over the 1995–2000 period. The observation data are indicative of a very strong spatiotemporal variability of the global annual SAT average (such a variability is a cause of why reliable enough assessment of SAT uncertainties is impossible). For example, the global annual SAT average change suggests that climate warming in the 20th century fell primarily in two time periods, namely, 1910–45 and 1976–2000. Compared to IPCC-2001, the new version of the global annual SAT average change in the Northern Hemisphere excludes the warming period of 900–1200 and the cooling-off period of 1550–1900. Only such a revolution in interpretation allows for the argument that climate warming in the Northern Hemisphere in the 20th century was the strongest over the last 1,000 years with the 1990–2000 period as the warmest decade and the year 1998 as the warmest year. Owing to specifically these arguments, an impression may be created that it is the marked increase in CO_2 concentration over the last three decades that is specifically responsible for such extreme temperatures. By excluding these (not simulated by numerical models) natural phenomena one can achieve certain adequacy between the modelled and observed temperatures.

IPCC-2001 makes no mention of the previously supposed climate warming enhancement in the high latitudes of the Northern Hemisphere as a characteristic feature of human-induced global warming. However, SAT analysis directly measured at the 'North Pole' stations over a 30-year period (Adamenko and Kondratyev, 1999) and the dendroclimatic proxy data for the last two to three centuries shows that there was no warming enhancement in the Arctic region. The temperature changes during the last century and recent decades exhibited major spatiotemporal non- homogeneity manifested as the simultaneous formation of climate warming and cooling-off regions in the Arctic region (see also Stafford *et al.*, 2000).

Since the 1950s, with a more or less adequate aerological observational network, the trends for the global annual SAT average and lower troposphere temperature have been almost identical, about 0.1°C/10 years (Angell, 1999, 2000a, 2000b). According to satellite-based microwave remote sensing data (over the period 1979–2000), the global annual average temperature of the lower tropospheric air increased by $\sim 0.036°C/10$ years, that is, significantly slower compared to SAT (0.24°C/10 years). This difference is even more appreciable if one excludes the influence of the sharp temperature rise in 1998 due to the severe El Niño phenomenon. This is evidenced by low trends for the 1979–97 period, namely, 0.012°C/10 years for the tropospheric air (that is, cooling according to the microwave remote sensing) and 0.158°C/10 years for SAT. There is no such trend difference in the global circulation models that are the main IPCC instruments for understandingy climate change and this casts serious doubts upon their capabilities. The controversial nature of existing temperature observation data has been convincingly analyzed by Wojick (2001).

Sonechkin (1998) has accomplished a study of self-similar and trend-like properties of hemispheric SAT time series as realizations of a fractional Brownian motion by means of the wavelet transform technique. The results obtained indicate the first evidence of a crossover scale that separated the obvious internally-induced, statistically stationary, chaotic oscillations from the substantially longer, trend-like SAT variations, which origin is not clear and may be assumed to be externally forced. The residual trend-like components reveal a single linear warming trend that was started at the beginning of the 20th century. The increment of this trend is equal to 0.59°C per 100 years for both hemispheres. Later on, Sonechkin *et al.* (1999) demonstrated a very strong coupling between long-term SAT variations and Southern oscillation dynamics. Datsenko and Shabalova (2001) have applied a similar approach to analyze seven seasonal early instrumental temperature series at various locations in Europe. They emphasized the crucial role of seasonality in the spatiotemporal structure of low-frequency SAT variability.

5.2.2.2 *Snow and ice cover*

These climate characteristics are treated by IPCC as important indirect global warming indicators. IPCC-2001 says that there was a −10% snow cover reduction since the late 1960s and a −2-week reduction of the annual lake and river ice cover in mid- and high-latitudes of the Northern Hemisphere in the 20th century. This was paralleled by the retreat of mountain glaciers in non-polar regions. Also, IPCC-2001 states that the sea-ice cover in the Northern Hemisphere in the spring and summer periods has reduced since the 1950s by 10–15%, with a much less significant thinning in the winter period. Since the 1970s, regular satellite observations have revealed no marked trends for ice cover either in the Arctic or Antarctic regions. Combined with insignificant temperature trends, this allows for the conclusion that the ice thickness in the Arctic region is markedly influenced by ocean currents and the North Atlantic Oscillation (NAO) dynamics.

5.2.2.3 *Sea surface level and heat content of the upper layer of the ocean*

These characterisitcs are also important for IPCC in view of the serious concerns about the global consequences of a sea-level rise in the World Ocean. Because of exclusively fragmentary, intermittent, and unreliable measurement results available, the IPCC-2001 actually presents the calculation-corrected data. It states that the World Ocean level has risen by 0.1–0.2 m in the 20th century owing to thermal expansion of sea water and land ice melting due to global warming. The report says that in the 20th century the World Ocean level rising rate exceeded almost tenfold that over the last 3,000 years. Since the 1950s, ocean surface temperature rising has caused an increase in the heat content of the upper ocean layer.

Levitus *et al.* (2001) analyzed the warming data for individual components of the climate system in the second half of the 20th century. These data were derived from the analysis of the increase in the heat content of the atmosphere and the ocean, as well as the heat consumed by melting of certain components of the cryosphere. The results suggested an increase in the heat content of the atmosphere and the ocean. Over the 1950–90 period, the increase in the heat content of the 3 km-thick upper ocean level exceeded at least by an order of magnitude those for other components of the climate system. The increase in the heat content of the ocean observed in the 1955–96 period was estimated at 18.2×10^{22} J, and that in the case of the atmosphere, at 6.6×10^{21} J only. As to the latent heat due to phase transformations of water, these were estimated at 8.1×10^{21} J (decrease in the mass of terrestrial glaciers); 3.2×10^{21} J (sea ice cover reduction in Antarctica); 1.1×10^{21} J (melting of mountain glaciers); 4.6×10^{19} J (snow cover reduction in the Northern Hemisphere); and 2.4×10^{19} J (melting of permanent ice cover in the Arctic region).

In addition, Levitus *et al.* (2001) compared the observation data with the results of numerical modelling by using an interactive model of the 'atmosphere-ocean' system developed at the Geophysical Fluid Dynamics Laboratory in the USA. This comparison took into account the radiation effects due to the observed GHG concentration growth, and changes in the sulphate aerosol in the atmosphere and extra-atmospheric insolation, as well as volcanic aerosol. It was found that the observed heat content variations in the ocean can be primarily attributed to GHG concentration growth in the atmosphere, although one has to take into account a major uncertainty in assessing the radiative forcing due to sulphate aerosol and volcanic eruptions. The latter fact makes the work by Levitus *et al.* (2001) of insufficient reliability as regards human-induced warming identification; but Levitus et al did mention the major variability of the heat content of the World Ocean from year to year. They emphasized that they partially attributed external warming of the World Ocean in the 1990s to the multidecadal warming of the Atlantic Ocean and the Indian Ocean, as well as to the positive polarity of the possible biannual heat content fluctuations of the Pacific Ocean. The observed variations of the heat content of the World Ocean can be related to the modes of hemispherical or global variability of the atmosphere from the ocean level to the stratosphere. Gaining insight into the nature of this kind of possible relationship is of major importance in understanding the mechanisms governing global climate.

As has been already mentioned, recent developments in identifying human-induced climate changes were confined for the most part to analyzing a comparatively long series of SAT data. Also, much more limited bodies of information on sea-ice cover variations, vertical temperature profiles (radiosonde data), and satellite microwave remote sensing results were analyzed. On the other hand, numerical modelling results suggest that in this context more representative than SAT would be the extremely scanty data on the amplitudes of the annual and daily temperature variations in winter.

Being the major component of the global climatic system, the World Ocean holds priority in variability analysis, especially after Levitus *et al.* (2001) revealed an increase in the heat content of the upper layer of all the world's oceans within the past 45 years. In this connection, Barnett *et al.* (2001) discussed the results obtained by comparing numerical modelling and observation data for the heat content of the 3 km-thick upper layer of different oceans. The calculations utilized the 'parallel' climate model (PCM) for the 'atmosphere-ocean' interactive system (no flux adjustment) and five scenarios of the growth patterns for GHG concentration and sulphate aerosol content in the atmosphere.

The comparison showed that the calculated heat content anomalies (that is, deviations from the data corresponding to the control integration) did not differ from the observed values (for the 1950–90 period), the statistical significance being at the level of 0.05. The only exception (in the global averaging case) was the data for the 1970s. In this case, the model does not simulate the heat content anomaly observed during this decade. On the whole, the probability of fact that the heat content anomalies are due to natural variability of the climatic system does not exceed 5%. This makes realistic detection of the anthropogenic signal of climate changes impossible.

It should be noted, however, that oceans differ substantially in the nature of warming. A typical feature of the Atlantic Ocean (especially of its southern area) is an intensive vertical mixing and rapid propagation of the warming deep into the ocean. In other oceans this process has a much slower rate. The results obtained allow an important conclusion: that is, climate models should simulate not only SAT but also ocean heat content variations. Barnett at al. (2001) showed certain weak points of the numerical modelling undertaken – in particular, that natural climate variability was assessed based on control numerical modelling data only.

Cai and Whetton (2000) called attention to the fact that the enhancing effect of 'greenhouse' warming on the sea surface temperature field in the Pacific Ocean tropics can affect significantly precipitation on a global scale in the future. Studies on these controversial problems, using both observation and numerical modelling data, yielded very different results. Climate warming during recent decades was similar in the spatial structure to the El Niño/Southern Oscillation (ENSO) phenomenon. In view of the lack of the data for such a structure over the entire century, it was suggested that the observed warming structure resulted from multidecadal natural climate variability rather than from greenhouse forcing-induced change.

The first results of numerical modelling with the use of interactive models of the 'atmosphere-ocean' system showed that the warming structure characterized by a zonal sea surface temperature field gradient in the equatorial zone should be similar to El Niño, in contrast to certain theoretical studies which point to a similarity with La Niña. To settle this controversy, Cai and Whetton (2000) utilized for climate numerical modelling the interactive model developed at the Australian National Scientific Centre, CSIRO MARK 2. They showed that the initially formed spatial structure of warming is similar to La Niña (the strongest warming in extra-tropical latitudes for weak La Niña which is similar to the structure in the tropics), which later on (after the 1960s) transformed to a structure similar to El Niño. Such results were yielded by three versions of numerical modelling (in addition to the control integration over 1,000 years) with GHG concentration growth in the atmosphere prescribed according to the observation data (1880–1990) and the IS92a scenario

(1990–2100). These simulations did not take into account the influence of aerosol on climate formation.

The above-mentioned transformation of the spatial structure of climate warming is due to warm extra-tropical waters that, after deep submerging, come through the sub-tropics and reach the tropical zone where upwelling arises. This is specifically the reason for climate change. These results can be interpreted as confirming the conclusion that warming having a characterisitc El Niño-like spatial structure, observed over recent decades, is at least partly attributable to human-induced enhancement of the atmospheric greenhouse effect. It was noted, however, that despite the similarity between observed and calculated warming structures, observations carried out before the 1950s are less reliable. Also, conditions similar to La Niña were observed once again recently (in 1995–96 and 1998–2000).

5.2.2.4 *Other climate parameters*

The observation data suggest a 0.5–1%/10 years precipitation enhancement in the 20th century in most terrestrial regions in the mid- and high-latitudes of the Northern Hemisphere. This was paralleled by approximately 0.3%/10 years precipitation reduction in most of the sub-tropical-latitude land regions which, however, weakened in more recent years. As to the World Ocean, the lack of adequate observation data prevented revealing reliable precipitation trends. Since the middle 1970s, the most stable and intensive phenomena have been those of ENSO; such dynamics were manifested in the features of the regional variations of SAT and precipitation in most tropical and sub-tropical zones. The observation data on the intensity and occurrence of tropical and extra-tropical cyclones, as well as local storms, remain inadequate which prevents revealing any clear trends (Grigoryev and Kondratyev, 2001).

5.2.2.5 *Greenhouse gas and anthropogenic aerosol concentration in the atmosphere*

IPCC-2001 notes that since 1750 carbon dioxide concentration in the atmosphere has increased by approximately one third and attained its highest level over the last 420,000 years (possibly over the last 20 million years), as evidenced by borehole data. About two thirds of the CO_2 concentration increase over the last 20 years is accounted for by fossil fuel burning emissions (the remainder resulted from deforestation and, to a lesser extent, from the cement industry). Interestingly, by the end of 1999, the CO_2 emissions in the USA exceeded the 1990 level by 12%, which figure will further increase by another 10% by the year 2008 (Victor, 2001). At the same time, the Kyoto Protocol requires a 7% emission reduction by 2008 relative to the 1990 level, which implies a total emission reduction by –25% (which is, certainly, absolutely impossible).

According to IPCC-2001, both the World Ocean and the land itself act now as CO_2 global sinks. In the ocean this action is due to chemical and biological processes and on the land to enhanced 'fertilization' of vegetation due to increasing CO_2 and nitrogen concentrations, as well as to changing land use patterns. The approach of IPCC-2001 to the carbon cycle remains virtually identical to those opinions adopted in the early 1990s, although it does devote a special chapter to the problem. From the very beginning, IPCC-2001 was wrong in estimating forest and land use processes, taking fossil fuel burning and deforestation as the main sources of CO_2 emissions. In attempts to close the carbon cycle by balance calculations IPCC-2001 introduces an undefined 'fertilization sink' term which obscures the carbon cycle problem. Emissions due to 'carbon burning' by humans when breathing were exclude as a CO_2 source. According to Wigley (1998, 1999), this totals about 135 kg of carbon a year; for six billion people this makes about 0.8 Gt, and this is much greater than cement industry emissions which is a 'must' in IPCC-2001. As shown by Kondratyev and Demirchian (2000), an additional, beyond 270 ppm, anthropogenic increase in CO_2 concentration in parts per million K_a linearly varies with global population size. According to this dependence, current CO_2 concentration in the atmosphere can be parameterized as $K \sim 270 + K_a = 270 + 15P$, where P is the population size (in billions). For example, the concentration K of the atmospheric carbon dioxide in the year 2000, when the population size was about 6.1 billion, is equal to $K \sim 270 + 180 = 450$ ppm. This value differs by only 2.2% from that obtained at the Mauna Loa observatory, namely, 369.7 ppm. This inaccuracy is comparable with those for P and K values. When related to the $K_a - P$ linear dependence, the carbon cycle model not only simulates more adequately the processes observed in the carbon cycle but also (which is more important) links more accurately the forecasted future CO_2 concentration to the only variable; that is, the population size. For example, based on this relationship, one can expect that carbon dioxide concentration in the atmosphere will rise by the year 2100 (with a predicted population size in 2100 of 12 billion) to $K_a = 15 * 12 = 180$ and $K \sim 270 + 180 = 450$ ppm.

The fact that the CO_2 concentration growth in the twenty-first century will be due mainly to human activity is beyond question. Based on the model data, Report-2001 states that the biosphere and ocean will be gradually losing their importance as the concentration rise barrier. In this connection, IPCC-2001 presents a probable range of CO_2 concentrations by the end of the century, namely, 540–970 ppm against the pre-industrial and present-day levels of 280 and 369.7 ppm, respectively. Carbon cycle models listed in IPCC-2001 assign the primary role in the carbon oceanic sink to diffusion and chemical processes occurring in water. These models do not take proper account of the processes due to carbon dioxide transport by ocean currents, which is especially relevant to cold water formation regions. Such factors make the carbon dioxide absorbing capability look more optimistic. This is evidenced by an increase in the oceanic sink as stated in IPCC-2001.

Land use, where pattern and intensity are governed by aggregate economic activity, can create sinks for carbon. Likewise, deforestation does not necessarily decrease the sink. Making some of the Amazonian large forests a region of zero sink-source carbon balance is an agricultural practice which can achieve some positive results. The same role was assigned to lumber industry complexes in the former USSR, where logging was paralleled by creating new forest plantations in which sink significantly increased in time. These and other features of economic activity make a decisive contribution to the established linear dependence between the increase in the human-induced CO_2 concentration in the atmosphere K_a and the population size. Observations and modelling show that per capita carbon emissions amounted to 1.3–1.38 TgC/year by the end of the past century, whereupon they began to decline, which is quite natural in the context of the world economic development situation. It should be noted that the emissions declined naturally, not responding to any administrative measures such as those recommended by the Kyoto Protocol. All this makes unlikely the forecasts of carbon dioxide concentration above 450–460 ppm.

Since 1750 (indirect data) atmospheric methane concentration has increased by a factor of 2.5 and today it has a tendency to increase. The annual growth rate of CH_4 concentration decreased virtually to zero, exhibiting a greater variability in the 1990s compared to the 1980s. Since 1750, nitrous oxide concentration has increased by 16%. Through implementing the Montreal Protocol recommendations and follow-up measures, concentrations of a number of halocarbons, acting both as GHG and ozone-depleting gases, either increased at a lower rate or began to decline; on the other hand, there was a rapid growth in concentration of their substitutes and of some other synthetic compounds, for example, perfluorocarbons (PFC) and sulfur hexafluoride (SF6). As to other greenhouse gases, their concentrations expected for the future widely vary. For example, some experts believe that tropospheric ozone as a greenhouse gas can become equal in contribution to methane and will be an important factor in deteriorating air quality over most of the Northern Hemisphere.

The estimated RF which characterizes enhancement of the greenhouse effect in the atmosphere and is due to concentration growth of the atmospheric mix of MGC proved to be 2.42 W/m^2, into which various minor gaseous components such as CO_2, CH_4, halocarbons, and N_2O contributed with 1.46, 0.48, 0.33, and 0.15 W/m^2, respectively. The decrease in total ozone concentration observed in the last two decades could result in a negative RF of $-0.15\,W/m^2$, which can decline to zero during the present century provided efficient ozone layer-protected measures are taken. The increase in tropospheric ozone concentration since 1750 (by approximately one third) could lead to a positive RF of about 0.33 W/m^2.

Since publication of IPCC-2001 in 1996, many changes have been made to assess RF due to not only the previously considered sulphate aerosol but also to other aerosol types. This concerns especially carbon (soot) aerosol strongly absorbing solar radiation, as well as organic, sea salt and mineral aerosol. Major spatiotemporal variability of the aerosol content in the atmosphere and of its properties significantly complicates assessing the aerosol impact on the climate (Kondratyev, 1999). An example of this can be found in the 'alternative strategy' proposed by Hansen *et al.* (2000), an assumption that approximate mutual compensation of climate warming via CO_2 concentration growth and cooling is due to anthropogenic sulphate aerosol. In this strategy, more importance is assigned to anthropogenic methane emissions (mainly due to rice paddies) and carbon (absorbing) aerosol. Being acceptable, at least for the reason of making unnecessary the Kyoto Protocol, this new strategy – unfortunately – creates a broad scope for political games, now conducted around methane and carbon aerosol.

Another climate-forming factor to be taken into account is extra-atmospheric solar radiation change. This contribution to RF over the period since 1750 could have been approximately 20% of that from CO_2, due mainly to enhancement insolation in the second half of the 20th century (it is important that the 11-year insolation cycle be taken into account). The fact that IPCC-2001 neglects the phase of Middle Ages warming and the Little Ice Age indicates that climate experts are still far from understanding the possible mechanisms of enhancement of solar activity impact on climate (Haigh, 2000; Kondratyev, 1998).

5.2.3 Results and reliability of numerical climate modelling

The problem of numerical modeling has been analyzed in detail by, among others, Kondratyev, 1998, 1999; Bengtsson, 1999; The Greenhouse Effect, 1999; Houghton, 2000, 2001; Schlesinger and Andronova, 2000, 2001. Here, we restrict ourselves to brief comments. Progress is beyond question in improving completeness of numerical models of climate which take into account interactively all the components of the climatic system 'atmosphere–hydrosphere–cryosphere–biosphere'. Interactive simulation of the global carbon cycle in climate models has been started at last. The extreme complexity of climate models and the numerous empirical parameterization schemes they utilize for parameterization of various processes hinder assessment of the adequacy of these models, especially from the viewpoint of future climate prediction. This was specifically responsible for very schematic, contradictory, and unconvincing attempts to compare numerical modelling results and observation data.

One is forced to accept that, despite all the improvements made, the models in question do not describe reality; the major drawback is that their adequacy is quantitatively unverifiable. Moreover, the more detailed the simulation of the reality, the more unattainable the task of overcoming this drawback in view of the stochastic nature of the processes and models. This makes unconvincing, for example, conclusions concerning the long-term change of the annual global average SAT over the recent 150 years. If, according to IPCC-2001, the observed and calculated (with CO_2 and sulphate aerosol concentration growth taken into account) SAT variations do agree, one has to attach more attention to methane and carbon aerosol, as proposed by Hansen *et al.* (2000). Unfortunately, in both cases, the conclusions rest upon arbitrary opinions, and the agreement between the models and the observations is achieved in reality by nothing more than forced fitting. Also, a meaningful comparison of the theory with observations requires consideration of regional climate changes (not only SAT) and of not only average values of climate parameters but also of their variability characterized by higher-order moments. An indication of a necessity to further improve coupled climate models are the controversial conclusions concerning the fate of thermohaline circulation (THC) under conditions of global warming. According to Gent (2001), there will be no THC changes under global warming conditions. An unsolved problem is the 'warming commitment' discussed by Wetherald *et al.* (2001).

According to Charlson *et al.* (2001), anthropogenic aerosols strongly affect cloud albedo, and assessment of global average forcing showed that these values are of the same order of magnitude (though with opposite signs) as those due to GHG. Recent studies show that aerosol forcing can even exceed predictions.

The Achilles heel of climate models is parameterization of the biosphere dynamics (see Kondratyev, 1998; Gorshkov *et al.*, 2000; Zhang *et al.*, 2001). In this context, a good number of numerical experiments have been conducted previously to elucidate how deforestation affects the Amazon river basin. These experiments showed that total deforestation (replacement of rainforest by grass cover) will result in the reduction of evaporation from the surface and precipitation, though in a simultaneous rise in the surface temperature. This will be responsible for a rise of SAT within 0.3–3°C. Such changes are due mainly to an increase in surface albedo and a decrease in soil moisture, causing a decrease in the energy and water vapour fluxes to the atmosphere, moisture convection decay and latent heat release which, in turn, will be responsible for a decline in atmosphere warming. This will have a dual effect on atmosphere circulation:

(1) Changes in the ascending and descending air flows in tropics and sub-tropics (Hadley circulation cells).
(2) Changes in planetary wave generation conditions (Rossby waves) propagating from the tropics to mid-latitudes.

To elucidate in more detail how deforestation affects atmosphere circulation and climate, Gedney and Valdes (2000) carried out numerical experiments on simulating present-day ('reference') climate and total deforestation conditions for the Amazonian region. To this end they took advantage of a 19-level spectral (T42) general atmosphere circulation model. Deforestation should change the following climate-forming parameters: albedo ($13.1 \rightarrow 17.7\%$); roughness ($2.65 \rightarrow 0.2$ m); vegetation cover share ($0.95 \rightarrow 0.85$); leaf area index ($4.9 \rightarrow 1.9$); minimal vegetation cover resistance (150–200 s/m). All this will change the soil type as well. Numerical modelling revealed statistically significant variations of precipitation in the winter period in the north-east area of the Atlantic Ocean which result from deforestation and propagate further eastwards, towards western Europe. Such variations are due to changes in large-scale atmosphere circulation in mid- and high-latitudes. Simulation of such variations using the simple model confirmed that they are due to planetary wave propagation. This suggests that the results showing the interrelation between the processes occurring in the deforestation region and in the North Atlantic and western Europe areas are independent of the model chosen, with the variation range corresponding to assessed human-induced climate changes due to an increase in CO_2 and aerosol concentration.

Zhang *et al.* (2001) carried out a significantly more extensive numerical modelling of the climate implications of tropical deforestation under progressing 'greenhouse' warming due to the CO_2 concentration doubling. Utilizing the global climate model CCM1-Oz, developed at the National Center for Atmospheric Research, their calculations suggest a major decline of evapotranspiration (by $\sim$180 mm/year) and precipitation (by $\sim$312 mm/year), as well as a SAT increase by

3.0 K in the Amazonian region. Similar, though less pronounced, changes are observed in south-east Asia (precipitation reduction by 172 mm/year and warming by 2.1 K). Even less pronounced changes are observed in Africa (precipitation increase by 25 mm/year). Energy balance assessments showed that climate warming is due not only to greenhouse effect enhancement but also to a deforestation-induced decrease in evapotranspiration. Statistically significant climate changes are observed in mid-latitudes as well.

The conclusion arrived at in IPCC-2001 was subject to heated debate: namely, that the balance of evidence suggests a discernible human impact on global climate. Also, IPCC-2001 states that the 'anthropogenic signal' is already manifested against the natural climate variability background and that detection and attribution studies indicate the presence of the anthropogenic signal in climate observation data for the last 35–55 years. The nature-induced impacts could play a role in the warming observed in the first half of the 20th century, but they cannot explain the warming in the second half of the century. However, IPCC-2001 also indicates that the reconstruction of the climate for the last 100 years, as well as model estimates, demonstrate low probability that observed climate changes in the second half of the 20th century could be of completely natural origin. In addition, IPCC-2001 high uncertainty of the quantitative estimates for human-induced warming, especially as regards contributions of various warming factors (this holds, above all, for clouds and atmospheric aerosol). The judgements and conclusions mentioned are so evidently conflicting and unconvincing that here we do not comment. Certainly, the leading role in substantiating future climate predictions should belong to integral models describing the dynamics of the interaction between socioeconomic development and nature (see The Atmospheric, 1998; Kondratyev, 1999; Prinn *et al.*, 1999; Victor, 2001). It remains to be seen, however, how realistic are forecasts based on such models with their extraordinary complexity and inadequate input information. It seems likely that, at least in the not so distant future, integral models will serve only for preparing very schematic scenarios. An important advantage of such models is that they show regional climate change pattern. However, this pattern (the adequacy of which is doubtful) can be obtained only after multiple simulation of climate changes on a global scale, which is very problematic.

New data suggest that with a 1% annual growth of CO_2 concentration under various scenarios of the economy, development and various aerosol compositions, the global annual SAT average increase during the period 1990–2100 will be 1.4–4.8°C against the 1–3.5°C in IPCC-2001. Symptomatic in this connection is that, with improving models and an increase in their number, the divergence of this process deepened rather than contracted. Importantly, the SAT values calculated in terms of various models for the same scenario of MGC emissions were virtually identical to those with the same model for various scenarios. As to regional climate predictions, they are still statistically unreliable. However, one can probably hold as reliable the conclusion that in many terrestrial regions the warming will be faster than on the global scale, especially in high latitudes during the cold half-year. The most substantial (~40% in excess of the global average value) was calculated climate warming in the northern region of North America, as well as

Table 5.3. Observed and predicted anomalous weather and climate change.

Phenomenon	Observations (second half of the 20th century	Forecast (2050–2100)
Abnormal temperature maxima and number of unusually hot days	Almost all terrestrial regions	Revealed by the majority of models
Enhanced heat index	Many of terrestrial regions	Revealed by the majority of models
Abnormally intensive precipitation	Many of northern hemisphere mid- and high latitude regions	Revealed by the majority of models
Abnormally high temperature minima and decrease in the number of cold days	Almost all terrestrial regions	Revealed by the majority of models
Decrease in the number of frost days	Almost all terrestrial regions Almost all models	Possible, taking into account increase in minimal temperatures
Decrease in the daily temperature amplitude	Many of terrestrial regions	Possible, taking into account increase in minimal temperatures
Summer continent dehydration	Selected regions	Possible, taking into account increase in minimal temperatures
Maximal wind increase in tropical cyclones	Not observed, but the number of case studies is scarce	Selected models
Increase of medium and maximal precipitation in tropical cyclones	Insufficient data	Selected models

in northern and central Asia. By contrast, summer warming in south Asia and in south-east Asia and in the southern regions of South America and winter warming should be weaker than the global average warming. Numerical modelling data are indicative of the forthcoming enhancement of the water content of the atmosphere as well as precipitation. In particular, there can be precipitation increase in the mid- and high-latitudes of the Northern Hemisphere, as well as in Antarctica in winter. This conclusion is of special importance in the glacier dynamics content. For low latitudes, depending on the MGC emission scenario chosen, both precipitation enhancement and reduction is probable.

Responding to the great interest in possible anomalous events, IPCC-2001 gives prognostic estimates as correlated with recent observation data (see Table 5.3). This problem has been discussed by Karl *et al.* (1999) and Grigoryev and Kondratyev (2001). The deficient observational data and unreliable results of numerical modelling make the conclusions in Table 5.3 fairly vague.

Calculations of human-induced ('greenhouse') climate changes evidence a possible weakening in the future of THC in the oceans of the Northern Hemisphere. However, even models revealing such a weakening still reflect preserved 'greenhouse' warming in Europe. It is still unclear whether THC collapse is irreversible and which threshold conditions correspond to this kind of collapse. None of the available models predicts complete cessation of THC within the next 60 years.

Numerical modelling of the 'global warming' process suggests that there must be further snow and sea ice cover reduction in the Northern Hemisphere. Further retreat of glaciers (except for the Greenland and Antarctic, including western Antarctic, ice sheets) is expected in the 20th century.

IPCC-2001 believes that the anthropogenic impact on the global climate will be preserved for a long time and that it will determine the following specific features of the corresponding processes:

- Carbon dioxide emissions have a lasting impact on atmospheric CO_2 concentration. Even after several non-emission centuries the fraction of carbon dioxide retained in the atmosphere can reach 20–30% of the total emission level.
- The supposed stabilization of atmospheric CO_2 concentration requires significant reduction of carbon dioxide emissions, as well as a more significant decrease in emissions of other greenhouse gases.
- After stabilization of the CO_2 concentration level, the global average SAT will also continue growing for hundreds of years owing to the giant thermal inertia of the ocean (this concerns only 15–25% of the whole increase).
- The ocean level and ice sheets will continue to respond to previous climate changes during the thousands of years after climate stabilization. Model calculations show that warming of 5.5°C can result in the World Ocean level rising (owing to the Greenland ice melting) by 3 m within 1,000 years. Modern dynamic models for the western Antarctic ice sheet suggest that melting can be responsible for the ocean level rising by a further 3 m during 1,000 years, but one should take into account the fact that the possible long-term dynamics of the western Antarctica cryosphere have not been properly studied.

Conclusions concerning currently observed, and especially the likely future of, climate change are very uncertain. This applies both to the diagnostics of the present-day dynamics of climate, its simulation models and numerical modelling results.

According to IPCC-2001, the priority should be assigned to developments in the following directions:

- To stop further degradation of the conventional meteorological observation network.
- To continue research on global climate diagnostics so as to obtain a long series of observation data with a higher spatiotemporal resolution.
- To attain a more adequate understanding of the interaction between components of the ocean climatic system (including deep-lying layers) in their interaction with the atmosphere.

- To reach a more realistic understanding of long-term climate variability.
- To apply the 'ensemble' approach in numerical modelling of global climate in the probabilistic estimation context.
- To develop the integral totality ('hierarchy') of global and regional models, with special attention paid to numerical modelling of regional impacts and extreme changes.
- To substantiate interactive physicobiological climate models and socioeconomic development models with the aim of analysing the environment–society interaction dynamics.

This should be supplemented, in particular, by the following:

- Essential for understanding the regular features of the present-day climate are paleoclimatic investigations, especially on sudden – that is, within comparatively short time periods – changes (Kondratyev, 1998, 1999; Kukla, 2000).
- Intensive satellite-based remote sensing studies have not yet provided the adequate global information required for climatic system diagnostics, because the existing satellite-based and conventional observation systems operate in a far-from-optimum regime. Despite much effort to, and major progress in, development of the Global Climate Observation System (GCOS), Global Ocean Observation System (GOOS), Global Terrestrial Observation System (GTOS) and (at a later time) of the Integrated Global Observation System (IGOS), the task of optimizing global observation systems still remains to be solved. We still should realize that, along with accumulation of long homogeneous observation data series required for climate system diagnostics, there is also a need for problem-driven ('focused') observational experiments. These latter will address problems such as global carbon cycle, anthropogenic impact on the stratospheric and tropospheric ozone, the dynamics of the processes occurring in the 'aerosol–clouds–radiation' system, and the biotic regulation of the environment, etc. (see Kondratyev and Cracknell, 1998; Kondratyev, 1999; Goody *et al.*, 2001).
- The IPCC documents vaguely estimate levels of anthropogenic impact on global climate. Reilly *et al.* (2001) are correct in their explanation of this situation, primarily by the lack of quantitative estimates for the uncertainty of the results obtained (this is the case, for example, in the expected rise of SAT by 1.4–5.8°C). Clearly, in such a situation the decision making on environmental policy (e.g. as in the Kyoto Protocol) relies on views lacking a serious scientific substantiation (Soon *et al.*, 2000). In this connection, a question arises as to whether predictions for the year 2100 make sense in view of the fact that prospects for global socio-economic development cannot be predicted. The only possible answer to this question suggests itself. One can only prepare entirely conditional scenarios, relying on which in political decision making would be unwise and even fraught with danger. This is even more pertinent to regional scenarios, which are of particular practical interest, rather than global average estimates which can be likened to the 'hospital average temperature'. Naturally, we cannot accept the attempts by Allen *et al.* (2001) to excuse the lack

of quantitative estimates of uncertainties – especially by pleading the fact that in 1990 the IPCC was pressed to make a statement attributing observed climate change to anthropogenic impact – on the grounds that otherwise it might have been done by someone else. The opinion advanced by Wigley and Raper (2001) defending the conclusions arrived at in IPCC-2001 also cannot be regarded as convincing evidence.

5.2.4 Energy production development and the environment

5.2.4.1 Introduction

The problem of energy production is a key one for the development of civilization both from the viewpoint of ensuring economic progress and the impact on the environment. Whereas the attention of the participants of the Second UN Conference on Environment and Development (in 1992) and a subsequent Special Session of the UN General Assembly, 'Rio + 5' (in 1997) was focused on the Framework Convention on Climate Change – where special attention was paid to the problem of greenhouse gases emissions to the atmosphere – the main cause for heated debate was the rapid development of energy production in the 20th century (Kondratyev *et al.*, 1996; BP Statistical Review, 1997; Hansen *et al.*, 1998; Kondratyev, 1998a, 1998b, 1998e; Rose, 1998; The World Bank, 1998). It is this very subject that stimulates the appearance of this publication. Naturally, the first problem requiring consideration is analysis of present-day needs in energy and the scale of its production, as well as certain consequences of the process, such as GHG emissions (and above all, those of carbon dioxide). Detailed relevant information can be found in IPCC-2001 (and see Houghton *et al.*, 1995). Similar studies have been made at the World Energy Council (WEC) (see O'Riordan and Jaege 1996; Parikh, 1998; The World Bank, 1998) and the International Institute of Applied Systems Analysis (IIASA) (see BP Statistical Review, 1997).

The energy development scenario worked out by WEC proceeds from the assumption that the world's population will increase to 10 billion by the year 2050 and will then reach about 12 billion by 2100. A parallel moderate economic growth is envisaged which takes into account, for instance, a slow restructuring in the countries of Eastern Europe and the former USSR. It is supposed that in OECD (Organization on Economic Cooperation and Development) countries the energy consumption growth rate will be 0.7% (up to the year 2020), and in developing countries it will reach 2.6% (these figures based on present-day conditions concerning economic policy and slow technological progress). What is ignored is that currently existing energy sources may produce limitations on a global scale and regional deficit is a highly probable perspective (oil prices can be especially dynamical). A characteristic peculiar feature for decades to come will still be the dominating role of fossil fuels (about two thirds of primary energy, out of which one half will be the use of coal). The main energy consumers are energy production enterprises and transport. Other energy consumers include industry, heating, domestic needs, etc. Further energy production development will bring about increasing emissions of SO_2 and nitrogen oxides unless large investments are made to advance the present technologies. By the year 2050 carbon dioxide emissions will increase up to 11 Gtons of C per year (today the level is 6.5 Gtons) (Kondratyev, 1992b; Gorshkov, 1995; Kondratyev *et al.*, 1996). The above said reveals the unpleasant fact that, under the terms of the scenario discussed for global energy production and consumption development, local, regional and global environmental problems will increase to unacceptable (and dangerous) levels.

5.2.4.2 Energy production and the environment: local and regional scales

So far, economic market mechanisms have proved inadequate for efficient prevention of the negative environmental impacts of energy production. Many plants and factories, domestic users of energy, transport, etc., mostly contaminate the environment but do not compensate for the harm they do in terms of economic and social losses. These latter include: human diseases (above all, respiratory ones), decrease of harvest, damage to ecosystems (lakes, forests, etc.), loss of soil fertility, damage to buildings, cultural and architectural monuments, etc. In many cities of the world ever-increasing air pollution has become the main threat to human health. According to conservative estimates available, aerosol pollution in combination with SO_2 emissions are the cause of no less than 0.5 million premature lethal cases, to say nothing about 4–5 million cases of bronchitis and numerous other serious diseases. The result is severe economic loss. The data of Table 5.4 illustrate the effects of atmospheric pollution due to fuel burning on human health (The World Bank, 1998).

Here the primary pollutants are components that are directly emitted to the atmosphere; the secondary pollutants are products of chemical (photochemical, gas-to-particle conversion) reactions in the atmosphere. $PM_{2.5}$ and PM_{10} are aerosol particles with a diameter less than 2.5–10 μm, respectively.

Two peculiar features are typical for the interrelationship between energy production development and negative impacts on the environment in developing,

Table 5.4. The effect of air pollution on human health ($PM_{2.5}$ or PM_{10}): minor particles as indirect indicators

Polluting components	Primary polluters	Secondary polluters	Products of diseases secondary polluters	Mortality
$PM_{2.5}$ or PM_{10}	+	+	+	+
SO_2	+	+	(PM_{10})	+
NO_x	+	+	(O_3 and PM_{10})	+
VOC	+	+	(O_3 and PM_{10})	+
Ozone		+		+
Lead	+		+	

Table 5.5. The harmful effect of polluted urban atmosphere on human health from the data for 1993 (in $ million).

City	Coal	Oil fuel	Diesel fuel	Petrol	Total
Mumbai (India)	850	126	125	19	1 120
Krakov (Poland)	200	1.5	35	8.7	245
Santiago (Chile)	no data	278	348	106	732
Shanghai (China)	3 479	24.5	45.5	8.5	3 558
Total	4 528	430	555	142	5 655

and some other, countries. One is the prevalence of coal as an energy carrier for industrial and domestic needs; its consequence naturally is a high level of atmospheric contamination in cities. This situation is characteristic for a number of countries, including China, India, Poland, and Turkey. The intensive development of transport also greatly contributes to the increase of city pollution (yielding such pollutants as PM_{10}, SO_2, NO_x, VOC, O_3, and lead) both in the megacities of developing countries, such as in Brazil, Indonesia, Mexico, and Thailand, and in the big cities of the advanced industrial countries.

The data of Table 5.5 illustrate the influence the atmospheric pollution of cities has on human health. In terms of material loss, such pollution can reach, for example in the case of Shanghai, US$3.5 billion per year. It can be seen that various kinds of economic activity contribute substantially to this pollution (which makes it difficult to fight) and the dominant role of coal, among other energy carriers, is obvious. The data of Table 5.6 reflect the contribution of various sectors of the economy to the pollution of the urban atmosphere (due to averaging, the sum of relative contributions may not be equal to 100%).

Analysis of these data makes it possible to draw the following two conclusions in the context of preventing atmospheric pollution:

(1) Significant results can be achieved only if the contributions of the transport and minor emissions sources are taken into account.
(2) Relative contributions of various pollution sources are very different in different cities, which makes necessary local specific measures to fight the pollution.

Essential ecological consequences of pollution show themselves in impacts which are not directly relevant to human health. One of the most important consequences of this kind is the formation of acid rain as a result of chemical reactions with the participation of SO_2 and NO_x and the subsequent long-distance transport of several thousand kilometres, which is completed as a process of dry and wet deposition. Other serious consequences are the effects of energy production and transport upon soils, water reservoirs and air quality. These often take the form of accidents at enterprises where oil and/or gas are processed and also where supertankers are involved in breakdowns or accidents resulting in huge oil spills.

Table 5.6. Contribution of various economy sectors to atmospheric pollution of big cities. City contribution of various economy sectors to missions due to fossil fuel burning (%).

	Electric power stations, heating	Major industry	Minor sources	Transport
PM_{10}				
Krakow	66	22	10	2
Mumbai	5	39	61	6
Santiago	2	22	45	31
Shanghai	6	64	30	0
SO_2				
Krakow	68	26	6	1
Mumbai	35	35	25	4
Santiago	5	59	18	18
Shanghai	29	55	16	1
NO_x				
Krakow	50	19	2	29
Mumbai	19	17	10	53
Santiago	1	14	6	79
Shanghai	35	45	8	14

5.2.5 Conclusion

The main conclusion following from our discussion of the IPCC-1996 and IPCC IPCC-2001, both of which have been used to substantiate the need to adopt and realize the Kyoto Protocol, is that they are unsuitable as a basis for decision making. Gaining insights into characteristic features of present-day climate system dynamics and, especially, assessing likely climate changes in the future are seriously complicated by the lack of reliable estimates of the contribution made by anthropogenic factors to present-day climate change. We clearly realize that human-induced enhancement of the atmospheric greenhouse effect due to GHG concentration growth in the atmosphere should be responsible for certain changes in the global climate. But using this realization to create panic and to solve political problems cannot be tolerated. Unfortunately, many of the leading figures in IPCC act in accordance with specifically this principle.

In understanding how realistic are climate predictions, verification of the model adequacy to currently observed climate changes and paleoclimate dynamics (based on indirect data) is of crucial importance. As to the use of modern observation data, the situation seems fairly paradoxical. Such verification is virtually confined to average temperatures, while it is evident that one needs many other kinds of information. The paradox consists in that satellite observations provide a giant excess of non-systematized data, which is paralleled by the already mentioned degradation of the conventional (*in situ*) observation network.

Verification of global climate models by comparing numerical modelling and observation data is an extremely complex task. It has been solved most often by comparing long series of data on the annual global average SAT, which virtually always yields the same conclusion (despite substantial – sometimes cardinal – distinctions in accounting for climate-forming processes): namely, that calculations agree on the whole with observations. Another feature characteristic of these developments is a conclusion – not adequately substantiated scientifically – of a major (or even dominating) contribution from anthropogenic factors (above all, greenhouse effect) to climate formation. Clearly, this approach to model verification cannot be taken seriously, because:

(1) Modern climate models are still imperfect from the viewpoint of interactive account of the biospheric processes, 'aerosol-clouds-radiation' interaction, and many other factors.
(2) The only long-term (100–150 year long) SAT observation series available is far from being adequate from the viewpoint of calculating annual global average SAT (not to mention other climate parameters).

Recent developments under GCOS, GOOS, GTOS, and IGOS programmes are undoubtedly useful, but they still do not substantiate the optimal global observation system, a question which has been discussed in detail (in monographs by Kondratyev, 1998; Kondratyev and Cracknell, 1998; and, in more recent years, by Goody *et al.*, 2001). This is primarily explained by imperfect climate models intended as the conceptual basis for observation planning to be further improved following the refinement of the models. In this connection we should emphasize the need in analyzing the divergences disclosing specific 'weak points' of models rather than in making illusory statements treating global climate models as fairly realistic. Clearly, the focus of consideration should be not only on SAT but on the totality of the climate parameters. The primary attention should be given to the model's simulation of climate change (including at least second-order moments).

Paleodata indicate that in the geological past there were strong and sometimes very rapid climate changes. For example, Alverson *et al.* (2000) noted that the ocean level changed by over 100 m, the stable changing rate being over 1 m, within 1,000 years. Such changes significantly exceed the supposed human-induced changes corresponding to the doubling of atmospheric CO_2 concentration, which shows that any concerns about anthropogenic impact on the climate are unjustified. The problem consists not so much in forecasting in detail the climate of the future as in analyzing how sensitive is modern society and its infrastructures to likely climate change. It is worth mentioning that for many countries, including Russia and the USA (see Kondratyev, 2001), the forecast warming means a positive phenomenon rather than danger. In this connection, paleodata are of a greater value for climate prediction than numerical modelling-based schematic scenarios.

As to climatic predictions and the atmospheric greenhouse emission reduction recommended by the Kyoto Protocol, the former clearly need to be interpreted as schematic scenarios and the latter, correpondingly, as having no real grounds. This suggests an acute need in revising the FCCC (before 'Rio + 10') and rejecting unjustified and non-realistic recommendations contained in the Kyoto Protocol that pose threats to socioeconomic development. The collapse of the Sixth Conference of the FCCC signatory-countries held in The Hague on November 13–24, 2000 suggests that these costly conferences are fruitless and the global climate change problem needs a serious scientific dicussion without the domination of the 'global warming' concept. The real situation is that greenhouse emissions continue to increase and that this trend will be preserved, in particular in the USA, and all the reasonings on the importance of 'flexible market mechanisms', 'emission trading', etc., are therefore purely rhetorical.

The current status of climate theory by no means receives only optimistic assessments. In this connection it is appropriate to mention opinions of American experts (see Shackley *et al.*, 1998). For example, G. North from Texas University finds it diffcult to judge whether climate models have improved over the recent five years and believes that the uncertainties are still as large as they were 20 years ago. P. Stone from Massachusetts Institute of Technology thinks that the major uncertainties in predictions of present-day climate changes have not decreased at all. R. Charlson from Washington State University believes that one would be wrong in thinking that one understands the climate. In R. Kerr's opinion, the information contained in IPCC-2001 substantially reduces uncertaintly in solving the problem of detecting the 'anthropogenic signal' for climate and supports the conclusion that global climate changes have anthropogenic origin but, at the same time, not only does not decrease but even extends the range of uncertainties in climate predictions. R. Gutzler from New Mexico State University says that it seems the solution to the 'anthropogenic signal' detection problem has almost been found. According to IPCC-2001, the warming observed during the recent 50 years is, probably (by 66–90%), due for the most part to GHG concentration growth. Here, it should be noted, the problem of quantitative estimation of what is 'almost' and what is 'for the most part' remains to be solved. These vague formulations clearly reflect the degree of uncertainty of climate numerical modelling results.

Very much to the point was the remark by Soros (2000) that, at the present time, CO_2 emissions in the USA are about 16% of those made in 1990, the emissions in the European Community member-countries are about 6% on average, those in Japan about 5%, and those in Australia about 24%. Thus, in the 1990s there was a rise, rather than a stabilization, of atmospheric emissions of carbon dioxide. Also, there is no evidence to suggest any serious emission reduction efforts – CO_2 emission reduction in Germany and the UK has nothing to do with the Kyoto recommendations. Soros was right in voicing a concern about the credibility gap in the Kyoto Protocol and the clear lack of prospect for its ratification by the leading industrial countries.

Tol (2000) cautioned, with good reason, against the misunderstanding that without fossil fuel the world will be like a paradise. Though attractive on a small scale, renewable energy sources have unclear prospects when viewed on a large scale: the limitations of, for example, hydro energy and wind energy become evident. All this reflects the undoubted truth that one needs to search for ways of further developing civilization and to substantiate an adequate environmental policy in the context of the dynamics of the 'society–nature' interactive system. Finding a solution to this problem will require unprecedented cooperative efforts by experts in natural and social sciences.

5.3 DEFORESTATION

One of the global problems of the present-day is the mass destruction of forests. Under human impact, forests have receded in all continents and in almost all countries. Thus, for 20 years, according to the data over the period 1966–68 to 1986–88 the global area of forests was reduced by 6% (United Nations, 1991).

5.3.1 Forest resources and tendencies to change

The total area of forests on the planet as at the year 1990 was estimated at 5120227×10^3 ha, which amounts to about 40% of the land surface. The term 'forest' includes natural forests and forest plantations as well as reforested lands. At first sight, forest resources are vast and possibly that is why forests have for so long been actively and carelessly used for man's various economic needs. Although in some individual countries forest cutting has been compensated for by plantings, such compensation has not been universal throughout the planet. Only recent studies (United Nations, 1991; Brown *et al.*, 1996; World Resources, 1996; Grigoryev and Kondratyev, 1998b) have shown that forest cover and its condition has an extremely important effect on the ecological situation of the earth's large regions and of the whole planet. Therefore, observation of the global forest condition and its tendencies has become important.

At present two-thirds of the planet's forest resources consist of natural forests ($3,444,369 \times 10^3$ ha) and one third includes reforested lands ($1,677,719 \times 10^3$ ha). For the 10 years, 1981–90, the area of forest cover was reduced by 0.19% every year. This reduction occurred differently in different regions of the earth and the extent of forests affected also varied within a wide range. Thus, for example, in Africa where almost one-fifth ($1,196,676 \times 10^3$ ha) of the planet's forests is situated, in the same 10 years the area of forests was reduced by 0.24% every year. And, taken together, the forests area of North, Central and South America – $2,009,606 \times 10^3$ ha – was reduced by 0.31% every year during the same period. The data on forest dynamics in Asia as a whole – the area of which (without the former USSR) is 657361×10^3 ha – are not available, though it is known that in Asia regions of forest are intensively reduced.

It is interesting that in the former USSR during 1981–90 the forest area increased every year by 0.01% (the total area is about $941,530 \times 10^3$ ha). This is probably explained by the general decline in the economy during these years and the subsequent reduction in cuttings, rather than by forest plantings. In Western Europe, where the area covered by forest ($174,340 \times 10^3$ ha) is more than five times smaller than in the former USSR, the forest area increased much more rapidly (by 0.11% per year). This increase was due to more zealous forestry management.

In Australia and Oceania, where the forest-covered area is slightly larger than it is in Europe ($200{,}971 \times 10^3$ ha), change is minimal, being reduced only by 0.02% per year.

On a global scale, the area of forests continue to decrease – according to the data as of 1990, by 2% for all of the planet's forests. The most marked decrease is in the tropical countries where for 10 years forests have decreased by 3.6%. The tropical forest area in 1990 was $2{,}727{,}999 \times 10^3$ ha and in such a landscape the forests are rich in species and economically valuable; this explains their intensive destruction, particularly in the countries of Latin America and the Caribbean where the decline is by 4.5%. Of importance is, of course, an acute need for the economies of developing countries to export their resources.

In states of extratropical or moderate belt the forest-covered area for the years 1981–90 even increased a little (by 0.1%), mainly due to forest planting; by 1990 it amounted to $2{,}392{,}230 \times 10^3$ ha, although it should be mentioned that sixth-sevenths ($2{,}063{,}569 \times 10^3$ ha) of these forests are situated in developed countries and for 10 years to all intents and purposes their area did not change.

Changes in the forest cover of developing countries (in mid-latitudes) are on the whole insignificant. Their area (as of 1990, $328{,}665 \times 10^3$ ha) increased every year by 0.8%, this occurring due to the forests located in Asia, Australia, Oceania (their area increased by 5.3% every year). On the other hand, in Africa and Latin America, the forest area was reduced catastrophically (by 7.2% and 5.3%, respectively).

It should be emphasized here that of particular concern is the destruction of ecologically valuable natural tropical forests, rather than tropical forests in general (including also reforested lands). For the period 1981–90, the area of natural forest decreased by 8.1% to become $1{,}761{,}228 \times 10^3$ ha. In this case the forests in Asia reduced most rapidly (by 11% for the decade).

The causes for the deforestation of the earth are various. One of them is wood felling for industrial purposes, widespread everywhere but especially so in the tropical belt. The owners of large forests in this belt tend to be located in mainly developing countries and the nature of such countries and their economics and the demands of international trade determine the high stress experienced by forest-covered areas.

Tropical forests are destroyed as a result of mass cuttings made mainly in the interests of foreign transnational monopolies, rather than those of the forest host countries. Forest raw materials are exported to other countries – to Japan, the USA, the United Kingdom, France, Germany and so on. Forests are also cut to make way for cities, villages, communication lines and soon rapidly disappear as a result of extending agricultural land. In many countries in the equatorial and tropical belts of the earth, the slash-and-burn system of farming is used – for example, in Indonesia this primitive method of farming is still in use, even now in the twenty-first century. In economically poor underdeveloped countries, including many developing states of Africa, in India and in Pakistan, wood is widely used as a fuel despite the well-known fact that this is a most wasteful method of using forest resources.

The most important factor in the destruction of the forests is population growth, which is particularly typical of developing countries. For example, the population in

Africa has doubled in the last 50 years and this demographic outburst is also observed elsewhere – for example, on another 'green' continent, South America; a bursting population growth increases abruptly the load carried by tropical forests on both these continents.

There is no doubt that, in future, reforestation on plantations will become of greater importance in preserving the forest cover of the planet. In 1990 the total area of forest plantations was $2,245 \times 10^3$ ha; this area increased by 7.77% every year during the decade, 1981–90.

In developing countries the area of plantations for the period 1980–90 increased by 88% (32,000,000 ha). In this case, in the tropical states of Asia, Africa and South America this value was much higher than in the countries of moderate belt (e.g. in Asia it reached 189%, in Africa 76%, in Latin America 75%).

Of course, the reduction of the planet's forest cover is only one component of change forests are undergoing. Over vast areas of Russia, Ukraine, Byelorussia, and in the USA and Canada, many planted (not original) forests have long existed, less rich in species variety than original forests (poorer in fauna as well) and less valuable economically. These forests have become degraded to some extent as a result of different kinds of pollution, including radioactive fallout.

Today there is great concern regarding the simultaneous impoverishment and disappearance of tropical ecosystems, the creation of which has taken millions of years, unlike other ecosystems. In some regions of the earth, particularly in Indonesia and Brazil, the consequences of destroying forests, especially tropical forests, could mean that extensive areas of forest will become waste lands in the near decades.

As is well known, tropical ecosystems have a very wide variety of species of animals and vegetation. Thus, for example, in the Amazon area 1.7 million species of plants and animals have so far been identified. Among them are 47,000 vertebrates, 750,000 insects, and 250,000 plants. It is also assumed that there are some 5 million species of animals and vegetation; some specialists assess this number as even higher – 30 million (Grigoryev and Kondratyev, 1998b).

Destruction of forests results in replacing one landscape with another. Thus, in the Ukraine, islands of oak groves have long existed on the steppes of the Odessa region (Grigoryev, 1985). Where there is a lack of forests natural growth like these oak-groves has a very high value. Consequently these forests are cut out instead of becoming reserves. One of the most famous in the region and under a present threat of destruction is Savranskaya Oak-Grove, a 9,000-ha area located on the bank of – the South Bug River.

In Siberia as a result of cutting, swamps have started to form in many lowland areas. The most valuable forests, pine and partly cedar, are cut out first and larch forests occupy a more and more extensive area. As a result, the forest cover becomes poorer everywhere.

Waste land will soon form at the sites of removed forest, especially in tropical areas, and the processes of erosion begin and a washout of soil is developed. Forest tracts are replaced by savannahs. This is true of, for example, some regions of Indonesia which have become a vast waste land.

The consequences of forest cutting can include a change of hydrological and climatic conditions. In territory that dries up, destructive droughts are observed, for example, in the southern areas of Brazil. This is not surprising: in one south Brazilian state, Rio Grandid-du-Sul, the forest covering was reduced from 35% to 2% within 50 years.

Two contradictory viewpoints of the condition of the Earth's forests are pertinent today among specialists. The majority view is that the forests are destroyed progressively, as a result of expanding economic use which can cause a number of different, including global ecological, consequnces such as a decrease in biodiversity, effect on climate, displacement of natural zones, etc. According to the other viewpoint (Sedjo, 1995), the situation is not disastrous. On the contrary, it is believed that the moderate belt – which accounts for 75% of all industrial production of forests – has within it many regions, for example in Canada, where stabilization and even extension (in Europe and the former USSR) of forest areas is observed. Industrial cuttings here are, as a rule, made either in planted forests, or on forest plantations. Indeed, the area of forest plantations continues to expand worldwide; in 1985, estimated areas for developed countries amounted to 60,000,000 ha (including 21,900,000 ha in the former USSR) and for developing countries 35,900,000 ha.

The basic reason for the removal of forests is to develop land for agriculture. For this purpose, tropical forests are the first to be cut down; as a general rule, mid-latitude forests are cut down for commercial purposes.

Unlike the forests of the temperate belt, the tropical forests are now drastically reducing in area (according to the 1990 data, at a rate of 0.6% every year) and the setting of forest plantations is practised to a much lesser extent than is the case in temperate latitudes. It is doubtful if the continuing reduction of the tropical forest area can be stopped.

5.3.2 Satellite information on forest cover destruction

The value of space observations in monitoring forest destruction consists, in particular, in showing a generalized pattern of actual distribution of preserved forests in one or another region or on the planet as a whole. In some cases, space information on tracts of forest is considerably more obvious and convincing than evidence in detailed maps. A typical example is that of the American specialists who, on viewing satellite images of the eastern part of the USA in the 1970s, were astonished to see how low was the forest coverage between Washington and Baltimore (Satellites, 1972). Moreover, they expressed their regret that these pictures had not been to hand some 25 years earlier. The specialists believed that, had space photography been available sooner, these particular forests would have suffered less vegetation damage and their planning and management would have been improved.

Analysis of forest vegetation using space images is made by spectral differences, taking into consideration the location of the types of forests studied (Isaev and Ryapolov, 1988; Investigation, 1987; Manual, 1983).

Even pictures of low space resolution (from meteorological satellites) are useful for the separation and estimation by area of seven zonal types of vegetation cover: namely, tropical forests; taiga; leaf-bearing forests; tundra; steppe; semi-desert; and desert. To this end, measurements of normalized vegetation index are obtained from an NOAA/AVHRR satellite.

Information from meteorological satellites at high altitudes enables an estimate to be made of the condition and change in forest cover on a continental scale. Compilation of a small-scale (1:1,000,000) map of South American forests by Stone *et al.* (1994) used images received from an *NOAA*/AVHRR satellite with a resolution of 1 km and from an *NOAA*/AVHRR GVI satellite with a resolution of 15 km.

Using spectral data, 39 classes of vegetation cover have been selected with the help of digital processing of images and the use of NDVI. Analyzed images have been obtained in different spectral intervals: in the visible spectrum region 0.58–0.68 μm, close infrared region 0.725–1.1 μm, middle infrared region 10.3–11.3 and 11.5–12.5 μm.

To study the dynamics of vegetation cover, images have been used of past years from *NOAA-9*, *10*, *11* and other satellites, as well as small-scale cartographic materials. Comparison of these images has shown that for 10 years, by 1988, 9% of tropical forests had disappeared, 22% of tropical forests turned out to be degraded or partially cut down, and 25% of savannahs and 18% of other forest land had also considerably degraded.

According to a vegetation map compiled in 1985 by the international organization FAO, the area of forests and forest lands in South America amounted to 9,200,000 km^2. According to satellite data (as of 1988), this same area was 9,630,000 km^2. There is reason to believe that the latter map is more accurate than the former one, since the routine mapping of vegetation cover in the interior regions of the South American continent by traditional techniques is very difficult.

Satellite data enabled effective estimation of the disappearance or degradation of forests in all individual states of South America. Thus, by 1988, 31% of all forests in Brazil had been degraded.

Images obtained from manned space vehicles, as well as from satellites of the *Landsat* and *SPOT* type, allow more thorough estimations to be made of forest cover and its changes. Such information proves helpful in assessing forestry management in one or another country, as well as the monitoring the balance between cut down and planted forests, introduction of afforestation practice, estimation of forest-protecting measures, and so on.

Satellite images, unlike more detailed aerial photographs, demonstrate groups of forest types, and some individual types. Their decoding is made on the basis of differences in spectral characteristics. Of substantial importance, too, is the estimation of the location of a group of forest types in a landscape. Such an approach, which includes a close relationship between forest and other components of the landscape, was worked out in Russia as long ago as the 1960s when the process of decoding forests in aerial photographs (Beresin *et al.*, 1969) was being developed. In decoding, a large integrability of forest cover image in space pictures determines a higher role should be given to the landscape approach (Isaev and Ryapolov, 1988; Investigation, 1987).

The use of space images turned out to be very promising in the study of the regional features of forests, especially due to their small-scale and middle-scale mapping. In Russia (and in the former USSR) small-scale mapping of forest cover was based on the materials of spectrozonal (multizonal) space photography. These images were made from satellites of the *Kosmos* series and the orbital station 'Salyut' (with resolution of 5–20 m), as well as scanned space images of high resolution (Isaev and Ryapolov, 1988; Investigation, 1987).

Mapping of forests in the different regions of Siberia uses information obtained from space images to produce maps at a scale of 1:200,000–1:1,000,000. One of the maps (1:1,000,000 scale) produced shows forest resources in Mongolia and a map of forest types (1:1,000,000 scale) was compiled for the territory to the west of the Baikal–Amur main railway line. Such general surveys showing the condition of forests of extensive areas are planned to be performed at intervals of 10–25 years.

The use of space images produces not only forest resource maps but also a series of other associated maps: forest cover, geobotanical information, forest pathology, typology, pyrology, location conditions, forest swamp, soil, hunting, and so on. Such a spectrum of small-scale and, mostly, associated maps, enables a complete estimation of forest resources, conditions and perpectives. Such a series of maps (1:1,000,000 scale) based on space images was compiled for the Angara–Yenisey region in 1980–85.

Mapping of forests in this way is carried out in a growing number of countries. In India, for example, the condition of large forests has been estimated using *IRS-IA* satellite images with a resolution of 36 m. The accuracy of maps prepared on this basis in some cases exceeded 80%, being verified by ground-based studies (Grigoryev and Kondratyev, 1993). Satellite images from *IRS-IA*, as well as from a *Landsat* satellite made possible compilation of maps (1:250,000 scale) of forest cover for the whole of India over the years 1972–91 (Rao, 1996).

The total forest area in India for 1972–91, although it changed, decreased only slightly – from 21.6% to 19.47% of the whole country. The area of dense large forests decreased appreciably from 14.1% to 10.9% (that is, by 25%) only for the period 1972–82. However, in subsequent years, due to steps taken, the decrease was not only arrested but the total forest area was actually increased, albeit slowly, to 11.72% in the years 1989–91.

The most common reason for forest destruction is, of course, the extension of agricultural lands but the long-term impact of big cities, different forms of transport and industrial constructions on their immediate surroundings has also led to significant damage of forests. Thus, in space images of Petersburg one can see that large forests are no longer found near the city as once they were; the forests have retreated some tens of kilometres, especially to the south, giving way to industrial constructions, populated areas, and ploughed lands.

The destruction of forest ecosystems is hastened by the use of wood for industry and as a fuel. In Russia, cuttings, both new and old, are well seen from space against a background of forest-covered, and not yet developed, zones which stand out due to their typical shape in plan (most often rectangles) and higher brightness. New cuttings, not shown on conventional maps, are clearly observed even in small-scale space images of low resolution (including in the low hundreds of metres). An example of such are the forests in the Komi Republic, and in the Kirov and Arkhangelsk regions.

Detailed space images with a resolution of 5–30 m also isolate other features in the developing forest, such as networks of dragging roads and lumber yards. In spectrozonal pictures, particularly, clear observations can be made of the consequences of cutting – for example, the phenomena of swamping to be seen in those locations that develop, generally on flat plains, over hardly penetrable rocks. This unfavorable process, as surveys from space show, follows on from cuttings made in Siberia and European Russia, especially in those cases where the necessary good husbandry of the forest fails to take place.

Special types of cutting result from the construction of buildings, particularly as communication lines – railways, roads, pipe lines, etc. – are developed. Space images show large forests already cut out along the biggest main lines of the continents, as for the Baikal–Amur railway in Asia and the transamerican highway in South America. Along these lines of continuous belts of cut forest are peculiar centres of forest destruction, clearly seen from space. These 'moustaches', linearly extended belts of cuttings pushed deep into taiga or jungle not yet touched, contain newly populated areas and consist of settlements and roads which, in their turn, allow further destruction. If such development continues without appropriate planning, and unregulated forest cutting is allowed, negative ecological consequences are possible.

Judging from space images of Brazilian selva, developments of this kind are resulting in vast tracts of forest being cut along the tributaries of the Amazon. These are developments which encroach on the water protection zone (see Stone and Woodwell, 1985; *La déforestation*, 1992; Manual, 1983; Scholes and Tucker, 1994).

These observations on destruction of Amazonia's rainforest and estimations of settlement expansion leading to selva destruction in Brazil were made possible by the use of Landsat satellites. It was established that, by 1977, Amazon forest was removed from an area of 208,000 km^2. Over ten years, up to 1988, 588,000 km^2 were deforested. Thus, in those years, 38,000 km^2 of Amasonia selva were deforested every year. This amounts to about 1% of the total selva area (4 million km^2) according to one survey (*La déforestation*, 1992).

It is possible to study forest areas when they are covered with clouds. Using radar survey allows observers 'to see' the target location through clouds, though less extensively than in normal circumstances. Data on the cutting of forests and their dynamics were obtained by radar survey of Amazonia selva in 1981 and 1984 from the *Space Shuttle*. These data differed from control ground-based information by not more than 10% (Stone and Woodwell, 1985).

Extensive degradation of forests takes place not only due to cutting but also because of fire, disease and damage done by pests. To estimate forest damage caused by insects, maps (at scales from 1:100,000 to 1:500,000) have been compiled from space satellite pictures. For example, in China, images from *Landsat* and China's

Fu-1 satellite (Grigoryev and Kondratyev, 1993) differentiated, taking into account spectral differences, those forests which were particularly heavily damaged by caterpillars.

Forest surveys from space also promise an estimation of the ecological consequences with respect to other components of the natural environment, even those on a regional scale. Surveys in microwave spectrum region have proved to be very interesting in the study of selva destruction and the drying up of Amazonia. In this region, as is well known, the features of surface moistening and the difference in moisture of the earth's covers (ecosystems) have been revealed. The survey of selva employed microwave sensors from Brazil and data obtained from USA satellites with an interval of five years (1968–73) and the results indicate a decrease of moistening in vast areas of Amazonia. These data were confirmed by ground-based observations (Grigoryev, 1985).

There are promising studies of forest biomass using satellite data; these studies are designed to estimate the dynamics of carbon in the 'forest–atmosphere' system. Thus, in India, studies conducted in Radjardji National Park, Uttar Pradesh state, found that for $1\,\mathrm{km}^2$ of forest area, 0.188×10^9 of carbon is absorbed every year. (Unni *et al.*, 1991; Grigoryev and Kondratyev, 1993).

Global estimations of biomass are now possible on the basis of mapping the vegetation index (taking into account the differences in brightness characteristics at different wavelengths), in particular, from *NOAA* satellite data. Thus, real possibilities are opened up for estimating in large regions, and on the whole planet, disturbances of gas exchange caused by forest destruction (Environmental, 1994).

Space images are also useful for observing reforestation of cleared spaces and burnt areas, as well as for monitoring forest ecosystems damaged by insects and pests. For making general estimations in Russia, space images with 5–30-m resolution have been used to survey damaged forests in some areas of Siberia. Maps of damaged forests, on a landscape basis, have also been prepared to show the process of restoration (see Investigation, 1987; Isaev and Ryapolov, 1988) and more detailed estimations of reforestation are made with the help of aerial surveys.

The information presented confirms that at a global and regional scale (that is, whole continents and individual large and small countries) space images can provide both operative and objective information on forest resources and condition, thereby advancing the development of deforestation studies. This information is the basis for assessing the slow action degradation of the forest cover of the planet and the related possibility of negative ecological consequences arising from it.

5.4 DESERTIFICATION

5.4.1 Development and consequences of desertification

In Figure 5.1 a schematic map of global desertification of the earth's ecosystems is presented. Desertification is one of most negative consequences of human activities

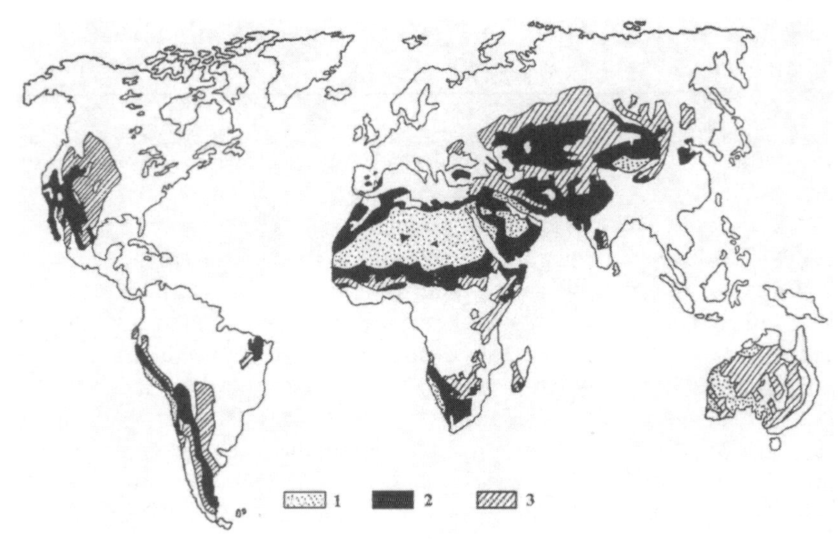

Figure 5.1. Schematic map of global desertification of the Earth's landscapes (ecosystems): 1 = natural deserts; 2 = landscapes subject to intensive desertification; 3 = landscapes subject to moderate desertification (*source:* Rao, 1996).

in arid and sub-arid regions and is generally defined as the degradation of ecosystems accompanied by the decrease of their bioproductivity. The phenomenon of desertification occurs both in the desert zone and, particularly, in the so-called marginal or border zones (Vinogradov, 1980; Babayev *et al.*, 1986; Zonn, 1986, 1997) best described as semi-deserts, dry steppes, and partially, savannahs.

Alhough this phenomenon can occur naturally (e.g. as a consequence of a decrease in the amount of precipitation), anthropogenic desertification has become widely distributed. It occurs in the regions developed by man and manifests itself in the increase of areas of arid, barren and low-yield lands.

However, in regions of intensive land development, natural and anthropogenic factors are very interrelated and the differences between these two types of desertification are extremely diffcult since their 'final' result, a degraded ecosystem, is the same.

An exaggerated human load on ecosystems due to drought in developed marginal landscapes of dry steppes and semi-deserts, especially in conditions of particular fragility and vulnerability, can trigger the phenomenon of location desertification.

In the second half of the 20th century, the planet suffered from large-scale desertification. This process is hastened by population growth and the ensuing load on the earth's resources and by the failure of humankind to take reasonable and rational techniques to deal with the phenomenon. The destruction of soil-vegetation cover, fragile and easily vulnerable in arid and sub-arid natural zones, is caused by many factors. Among them are excessive pasturing of cattle, careless use of ground waters, construction of buildings, and the use of trees and bushes for fuel.

Desertification would have received less attention had it not touched upon the vital interests of mankind. The desertification of a given location results in a change of soil–vegetation, hydrological and even climatic conditions which are extremely unfavourable for vital human activity. Symptoms of desertification include the thinning out, or complete destruction, of vegetation, destruction of the upper rich soil layer, soil blowing away, and the decrease of ground water level. This process occurs in those zones inhabited by the majority of mankind on the earth and in land areas most favourable for farming. As a result, large areas of rich land can turn into barren land and become unfit for agriculture.

When the possibility of using land for pasture or growing crops is lost, population has either to starve or to migrate. According to UN estimates for 1986, in the early 1980s as many as 150 million peasants had to give up farming in China, India, and the countries of south-east Asia because their land had become barren. In 1983–84, African regions subject to desertification saw tens of millions of people starve.

Desertification in the 20th century was a widespread phenomenon and was experienced by several tens of states and covered large regions of the earth.

On a global scale, desertification in Africa attracted the most world attention during the catastrophic years of 1968–73. It caused serious damage to vital human activity not only in 15 countries of Africa but also adversely affected the economy of the world community. In 1977, the Nairobi Conference was convened by the United Nations as a response to the extent of the desertification and its disastrous consequences.

According to approximate estimations made at the time, desertification affected almost 80 states with a combined population above 600 million and covered an area of about 9 million sq km. As a result of this phenomenon, 5–7 million ha of productive lands throughout the world degraded every year and the cost of the damage done by desertification was estimated at US$10 billion (Tolba, 1978).

At present (from the data of World Atlas on desertification, 1992 Paul, 1980) desertification covered the territories of above 100 states. The population of the degraded lands is 250 million. The total area of arid lands on Earth is 6,100,000 ha and, by 1992, desertification threatened 3,600,000 ha of this total. The danger of land degrading exists for 70% of the area of all arid lands on the planet (Zonn, 1986). According to a later estimation (UNEP, 1997) the risk of land desertification remains for 110 states of the world with a combined population of 1 billion.

Such disasters as desertification, droughts and dust storms are interrelated in many respects. The closest relationship between them exists in arid and sub-arid regions. In these regions, population calamities causing hunger, disease, death, migration of refugees, etc., are the result of total adverse effect on the population of all three types of the phenomena. To a certain extent, desertification plays an integral role in this triad; it reflects a negative change of landscape, unfavourable and unhealthy for man and other living organisms. The degradation of these marginal lands encircling deserts produces ecological disasters, especially where they support human populations. In the 20th century, disasters occurred in the American Midwest in the 1930s and in the Sahel in West Africa in 1968–73. At the beginning of the new twenty-first century, the problem of desertification continues to develop over the whole planet, including in Russia (particularly in Kalmykia) and in Central Asia around the Aral Sea.

On the fringes of the Sahara in West Africa, desertification is now on a large scale. Since the late 1960s, the desert has expanded; in its southerly movement it has encroached on the Sahel, a transition zone from semi-deserts to savannahs in Senegal and neighbouring countries. The encroachment between desert and savannah advanced by about 1–10 km, averaging some 4.8 km per year. Regions become 'joined' to the desert, sources dry out, wells dry up, pastures become barren, and in some places sands start moving so they begin to cover land which was developed earlier. In the Sahel during the 1970s, moving sands came close to the capital of Mauritania, the city of Nouakchott, which was founded in the late 1950s. Some city blocks were buried under sand. Elsewhere, the moving sands of the Sahara encroached upon the oases in the Nile Valley.

In the Sahel, desertification occurs in the belt which extends from west to east over thousands of kilometres, almost across the whole continent of Africa. In recent years, and particularly in 1968–73, drought here has continued almost without end. Land desertification and drought resulted in a sudden deterioration of ecological conditions, causing loss of cattle, famine, dangerous diseases, population migration and the death of people in large numbers. It would be incorrect to consider drought, and not man, to be the main cause of the degradation of the Sahel – although this is the opinion of some specialists. Actually, droughts only make an ecological situation worse. Vulnerable landscapes, like the Sahel bordering on the Sahara in a condition of drought, require a lowering of the anthropogenic load on themselves. However this did not take place.

In Russia and its neighbours – Kazakhstan, Turkmenistan, Mongolia, as well as Uzbekistan – areas of desertification cover tens of thousands of square kilometres. The problem is widespread in the Aral Sea area, around the Kara Kum Canal, near the oases along the valleys of the rivers Amu Darya and Syr Darya, Tedzhen and Murghab. In Kara Kum and on the Caspian Sea coastal plain pictures taken from space have revealed hundreds of 'centres' of desertification: bright, light spots of degraded land with a diameter of up to several kilometres which have appeared around wells, settlements, and farms.

During 1997, the process of desertification in Russia embraced about 50 million ha. It has become most intensively developed in Dagestan and the Kalmykia Republics, in Astrakhan, Rostov, Volgograd Regions, in Stavropol and Altai Territories, and in Tuva Republic.

For the last three decades in Russia, the Black Lands of the Kalmyk Republic have become a zone of ecological disaster due to land degradation (Grigoryev, 1985). During a long time, until the mid-1950s of the last century, these Black Lands were used as pastures and were ecosystems in an unstable state of equilibrium. The degradation of the region's ecosystems (an area of about 2 million ha) is due to two basic causes. First, lands were being ploughed which were too fragile. This was not the action of the local people – the Kalmyks. Second, animal husbandry culture was 'forgotten' here. Fodder supply on the pastures of the Black Lands amounted to about 260,000 tonnes of fodder units in 1997 and was intended to maintain 744,000 sheep. In fact, on the steppe there were almost 1,653,000 sheep and above 200,000 saigas. The Black Lands pastures became overburdened by a factor of 2.5. Out of 3 million ha of pastures 1,195,000 ha was threatened and 5,343,000 ha 'extremely threatened, and 665,000 ha of former pastures have now degenerated into moving sands and solonchaks. The total area of Kalmykia territory was 7,600,000 ha; by 1951 about 13% had been destroyed (Zaletayev, 1997). In the second half of the 20th century, the steppe of the Kalmyk Republic has become a barren desert. The adverse effects on the local economy are many and include not enough fodder for cattle, resulting in the cattle starving or cash having to be spent to bring fodder from other areas (Vinogradov *et al.*, 1990; Zonn, 1986).

A further aspect of desertification, and one that can be connected to the activities of man, are dust storms. They are widespread in nature, but there is evidence that lands degraded due to man's mismanagement can trigger them. For example, the frequent dust storms in the Sahel remove the thin rich soil layer from those lands threatened with desertification.

Dust storms developing in sub-arid areas on land intensively and unwisely farmed by man, can serve as indicators of the location of likely desertification. In the 1930s, the Great Plains of the American Midwest were threatened by a huge, so-called, 'Dust Bowl'. It appeared due to the desertification of prairie land – the 'granary' of America – as a result of the activities of farmers on the land which had caused its degradation. The outcome was decreasing crops, ruined farmers, famine, and an adverse effect on the local and national economy.

Similar events to this American disaster of national scale broke out between 1960 and 1970, in North Kazakhstan, then part of the former USSR. The so-called Virgin Lands of the region had been cultivated without due regard for the features of their ecosystems.

Those specialists gathered in 1977 at the Nairobi Conference were convinced that the advance of the desert could soon be stopped if special steps were adopted. Indeed, the Executive Director of UNEP, M. Tolba, believed that 'desertification' on the planet would be stopped by 2000 (Tolba, 1978). However, his optimistic forecast has proved to be unsuccessful.

5.4.2 Satellite observations of desertification

The rapid development of desertification and the large land areas affected make difficult the use of traditional methods for studying it. Space images of the earth provide various information, including that on desertification and its most widespread anthropogenic variety (Kondratyev *et al.*, 1983; Grigoryev, 1985). The value of space images is in the information obtained on the meso- and macro-characteristics of forms of anthropogenic desertification rather than in details. Such features are revealed only with difficulty when using traditional methods of observation.

One of the forms of anthropogenic desertification is related to the destruction of vegetation around wells, farms and populated settlements. This phenomenon is widespread in arid and sub-arid zones in deserts, semi-deserts and steppes. The development of the desertification process in deserts means a change in natural complexes which results in worsening ecological conditions for vital human activity. Space images express this form of desertification as round and irregularly round spots. They are, as a rule, found not as single images but in groups and cause a unique pattern of dot-and-spot.

Around wells, these 'spots' form due to cattle clearing already scanty and thin desert vegetation. Thus, the top soil layer is broken as the vegetation is cleared. As a result, such areas turn to a kind of 'ulcer' of deflating (blowing away) grounds, which begins the process of dust formation. The reflective properties of the surface landscape change greatly owing to this development and is the reason for the space images to record bright (light) slightly dim spots. They are clearly seen against the relatively darker background of the surrounding landscape image. The sizes of such spots differ: in Central Asia and Kazakhstan their diameter varies from 0.5–0.6 km to 5–6 km (most often they are of about 1–2 km diameter).

The phenomena of anthropogenic desertification of the form and type related to intensive cattle breeding is seen in the presence of natural conditions suitable for pastures. Therefore, this unusual pattern of space image is far from general. In some regions of Central Asia and Kazakhstan, for example, the spottiness is expressed in a stronger image, in other regions the pictures are weaker, and in many cases disappear altogether. In the sand deserts of the south and south-east areas of Turkmenia more than 150 spots are distinguished in space images. They are much stronger in images taken over the Caspian Sea coastal plain, particularly between the lower parts of the rivers Volga and Ural; in approximately the same area, as in the south-east of Turkmenia, about 900 spots were found, which are large centres of deflation (wind blowing out) having a total area of about 220,000 ha (Kulik *et al.*, 1980). This is indicative of much greater anthropogenic changes in the landscape, including in the vegetation cover of the Caspian Sea coastal territory. In some places the spots revealed by space images are found here so frequently that the total area of desertification amounts to as much as 50%.

The destruction of soil-vegetation cover occurs very frequently also around populated settlements. In these areas of arid and sub-arid zones, desertification takes place under the impact of different factors, in particular cattle pasturing, building, as well as cutting of bushes for fuel. In space images the desertification spots around small populated villages are revealed in similar fashion – such as around wells.

It should be emphasized that, though the above-mentioned type of desertification in the form of round spots was well known, there were very few particular data on its spread, development area, or the frequency of occurrence. Now, space images have become an inportant source of information, allowing us to specify and improve the existing small-scale maps of different regions of the earth.

Desertification actively develops along lines of communications. According to space images, there is a widespread type of line-oriented damage to vegetation cover occurring as belts extending along communication lines. Most roads in deserts and semi-deserts are built on stony and clay surfaces. The belts of degraded soil-vegetation cover particularly form along the lines of the temporary roads built in the period of geological exploration and other surveys and then along the roads themselves as they are constructed.

From the air, at an altitude of about 1–1.5 km, hundreds of such linear anomalies can be seen on the Ustyurt Plateau, a desert upland in south-west Russia. Particularly in the western and eastern parts of the plateau, only some of the lines connect up the isolated populated settlements and therefore are shown on maps. In space images (as compared to aerial photographs) a smaller number of such linear formations is observed. However, space images do enable the estimating of large-scale development of desertification of the plateau landscape by measuring its extent (Grigoryev, 1985).

One of the important factors contributing to rapid disturbance of fragile ecosystems in arid and semi-arid zones, is the mechanical destruction of the topsoil layer. This comes about when agriculturists employ all kinds of equipment, especially heavy equipment. Using tractors, multi-disk ploughs and other machines is, of course, a highly profitable practice in farming, but at the same time it destroys soil. The use of mechanized farming, in particular, has contributed to the wide development of desertification in Tunisia. In the south of the country, in the landscapes west of Gabes, mechanized farming began to be used in the 1960s and 1970s. As a result, desertification on a large-scale was observed from space over the period 1975–78. The resource satellite *Landsat* (Rapp, 1974; Rapp and Helden, 1979) revealed not only the above-mentioned 'dot-and-spot pattern but also linear forms of soil and vegetation destruction. Of particular interest in the use of space images, is that areas changed by human pressure on vegetation and soil cover can be determined. Such areas are shown as a lighter shade of the natural landscape image and usually observed around intensively developed river valleys, oases, etc. Thus, from Landsat satellite images, a digression of pastures has been established in the steppes of Ubsu Nur in Mongolia (Grigoryev, 1991). A desertification map (1:75000 scale) was compiled for the region using space images. Five desertification levels were distinguished, explained by the excessive pasturing of cattle and the removal of forest for fuel.

In a number of arid zones, large anthropogenically disturbed natural landscapes have been revealed within which desertification has occurred. These can be as extensive as hundreds, and even thousands, of square kilometres in area. One such region, situated in Afghanistan near its border with Turkmenistan, was discovered in 1975 with the help of a space survey.

The desert to the south of the border, on the territory of Afghanistan, is shown in the images by a much lighter colour than that located to the north of the border. The analysis of pictures and ground-based studies has shown that such a photo-anomaly in depicting the same landscape is explained by only one cause – anthropogenic activity. The lands on both sides of the border are used as pastures in both cases. The different methods of pasture husbandry in Turkmenistan and Afghanistan have resulted in the differences seen in the landscape. In Turkmenistan pasture is regulated and therefore the desert landscape has proved to be less disturbed. Changes have occurred mainly around wells, as well as in some small settlements, and are shown in the space images as bright spots and dots. Between these local disturbances of soil-vegetation cover the landscape (first of all, vegetation and soils) has been changed to only a smaller extent.

An absolutely different picture can be seen on the other side of the border. The colour of the landscape surface image is much lighter here, caused by a strong disturbance of the natural soil-vegetation cover. The disturbance is explained by excessive pasturing of cattle. In Afghanistan, unlike Turkmenistan, the landscape has formed under the impact of irrational running of pasture husbandry. The space images show, against a general light background of pastures in North Afghanistan, some bright spots of irregular round shape. These spots are analogous in origin to similar spots fixed in Turkmenia territory.

Thus, only with the aid of a space survey has the actual condition of soil-vegetation cover been established within the whole region. Two kinds of desertfication are revealed – 'dot' and area. The latter, area, is typical of the landscapes of North Afganistan and has resulted in serious disturbance to ecosystems. In Turkmenia, the 'dot' type of desertification develops primarily and in the region at that time (1975) was not widespread (it amounted to 10–15% of the total area) and had not become a serious threat to ecosystems.

Anthropogenic desertification of whole regions, such as is the case in Turkmenia and Afghanistan, caused by excessive pasturing of cattle is also observed in space images for other regions of the world.

One such region is on the east coast of the Mediterranean, at the international border between Egypt and Israel. At the same time, this is the border between two landscape areas differing in albedo value. This difference, due to desertification (see Figure 5.2), was revealed when, for the first time in this area, space observations were utilized.

On both sides of the border between Israel and Egypt the natural conditions of the desert are similar. However, different methods of farming and agricultural development of lands has caused a difference to be seen within the same landscape. It separates irrigated and intensively cultivated lands to the north-east (Israel) from non-irrigated areas to the south-west (Egypt) which are used, basically at random, for pasture.

According to *Landsat* measurements, the albedo of the Sinai surface (south of the border) was 0.443, the albedo of the Negev desert surface (north of the border) was considerably lower and was on the average 0.317, though in some places (areas with denser vegetation) it amounted to 0.271 (Otterman, 1971).

As a result of radiometric measurements, the presence of contrast has been established between both areas (north and south of the border), which amount to 40% and higher. The contrast remained in space images obtained in different spectrum intervals. Both ground-based and aircraft investigations have revealed that to the south of the border, in the area with higher albedo, almost all vegetation was destroyed as a result of excessive grazing on the pastures and dried and dead bushes prevail. North of the border, in the Negev Desert where vegetation has not been trampled down, evergreen species prevail, which causes low seasonal variability in the landscape image and lower albedo. According to these space

Figure 5.2. A Landsat space image of the Mediterranean Sea east coast obtained from an altitude of about 900 km. The straight-line border (perpendicular to the coast) of the states of Israel and Egypt is clearly seen; it also correponds to the development of two different types of the same landscape (sub-tropical desert). Note the strongly degraded pasture, due to excess 'overload' and desertification in Egypt.

images, the zone of desertification extends not only near the international border between Egypt and Israel, but also has a regional extension. Anthropogenic desertification covers almost the whole northern plain, a part of Sinai with an area above 10,000 km^2.

The occurrence of anthropogenic desertification is particularly extensive to the south of Sahara, where it extends over a territory of tens of thousands of square kilometres. Space images demonstrated that in the area of the Sahel affected by drought (in 1968–73) an extensive area actually appeared characterized by relatively higher reflective properties than was recorded earlier, in the period before the drought.

Ground-based expedition studies presented by specialists of different countries, including those from FAO, established that in the Sahel as a result of unsystematic and unplanned management of animal husbandry and, importantly, excessive pasturing of cattle, the pastures experienced an extreme load. The consequence was the exhaustion of thin and scanty vegetation, intensive loosening of the surface soil layer which was then bared to the wind and subsequently blown away.

Long-term observations of this process were conducted to monitor the anthropogenic desertification of the Sahel landscape (McLeod, 1974a). Control observations, aiming at a comparison, were made in special reservations – a cattle breeding farm was fenced off from other territory. In this farm (which is a polygon with a length of 113 km along the perimeter and some 11,000 ha in area) the land was also used for pasture. But, unlike the land located on the other side of fence, on the farm it was worked on a reasonably planned basis, without constant 'load' on the whole territory. Animals grazed and were then moved on the farm territory according to a plan. As a result, vegetation was grazed slightly and then rapidly revived. Outside the ranch, indiscriminate cattle grazing resulted in loss of vegetation and soils.

The special farm was a control observation organized in 1968 and the space images were obtained five years later, in 1973. Unlike the surrounding lands, the 'farm did not show any noticeable signs of desertification. The regulated management of pasture husbandry had yielded its results and these were fixed in space by *Landsat-l* images. The territory in the zone of reservation, rhomboid in plan, appeared in the images as darker when compared to the neighbouring areas.

These comprehensive ground-based studies were made to explain the differences in space images of the reservation territories and those outside it (McLeod, 1974b). The specific features of relief structure and climate in the reservation and outside it are similar. As the analysis of 1968 aerial photography has shown, the nature of landscape image in the farm zone location and outside it was also similar at that time. However, in five years, the sandy ground within the reservation was more stable than that beyond the farm border where the sandy soil proved to be well broken up. The microforms of aeolian relief were found far more often. The latter indicated wind blowing out soils and ground. The vegetation cover on the farm was also much more dense, explaining the difference between the land image within the reservation and that outside it.

In the southern areas of the Sahel, where farming is most highly developed, the causes of desertification of the location are excessive ploughing (without necessary land rest), irrational management of farming, errors in irrigation, an unreasonable use of water resulting in the decrease of ground water level, and the burning off of vegetation.

This advance of the desert into farming areas is clearly observed from aircraft and from space. For example, in 1968, the use of aerial photographs in the Sudan led to the discovery of a field of moving dunes extending to over 4 km in the area of El Fasher (Rapp and Helden, 1979). In the 1954 aerial photographs of the same territory the dunes were not observed. The field of dunes formed on ploughed land where excessive cultivation had taken place.

Had the cultivated lands in different areas of the Sahel been rested in earlier years – perhaps allowed to lie fallow for 15 to 25 years – by the early 1970s farmers could have reduced this resting time to 1–5 years and gone on to produce a larger amount of agricultural products to meet the conditions of a rapidly growing population (Schneider and Mesirow, 1997).

According to the data of space imagery (Paul, 1980), the continued advance of dunes on to agricultural lands is a natural event. The ecosystems and the nature of border zones, like the Sahel, are very fragile. It is clear that in drought years, if 'pressure' is applied to nature by means of, say, intensive ploughing, natural complexes are broken, soil is loosened up, rougher sand particles form dunes, finer ones are blown out by wind. In some areas of the Sahel the old overgrown dunes started moving. This, in particular, was established by comparing the results of aerial photography of 1957–58 with those of *Landsat* images obtained in 1975 at 1:1,000,000 scale for the region of Niger (Mainguet *et al.*, 1979).

One more type of local population activity contributes to the Sahel landscape desertification – stubble burning. The past year's dry grass is burnt every year with the aim of clearing out stubble and killing off different insects. This practice of destroying old vegetation is used in many regions of the world. In the Sahel it would have only positive consequences – if it were done directly before the onset of the rainy season. However, according to the observations of orbital station *Skylab* (Muehiberger and Wilmarth, 1977), and contrary to the claims of agriculture experts, most grass burning activity takes place after the rains, and not before. A convincing proof of such practice is the numerous smoke plumes from artificially induced fires in the Sahel observed in space images. Burning-out of vegetation before the onset of drought-bared soil, facilitated the drying of it, the break up of the upper layer, and more intensive exhaustion of the soil by animals. All this resulted in the process of the loosened ground blowing out; the strengthening of winds blowing from the Sahara in the dry period contributed to this.

In addition to the form of desertification experienced in the Sahel, as well as in other regions of the earth, other forms are also known – linear and dot forms. Particularly developed among them are dot forms around wells, watering places and populated settlements (McLeod, 1974a). They are caused by soil and vegetation exhaustion due to overuse by cattle, as well as the destroying of trees and bushes for use as fuel.

These desertification forms appear in space images as small-scale pictures – as bright light dots or round spots. The dots and spots correspond at a location, round or irregularly round in plan areas of exhausted vegetation having a diameter of hundreds of metres, even up to 1 kilometre in diameter. Such desertification forms are found over the whole of the Sahel and their lesser or wider distribution can be tracked by space images. According to French investigators, this form of desert degradation is also widely developed, in particular, in the region of Zinder (Niger) (Mainguet *et al.*, 1979) where most instances can be seen in space images.

Around some populated settlements, desertification occurrs due to destruction of trees and bushes mainly used as fuel. Thus, several decades ago in the Sahel, the zone of distribution of tree-bush vegetation extended as far as 5–10 km from the outskirts of Khartoum. By the late 1970s it had receded by 100 km and more (Rapp and Helden, 1979).

One of the distinguishing features of territory desertification is deflation of the surface soil layer. The development of this process is obtained from space images. In fact, any area in the Sahel where for some reason the vegetation cover, or the soil cover, is disturbed, it is sure to raise dust given the conditions of this region.Therefore, all the images of showing up different forms of desertification discussed above are at the same time revealing examples of deflation centres.

The images from space allow us to observe the consequences of the Sahel desertification as dust transfers. *Landsat* images (McLeod, 1974a) followed the transfer of dust and sand in the atmosphere over the Sahel – near to the fenced-off farm used as a control observation. The length of dust transfers usually exceeded 150 km and their total area amounted often to many thousands of square kilometres. It is important to mention that the observed dust transfers were non-transit; their centres are located in the Sahel and are the areas with the most heavily exhausted vegetation and soil.

According to space images, apart from dust storms, dust haze very often spreads over the Sahel. It decreases the contrasts of the earth's surface details which are usually seen much better on days without haze. The areas with a most frequent turbid atmosphere, best followed only from space, are related to the territories of the Sahel which undergo the highest desertification. They look like bright indistinct spots in the pictures and, related with the turbid atmosphere, were fixed more than once by the images in different areas of the Sahel, in particular, to the north-west of Lake Chad, and to the north-west of the River Niger bend. The dust storms starting in the Sahel lead to blowing out of the upper rich soil layer. According to some investigators, dust storms have resulted in a considerable decrease of productivity in the Sahel zone.

All the above-examined factors contribute to one thing – desertification and, first of all, the disappearance of zonal vegetation and breaking up of the soil surface. In many areas in the Sahel, desertification occurs in conditions of strong wind activity and this assists in producing the aeolian formations, moving sands. In some cases, in the 1970s, occurrences were noted just before the drought started.

Space images helped greatly in determining the displacement of the Sahel's southern vegetation boundary in the zone of deserts. While field studies, as well as aerial photographing, can give only local information for small areas of studies, space images give general information. In the Sudan, on the basis of analyzing space images and control ground-based studies, it has been established that the desert extended over distances of 30–120 km for only 17 years (Rapp and Helden, 1979). The space data agree in many respects with the results of ground-based and aerial observations of this whole region. Approximately the same information, that is, about a 100-km shift of the Sahara's southern boundary southward, was obtained for other areas both in the Libyan Desert and in the Nubian Desert.

Various viewpoints are held concerning the occurrence of natural disaster in the Sahel. Swedish specialists from the University of Lund assume that well-known ideas on the extending Sahara in 1960–70, at the rate of 5–6 km per year as well as the subsequent dynamics of desertification, are not at all related with the phenomenon of anthropogenic pressure. The satellite and ground-based data (according to these specialists) on the dynamics of vegetation cover productivity in this region correlate well with changes in precipitation (Helden, 1994).

The data obtained convinced Swedish specialists that the cause of disaster in the Sahel was a natural increase in climate dryness, a long-term drought. The natural component data on desertification (precipitation decrease and climate dryness increase) are not in doubt. At the same time one can see from these Swedish studies that they neglect the manifestations of anthropogenic degradation of the Sahel landscapes clearly recognized in space images and, thus, the anthropogenic component of desertification.

Study of the Sahel space images has revealed not only the extension of the Sahara zone to the south but also – and this is a principal finding – that the desertification occurs from inside the Sahel due to the appearance and expansion of desertification centres located around settlements and wells, rather than due to the extension of the Sahara zone. These centres reach the border of the desert zone in the Sahara (Mabbut, 1981) and give weight to the data on the anthropogenic factor in the Sahel disaster.

A considerable (though not prevailing) role of the anthropogenic factor in the degradation of the Sahel has also been shown by studies of tendencies in desertification over a long period of time (starting from 1973). These studies were carried out in the north-east part of Mali. *Landsat* images were used to serve as the initial material for analysis (Babayev *et al.*, 1986; Vinogradov *et al.*, 1990; Grigoryev, 1991) of the earth's surface albedo and NDVI and were corrected taking into account atmospheric effects.

The variability of the above two parameters of the earth's surface has demonstrated a steady tendency for deserts to form, particularly in the southern part of Mali under study. Thus, for the period 1973–86, the albedo of the surface ot the southern part of the region increased by 15%, and NDVI decreased by 17%. It was also found that the degradation of ecosystems occurs in the Sahel against a general background of drought growth for the recent 30 years with a considerable rise in anthropogenic activity in the region being studied.

Here, it should be emphasized that space images fix not only negative changes related to non-optimal economic activity in ecologically fragile, easily disturbed landscapes of arid and sub-arid zones. In some steppe and desert areas of the world, despite the growth of farming, active desertification is not observed. Thus, in China, it has been observed on the southern margin of the Takla-Makan desert (China) where the *Landsat* image has not fixed a distinct increase in the desertification process for the period 1973–88. At the same time, an increase was revealed in the area covered by vegetation due to the activities of farmers (Sugihara *et al.*, 1994).

On the whole, using space observations to monitor desertification makes it possible to reveal the variety, types and forms it takes in one or another region and, at the same time, to estimate the large-scale nature of the process and cover many areas at the same time to assess regional, and even subglobal, disasters. Thus, possibilities are opened up for regional and global monitoring of the development of the prerequisites that attend these disasters.

5.5 BIODIVERSITY

5.5.1 Modern tendencies in biodiversity state on the planet

At the present time, about 1.7 million species of different living organisms have been identified (World Resources, 1996). At the same time, new discoveries of unknown species allow for the possibility that the actual number of species of living organisms reach almost 14 million. However, according to some data, the figure may vary from 3 to 11 million, the majority of species being insects – from 2 to 110 million species.

Up until the present day, the Earth's ecosystems have been difficult to reach and have remained very little investigated (humid tropical forests, the bottoms of the oceans, high mountains, etc.). The most numerous of new species discoveries, the insects, are expected to be found in humid tropical forests. For many reasons, a number of species of living organisms are constantly subject to the threat of extinction and disappearance. Table 5.7 presents selected data for some countries where the risk of disappearance of animal species exceeds 10%. It is assumed that for each decade during the period of 1975–2015, between 1–11% of all existing species have been, or will be, extinguished (World Resources, 1996). It is supposed that 39% of all disappeared species have died due to the introduction of non-local species, 36% due to destruction of habitat, 22% as a result of hunting, and 2% for other reasons (including poisoning) (Brown *et al.*, 1996).

The number of species that disappeared between 1600 and 1994 total 626. Most of all, species of birds (36) disappeared in Oceania, reptiles (12) in Africa, fishes (30), mammals (37), invertebrates (162), and amphibians (2) in North and Central America. This unprecedented extinction of living organisms on the planet arouses alarm. It is known that all species live in interrelation with their habitat and with each other, and not in isolation. The disappearance of one or another species results in a breaking of ecological bonds, the global network of which supports the stability of the Earth's biosphere (Kondratyev, 1993b; Kondratyev, 1996; Kondratyev, 1997b). The question arises as to whether the reduction of biodiversity will threaten human life. Some possible scenarios have been 'calculated': in particular, it has been shown that the destruction of tropical forests in Amazonia can disturb the climatic balance in the region, influence the sea currents surrounding the continent and, in the end, form a different natural zone (where once there were tropical forests) which is unsuitable for aboriginal life.

The destruction of living organisms causes alarm also because, of course, some organisms are an important source of food and fuel for the inhabitants of a number of states. Thus, for example, in Niger about 80% of energy consumed by rural dwellers is provided by 'wild' trees (Brown *et al.*, 1996).

Table 5.7. The diversity and modern condition of mammals: selected data from 32 states showing those of them for which the risk to species exceeds 10% of all of them (Ourgensen and Nuhr, 1996).

Region/Country	Number of known species	Including those in risk state	Region/Country	Number of known species	Including those in risk state
Africa			**Europe**		
Djibouti	22	6	Austria	83	38
Egypt	105	12	Denmark	49	14
Guinea-Bissau	109	11	Finland	62	7
Libya	76	9	France	113	59
Mauritania	61	10	Germany	94	44
Morocco	108	113	Ireland	31	5
Nigeria	274	57	Italy	97	13
Tunis	77	14	Netherlands	60	39
West Sahara	15	7	Portugal	56	25
North America			Spain	135	17
The Bahamas	17	2	Sweden	65	10
Cuba	39	9	UK	77	24
Greenland	26	7	Former USSR	357	78
USA	466	49	**Oceania**		
South America			Australia	320	43
Argentina	255	26	New Zealand	69	14
Brazil	394	42			
Chile	90	10			

In many developing countries, a part of the population is still very dependent on the state of vegetation and the animal world.

There are different viewpoints on the problem of preserving biodiversity: some specialists consider that living natural resources should be urgently protected against any use. Other specialists believe that resources should be protected in such a way that they can used. It is evident that the latter viewpoint is more reasonable and clear that, for the time being, mankind cannot live without the use of biospheric resources. Figure 5.3 shows a map of natural landscapes on the planet; these should be regulated within reasonable limits so as not to cause disappearance (exhaustion of species and genetic diversity). The principle of biotic regulation of the environment, proved by V. G. Gorshkov, is of basic importance (Gorshkov, 1995) here.

The most important aspects of the strategy of using biological natural resources are, undoubtfully, determining their condition, finding out what species are threatened or in the process of being destroyed. According to the data, as of 1990, about 4,500 such species were identified among animals; and in 1993 their number was 5,929, with a different degree of threat to their existence (Brown *et al.*, 1996). In 1993, also, the threat of disappearance arose in particular for 741 species of mammals, 970 species of reptiles, 169 species of amphibians and, for some of

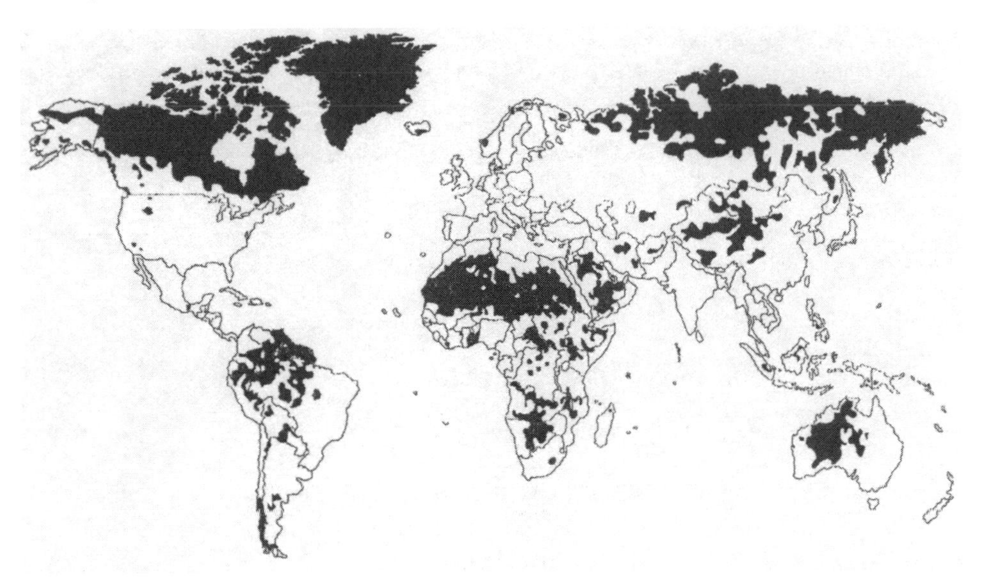

Figure 5.3. Map of the global distribution of natural (unchanged by man) landscapes as of August 1988.

them – 177, 188, and 47 species, respectively – the threat was extreme. There has long existed such forms of protecting 'wild' flora and fauna as that of entering rare, degrading and disappearing species in the *Red Data Book* and/or taking steps to prevent their further unreasonable use. The efficiency of such an approach became less viable due to insufficient account being taken of the fact that protection of an individual species is ineffective without simultaneous protection of the whole ecosystem.

One of the directions the preservation of the planet's biodiversity is taking is the organization of protected territories. In this case there are several types:

(1) fully protected territories, including:
 (a) research reservations,
 (b) national and local parks,
 (c) natural and local parks,
 (d) natural monuments; and

(2) partly protected territories, including:
 (a) natural reservations protected for highly specialized purposes and
 (b) protected landscapes (some of them can include cultural elements as well).

In all, there are 9,793 protected territories on Earth, each having an area of not less than 1,000 ha. Their total area amounts to $959{,}568 \times 10^3$ ha (7.1% of the land area). Different regions of the earth are characterized by different number and area of such territories. The most numerous 2,923) are in Europe, including Greenland, as well as in North and Central America (2,549). In Asia, Australia and Oceania, Africa and South America they are much less in number (1,174, 1,087, 727 and 706, respectively). However, according to the relative area of the protected territories, the first are Australia and Oceania (11%), as well as North and Central America (10.2%), followed by Europe (8.9%), Africa (7.1%), South America (6.3%) and Asia (4.4%).

These figures involve a fundamentally different specific nature of biodiversity in the countries of different regions of the earth. In Europe, for example, there is the highest level of nature degradation and (realizing the acuteness of the problem) the greatest number of protected territories. According to the level of nature protection in the world the first place belongs to North and Central America being much less populated (on the average) and less degraded (in respect of nature) than Europe. The area of protected lands in the regions of North and Central America is $230{,}199 \times 10^3$ ha and $223{,}905 \times 10^3$ ha respectively.

The greatest relative area of protected territories (that is, almost three times or more as great as the world average) exists in only 10 countries: New Zealand (22.4% of the total area of the country), Bhutan (20.6%), Venezuela (28.9%), Ecuador (39.2%), Dominican Republic (21.5%), UK (20.9%), Slovakia (20.7%), Germany (25.8%), Denmark, including Greenland (44.9%), and Austria (24.8%).

The most extensive areas of protected territories are located in the USA ($130{,}209 \times 10^3$ ha), Australia ($94{,}077 \times 10^3$ ha), Denmark ($99{,}618 \times 10^3$ ha), Canada ($82{,}358 \times 10^3$ ha), and in Russia ($70{,}536 \times 10^3$ ha). In the first four of the above-listed countries the portion of protected lands, with respect to the total area of the country, is larger – and in Russia is almost two times smaller – than the world average (13.9%, 12.2%, 44.9%, 8.3% and 4.1%, respectively). Among the largest countries, only China (6.1%) and India (4.4%) have a lower level of nature protection (according to this index), as in Russia. However, eight countries (including Yemen, Syria, and Laos) have no protected lands at all.

Some protected territories in individual countries have achieved international status in three categories: biospheric reserves, places of natural and natural-cultural inheritance, and wetlands. The largest area of lands occupied by biospheric reserves are in Denmark, including Greenland ($70{,}000 \times 10^3$ ha), Brazil ($29{,}940 \times 10^3$ ha), Russia ($9{,}561 \times 10^3$ ha) and Algeria ($7{,}276 \times 10^3$ ha).

It is clear now that neither protection of individual species of organisms, nor selection of complex protected territories can solve the problem of biological resources degradation. To replace the abstract concept of humanization of their use aiming at nature protection, a new concept is emerging – reasonable use in conditions of steady development or the steady preservation of biodiversity (Edwards, 1995).

It is not doubted that in this case one should take into consideration the ecological, economical and social needs of the different societies in different countries. Such an approach does not contradict the urgent requirements of providing for vital activity in a number of groups of population in underdeveloped countries and would enable the biodiversity problem to be solved and not only on local protected territories.

5.5.2 Using satellite information for studying biodiversity

Remote, including space, sensing of the earth is suitable for differentiation of various types of bioresources. The data on natural complexes identified in space images can be based, first of all, on space information on biodiversity. This information includes such important data as spatial and temporal images of landscapes, and information on their structure and composition.

Remote sensing assists in determining different types of both natural and anthropogenic effects on one or another landscape and in the obtaining of information on sources, direction and scale of such effects.

The most general information on biodiversity which can be obtained with the help of space images is information on the large-scale changes of natural zones and their vegetation cover. The largest biodiversity reduction (both vegetation and animal) on the planet occurs as a result of the destruction of large forests, either naturally (fires) or for anthropogenic reasons (cutting). Such changes to biodiversity occur also in those regions subject to desertification: in deserts and, particularly, in the so-called marginal areas such as semi-deserts, steppes, and savannahs. Space images enable observations of these events of deforestation and desertification thereby providing evidence of the change to biodiversity.

Special estimations of biodiversity using space information have been accumulated for various landscapes and regarding different aspects of vegetation and animals. For example, satellite images were successfully used for estimating biodiversity in the Western Ghats in India (Rao, 1996). This region was studied using satellite *IRS* images characterized by a wide variety and abundance of vegetation cover (1,700–2,000 species). Space images were fixed for the existing patchiness of forest cover species varying from coastal mangroves to evergreen tropical forests. Within the forest-covered territory space images differentiated types and complexes of forests, the extension of which depends on a set of factors (relief, soils, rocks, climatic conditions).

Indian specialists select in the images the habitats of one or another species of animals; for example, those related to the hilly areas of the Western Ghats covered with mixed highly varied forests. Information ‘is also obtained from ground-based estimations. However, specific data, ‘related to habitats and the natural complexes identified in space images can be extrapolated over an extensive area with similar natural conditions.

As animals migrate in their search for water, food, and favourable living conditions in general, some will pass from one natural region to another. These migrations provide an opportunity to model their habitats and ecosystems in work done in such establishments as the Indian Zoological Institute.

Wetlands (waterlogged or swampy lands) are one of the most important types of habitat for living organisms. Space imagery can provide various information on the type, qualitative and quantitative characteristics of, distribution and changes in, water formations and are useful in the monitoring and control of wetland ecosystems (Gill, 1993).

Biodiversity estimation is effectively realized on the basis of joint use of satellite telemetry, the data of satellite surveys and geoinformation systems (GIS). It is generally made on a regional level. Thus, for example, satellite data (telemetry and space images) are used for obtaining information on the distribution and behavior of caribou in the north of Alaska (Dau, 1986). Using GIS, digital map-schemes have been prepared showing disturbed land around routes of roads and oil pipelines.

Satellite telemetry and similar schemes have helped to establish that caribou species are much less numerous in places of insect concentration. The potential mass concentrations of insects have also been determined from digital map of vegetation compiled from *Landsat* imagery.

To estimate biodiversity, the various maps used are prepared on the basis of space images of hydrological network, snow cover, soils, fodder resources, etc. However, the vegetation map is the basic tool, compiled using *Landsat*, *ERS*, and AVHRR images. In its turn, this vegetation map can be the basis for preparing the dot map showing distribution of vertebrate and invertebrate animals (with consideration made for the indicator types of biodiversity), species abundance and their distribution. In this case, selection of outlines is possible identifying rare and disappearing species of plants and animals and, vice versa, showing maximum species abundance.

The data obtained are helpful for determining those communities which should be, first of all, protected since the maximum biodiversity is related to them. The same information can be used for the organization of territory protection, determining areas for protected sites, buffer zones, and so on.

Space images in tandem with other observation data (aerovisual, space biotelemetry) have been used in estimating populations of big mammals (walrus, polar bear) in the Arctic (Belchansky *et al.*, 1992). Satellite *Okean* radar images were used and provided information on habitat and the distribution of ice cover types, razvodiya (open water in ice), ice cracks, water temperature, etc.

Aircraft observations helped to obtain more detailed information on ice types, its border with water, groups and individual animals. Data obtained from the correlation of relations between number of animals and habitat factors were also used. Probable habitats, in particular of walrus populations, were estimated on the basis of long-term data on the maximum density of the distribution of animal groups related to ice with its total concentration in the interval 1–5 (Belchansky *et al.*, 1992). The same method was used for calculating probability of occurrence of animals in areas with different compactness of other ice classes selected (of four classes) from radar images.

Thus, the data obtained from space images on habitats, boundaries and types of ice cover are one of the elements of an observation system for estimating the critical levels of diversity of big mammals in the Arctic.

In the Sahel, two types of satellite images were used for estimating. From the high-resolution images obtained with the help of a *Landsat* thematic cartographer, the diversity of landscapes was determined. The annual value of the biomass was calculated from the *NOAA* satellite NDVI data. These data (from images) were

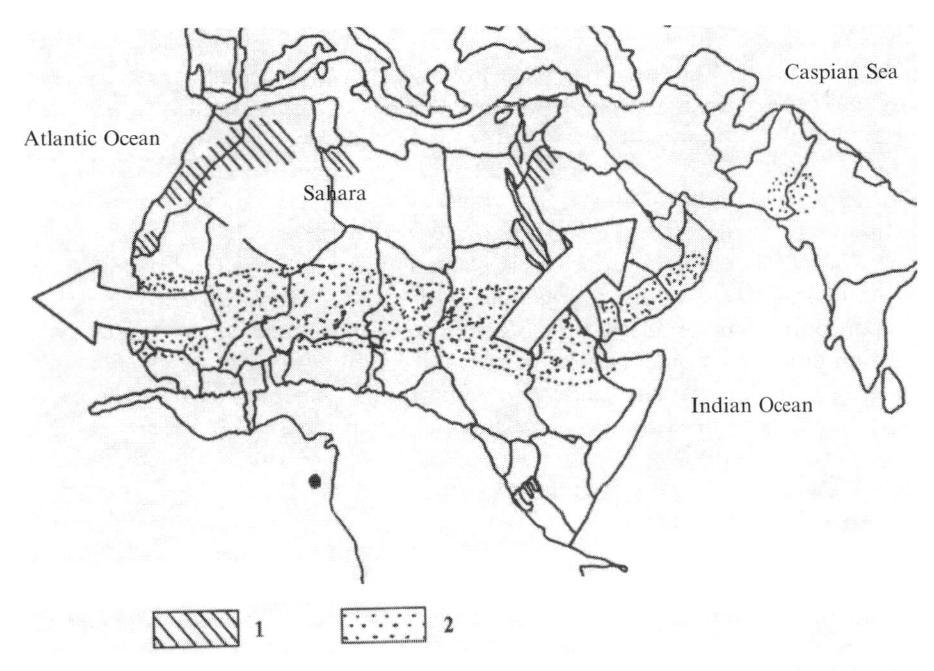

Figure 5.4. Areas of locust reproduction in 1988: 1 = spring (March–May) in the Atlantic and Mediterranean regions of North Africa and the Red Sea region (in deserts and desert low mountains); 2 = summer, south of the Sahara (in the Sahel) savannahs as well as in desert low mountains on the Arabian Peninsula. The arrows show the direction of locust movement, including that over the Atlantic Ocean (Rao, 1996).

compared with the bird diversity (as the indicator of biological diversity). The constructed model, based on three types of information, allowed an explanation of 40–50% of variations in the population of birds in the winters of 1991–92 and 1992–93 winters (Ourgensen and Nuhr, 1996).

The problem of biodivergency on the planet becomes important for mankind not only because of a reduction in the species of living organisms. The question is about the not fully understood consequences of the forthcoming disaster. At the same time a real threat to the vital activity of people in the present-day, and not in the future, is the mass reproduction of some species of living organisms.

These outbursts of vital activity of some organisms have long been known though they cannot always be explained. Among them is, for example, the phenomenon of reproduction of sea microorganisms forming so-called red tides at some sea coasts.

Among such phenomena is that of mass reproduction and invasion of locust to different regions (see Figure 5.4). According to approximate estimations, about 30 million sq kn of land is invaded by locust. These are exclusively agricultural lands – cultivated fields and pastures. Such locust invasions are accompanied by total devastation of the lands in question, causing famine among their population. At present, about 55 countries of the world are periodically subject to this type of biological disaster.

Satellite imagery can be used to determine those areas with conditions most favourable to mass reproduction of locust. It is possible to monitor from satellites the soil moisture condition and development of the vegetation cover which serves as fodder for locust. The data on these characteristics of the locust habitat (reproduction) can be obtained using, in particular, *NOAA*/AVHRR. The biomass is estimated using the normalized vegetation index.

The use of satellite images is particularly helpful in obtaining information on the centres of possible development of locust and likely devastation caused in remote and not easily accessible regions where ground-based monitoring would prove difficult. Examples of such landscapes are the deserts of north and east Africa, in the Near East, and in South-East Asia. The total area of such regions periodically subject to locust onslaughts amounts to 16 million km^2 (Rao, 1996).

Space images have proved to be useful in assessing regions of locust reproduction risk. Soon after, Africa and the Near East were subject to devastating locust invasions in 1986–88, the possibilities of using space techniques for future observations were investigated and tested by the German Society of Technical Cooperation (Dreiser, 1993). Experimental studies were carried out on the Red Sea coast in Sudan in the Tokar River delta, one of the basic regions of locust reproduction. Both *Landsat* TM images and confirming ground-based data were used. The areas covered with vegetation (the expected places for locust reproduction) were located with the help of NDVI. These areas were then masked, classified by the method of maximum likelihood and, later, equilization and comparison with ground-based data were made and a map (1:150,000 scale) of the area compiled.

The studies carried out (they were also made in Mali, Mauritania, and Nigeria) made it possible to locate successfully and promptly areas of mass locust reproduction and to orient resources to destroy them.

In Australia in 1991, *Landsat-5* images were used to forecast the development of locust and reveal their places of reproduction. It was found that the most acceptable of such places is the vegetation index. Besides, additional information on such places can be obtained with the help of data on soil cover and geological features of the location determined from space images (Bryceson, 1991).

Landsat TM and *SPOT* space information were used to locate the areas of distribution of the Aedes mosquito – a Rift Valley Fever carrier – on the east coast of Africa in 1989. The vegetation cover features were decoded and normalized and difference vegetation indexes were calculated from the images obtained (Ambrosia *et al.*, 1989). It was established that the high index values proved to be related to the areas of distribution of mosquito, the carrier of the dangerous disease. The highest index values were determined for the periods of the highest mosquito number, both for the wet (January) and dry (August) season.

In vast areas of forest in some regions of the world, where there are sudden outbursts of pest reproduction, the laborious process of using traditional ground-based techniques to monitor the phenomenon has encouraged specialists to use remote monitoring techniques.

In the Siberian forests, great damage is done to trees and forests by pests. For example, in the taiga forests of Krasnoyarsk Territory, outbursts of mass reproduction of Siberian silkworm, black longhorn beetle, and Pine Geometrid occur periodically. The areas of destroyed forests infested with such pests are usually distinguished by shade or colour from space images of 1:200,000–1:1,000,000 scale (Isayev and Ryapolov, 1988).

For more detailed investigations of forest damage done by pests, remote sensing images of 1:50,000–1:15,000 and larger scale are employed.

In the Siberian forests a technique was tested for using space and aeroimages to forecast and map potential centres of forest entomopests (Malysheva and Sukhikh, 1991). To this end, preliminary investigations for forest pathology monitoring were made in forests of a differing extent which had suffered damage by strong winds. These locations were identified using space images and ground-based data. Space pictures from *Kosmos* satellites were used. The maps prepared from space images of weakened (in this Siberian case by wind) forests orient an investigator to the areas of the highest probability of trees damaged by pests.

To locate centres of probable Siberian silkworm appearance, a method was devised to determine the fodder base of the pest using space images. It was revealed from ground-based data that the fodder base is usually situated in plantations with prevailing fir on plakors (flats, interfluves) growing on slopes angled at a low steepness of some 3°–5° and at heights with absolute values not exceeding 300 m. These data can be obtained by a combined use of space images, topographic maps, and ground-based observations.

The role of landscape-key method for monitoring the population of basic pests was developed at the Krasnoyarsk Forest Institute of the Russian Academy of Sciences. The Institute has considerable expertise in forecasting forest infestations of pests (Isayev *et al.*, 1991) The method is based on consideration of the confinement of certain species of pests to one or another natural location, landscape and their components. Space images are used for landscape mapping (1:300,000–1:1,000,000 scale) of territory and for compilation of maps showing potential centres of forest pests.

The technique developed for aerospace monitoring of forests allows a two to three times decrease in the volume of expensive and laborious ground-based work. It was used in Russia to establish the total area of dark coniferous forests subject to pests in the taiga forests of the central region of Krasnoyarsk Territory. These forests, on an area of 1,500,000 ha, are suitable for the mass reproduction of insect pests. The weakening of their reproduction outbursts is expected over an area of 1,000,000 ha. The forests in the rest of Krasnoyarsk Territory, some 2,500,000 ha in area, are not suitable for insect pests development.

The role of aerospace observation is very important for observing different kinds of protected territories – that is, territories protected from human activity and natural calamities. Of course, satellite images cannot provide complete and necessary information on biodiversity within such territories. However, they do enable the observation of its many characteristics; for instance, the degradation of vegetation cover as a result of fires, drought, lowering of ground water level, infestation with pests. This information is also useful in estimating macrochanges of habitat as an indirect judging of biodiversity changes.

In India, satellite images have been successfully used for observing the condition of a number of reserves, such as the National Park Kanha, the wild animal reserve at Chandana in Orissa and the Little Ran of Kutch reserve in Gujarat (Rao, 1996). In all these reserves, space images have been used to estimate numbers of wild animals.

Thus, despite many restrictions, satellite images, particularly detailed ones, are undoubtfully useful for making estimations of the condition and change of habitats and monitoring planet biodiversity. These estimations are based, first of all, on exclusive possibilities for the differentiation of space images of natural complexes of different scale. Quite definite habitats of different species (and their combinations) of vegetation and the animal world correspond to these complexes. The important components of such estimations are other types of information which can be obtained as a result of larger-scale aerial photography, aerovisual observations, selective (at key sites) ground-based investigations and satellite telemetry data.

5.6 DROUGHTS

5.6.1 Droughts as natural and man-made disasters.

Drought is one of the most threatening of disasters. Its consequences are considerable in the effect it has on population and environment. Droughts cause famine, mass disease and death and are a result of landscape degradation and location desertification, hence the unripening of crops, loss of cattle, lack of fodder and water (see Figure 5.5).

Droughts occur for different reasons (Buchinsky, 1976, Disasters, 1978, Rao 1996): lack of precipitation (climatic drought); unusual distribution of precipitation by seasons of the year (agricultural drought); landscape change due to erosion; anthropogenic reasons – in particular, location deforestation, ground water level decrease, soil moisture deficit (hydrological drought).

Unlike other disasters, droughts are characterized by their long-time duration, which can last from several weeks to several years – for example, the droughts in the Sahel in 1968–73 and in the mid-1980s. And Heatcot (1978) found that Australian farmers believed they could, with difficulty, survive a one-season (that is, lasting one year) drought but a recurrent two- or three-year drought would lead to their ruin.

In 1972 in European Russia one of the worst and longest-term droughts occurred (Buchinsky, 1976). An unusually hot summer affected an area from the Kola Peninsula to the Caucasus. On Kola Peninsula itself the average July temperature was twice as high as normal. In Moscow on some days the air temperature amounted to 36°C. On the White Sea it was hotter than in the Crimea. The cause of this drought was the formation of an extensive summer anticyclone after a

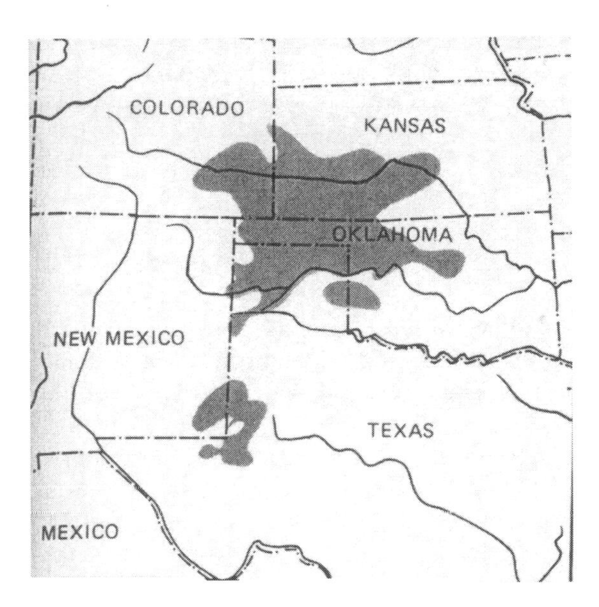

Figure 5.5. An area of desertification known as the 'Dust Bowl', which appeared in the American Midwest in the 1930s during a period of drought and simultaneous 'pressure' of farmers on prairie landscapes in the Great Plains (*source:* Ramade, 1987).

cold front. Cold air from the northern regions swept European Russia but little moisture was created and most that was dried up.

In the southern areas of Russia, and in the Ukraine where the effects were most severe, the drought affected the development of cereals and other crops and caused pasturc to burn out.

One of the consequences of this drought was an outbreak of numerous forest and bog fires in central and northern Russia which continued during the summer and early autumn. Smoke from the fires spread throughout European Russia and even reached as far as Lake Balkhash in Kazakhstan, Turkey and into Germany; in the Moscow region, and in Ryasan Polesye, the smoke caused some people to migrate in search of clean air. The damage done to the Russian economy extended beyond the forestry industry. Haze reduced the transparency of the atmosphere and had a negative effect on ripening crops. This drought, as with so many, covered a huge area extending over almost the whole (except for the western territories) of the eastern European plain from Murmansk to Krasnodar.

A drought in Australia in 1963 covered almost 50% of the country. In Africa, the 1968–73 drought which struck the Sahel also affected the whole continent from the Atlantic Ocean to the Indian Ocean.

Before the Sahel disaster, drought had not attracted particular attention in the world community; at that time, drought was not considered a 'serious' disaster. The reason may have been the long-term nature of the development of drought.

In recent decades, droughts have been seen as among the most serious of natural and anthropogenic disasters and the number of deaths and victims affected bear this out. Thus, for the period 1968–92, every year (that is, on the average) droughts and subsequent famine were the cause of 50% of all disaster deaths (73,606 persons of 140,315) and victims affected (58,973,495 persons of 113,029,728 (Smith, 1997).

In addition to death and famine, material damage caused by droughts is substantial. Thus, the consequence of the 1963 drought in Australia was the loss or emaciation of 64% of the sheep herd (out of an estimated stock of 137 million) and 60% of cattle (out of 24 million) at a cost of US$1 billion. The 1987 drought in India caused damage to 58,600,000 ha of sown land (Rao, 1996); some 285 million people were affected, a large number becoming victims of suffering, hunger, disease, and ruin.

Of particularly disastrous nature are droughts in arid and sub-arid areas. Here desertification and dust storms are their constant companions. All three events (droughts, dust storms, and desertification) produce by 'joint efforts' an extremely unfavourable ecological situation. In different conditions each of these events can play sometimes a main (or greater) role in worsening the conditions of the population. Thus, for example, in the 1930s in the USA during the 'Dust Bowl' drought which affected the Great Plains, a huge role was played by dust storms. During the development of dust storms the rich soil layer was blown out by wind and destroyed and crops were damaged.

The consequences of such disastrous droughts as these affect human populations differently. In developing regions of the world, as in the Sahel, the large-scale consequences of drought are famine, mass disease, and death from malnutrition. In developed countries the death of people in such situations is normally kept to a minimum.

Material damage is also greater in developing countries. Thus, the annual damage caused by drought in Tanzania in the 1970s was 1.8% of the annual national income. In Australia, even during the worst droughts, annual damage does not exceed 0.1%. The same is true, even more so, in the USA. The 1988 drought resulted in a US$39 billion loss, the greatest economic loss in America during the 20th century (Riebsame, 1991). However, the damage did not exceed 0.1% of the annual national income of the USA. In our example, these statistics indicate that both for Australia and for the USA material damage was 10 times less tangible than it was for Tanzania.

In recent years drought and desertification visited on the poorest countries of the world have been disastrous for their populations. A set of remedial steps have not yet been taken. This concerns especially the states of the sub-Sahara area where 29 of the 36 poorest countries in the world are located. The population here grows particularly rapidly – by 3.11% per year, while food output increases only by 1.6% (Smith, 1997). According to some data, the population of the sub-Saharan area will, at least, double by 2005 – as compared to the 1980 population of 385 million people.

The rich and highly-developed countries of the world, such as the USA and the European states, are able to overcome drought by means of government support, the availability of subsidies, insurance, and a wide range of agricultural measures. For example, the consequences of the Great Plains drought in 1952–56 were much more severe than the drought in the same area during the 1930s, yet were eased in many respects due to government support of farmers (Visvader and Burton, 1978).

In poor countries, in contrast, the consequences of drought remain devastating; in 1968–73 the Sahel drought attracted the attention of many international organizations, including the United Nations. In 1983–86, again, the Sahel experienced a disastrous drought with many millions of inhabitants suffering at its height (in early 1985). In some countries of this same area of Africa, the number of victims was 4.7 million in Mozambique, 1.1 million in Senegal, 2.5 million in Mali, 1 million in Angola, and 5.2 million in Ethiopia (Courier, 1985).

The minimization of losses as a result of drought is achieved through development of two types of risk reduction. First, early warning of the drought with the help of space observations predicting its likely development. Second, longer term planning for overcoming the disaster, including measures preparing for it in the locality, especially water conservation, irrigation development, and optimization of agricultural land use.

However, study of the strategic situation is the requirement; that is, the relationship between organizing the whole economy and social policy of drought-affected countries including reduction of population growth, re-orienting the economy to avoid industries that use large volumes of water, etc. In other words, organization of the whole population in a condition of constant drought risk.

Current developments in drought forecasting are various, including those based on analyzing the heat anomalies in the Atlantic Ocean – and even in the Pacific Ocean (e.g. effects of the El Niño phenomenon). Space observations become more and more important in such forecasting.

5.6.2 Satellite observations of droughts

Information on different manifestations of the drought phenomenon can be obtained from space observations. Satellites provide information on precipitation, the temperature and humidity of the earth's surface and important detail on vegetation cover and its condition.

The main parameter characterizing drought development and information obtained from satellites is the help of NDVI – the normalized difference vegetation index. The index enables best estimation, in real time, of an ecological situation, a 'drought course', and thereby orients land owners and administrators – in particular, those in charge of agricultural policy – who have to make adequate decisions.

Using Landsat images, the NDVI was calculated for the Sahel during the period of drought and active desertification, in particular for the effects it had on Mali. For the period 1973–86, when the gradual growth of the drought was taking hold, the NDVI decreased by 17% (Goita and Royer, 1993). The dynamics of the index reflected the real picure of a worsening ecological situation in this part of the Sahel in relation to the development of drought processes.

In the USA, the vegetation indexes derived from satellite imagery for the desert in the southern part of the state of New Mexico were used to estimate a number of vegetation parameters during compilation of growth maps. The results of spectral photography showed an increase in dryness in the territory – indicated by information on the reduction of grass cover area and the increase of bush cover area (Peters and Eve, 1995).

An important indicator of the consequence of drought developing in arid and sub-arid areas is the change of surface albedo determined from satellite data. In the Sahel, after the destructive drought of 1968–73, albedo was measured from Landsat. The general decrease of albedo in different earth covers was fixed. According to the albedo values, the surface was divided into five classes of different natural vegetation. For the period 1973–79, according to satellite observations, the vegetation became more abundant, the portion of bare surface areas decreasing from 9.1% to 1.4%. This decrease was caused by precipitation amount increase (Grigoryev, 1991).

Small-scale images obtained using Meteosat can be used in covering extensive areas and measuring albedo on them. In such a way, albedo maps were prepared for West Africa (Pinty *et al.*, 1984). In addition to satellite information, ground-based radiation measurements were also used. The investigation procedure involves several measurements of the same territory for a day. Meteosat photography enables almost continuous observation of the territory. Among the procedure restrictions there is a necessity for at least one control point for ground-based measurements, as well as cloudless conditions for photography.

The map for dry and wet seasons has revealed a considerable change in albedo from February to June for the territory situated westward of 9°W and southward of 18°N. The most appreciable changes of albedo (about 25%) were found near 14°W in the territory of Senegal and West Mauritania.

In regions subject to drought it is possible to observe the content of soil moisture in surface horizons. The spectral data on desert surfaces, their relation in different spectrum regions and the dynamics indicate the presence of moisture in soils and, thus, a drought event.

Space images of the Earth's surface in the heat infrared spectrum region (10.50–12.5 μ) are also used for estimating soil moisture. Experimental determinations of soil moisture have been made from geostationary *Meteosat-2* images for the south part of Niger in the Sahel. The relationship has been established between its content in the upper 0.5 m soil layer (from the data obtained by ground-based investigations on test sites) and the day-time maximum radiobrightness temperature determined from satellite (Wilkinson *et al.*, 1983). As a result, a mathematical model has been developed for the relationship between vertical distribution of soil moisture and day-time variations of surface temperature, the correctness of which was confirmed at test sites.

Another approach is to obtain data on soil moisture – on the basis of measurements from satellite radiobrightness fields of temperature, clouds carrying rains, and finding the relationships between these data and the amount of precipitation fall.

Dust storms are a reliable indicator of drought development in arid and sub-arid regions, when monitoring with the help of space observations. The more frequent occurrence of dust storms is an indicator of dryness increase and sudden dust storms in one or another region can indicate the start of drought and the related location desertification. Thus, during the 1968–73 drought in the Sahel, far more frequent dust storms were observed over the region – and over the Atlantic Ocean – from satellite images (Kondratyev *et al.*, 1983). An alarming symptom of drought development was also the appearance of an extensive dust cloud in the west part of the Great Plains in the American Midwest in February 1977. Detected by *GOES-I* satellite (McCauley *et al.*, 1981) it was formed by dust transported from two locations; one located at the boundary of Colorado and Kansas, and the other at the boundary between New Mexico and Texas. New Mexico had already functioned – 'raised dust' – in the disastrous drought of the 1930s and was to do so again in the 'dry 1950s'. The dust cloud, tracked by satellite, crossed the USA and reached the Atlantic. Its appearance was evidence that soil-grounds had dried up in the prairies of the western Great Plains.

As is well known, satellites can provide information on the different features of landscape structure: potential drought areas and related soil cover, erosion separation, geomorphological features of the structure, water-resource potential, as well as the specifics of territory development by man and particular land use. All these data, together with data of an economic-geographical nature, information on harvests and the natural situation in the past years are utilized in forecasting droughts.

Of course, in each region the leading natural and socio-economic factors, the prerequisites of the appearance of droughts disastrous for population, are different.

It is important in preparing to repel drought to have satellite information on air currents, cyclones, monsoons, areas of high or low pressure, and anomalies in precipitation. In India, for example, information on the monsoon is paramount. Various data (including the parameters of monsoon rainfall) is obtained from satellites – in particular from the Indian satellite *IRS-lc* (Rao, 1996). At the same time, other satellite data are also used – the temperature of the Indian Ocean surface, land albedo, the dynamics of cloud cover, the activity of thunderstorm winds in the upper atmosphere, humidity profiles, thermal variability of land surface, snow cover in the mountains. Satellite information and the results of modelling natural situations are a component of a special programme on drought warning in India. According to this programme, a detailed monthly estimation is made of possible drought development and its likely effect on agricultural products yield.

Information on forthcoming drought obtained from satellites is becoming standard practice. Thus, satellite data obtained in the summer of 1984 in the Sahel alerted the farmers of Nigeria to a possible 5% loss of crops (Jasinski and Karnovitz, 1985). Such information was obtained with the help of both a polar-orbital *NOAA-9* satellite, with a period of rotation of about 90 minutes, and the European geostationary Meteosat which retransmitted continuously images to the given territory. On the basis of satellite data, cloud cover indexes were calculated, rain clouds were decoded, and a possible distribution of dryness index was forecasted. As current 'supporting' data, results of observations at meteorological stations in the Republic of Chad, the Sudan, and Ethiopia were also made available.

5.7 SLOW OSCILLATIONS OF INLAND WATER LEVELS

In the last decades of the 20th century disaster possibilities were actively discussed as a result of the World Ocean level rise – this latter due to climate warming and the melting of glaciers. A global disaster involving such a sea level rise is a real possibility for inhabitants of ocean coastal areas but of low probability in the early decades of this new twenty-first century. However, some parts of oceanic or sea coast are subject to slow flooding related to other factors, or to tectonic shifts or anthropogenic effect – for example, the pumping out of underground waters and a subsequent lowering of the location affected.

Population is much more sensible to the oscillations of inland reservoir levels. Such events are quite typical for lakes and inland seas in arid and sub-arid regions. Occurring usually in sight of man with a periodicity, in particular, of 10 years, these oscillations of reservoir levels sometimes acquire a great importance: they can reach several metres, the coastline shifting over many – and in some cases – tens of kilometres. It is in these situations that disasters might occur on intensively developed coastlines and at some reservoirs.

5.7.1 Manifestations of disasters and their scales

In the second half of the 20th century, the oscillations of reservoir levels in different regions of the Earth were the cause of ecological disasters in Africa, on Lake Chad, and in the USA, on the Great Salt Lake, and in the Caspian and Aral Seas in Central Asia. In the case of both the Caspian Sea and the Great Salt Lake the disaster was transgression of their waters, and in the Aral Sea and on Lake Chad the contrary, regressions. In all cases these events were unexpected, and proved to be very destructive for the populations involved; the features that were specific depended on the natural location of each disaster and the welfare level of the population before the disaster struck.

5.7.1.1 Caspian Sea

The Caspian Sea level has risen gradually in two recent 10-year periods, causing flooding and partial flooding of the coastline. The effect on the vital activity of the population of the coastal areas of a number of states was appreciable, particularly in Russia, Turkmenia, Kazakhstan and, to a lesser extent, Azerbaijan and Iran. The rise of the Caspian Sea level began in 1978 and was unexpected; both the administrative authorities in the coastal regions and the majority of specialists were caught unawares. For the period 1978–98, the sea level rose by 2.5 m and reached an absolute level of 26.5 m (*Geology of the Near Caspian*, 1997). The transgression changed twice to insignificant regression, for a short period of time.

In the first years the unexpected rise of the sea level was considered as a positive event: since the long-term preceding period of its drop had resulted in unfavourable consequences both for the whole marine ecosystem and the human population and economy of the coastal regions. In the former USSR, the practical problem of 'saving the sea' was considered on a national responsibility. However, the sea level rise started in 1978 and proved to be long term, extending over many years and rising to an unexpected level. On average, for the transgression time period the sea level rose by 14 cm every year.

The extensive territories of Dagestan, Kalmyk and the Astrakhan region in Russia were flooded or partially flooded. The movement of the sea inland occurred at the rate of 1–2 km per year. The surge events with amplitude 2–3 m penetrated the coast over a distance of 20 km and more.

Agricultural lands were damaged as a result of flooding or partial flooding, 320,000 ha of valuable lands being destroyed. Many lines of communication were also destroyed (including roads and power supply lines), some dams were ruined, and the buildings of many industrial enterprises were damaged. Tens of villages, and some cities, suffered from the sea level rise among them Kaspiysk, Derbent, Sulak, and Makhachkala. In Kaspiysk the port structures were damaged from the intensive abrasion of the coastline (Rychagov, 1996).

As a result of the destruction of dams and irrigation systems and the flooding of cities with sea water, drinking water was contaminated; the medical-ecological situation became worse, in particular on the Russian coast of the Caspian where infectious intestinal and parasitic diseases took hold (Elpiner, 1996).

The sea level rise also had an adverse effect on other countries, including Turkmenia. Here, in the open sea, some 1.5–2 km off the eastern sea coast were oil derricks, boundary towers, and power supply lines. In Krasnovodsk, the waves washed away railroad banks in the city of Cheleken and tens of residential dwellings were partially flooded, as well as number of buildings in industrial enterprise zones.

In Kazakhstan, in the new shallow waters between the deltas of the River Volga and River Elba, about 1–1.5 million birds died in 1980–85 from botulism, brought about by the changed ecological conditions (Zaletayev, 1996).

The Caspian Sea level rise resulted in ecological disaster and many consequences (including sanitary ones) arose from it; in Russia alone, the direct economic damage amounted to US$5 billion by late 1993 (International, 1998). Undoubtedly, the cause of the disaster was natural. Specialists discussed different versions of why this should be and, in particular, hydrogeological (run-off in depth from the Aral basin) and tectonic (rise of the sea bottom) causes were thought to be significant. Among these and other hypotheses, the hydrologic climatic is the one preferred – an increase of river run-off due to precipitation increase in the Caspian Sea basin and first of all in the Volga basin.

Another, though not less considerable, theory is the anthropogenic factor. The unplanned development of low areas of the sea coast – in what was well known (from historical documents) to be a location subject to periodic flooding – and the construction in the risk zone of port and industrial structures, city buildings, communication lines, infrastructure developments related to the reconnaissance and production of mineral resources (particularly in Kazakhstan) and the agricultural development of the territory, had caused disaster.

5.7.1.2 Great Salt Lake

The sudden transgression of the waters of a lake, the biggest in the American West, also caused a disaster though on a smaller scale than that which occurred on the Caspian Sea.

After a long-term drying out through the 1960s and 1970s, the level of the Great Salt Lake waters rose suddenly in the 1980s. Earlier, in 1963, the water had been at a height of 1,257 m above sea level and the lake water plane amounted to 1,449 km^2. Specialists and the administrative authorities had given assurances that the process of lake drying would not stop, resulting in speculative building and development in the drying coastal belt surrounding the lake.

Roads and other lines of communication were built along the Great Salt Lake bottom, buildings were constructed, vast evaporation ponds were built for the production of sodium carbonate and potash sulphate, the areas fed by the fresh waters of mountain rivers were enclosed with dams – which became stopping and resting places for migrant birds (Gore, 1985).

The lake level rise was caused by abundant rainfalls in the mountains during September 1982 (de Blij *et al.*, 1997). By 1983–86 the annual income of river waters to the lake proved to be 200% higher than the usual amount. In 1984 the water level rose to 1,264 m (0.9 m higher than the record level of 1873). As a result, the lake area increased sharply and amounted to 3,703 km^2. This led to flooding the important lines of communication and swamped communities – including the places of rest for the millions of birds which were forced to alter their migration pattern. In 1982 the bed of railroad had to be lifted urgently, having originally been built in 1969 on the bank of the lake which it crossed from east to west.

At the same time landslides and mud streams were set off in the surrounding Wasatch Range, causing damage to developments built by the people of the sub-mountain region. It is supposed that more frequent underground shocks contributed to the development of the landslides – or they could be due to the increase of water pressure on the thickness of rocks, the increase of seismic activity along the well-known active tectonic plate breaking over the niche of the lake basin.

The damage done by the rising waters of the lake – from 1982 up to 1986 it rose by 360 cm – amounted to several millions of dollars. There was a danger of the international airport and parts of Salt Lake City itself flooding. In 1987, the city administration started implementing a project to transfer surplus water to a neighbouring dry hollow; as the station for water transfer was readied, the Great Salt Lake water level started suddenly to decrease.

5.7.1.3 Lake Chad

A disastrous situation arose during the 1970s and 1980s on the coast of Lake Chad, the largest reservoir in the west of North Africa. Located in an area of desertified savannahs, the lake had long attracted different peoples of Africa. Its waters are usually characterized by slight salinity and near the coast, in the places where rivers flow into the lake, the water is practically fresh. The lake is abundant with fish and its coastal lands in areas of seasonal river flooding are widely used for agriculture (Pouyaud and Colombani, 1989).

The lake and its coast are parts of four nation states: Niger, Chad, Nigeria, and Cameroon. Many hundreds of thousands of people live in the environs of the lake which began drying out in 1962 when the water level rapidly decreased. By 1985 its level had dropped from 283.4 m to 275.4 m above sea level; its area decreased almost by the factor of 9, from 26,000 km^2 to 3,000 km^2. In some places the lake coast receded by 50 km. As the lake became more shallow, its mean depth decreased from 5.8 m to 3.5 m.

The damage caused to the fishing industry was substantial, the tonnage of catches reducing from 120,000 to 80,000; the lands in the river deltas were subject to degradation due to the great decrease of ground waters, causing rivers to shoal. As a result, stock breeders and farmers suffered. On the whole, the effects of Lake Chad drying out threatened the vital activity of about 10 million people.

The basic cause of Chad's aridification was the drought that was particularly strong in the Sahel in 1971–73 and 1982–84. At that time the south isohyet up to 300 mm per year shifted far to the south by 200–300 km. As a result the water volume of rivers feeding the reservoir greatly decreased. Their annual run-off for the period from 1961–62 to 1984–85 decreased from 53.3 km^3 to 6.7 km^3.

Some contribution to worsening the ecological situation in the region, on the lake coast, in the deltas and low reaches of the rivers flowing into the reservoir was made by the unconsidered actions of man. Among them were the construction of hydrosystems and dams which, though useful in some cases, reduced river floods in their low reaches, which tended to undermine the fodder base of stock breeders on pastures; large water amounts taken for such water-loving cultures as rice and cotton; population migration due to war actions and the subsequent increase of load on the natural resources of those regions where migrants accumulated (Levintanus, 1991).

However, unlike the Aral Sea disaster, water discharges (for irrigation and other needs) from the rivers in this region are much less (5–10% of river run-off) so the main cause of the disaster was drought. At the same time, there is no doubt that human activity in the adjoining areas of Lake Chad was insufficiently adequate to the ecological situation. Intensive forest cutting in the lake basin, the resulting soil cover erosion, and unreasoned installation and use of hydrotechnical structures contributed to the deterioration.

5.7.1.4 Aral Sea

One of biggest ecological disasters of the 20th century occurred in the Aral Sea, caused by a drop in sea level. The area affected included not only the Aral Sea coast but also the extensive territories in its basin and affected the vital activity of many millions of people, mainly the inhabitants of Uzbekistan, Kazakhstan, and part of Turkmenia.

From 1961, the rate of river run-off into the sea-lake began to suffer an abrupt decline. By the mid-1970s the flow had decreased to 7–11 km^3 as compared to an average 56 km^3 in the period 1911–1960 (in the area of recharge it reached at that time 110–120 km^3 per year) (Sevastyanov, 1991). In some years of the mid-1980s, river run-off was decreasin to 3–4 km^3 and even stopped altogether. By 1988, the sea level had fallen by 13.5 m, and in, the mid-1990s an almost 15 km fall was recorded. The sea coast (in the eastern and southern parts of the reservoir) receded by many tens of kilometres (in some places up to 150 km).

The basic cause of the disturbance in the Aral Sea was the drop in sea level, due to a sharp decrease in the water run-off from the only two rivers, Amu Darya and Syr Darya, which flowed into the sea. The lack of water is explained by the excessive taking of water from the rivers, mainly for irrigation. This draining was historical; until the mid-1950s, 30% of river resources were taken for economic needs (Sevastyanov, 1991), whereas in the following decades the removal of water increased considerably. This had its origins in the extending of the area of irrigated lands and in irrigating more and more new pastures in areas adjoining valleys – in the middle and upper rivers flows – rather than in the deltas.

The development of new territories was, of course, highly advantageous to the economy. At the same time, the excessive taking of water from rivers, mainly upstream and far from the Aral Sea itself and its environs, resulted in sudden consequences for the lower reaches of the two main rivers of the area.

The annual (averaged) water used for irrigation (from the rivers flowing into Aral Sea) amounted in the period 1951–60 (that is, on the eve of the sea level fall) to 37.6 km^3 per year. It increased drastically in the years 1961 to 1965, to 64.1 km^3 per year, and even more, by a factor of 3 to 111.1 km^3 per year in 1965–85 (Sharonenko, 1998). It should be mentioned that from 1975, when the situation in the Aral Sea began to be actively discussed, there was no observed sharp increase in water use for irrigation. For the years 1975–78, 1980, 1985–89 it was 107.9, 110.3, and 111.1 km^3 per year respectively.

The drastic reduction of river run-off and the drop of sea level caused a corresponding decrease of ground water level in the environs of the Aral Sea, particularly in the deltas, drying-up of oases in the deltas and river valleys, salinization of soils, and an occurrence of dust-salt transfers (on the former reservoir bottom in the drying belt) destructive to plants. These changes to the ecological balance in the Aral Sea area proved to be destructive for its inhabitants: water salinity meant that fish almost disappeared, thus fishing stopped and fish-processing plants were closed, and the sea level drop prevented navigation with subsequent unemployment in the shipping and related industries.

Due to the oasis land salinization, location desertification, soil and water contamination by run-offs, agricultural production decreased drastically. The population sick rate increased, including cases of esophagus cancer and chronic gastritis, as did the death rate, particularly that of children. In the 1980s, infant mortality in the area of the Aral Sea amounted to 110 per 1,000 births – much higher than, for instance, in the Sudan (respectively, 81 per 1,000 births), Syria (973), Mexico (82) and a number of other countries (Glazovsky, 1997).

5.7.1.5 Conclusion

The question arises as to why the transgressions and regressions of all four reservoirs – Lake Chad, the Great Salt Lake, the Caspian Sea, and the Aral Sea – turned out to be unexpected and caused disasters.

These events should have been far from being unexpected. The frequency of rise and drop in the water levels of these lakes had been established years before. The Great Salt Lake, that relict of a huge post-glacial reservoir, had become its present size about 8,000 years ago and since then had partly dried more than once, again revived, and two more times had flooded what was to later be the the location of modern Salt Lake City.

The history of periodical fluctuations of the Caspian Sea was well known for several centuries and its rhythmic fluctuations established. However, these rhythms are not distinct and therefore were difficult to observe: inadequate forecasting of sea level fluctuations based on probabilistic methods resulted therefore in the incorrect estimations of the Caspian Sea evolution in the 1970s. Nevertheless, the transgression was forecasted by a number of investigators, including S. Z. Antonov and B. A. Appolov and others, as long ago as 1963. Their forecast, made independently of each other, had as its basis solar–terrestrial relationships.

Similarly, the large regressions of Lake Chad and the Aral Sea were predictable. It was well known that the level of Lake Chad in the previous 1,000 years had greatly decreased and an 80–90-year cycle in its fluctuations had been observed. The large regressions of the lake were observed in the middle of the fifteenth, sixteenth, eighteenth and nineteenth centuries – and in the 20th century a considerable decrease in lake level occurred twice (the first time in the early part of the century, in 1905–08).

The second large decrease of the Lake Chad water level in the 1970s and 1980s was forecast by only one investigator, General J. Tillo, as long ago as 1928. He twice visited Lake Chad in the periods of the rise and drop of its level in the early century and made a forecast of strong lake regression based on the analysis of solar-terrestrial relationships and comparing them with the movement of glaciers in the Alps (Pouyaud and Colombani, 1989).

The regression on the Aral Sea could also have been forecast, though on another basis – taking into consideration abruptly increased water use. However, the viewpoint of climatologist A. I. Voeikov prevailed among specialists at that time, the early 20th century. As expressed by Voeikov, it was considered that the Aral Sea was a huge evaporator of water and that people needed to take the river water flowing into it for their oases and irrigation of crops.

The basic lesson in all these events is that unconsidered development of the coasts and basins of lakes without a due regard for possible – and drastic – water level fluctuations leads to disaster. The projects put in place for 'saving' the regressing resevoirs of Lake Chad, the Aral Sea, and the Caspian Sea (in the period of its drying until 1978) by transferring river run-off can lead to unexpected ecological consequences and are costly. Planning of the whole economy seems most realistic, in particular the water industry and developments on coasts and basins. To overcome the ecological crisis in the Aral Sea area, the structural reconstruction of the economy is required, adjusted to changing natural conditions. Such a proposal excludes the suggestion that the inhabitants of the disaster zone be moved – for example, from Kara-Kalpak as proposed by some specialists in Moscow. This according to the principle: if there is no man, there is no problem.

The above discussion will also refer to disasters related to reservoir transgression. For example, it has been predicted that the Caspian Sea level will continue to rise until around the years 2020–25 (Maximov, 1992; Rychagov, 1996; Geology of the Near Caspian, 1997). If this prediction is disregarded, then, by 2010, the economic damage in Russia will have reached US$20 billion (International,1998).

5.7.2 The results of satellite observations of ecological disaster regions on lake coasts

To monitor the ecological situation in the after events of these various disasters, space information was widely used, especially in the regions of the Aral Sea and the Caspian Sea as well as Lake Chad. One of the essential advantages of using space images is being able to obtain information immediately about a whole lake, including the dynamics of changes in reservoir boundaries. To this end, small-scale images from meteorological satellites were employed.

The space information became the basic source of data on the relief, vegetation cover, other landscape features forming in the lake regression or transgression area. As a rule, such landscapes – in particular, those on the coasts of the Aral and Caspian Seas – had been developing for a long period of time. Thanks to the simultaneous coverage of space images of the lake and its basin, observing the relationships of events occurring in its different parts was easier than using ground-based data.

We now discuss the various landscapes under observation.

Caspian Sea

The space images of the Caspian Sea enabled study of the dynamics of sea coast flooding. For example, the water level rise which had occurred in 1977 in the area between the Volga delta and the Buzachi Penisula was studied in 1992 (Kravtsova, 1993). In this case multi-zonal images were used, a 1:500,000 scale with resolution of 15 m and a more detailed survey with 5 m resolution on a 1:200,000 scale.

This revealed flooding of the coast up to 15 km wide and degradation of reed thickets between the River Ural delta and Buzachi Peninsula. An expansion of the reed thicket belt was also revealed between the deltas of the Volga and Ural rivers. Other evidence determined the effects the Caspian Sea level rise caused: changing the outline of saline marshes, creation of saline lakes, overgrowing of moving sands, appearance of areas with hydromorphic vegetation, marsh forming, etc.

Changes, no less appreciable, in the natural environment were also fixed by space observations on the western sea coast. The transformation of accumulative beaches to lagoon coasts was observed on the Dagestan coast (Geology of the Near

Caspian, 1997). This phenomena of new formations were also observed in other areas, notably to the north of Derbent city in the area of Adzhi Lake, in the city of Kaspiisk, on the Agrakhan Peninsula between Kaspiisk and Makhachkala city.

The sea level rise has changed the formation of the deltas on the Dagestan sea coast – in particular, that of the River Sulak and the 'new delta' of the River Terek. The images show the slower accretion of deltas – their movement to the sea (due to partial flooding and reduced arrival of rock waste). The partial destruction of the River Sulak deltas was also observed. The flooding of the 'new delta' on the River Terek can cause an outflow of a part of the river water to the north part of Agrakhan Bay. It has been established from space observations that the extension of abrasion coasts for the period from 1977 up to 1991 has increased and that they now amount to 30% of the whole length of the Dagestan coastline (as compared to 10% in 1977) (Kravtsova, 1993).

Great Salt Lake

Space observations have fixed the large-scale changes of the Great Salt Lake coastline due to its sudden transgression. The area of low coast flooding, detected from space images, in the development regions (existing lines of communication, city structures for industrial and agricultural use, etc.) has oriented the administrative authorities to more optimal ways of planning coastal development and the necessity of using lake level forecasting (Gore, 1985).

Aral Sea

A study of the largest transformations of the natural environment of the Aral Sea area observed from space (Grigoryev, 1985, Dech and Ressl, 1993) showed a considerable decrease in sea level. The space images fix the changes of coastline, an increase in the size of islands, including among others Vozrozhdenia and Barsakelmes, and the connection of some of the islands to the Aral coast itself. Formation of a drying belt at the former sea bottom (see Figure 5.6) was also observed.

All the above-mentioned changes in the configuration of the coast and the size of the islands became apparent after 1961 when, as revealed by space images, a steady decrease in the sea-lake level began (Figure 5.7).

The space images also revealed a number of other changes in the Aral Sea area. Here we discuss only the largest and essential changes. The first, and most considerable of these occurred in the deltas of the Amu Darya and Syr Darya rivers. They are related to the process of desertification due to a drastic reduction in the supply of water brought by rivers. Comparison of successive space images obtained for different years, as well as of aerial photographs and maps, shows the drying of river channels and a reduction in the size – and even the disappearance – of many lakes to be current in both the deltas. Even large lakes dry considerably, in particular Sudochie Lake in the delta of Amu Darya.

One of the main processes of landscape formation in the deltas of the Syr Darya and Amu Darya has been the recent formation of bogs. Drastic reduction in river

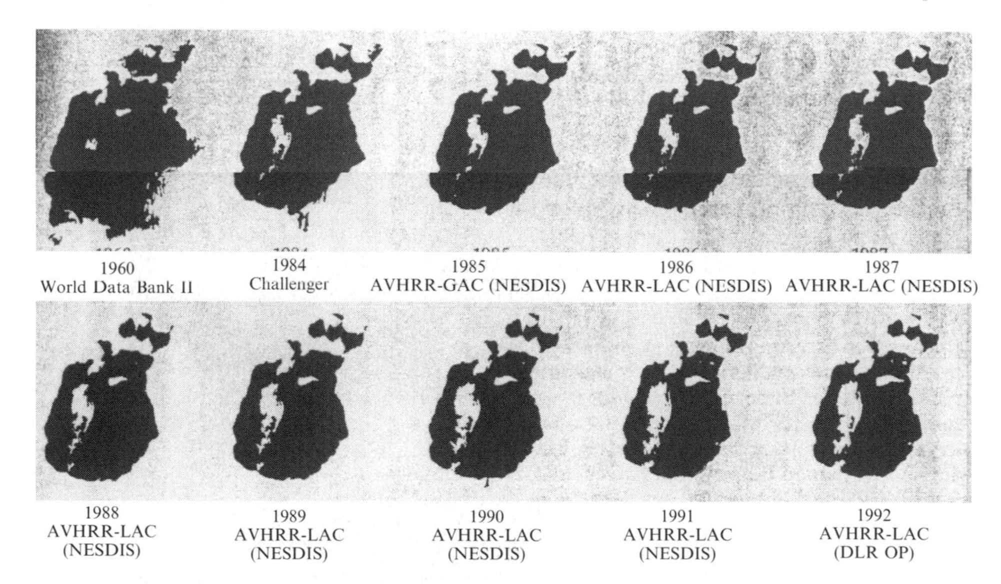

Figure 5.6. Successive images of the Aral Sea. The first is a cartographic image obtained in 1960; the others were taken from *Challenger* in 1984 and from *NOAA*/AVHRR meteorological satellites. All the images have been reduced to one scale. The reduction of the area of water and the change to the coastline for the period of 1960–92 were due to the disastrous drop in the sea level.

floods, cause the swampy low lands to start drying. The ground water level in the low parts of the Amu Darya delta in which bog soils formed earlier has now decreased to only several meters (Rafikov and Tetyukhin, 1981). The soils of such low lands became saline and in some places solonchaks form. Due to high albedo, the solonchaks are distinguished as bright spots and are clearly depicted in the space images.

As a consequence of the desertification of these deltas, tugai vegetation started disappearing which, before 1961, was widespread here. Tugai forests grow along the beds and channels of formed by near-river belts of trees and bushes and can be up to 56 km wide. Such belts of dark shade stand out well in space images. The disappearance of tugai forests occurs along channels of highly reduced or discontinued run-off waters (Rafikov and Tetyukhin, 1981), causing a transformation as the complete or partial discontinuance of flooding results in a considerable drop of ground water level along the coastal belt.

The development of the above processes is depicted on space images by a lighter shade as tugai forests disappear. This same decrease in the water supply to deltas occurred not only as a result of drying of lakes, channels, bogs, but also because of the overall decrease in the ground water level due caused by the sea level drop. The latter leads to aridization of natural complexes, and the drying of vegetation cover on the delta areas between channels. Disappearing vegetation intensifies the process of soil blowing out and the scattering of sand. Such events, revealed by lighter images, show up on even more ancient elevated delta areas – here,

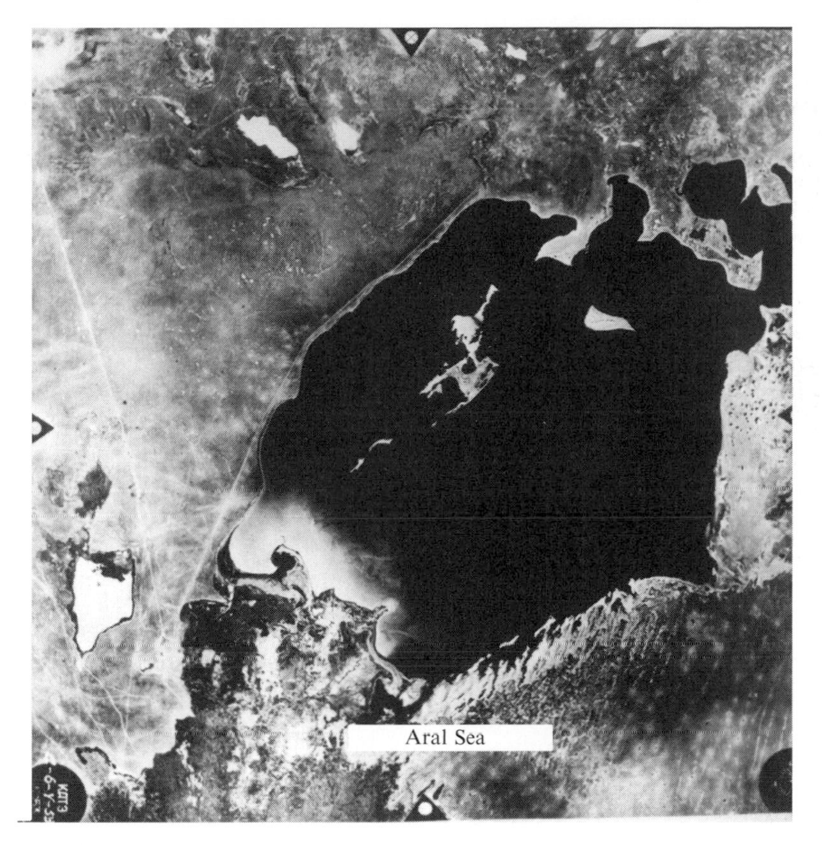

Figure 5.7. A space image of the Aral Sea obtained by the Soviet orbital station *Salyut* from a height of about 250 km. The sea is shown in its period of disastrous drop in 1997; on the east coast there is an extensive young drying-out belt (light areas). In the north, the 'small sea' is almost isolated from the 'big sea'.

in the eastern part of Amu Darya delta and in the western part of Syr Darya, where the ground waters always lay deeper than in younger deltas. This decrease in ground water level also occurs on the plains outside the Aral Sea deltas, to more extent in the south, to less extent in the north. In the south, in particular, the formation and advance of mass moving sands, is related to takyrs. This process can be seen in space images due to the appearance in takyr images, earlier homogeneous in shade, of a specific patchiness of the picture (Sadov, 1981).

Outside the river deltas, other changes fixed by space images include the formation of lakes – one of the largest is Sarykamyshskoe Lake – as a result of disposal of waste waters. The waters of these lakes, which used to flow into the Aral Sea now disappear due to evaporation.

In some cases, space images provide the most important information on processes occurring in the Aral Sea. For example, in the dried belt, monitoring processes of relief formation actively developing and new natural complexes as they formed: residual small lakes, aeolian forms, solonchaks. Dust storms also originate here. In the spring of 1975 space images tracked for the first time the movement of a large dust cloud over the Aral Sea, determining the distribution of powerful dust formations (Grigoryev and Lipatov, 1977, 1979, 1982). Thus, the area of salt receiving the dust fall was revealed, as well as its possible effect on agricultural lands (Figure 5.8).

By analysing everyday space information (many hundreds of images) a map-scheme of dust formations was compiled for the Aral Sea area for the period of 1975–81 (Grigoryev and Lipatov, 1979). The main direction of dust movement (60% in most cases) is to the south-west, to the oases of the Amu Darya delta, at a distance of up to 200 km. In 25% of the cases the dust flows move in a westerly direction to Ustyurt Plateau; in the rest, movement is in other directions, sometimes according to the images, over the oases of the Syr Darya Valley.

The areas of salt-containing dust fall correspond to a considerable extent to the areas of dust clouds distribution seen in the images. The dust precipitates mainly on the water area surface. It also precipitates, though to a smaller extent, on the Aral Sea coast, relatively more often in the south and south-west part. The above-mentioned map-scheme shows that the area of dust distribution is 200,000 sq km.

Space observations confirm that dust flows often reach the oases in the Amu Darya delta. According to approximate estimations, an average of 1.5 million tonnes of salt-containing dust fell there during dust storms – the maximum recorded in one powerful storm being up to 3 million tonnes. For a year (in the observation period from 1975 up to 1990) some 15–90 million tonness of dust was blown out (in different directions) from the dry Aral Sea bottom (Grigoryev and Lipatov, 1977, 1982; Grigoryev and Zhogova, 1992).

Thus, from space images five large zones of the Aral Sea area can be selected, according to a set of features in which – beginning from 1961 – the changes of natural environment occurred because of the impact of human activities (Grigoryev, 1985).

- The first zone is the water area of the Aral Sea itself; images observe the reduction of the water area and the change of the coastline configuration, the appearance of new – and disappearance of some old – islands, and map the shoals. All this information is very important for human activities.
- The second zone is the drying zone. Space images are effective for investigating various transformations of natural complexes. Almost all diverse phenomena occurring here, on this newly formed land area do not affect directly human population activities (the zone has not yet been used). The exception is dust storms transporting the salt dust to other regions.
- The third zone includes the deltas (including oases) of both the largest rivers in Central Asia. The space images show the basic ecological changes occurring there include drying of lake beds and appearance of new reservoirs (from disposal of waste waters), disappearance of bogs, appearance of solonchaks, the destruction and change of vegetation, and the blowing out of sands.

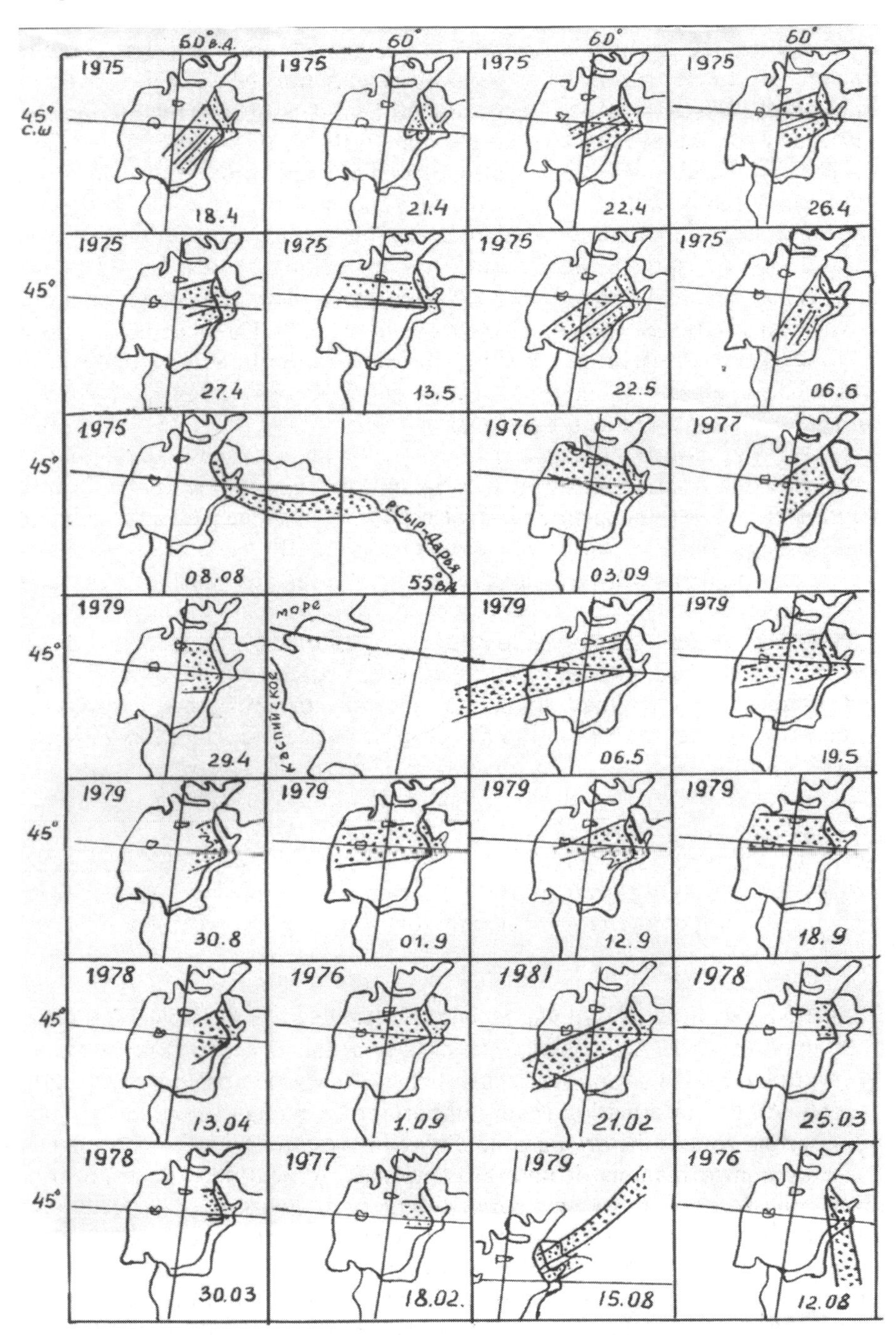

Figure 5.8. Maps of the distribution of a powerful dust cloud tracked over the Aral Sea, compiled by analysing images from different meteorological satellites for the period 1975–79 (*source:* Grigoryev and Lipatov, 1982).

Observation of these processes and phenomena from space is difficult to overestimate, since densely populated cities and agricultural lands, where the above nature disturbances are most appreciable, are related to the deltas.

- The fourth zone is the deserts around the Aral Sea, outside the delta areas. The processes of aridization of natural complexes due to ground water level decrease are less distinct here than in the preceding zones. They can be found primarily because the space images are highly detailed. Observation of the development of aridization processes is important since damage to pasture husbandry is widespread in this zone. In small-scale poorly-detailed space images the zone boundaries would not yet be distinct.
- The fifth zone of anthropogenic changes of the natural environment is the largest. It correponds to a vast region of distribution of powerful dust blowing out from the north-east a recently dried part of the Aral Sea bottom. Unlike all preceding zones, each of which is individual and is separated from each other, this zone includes the rest and, moreover, extends far from their limits. The fifth zone is distinguished only on the basis of analyzing space images. The study of space-time regularities of distributing dust in this zone is of considerable interest: the precipitation of salt dust can be destructive to farming, gardening and pasture husbandry. The space observations of the Aral Sea confirm that the dried sea bottom can become a huge sand solonchak desert, the centre of salt-dust blowing out, degradation and destruction of oases due to desertification. They also point authorities, charged with managing such a situation, to policies that regulate the ecological situation.

Lake Chad

To control the unfavourable situation which arose as a result of the great decrease in the Lake Chad water level, detailed images from piloted spaceships and those from meteorological satellites were utilized. Changes in the lake area were estimated for the period of 1965–84 and its steady decrease in size was monitored up to the year 1984 (Helfert and Holz, 1985).

These observations enabled the establishment of a sequence of ground water level decrease, based on sea bottom colour changes along the dried coast.

The most success in monitoring the current situation in the Lake Chad region has been gained from *Landsat* and *NOAA* satellite information. More detailed images of the region, obtained with the help of multi-zonal scanning device MSS (on board the *Landsat* satellite), proved to be useful in studying change in the lake table and the condition of its basin. The images made with the help of a radiometer with high resolution AVHRR (on board the *NOAA* satellite) turned out to be more suitable for routine observation of the reservoir coastline, as well as the basin parts with distinct vegetation cover (Scheider *et al.*, 1985).

5.8 CONCLUSION

Space images have proved to be informative for estimating the ecological situation due to the disastrous fluctuation of water level in all four lakes in our example.

The images can be an important component in forecasting disasters since they allow simultaneous observation of significant changes of the natural environment on extensive territories in the basins of reservoirs, some of which are situated in natural zones that are different from that of the reservoir.

5.9 BIBLIOGRAPHY

Abrahamson, D. E. (ed.) (1989) *The Challenge of Global Warming*. Island Press, Washington, DC, 356 pp.

Abstracts of the European Conference on Atmospheric UV Radiation. Helsinki, 29 June to 2 July 1998. Finnish Meteorological Institute, 127 pp.

Adamenko, V. N. and Kondratyev, K. Ya. (1999) Global climate changes and their empirical diagnostics. In: *Anthropogenic impact on the nature of the North and its ecological implications* (eds: Yu. A. Izrael, G. V. Kalabin, and V. V. Nikonov). Kola Scientific Center, Russian Academy of Sciences, Apatity, pp. 17–34 (in Russian).

Adamenko V. N. and Kondratyev, K. Ya. (2002). *In situ observations and proxy data on climate change in the Arctic during the last few decades* (in print)

Adams, P. J., Seinfeld, L. H., Koch, D., Mickley, L. and Jacob, D. (2001) General circulation model assessment of direct radiative forcing by sulphate-nitrate-ammonium-water inorganic aerosol system. *J. Geophys. Res.* **106**, 1097–111.

Adem, J. and Garduño, R. (1998) Feedback effects of atmospheric CO_2-induced warming. *Geofis. Int.* **37**(2), 55–70.

Adequacy of Climate Observing Systems (1999) National Academy Press, Washington, DC, 51 pp.

Adger, W. N. and Brown, K. (1994) *Land Use and the Causes of Global Warming*. John Wiley & Sons, Chichester, UK, 271 pp.

Alcamo, J. (1995) *Integrated Modeling of Global Climate Change*. Kluwer Acad., Dordrecht, 328 pp.

Alexandris, D., Varotsos, C., Kondratyev, K. Ya., and Chronopoulus, G. (1999) On the altitude dependence of solar effective UV. *Physics and Chemistry of the Earth* **24**(5), 515–17.

Allan, R., Slingo, J., and Wielicki, B. (2001) Changes in tropical OLR–a missing mode of variability in climate models? *Abstracts 8th Scientific Assembly IAMAS, Innsbruck, 10–18 July 2001.*

Allen, M., Raper, S. and Mitchell, J.(2001) Uncertainty in the IPCC's Third Assessment Report. *Science* **293**, 430, 433.

Alverson, K. D., Oldfield, F., and Bradely, R. S. (eds) (2000). *Past Global Changes and Their Significance for the Future*. Pergamon Press. London, 479 pp.

Ambrosia, V. G., Linthicum, K. G., Bailey, C. L. *et al.* (1989) Modelling Rift Valley Fever (RVF) decease vector habitats using active and passive remote sensing system. *IGARSS '89: Remote Sensing: Econ. Tool Nineties (and) 12th Canadian Symp. Remote Sensing, Vancouver*, pp. 2758-2760.

Anisimov, V. N. and Polyakov, V. Yu. (1999) On the forecast of air temperature variations for the first quater of the XXI century. *Meteorol. Hydrol.* **2**, 25–31.

Andronache, C. and Chameides, W. L. (1995) Interactions between sulphur and global ozone dynamics. *Int. J. Remote Sensing* **16**(10), 1887–95.

Andronova, N. and Schlesinger, M. E. (2000) Causes of global temperature changes during the 19th and 20th centuries. *Geophys. Res. Lett.* **27**, 2137–40.

Angell, J. K. (1999). Comparison of surface and tropospheric temperature trends estimated from a 63-station radiosonde network, 1958–1968. *Geophys. Res. Lett.* **26**, 2761–4.

Angell, J. K. (2000a) Tropospheric temperature variations adjusted for El Niño, 1958–1998. *J. Geophys. Res.* **105**, 11841–9.

Angell, J. K. (2000b). Difference in radiosonde temperature trends for the period 1979–1998 of MSU data and the period 1959–1998 twice as long. *Geophys. Res. Lett.* **27**, 2177–80.

Annual Review of the World Climate Research Programme and Report of the Twentieth Session of the Joint Scientific Committee (1999), Kiel, Germany, 15–19 March 1999. WMO/TD, No. 976. Geneva, 110 pp.

Appendices. A Plan for a Research Program on Aerosol Radiative Forcing and Climatic Change (1996) National Research Council, National Academy Press, Washington, DC, 161 pp.

Ariya, P. A., Jobson, B. T., Sander, R., Niki, H., Harris, G. W., Hopper, J. F. and Anlauf, K. G. (1994) Measurerments of C_2–C_7 hydrocarbons during the Polar Sunrise Experiment.

Arking, A. (1996) Absorption of solar energy in the atmosphere: discrepancy between model and observations. *Science* **273**(5276), 779–82.

Arnone, R. A., Ladner, Sh., La Violette, P. E., Brock, J. C., and Rochford, P. A. (1998) Seasonal and interannual variability of surface photosynthetically available radiation in the Arabian Sea. *J. Geophys. Res.* **103**(C4), 7735–48.

Asharonenko, S. I. (1998) Aral crisis: the 20th century. *Ecology* **4**, 11–16 (in Russian).

Assessing Climate Change: Results from the Model Evaluation Consortium for Climate Assessment (1997) Gordon and Breach Science, Amsterdam, 418 pp.

Austin, J., Knight, J., and Butchart, N. (1999) Three-dimensional chemical model simulations of the ozone layer: 1979–2015. *Quart. J. Roy. Meteorol. Soc.* **125**, Part B, 561.

Babayev, A. G., Nechayeva, N. T., and Orlovaky, N. S. (1986) The problem of deserts, some results and new requirements. *Problems of Desert Development* **3**, 5–12 (in Russian)

Bacmeister, J. T., Kuell, V., Offermann, D., Riese, M., and Elkins, J. W. (1999) Intercomparison of satellite and aircraft observations of ozone, CFC-II, and NO_x using trajectory mapping. *J. Geophys. Res.* **104**(D13), 16379–90.

Baliunas, S. L. and Glassman, J. K. (2001) Bush is right on global warming. *Weekly Standard Magazine* **6**(39).

Baliunas, S. L. and Soon, W. H. (1997) *An assessmant of the sun–climate relation on timescales of decades to centuries: The possibility of total irradiance variations.* Harvard-Smithsonian Center for Astrophysics. Preprint Series No. 4565, Cambridge, MA, 10 pp.

Balling, R. C. Jr and Christy, J. R. (1996) Analysis of satellite based estimates of tropospheric diurnal temperature range. *J. Geophys. Res.* **101**(D8), 12827–32.

Baranka, G. (1999) Near surface ozone concentration evaluation and prediction in Budapest. *Időjárás* **103**(2), 107–21.

Barnett, T.P., Pierce, D.W., and Schnur, R. (2001). Detection of anthropogenic climate change in the world's oceans. *Science* **292**, 270–4.

Barrie, L. A. and Platt, U. (1997) Arctic tropospheric chemistry: Overview to Tellus special issue. *Tellus* **49B**, 450–4.

Barrie, L. A., Platt, U., and Shepson, P. (1998) Surface ozone depletion at the polar sunrise fueled by sea-salt galogens. *IG Acivities Newsletter* **14**, 4–7.

Bates, N. R., Knap, A. H., and Michaels, A. E. (1998a) Contribution of hurricanes to local and global estimates of air-sea exchange of CO_2. *Nature* **395**(6697), 58–61.

Bates, T. S., Huebert, B. S., Gras, J. L., Griffith, F. R., and Durkee, P. A. (1998b) International Global Atmospheric Chemistry (IGAC) Project's First Aerosol Characterization Experiment (ACE-1) Overview. *J. Geophys. Res.* **103**(D13), 16297–318.

Baumgardner, D., Miake-Lye, R. C., Anderson, M. R., and Brown, R. C. (1998) An evaluation of the temperature, water vapor, and vertical velocity structure of aircraft contrails. *J. Geophys. Res.* **103**(D8), 8727–36.

Belan, B. D., Kovalevsky, V. K., Plotikov, A. P., and Sklyadneva, T. K. (1998) Timal changes of ozone and nitrogen oxides in the near ground atmospheric layer in the area of Tomsk. *Optics of the Atmosphere and Ocean* **11**(12), 1325–7.

Belan, B. D. and Sklyadneva, T. K. (1999) Tropospgeric ozone concentration change in dependence on solar radidance. *Optics of the Atmos. and Ocean* **12**(8), 725–9.

Belchansky, G. I., Ovchinnikov, G. K., Petrosyan, V. G., Penk, L. F., and Douglas, D. S. (1992) Satellite monitoring data processing for the study of big mammals in the Arctic environment. *Studying the Earth from Space* **2**, 75–81.

Benestad, R. E. (2001). The cause of warming over Norway in the ECHAM4/OPYC3 GHG integration. *Int. J. Climatol.* **21**, 371–87.

Bengtsson, L. (1999) *Climate modeling and prediction – achievements and challenges.* WCRP/WMO, No. 954, pp. 59–73.

Bengtsson, L. A. (1997) A numerical simulation of anthropogenic climate change. *Ambio* **26**(1), 58–65.

Beniston, M. (1998) *From Turbulence to Climate.* Springer for Science, The Netherlands. 328 pp.

Beresin, A. M., Vavilov E. G., and Grigoryev A. Al. (1969) *Indicative role of forest plants for decoding soils and quaternary deposits.* Nauka, Leningrad, 128 pp. (in Russian).

Berger, A. and Loutre M.-E. (1997) Paleoclimate sensitivity to CO_2 and insolation. *Ambio* **26**(L), 32–7.

Berger, A. and Loutre M.-E. (2001) An exceptionally long interglacial ahead?. In: *First Int. Conf. on Global Warming and the Next Ice Age, Dalhousie University, 19–24 August*, pp. 245–8. Vancouver.

Bhatt, P. P., Remsberg, E. E., Gordley, L. L., Melnemey, J. M., Brackett, V. G., and Russel, J. M. III (1999) An evaluation of the quality of Halogen Occultation Experiment ozone profiles in the lower stratosphere. *J. Geophys. Res.* **104**(D8), 9261–76.

Bigelow, D. S., Slusser, J. R., Beaubien, A. E., and Gibson J. H. (1998) The USDA ultraviolet radiation monitoring program. *Bull. Amer. Meteorol. Soc.* **79**(4), 601–15.

Biggs R. H. and Joyner M. E. B. (1994) *Stratospheric Ozone Depletion/UV-B Radiation in the Biosphere.* NATO ASI Series, Global Environmental Change. Springer-Verlag, Berlin, 358 pp.

Boehmer-Cristiansen, S. (1994a) Global climate protection policy: the limits of scientific advice, Part 1. *Glob. Environm. Change* **4**, 140–59.

Boehmer-Cristiansen, S. (1994b) Global climate protection policy: the limits of scientific advice, Part 2. *Glob. Environm. Change* **4**, 185–200.

Boehmer-Cristiansen, S. (1996) Political pressure in the limits of scientific consensus. In: J. Emsley (ed.) *The Global Warming Debate.* Bourne Press, Bournemouth, pp. 234–48.

Boehmer-Cristiansen, S. (1997) Essay review. *Energy & Environment* **8**(4), 323–6.

Boehmer-Cristiansen, S. (1999) Climate change and the World Bank: Opportunity for global governance? *Energy and Environ.* **10**, 27–50.

Boehmer-Cristiansen, S. (2000) Who and how determines the climate change-concerning policy? *Izv. Rus. Geogr. Soc.* **132**, 6–22 (in Russian).

Bojkov, R. D. (1986a) The 1979–1985 ozone decline in the Antarctic as reflected in ground based observations. *Geophys. Res. Lett.* **13**(12), 1236–9.

Bojkov, R. D. (1986b) Surface ozone during the second half of nineteenth century. *J. Appl. Meteorol.* **25**, 343–52.

Bojkov, R. D. (1988) Ozone changes at the surface and in the free troposphere. In: I. S. A. Isaksen (ed.) *Tropospheric Ozone*. D. Reidel, Amsterdam, pp. 83–96.

Bojkov, R. D. (1998) International Assessment of Ozone Depletion. *WMO Bull.* **48**, 35–44.

Bojkov, R. D. and Fioletov, V. E. (1995) Estimating the global ozone characteristics during the last 30 years. *J. Geophys. Res.* **100**, 16537–51.

Bojkov, R. D. and Reinsel, G. C. (1985) Trends in tropospheric ozone concentrations. *Atmospheric Ozone*. D. Reidel, Dordrecht, pp. 775-781.

Bojkov, R. D. and Visconti, G. (1998) Atmospheric ozone. *Proc. XVIII Quadrannial Ozone Symposium. L'Aquila, Italy, 12–21 September 1996*. Parco Scientifico Technologico d'Abruzzo, pp. 1, 2.

Bojkov R. D., Mateer C. L., and Hansson A. L. (1988) Comparison of ground-based and total ozone mapping spectrometer using in assessing the performance of the Global Ozone Observing System. *J. Geophys. Res.* **93**(D8), 9525–33.

Bojkov R. D., Zerefos C. S., Balis D. S., Ziomas I. C., and Bais A. F. (1993) Record low total ozone during northern winter of 1992 and 1993. *Geophys. Res. Lett.* **20**, 1351–4.

Bolin, B. (1998a) The Kyoto negotiations on climate change: A science perspective. *Science* **279**(5349), 330–42

Bolin, B. (1998b) The Kyoto negotiations on climate change: A science perspective. *The Globe* **1**, 3, 4.

Bolin, B. (1998c) The WCRP and IPCC: Research inputs to IPCC Assessments and needs. *WCRP/WMO* **904**, 27–36.

Borell, P. M. and Borell, P. (1999) *EUROTRAC '98. Transport and Chemical Transformation in the Troposphere*. WIT Press, Southampton, UK, Vol. 1, 856 pp., Vol. II., 952 pp.

Bovensman, H., Burrows, J. P., Buchwitz, M., Frederick, J., Noél, S., Rozanov, V. V., Chance, K. V., and Goede A. P. H. (1999) Sciamachy: Mission objectives and measurement modes. *J. Atmos. Sci.* **56**(2), 127–50.

BP Statistical Review of World Energy (1997) The British Petroleum, London, 41 pp.

Bradely, R. S. (2001). Many citations support global warming trend. *Science* **292**, 2011.

Brasseur, G. P. (ed.) (1997) *The Stratosphere and its Role in the Climate System*. Springer-Verlag, Heidelberg, 368 pp.

Brasseur, G. P, Amanatidis, G.T., and Angeletti, G. (eds) (1998a) European Scientific Assessment of the Atmospheric Effects of Aircraft Emissions. *Atmos. Environ.* **32**(13), 2327–422.

Brasseur, G. P., Cox, R. A., Hauglustaine, D., Isaksen, L., Leiieveld, J., Lister, D. H., Sausen, D., Schumann, L. J., Wahner, A., and Wiesen, P. (1998b) European scientific assessment of the atmospheric effects of aircraft emissions. *Atmos. Environ.* **13**(32), 2329–418.

Brasseur, G. P., Hauglustaine, D. A., Walters, S., Rasch, P. J., Müller, J.-F., Granier, C., and Tie, X. X. (1998c) MOZART, a global chemical transport model for ozone and related chemical tracers. 1. Model description. *J. Geophys. Res.* **103**(D21), 28265–90.

Brekke, P. (2001). The Sun's role in climate change. *First Int. Conf. on Global Warming and the Next Ice Age, Dalhousie University, 19–24 August 2001*. Vancouver, pp. 241–4.

Brown, D. A. (1998) Making CSD work. *Linkages J.* **3**, 2–6.

Brown, L. R., Flavin, C., and French, H. (2001) *State of the World*. Earthscan, London, 275 pp.

Bruce, J. P., Lee, H., and Haites, E. F. (eds) (1996) *Climate Change 1995. Economic and Social Dimensions of Climate Change*. Cambridge University Press, 448 pp.

Bryant, E. (1997) *Climate Processes and Change*. Cambridge University Press, 209 pp.

Bryceson, K. P. (1991) Likely locust infestation areas in Western New South Wales, Australia, located by satellite. *Geocarto Int.* **6**(4), 21–37.

Buchinsky, I. E. (1976) *Draughts and Hot Dry wind*. Hydrometeoizdat, Leningrad, 215 pp. (in Russian)

Burrow, J. P., Weber, M., Buchwitz, M., Rozanov, V., Ladstätter-Weisenmayer, A., Richter, A., DeBeek, R., Hoogen, R., Bramstedt, K., Eichmann, K-U., Eisinger, M., and Peruer, D. (1999) The Global Ozone Monitoring Experiment (GOME): Mission concept and first scientific results. *J. Atmos. Sci.* **56**(2), 151–75.

Cai, W. and Whetton, P. H. (2000). Evidence for a time-varying pattern of greenhouse warming in the Pacific Ocean. *Geophys. Res. Lett.* **27**.

Caldeira, K., Ran, G. H., and Duffy, P.B. (1998) Predicted net efflux of radiocarbon from the ocean and increase in atmospheric radiocarbon content. *Geophys. Res. Lett.* **25**(20), 3813–14.

Campos, T. L., Weingeimer, A. S., Zheng, J., Montzka, O. D., Walega, J. G., Granck, F. E., Vay, S. A., Colins, J. E. Jr, Sachse, G. W., Anderson, B. E., Brune, W. H., Tan, D., Faloona, I., Baughcam, S. L., and Radley, B.A. (1998) Measurements of NO and NO_x emission indices during SUCCESS. *Geophys. Res. Lett* **25**(10), 1713–16.

Capellani, F. and Kochler, C. (1999) Ozone and UV-B variations at Ispra from 1993 to 1997. *Atmos. Environ.* **33**(23), 3787–90.

Carson, D. J. (1999) Climate modelling: Achievements and prospects. *Quart. J. Roy. Meteorol. Soc.* **125**(33), 1–27.

Cavalieri, D. J., Gloerson, P., Parkinson, C. L., Comiso, J. C., and Zwally, H. J. (1997) Observed hemispheric asymmetry in global sea ice changes. *Science* **278**, 1104–6.

Cess, R. D., Zhang, M. H., Ingram, W. J. (1996) Cloud feedback in atmospheric general circulation models. *J. Geophys. Res.* **101**(D8), 12791–4.

Chandra, S., Varotsos, C., and Flynn, L. E. (1996). The mid-latitude ozone trends in the Northern Hemisphere, *Geophys. Res. Lett.* **23**(5), 555–8.

Chanin, M. L. and Geller, M. (1998) Report of the WCRP/JSC Meeting. *SPARC Newsletter* **11**, 1–2.

Charlson, J. and J. Heintzenberg, J. (eds) (1995) *Aerosol Forcing of Climate*. John Wiley & Sons, Chichester, UK.

Charlson, R. J. and Wigley, T. M. L. (1994) Sulfate aerosol and climatic change. *Scientific American* **1270**, 28–35.

Charlson, R. J., Seinfeld, J. H., Nunes, A., Kulmala, M., Laaksonen, A., and Facchini, M. C. (2001) Reshaping the theory of cloud formation. *Science* **292**, 2025–6.

Chase, T. N., Pielke, R. A., Knaff, J. A., Kittel, T. G. F., and Eastman, J. L. (2000) A comparison of regional trends in 1979–1997 depth-averaged tropospheric temperatures. *Int. J. Climatol.* **20**, 503–18.

Chernikov, A. A., Borisov, Yu. A., Zvyagintsev, A. M., Kruchenitsky, R. M., Perov, S. P., Sidorenkov, N. S., and Stasyuk, O. V. (1998) El Niño impact on the Earth's ozone layer in 1997–1998. *Meteorol. Hydrol.* **3**, 104–10.

Chernikov, A. A., Kruchenitsky, R. M., Zvyagintsev, A. M., Ivanova, N. S., and Borisov, Yu. A. (1999) Ozone content over Russia and adjacent territories in the second quarter of 1999. *Meteorol. Hydrol.* **9**, 118–24.

Chopey, N. P., Cooper, C., Crabb, C., and Ondrey, G. (1999) Technology to cool down global warming. *Chem. Eng. (USA)* **106**(1), 37, 39, 41.

Christiansen, B. (1995) Radiative forcing and climate sensitivity: The 0zone experience. *Quart. J. Roy. Meteorol. Soc.* **125**(560), Part B., 3011–38.

Christy, J. R., Spencer, R. W., and McNider, R. T. (1995) Reducing noise in daily MSU lower tropospheric temperature data set. *J. Climate* **8**, 888–96.

Christy, J. R., Spencer, R. W., and Lobl, E. S. (1998) Analysis of the merging procedure for the MSU daily temperature time series. *J. Climate* **11**, 2016–2041.

Ciasis, P. (1999) Restless carbon pools. *Nature* **398**(6723), 111–12.

Clark, J. S., Carpenter, S. R., Barber, M. M., Collins, S., Dobson, A., Foley, J. A., Lodge, D. M., Pascual, M., Pielke, R. Jr, Pizer, W., Pringle, C., Reid, W. V., Rose, K. A., Sala, O., Schlesinger, W. H., Wall, D. H., and Wear, D. (2001) Ecological forecasts: An emerging initiative. *Science* **293**, 657–60.

Climate Change 1995 (1996a) *Impacts, Adaptation and Mitigation of Climate Change: Scientific-Technical Analysis* (ed. by R. T. Watson, M. C. Zynowera, R. H. Moss, and D. J. Dokker). Cambridge University Press, Cambridge, 878 pp.

Climate Change 1995 (1996b) *Economic and Social Dimensions of Climate Change* (ed. by J. P. Bruce, H. Lee, and E. F. Haites). Cambridge University Press, Cambridge, 488 pp.

Climate Change (1997) *Scientific Certainties and Uncertainties.* Natural Environment Research Council, Swindon, UK, 6 pp.

Climate Change (2001) *The Scientific Basis, 2001* (eds J. T. Houghton *et al.*) Contribution of the WGI to the Third Assessment Report of the IPCC. Cambridge University Press, 881 pp.

Climate Change and Its Impacts (1999) *Stabilisation of CO_2 in the Atmosphere.* The Met Office, Bracknell, UK, 28 pp.

Collatz, G. J., Bounoua, L., Los, S. O., Randell, D. A., Fung, I. F., and Sellers, P. J. (2000) A mechanism for the influence of vegetation on the response of the diurnal temperature range to changing climate. *Geophys. Res. Lett.* **27**, 3381–4.

Common Questions about Global Change (1997) UNEP/WMO, Nairobi, Kenya, 24 pp.

Considine, D. B., Stolarski, R. S., Hollandsworth, S. M., Jackman, C. H., Fleming, and F. L. (1999) A Monte Carlo uncertainty analysis of ozone trend predictions in a two-dimensional model. *J. Geophys. Res.* **104**(D1), 1749–66.

Courel, M. F. and Habif, M. (1982) Measurement of changes in Sahelian surface cover using Landsat albedo images. *Adv. Space Res.* **2**(8), 37–44.

Courier (UNESCO) (1985) **2**, 10–11.

Covey, C., Sloan, L. C., and Hhoffer, M. I. (1996) Paleoclimatic data constraints on climate sensitivity: The paleocalibration method. *Clim. Change* **32**(2), 165–84.

Cox, R. A. (1999) Ozone and peroxy radical budgets in the marine boundary layer: Modeling the effect of NO_x. *J. Geophys. Res.* **104**(D7), 8047–56.

Craig, S. G., Holmen, K. J., Bonan, G. B., and Rasch, P. J. (1998) Atmospheric CO_2 simulated by the National Center for Atmosphere Research Community Climate Model. 1. Mean fields and seasonal cycles. *J. Geophys. Res.* **103**(D11), 13213—35.

Crowley, T. J. (2000) Causes of climate change over the past 1000 years. *Science* **289**, 270–7.

Crutzen, P. J. (1970) The influence of nitrogen oxides on the atmospheric ozone content. *Quart. J. Roy. Meteorol. Soc.* **96**, 320–25.

Crutzen, P. J. (1998) The Bulletin interview. *WMO Bull.* **47**(2), 111–23.

Cubasch U. (ed.) (1997) *Anthropogenic climate change.* European Commission, Brussels, EUR 17466EN, 73 pp.

Cubasch, U., Voss, R., Hegerl, G. C., Waszkewitz, and Crowley, T. J. (1997) Simulation of the influence of solar radiation variations on the global climate with an ocean-atmosphere general circulation model. *Climate Dynamics* **13**(11), 757–67.

Daggett, D. L., Sutkus, D. J., Du Bois, D. P., and Daughcum, S. L. (1999) *An Evaluation of Aircraft Emissions Inventory Methodology by Comparison with Reported Airline Data.* NASA/CR-19990209480, GSFC, Greenbelt, MD, September 1999, 75 pp.

Dameris, M., Grewe, V., Kohler, L., Sausen, R., Bruhl, C., Groos, J.-U., and Steil, B. (1998) Impact of aircraft NO_x emissions on tropospheric and stratospheric ozone. Part II: 3-D model results. *Atmos. Environ.* **32**, 3185–200.

D'Arrigo, R., Jacoby, G., Free, M., and Robock, A. (1999) Northern Hemisphere temperature variability for the past three centuries: Tree-ring and model estimates. *Clim. Change* **42**, 663–75

Datsenko, N. M. and Shabalova, D. M. (2001) Seasonality of multidecadal and centennial variability in European temperatures: The wavelet approach. *J. Geophys. Res.* **106**, 12449–61.

Dau, J. (1986) Distribution and behavior of barren-ground caribou in relation to weather and parasitic insects. M.S. Thesis, University Alaska, Fairbanks, 149 pp.

Davis, W. E., Vaughan, G., and O'Connor, F. M. (1998) Observation of the near-zero ozone concentration in the upper troposphere in midlatitudes. *Geophys. Res. Lett.* **25**(8), 1173–6.

Day, T. A., Ruhland, C. T., and Xiong, F. (1997) Impacts of ultraviolet-B radiation and regional warming on antarctic vascular plants. *Antarct. J. (US)* **32**(5), 155–6.

de Blij, H. J., Glantz M. H., Harris S. L. *et al.* (1997) *Restless Earth.* National Geographical Society, Washington, DC, 215 pp.

Dech, S. W. and Ressl, R. (1993) Die Verlandung des Aralsees. Eine Bestandsaufnahme durch satellitenfernkundung. *Geographische Rundschau*, **45**(6), 345–52.

DeFreitas, E. (2001) Is it reasonable for climate scientists to continue to claim that the observed changes in concentration of greenhouse gases in the atmosphere are dangerous? *First Int. Conf. on Global Warming and the Next Ice Age, Dalhousie University, 19–24 August 2001.* Vancouver, p. 172.

de Laat, A. T. J., Zachariasse, M., Roelofs, G. J., van Velthofen, P., Dickerson, R. R., Rhoads, K. P., Oltmans, S. J., and Lelieveld, J. (1999) Tropospheric O_3 distribution over the Indian Ocean during spring 1995 evaluated with a chemistry climate model. *J. Geophys. Res.* **104**(D11), 13881–95.

Delworth, T. L., Manabe, S., and Stouffer, R. J. (1997) Multidecadal climate variability in the Greenland Sea and surrounding regions. A coupled model simulation. *Geophys. Res. Lett.* **24**, 257–60.

Demirchian, K. S. (1992a) Global and local problems of ecology and development of power generation of the country. *Izv. Russian Acad. Sci. Energy* **3**, 3–17 (in Russian).

Demirchian, K. S. (1992b) Evaluation of natural and anthropogenic factors impact on the temperature of the planet. *Izv. Russian Acad. Sci. Energy* **5**, 3–15 (in Russian).

Demirchian, K. K. (1994a) Methods to identify contribution of electric power production to increase in global temperature. Development of the problem. *Izv. Russian Acad. Sci. Energy* **1**, 92–98 (in Russian).

Demirchian, K. K. (1994b) Methods to identify contribution of electric power production to increase in global temperature. *Izv. Russian Acad. Sci. Energy* **5**, 118–22 (in Russian).

Demirchian, K. S. and Demirchian, K. K. (1993) Phenomenological model for determination of forced change in global temperature. *Izv. Russian Acad. Sci.. Energy* **5**, 15–26 (in Russian).

Demirchian, K. K. and Demirchian, K. S. (1994) Determination of anthropogenic emissions sinks from the atmosphere. *Izv. Russian Acad. Sci. Energy* **5**, 19–26 (in Russian).

Demirchian, K. S. and Kondratyev, K. Ya. (1999) Development of power generation and the environment. *Izv. Russian Acad. Sci. Energy* **6**, 3–46 (in Russian).

Deque, M., Marquet, P., and Jones, R. G. (1998) Simulation of climate change over Europe using a global variable resolution general circulation model. *Climate Dynamics* **14**(3), 173–89.

de Rosnay, P., Bruen, M., and Polcher, J. (2000) Sensitivity of surface fluxes to the number of layers in the soil model used in GCMs. *Geophys. Res. Lett.* **27**, 3329–32.

Derwent, R. G., Collins, W. J., Johnson, C. E., and Stevenson, D. S. (2001) Transient behavior of tropospheric ozone precursors in a global 3-D CTM and their indirect greenhouse effects. *Clim. Change* **49**, 463–87.

Doerell, P. E. (1999) Anti–global warming measures would hit mining hard. *Eng. Mining J.* **200**(2), 24–7.

Dreiser, U. (1993) Mapping of desert locust habitats using remote sensing techniques. *Int. Symp. Operational Remote Sensing, Enschede*. ESA, Enschede, pp. 63–9.

Dubey, M. K., Smith, G. P., Hartley, W. S., Kinnson, D. E., and Corull, P. S. (1997) Rate parameter uncertainty effects in assessing stratospheric ozone depletion by supersonic aviation. *Geophys. Res. Lett.* **24**(22), 2737–40.

Duffy, P. B., Doutriaux, C., Santer, B. D., and Fodor, I. K. (2001) Effect of missing data on estimates of near-surface temperature change since 1900. *J. Climate* **14**, 2809–14.

Dufresne, J. L., Fredlingstein, P., Fairhead, L., Le Treut, H., Ciais, P., and Monfray, P. (2001) Feedback process between carbon cycle and climate in the IPSLM coupled model. *Abstracts 8th Scientific Assembly of IAMAS, Innsbruck, 10–18 July 2001*, p. 10.

Ebel, A., Friedrich, F., and Rodhe, H. (1997) *Tropospheric Modeling and Emission Estimates*. Springer–.Verlag, Heidelberg, 450 pp.

Eddy, J. A. and Oeschager, H. (1993) *Global Change in the Perspective of the Past*. John Wiley & Sons, Chichester, UK, 383 pp.

Edmonds, J. A. (1999) Beyond Kyoto: Toward a technology greenhouse strategy. *Consequences* **5**(1), 17–28.

Edwards, S. R. (1995) Conserving biodiversity resources for our future. In: R. Balies *The State of the Planet*. Free Press, New York, pp. 211–65.

Ehhalt, D. H. (1998) Radical ideas. *Science* **279**, 1002–3.

Eichmann, K.-U., Bramstedt, K., Weber, M., Rozanov, V. V., Hoogen, R., and Burrows, J. P. (1999) O_3 profiles from GOME satellite data. II. Observations in the Arctic spring 1997 and 1998. *Phys. Chem. Earth (C)* **24**(5), 453–7.

Ellsaesser, H. W. (ed.) (1994) The unheard arguments: A rational view of stratospheric ozone. *21st Century Science and Technology* **7**(23), 37–45.

Ellsaesser, H. W. (ed.) (1999) *Global 2000 Revisited. Mankind's impact on spaceship earth*. Paragon House, New York, 436 pp.

Ellsaesser, H. W. (2001) *The current status of global warming*. Marshall Institute, Washington, DC, 5pp.

Elpiner, L. I. (1996) Medical and ecological aspects of the Caspian region. *Arid Ecosystems* **2**(2–3), 81–4 (in Russian).

Emsley, L. (1996) Global warming: The Report of the European Science and Environment Forum. Introduction. In: J. Emsley (ed.) *The Global Warming Debate*. The Report of the European science and environment Forum, Bourne Press, Bournemouth, UK, pp. 22–7.

Environmental Remote Sensing from Regional to Global Scales (1994) John Wiley and Sons, Chichester, UK, 238 pp.

Erlick, C., Russel, L. M., and Ramaswamy, V. (2001). A microphysics-based investigation of the radiative effects of aerosol-cloud interactions for two MAST Experiment case studies. *J. Geophys. Res.* **106**, 1245–69.

Ernst, W. G. (ed.) (2000) *Earth Systems. Processes and Issues*. Cambridge University Press, 576 pp.

Essenhigh, R. H. (2001) Does CO_2 really drive global warming? *Chemical Innovation* **31**, 44–6.

European Commission (2001) *European Research in the Stratosphere 1996–2000. Advances in our understanding of the ozone layer during THESEO*. Office for Official Publications of the European Communities, Luxembourg, 378 pp.

EUROTRAC-2 TOR-2 Tropospheric Ozone (1999) Research, Annual Report 1998, International Scientific Secretariat, Munich, 163 pp.

Evans, W. F. J. and Puckrin, E. (1999) Remote sensing measurements of tropospheric Ozone by ground-based thermal emission spectroscopy. *J. Atmos. Sci.* **56**(2), 311–18.

Extended and updated (1751–1995) CO_2 emission time series available (1998) CDIAC Communications, Los Angeles, No. 24, 18 pp.

Fabian, C. P. and Singh, O. N. (eds) (1999) *Reactive Halogen Compounds in the Atmosphere*. Springer, Berlin, 227 pp.

Feister, U. and Grewe, R. (1995) Higher UV radiation inferred from low ozone levels at northern mid-latitudes in 1992 and 1993. *Global and Planetary Change* **11**(1), 25–34.

Fenn, N. A., Browell, E. V., and Rutler, C. F. (1999) Ozone and aerosol distributions and air mass characteristics over the South Pacific during the burning season. *J. Geophys. Res.* **104**(D13), 16197–212.

Fioletov, V. E., Griffen, E., Kerr, J.B., Wardle, D. I., and Uchino, O. (1998). An assessment of volcanic sulphur dioxide on spectral irradiance as measured by Brewer spectrophotometers. *Geophys. Res. Lett.* **25**(10), 1655–8.

Fioletov, V. E., Kerr, J. B., Hare, E. W., Labow, G. J., and McPeters, R. D. (1999) An assessments of the world ground-based total ozone network performance from the comparison with satellite data. *J. Geophys. Res.* **104**(D1), 1737–48.

Fiore, A. M., Jacob, D. S., Logan, J. A., and Yin, J. H. (1998) Long-term trends in ground level ozone over the contiguous United States, 1980–1995. *J. Geophys. Res.* **103**(DI), 1471–80.

Fisher, H., Nikitas, C., Parchatka, K., Zenker, T., Harris, G. W., Matuska, P., Schmitt, R., Mihelcic, D., Muesgen, D., Paetz, H.-W., Schultz, M., and Volz-Thomas, A. (1998) Trace gas measurements during the Oxidizing Capacity of the Tropospheric Atmosphere campaign 1993 at Izaña. *J. Geophys. Res.* **103**(DI1), 13505–18.

Fishman, J., Watson, C. E., Larsen, J. C., and Logan, J. A. (1999) Distribution of tropospheric ozone determined from satellite data. *J. Geophys. Res.* **95**(D4), 3599–617.

Folland, C. K, Sexton, D. M. H., Karoly, D. J., Johnson, C. E., Rowell, D. P., and Parker, D. E. (1998) Influences of anthropogenic and ocean forcing on recent climate change. *Geophys. Res. Lett.* **25**(3), 353–6.

Folland, C. K, Frich, P., Basnett, T., Rayner, N., Parker, D., and Horton, B. (2000).Uncertainties in climate datasets – a challange for WMO. *WMO Bull.* **48**(1), 59–68.

Fortuin, J. P. F. and Kelder, H. (1998) An ozone climatology based on ozonesonde and satellite measurements. *J. Geophys. Res.* **103**(D24), 31709–34.

Fraedrich, K., Gerstengarbe, F.-W., and Werner, P. C. (2001) Climate shifts during the last century. *Climate Change* **50**, 405–17.

Fraser, P. J. and Prather, M. J. (1999) Atmospheric chemistry: Uncertain road to ozone recovery. *Nature* **398**(6729), 663–4.

Freeman, P., Martin, L., Machler, R., and Warner, K. (1999) Natural catastrophes, infrastructure, and poverty in developing countries. *Options* **Fall/Winter**, 6–16.

Friedl, R. R. (1999) Unraveling aircraft impacts. *Science* **286**, 57–8.

From the Candidates (2000) Gore and Bush address key environmental issues. *Resources* **Fall**(141), 5–8.

Fuglestvedt, J. S., Bernsten, T. K., Isaksen, I. S. A., Mao, H., Liang, X. Z., and Wang, W. (1959) Climatic forcing of nitrogen oxides through changes in tropospheric ozone and methane: Global 3D model studies. *Atmos. Environ.* **33**, 961–77.

Fung, K. K. and Ramaswamy, V. (1999) On shortwave radiaton absorption in overcast atmospheres. *J. Geophys. Res.* **104**, 22233–42.

Fusco, A. C. and Salby, M. L. (1999) Interrannual variations of total ozone and their relationship to variations of planetary wave activity. *J. Climate* **12**(6), 1619–29.

Gaffen, D. J., Santer, B. D., Boyle, J. S., Christy, J. R., Graham, N. E., and Ross, R. J. (2000) Multidecadal changes in the vertical temperature structure of the tropical troposphere. *Science* **287**, 1242–5.

Gates, W. L. (1996) *Review of main development since the sixteenth session of the JSC XVII Doc. 4, Toulouse, France, 11–16 March 1996*, 6 pp.

Gedney, N. and Valdes, P. J. (2000) The effect of Amazon deforestation on the Northern Hemisphere circulation and climate. *Geophys. Res. Lett.* **27**, 3053–6.

Gent, P. R. (2001) Will the North Atlantic Ocean thermocline circulation weaken during the 21st century? *Geophys. Res. Lett.* **28**, pp. 1023–6.

Geophysical Research Letters, November Supplement (1986) **13**(12), 1191–362.

Gerholm, T. R. (ed.) (1999) *Climate Policy after Kyoto*. Multi-science, Brentwood, UK, 170 pp.

Gierens, K. and Jensen, E. (1998) A numerical study of the contrail-to-cirrus transition. *Geophys. Res. Lett.* **22**(15), 4341–4.

Gill, G. S. (1993) Landuse mapping of Harike wetland and environs. *Spectrum* **3**, 4.

Glazovsky, N. F. (1997) The concept of a way out from the Aral crisis. *Izv. USSR Acad. Sci. Ser. Geogr.* **4**, 28–41 (in Russian).

Glavas, S. (1999) Surface ozone and NO_x concentrations at a high altitude Mediterranian site, Greece. *Atmos. Environ.* **33**(23), 3813–20.

Global Aspects of Atmospheric Chemistry (1999) Springer, Berlin, 334 pp.

Gloerson, P. and Mollo-Christensen, E. (1996) Oscillatory behavior in Arctic sea-ice concentrations. *J. Geophys. Res.* **101**(C3), 6641–50.

Gloerson, P., Parkinson, C. L., Cavalieri, D. J., Comiso, J. C., and Zwally, H. J. (1999) Spatial distribution of trends and seasonality in the hemispheric ice covers: 1978–1996. *J. Geophys. Res.* **104**, 20827–35.

Godzik, B. (1998) Ground level ozone concentration in the Krakow region, southern Poland. *Environ. Pollut.* **98**(3), 273–80

Goita, K. and Royer, A. (1993) Land surface climatology and land cover change monitoring since 1973 over a North Sahelian zone (Ansong-Mali) using Landsat data. *Geocarto Int.* **8**(2), 15–28.

Goldenberg, S. B., Landsea, C. W., Mestas-Nuñez, A. M., and Grey, W. M. (2001) The recent increase in Atlantic hurricane activity: Causes and implications. *Science* **293** 474–9.

Goody, R. (2001). Climate benchmarks: Data to test climate models. *Studying the Earth from Space* **6**.

Goody, R., Haskins, R., Abdou, W., and Chen, L. (1995) Detection of climate forcing using emission spectra. *Studying the Earth from Space* **5**, 22–33.

Goody, R., Anderson, J., and North, G. (1998).Testing climate models: An approach. *Bull. Am. Meteorol. Soc.* **79**, 2541–9.

Goody R., Anderson J., Karl T., Miller R., North G., Simpson J., Stephensens C., and Washington, W. (2001) Why monitor the climate? *Bull. Am. Meteorol. Soc.* (in press).

Gore, R. (1985) No way to run a desert. *Nat. Geogr.* **167**(6), 694–723.

Gorshkov, V. G. (1990) *Biosphere Energetics and Environmental Stability*. VINITI, Moscow, 238 pp. (in Russian).

Gorshkov, V. G. (1995) *Physical and Biological Basis of Life Stability*. Springer-Verlag, Heidelberg, 470 pp.

Gorshkov, V. G., Gorshkov, V. V., and Makarieva, A. M. (2000) *Biotic Regulation of the Environment. Key Issues of Global Change*. Springer/Praxis, Chichester, UK, 367 pp.

Grabb, M, Vrolijk, C., and Brack, D. (1999). *The Kyoto Protocol. A Guide and Assessment*. Royal Institute of International Affairs, London, 342 pp.

Graedel, T. E. and Crutzen, P. J. (1997) *Atmosphere, Climate, and Global Change*. Scientific American Library, New York, 196 pp.

Graf, H.-F., Kirchner, L., and Perlwitz, J. (1998) Changing lower stratospheric circulation: The role of ozone and greenhouse gase. *J. Geophys. Res.* **103**(D10), 11251–62.

Granier, C. (1999) Impact of stratospheric ozone changes on the distribution of tropospheric species. *SPARC Newsletter* **13**, 10–103.

Greenhouse forecasting still cloudy. Model gets it right – without fudge factors (1997) *Science* **276**(5315), 1040–2.

Grewe, V., Dameris, M., Hein, R., Kohler, L., and Sausen, R. (1999) Impact of future subsonic aircraft NO_x emissions on the atmospheric composition. *Geophys. Res. Lett.* **26**, 47–50.

Grigoryev, Al. A. (1985) *Anthropogenic impacts on the environment from satellite observations*. Nauka, Leningrad, 240 pp. (in Russian).

Grigoryev, Al. A. (1991) *Ecological lessons of the historical past and modern times*. Nauka, Leningrad, 250 pp. (in Russian).

Grigoryev, Al. A. and Kondratyev, K. Ya. (1993a) Remote sensing of the environment and natural resources in India. *Studying the Earth from Space* **3**, 118–22.

Grigoryev, Al. A. and Kondratyev, K. Ya. (1993b) Aerospace natural resources and ecological studies in China. *Studying the Earth from Space* **2**, 119–23.

Grigoryev, Al. A. and Kondratyev, K. Ya. (1998a) Natural and anthropogenic disasters: problems of risk. *Izv. Russian Geogr. Soc.* **130**(3), 13–24 (in Russian).

Grigoryev, Al. A. and Kondratyev, K. Ya. (1998b) Global natural resources. *Izv. Russian. Geogr. Soc.* **130**(1), 5–16 (in Russian).

Grigoryev, Al. A. and Kondratyev, K. Ya. (2001) *Environmental Disasters*. Scientific Centre, Russian Academy of Sciences, St Petersburg, 691 pp. (in Russian).

Grigoryev, Al. A. and Lipatov, V. B. (1977) Dust storms in the coast regions of the Aral Sea from space imagery. *Remote Sensing of Earth Resources (USA)* **6**, 575–99.

Grigoryev, Al. A. and Lipatov, V. B. (1979) Dust storms in Priaralje from satellite surveying data. *Development and Transformation of the Environment*. LSPI, Leningrad, pp. 94–103 in Russian).

Grigoryev, Al. A. and Lipatov, V. B. (1982) Distribution of dust pollutants in Priaralje from satellite observations. *Izv. USSR Acad. Sci. Ser. Geogr.* **6**, 13–21 (in Russian).

Grigoryev, Al. A. and Zhogova, M. L. (1992) Heavy dust outbreaks in Priaralye. *Russian Acad. Sci.* **324**(3), 672–5 (in Russian).

Groos, J.-U., Muller, R., Becker, G., McKenna,D. S., and Crutzen, P. J. (1999) The upper stratospheric ozone budget: An update of calculations based on HALOE data. *J. Atmos. Chem.* **34**(2), 171–83.

Grübler, A. and Nakicenovic, N. (2001) Identifying dangers in an uncertain climate. *Nature* **412**, 15.

Guirlet, M. and Harris, N. (1999) Third European Stratospheric Experiment on Ozone: Second year of activities. *SPARC Newsletter* **13**, 16–17.

Guirlet, M., Kibane-Dawe, I., and Harris, N. (1998). Third European Stratospheric Experiment on Ozone (THESEO). *SPARC Newsletter* **10**, 24.

Gushchin, G.P. (1999) Forty-year observations on total ozone by the network of Russia and CIS states. *Meteorol. Hydrol.* **6**, 37–42.

Guyot, G. (1997) *Physics of the Environment and Climate.* John Wiley & Sons, Chichester, UK, 256 pp.

Haigh, J. D. (1996) The impact of solar variability on climate. *Science* **272**, 981–4.

Haigh, J. D. (2000) Solar variability and climate. *Weather* **55**, 399–407.

Hales, J. (1996) Scientific background for AMS Policy Statement on Atmospheric Ozone. *Bull. Amer. Meteorol. Soc.* **77**, 1245–53.

Hall, A., Manabe, S. (2000) Effect of water feedback on internal and anthropogenic variations of the global hydrologic cycle. *J. Geophys. Res.* **105**, 6935–44.

Han, Q., Rossow, W. B., Chou J., and Welch, R. M. (2000) ISCCP data used to address a key IPCC climate issue: An approach for estimating the aerosol indirect effect globally. *GEWEX News* **10**, 3–5.

Hanna, E. (1996) The role of Antarctic sea ice in global climate change. *Progress in Physical Geography* **20**(4), 371–401.

Hanna, E. (1998) Solar-driven global warming? *J. Meteorol.* **23**, 131–7.

Hansen, G. and Chipperfield, M. P. (1999) Ozone depletion at the edge of the Arctic polar vortex 1996/1997. *J. Geophys. Res.* **104**(D1), 1837–46.

Hansen, J., Sato, M., and Ruedy, R. (1997) Radiative forcing and climate response. *J. Geophys. Res.* **102**(D6), 6831–64.

Hansen, J., Sato, M., Lacis, A., Ruedy, R., Tegen, I., and Matthews, E. (1998) Climate forcing in the Industrial era. *Proc. Natl Acad. Sci. USA* **95**, 12753–8.

Hansen, J., Sato, M., Ruedy, R., Lacis, A., and Oinas, V. (2000). Global warming in the twenty-first century: An alternative scenario. *Proc. Natl Acad. Sci. USA* **97**, 9875–80.

Harries, J. E. (1996) The greenhouse earth: A view from space. *Quart. J.Roy. Meteorol. Soc.* **122**(532), Part B, 799–818.

Harris, N. and Hudson B. (1998).Summary of the SPARC-IOC Assessment of Trends in the Vertical Distribution of Ozone. *SPARC Newsletter* **10**, 11–14.

Harris, N. R. P., Kibane-Dawe I., and Amanatidis, G. T. (eds) (1998) Air pollution research report 66–Polar stratospheric ozone 1997. *Proceedings of the fourth European Symposium. 22–26 September 1997, Schliersee, Bavaria, Germany.* European Communities (EUR 18032) EN, Brussels, 772 pp.

Harshvardhan, (1993) Aerosol-climate interactions. In: *Aerosol–Cloud Climate Interactions* (ed. by V. Hobbs). Academic Press, San Diego, pp. 76–96.

Haskins, R., Goody, R., and Chen, L. (1999) Radiance covariance and climate models. *J. Climate* **12**(5), Part 2, 1409–22.

Hasselman, K. (1997a) Multi-pattern fingerprint method for detection and attribution of climate change. *Climate Dynamics* **13**(9), 601–12.

Hasselman, K. (1997b) Are we seeing global warming? *Science* **276**(5315), 914–16.

Hasselman, K. (1999) Intertemporal accounting of climate changes harmonizing economic efficency and climate stewardship. *Clim. Change* **41**(3), 333–50.

Hauglustaine, D. A., Brasseur, G. P., Walters, S., Rasch, P. J., Mailer, J.-F., Emmons, L. K., and Caroll, M. A. (1998) MOZART, A global chemical transport model for ozone and related chemical tracers. 2. Model results and evaluation. *J. Geophys. Res.* **103**(D21), 28291–335.

Haywood, J.M., Stouffer, R.J., Wetherald, R.T., Manabe, S., and Ramasawamy, V. (1997) Transient response of a coupled model to estimated changes in greenhouse gas and sulfate concentrations. *Geophys. Res. Lett.* **24**(11), 1335–8.

Heatcot, R. L. (1978) Drought in Southern Australia. *Natural Calamities: The Study and Methods to Control* (translated from English). Progress, Moscow, pp. 118–35 (in Russian).

Hegerl, G. (1998) Climate change: The past as guide to the future. *Nature* **392**(6678), 758–9.

Hegerl, G., von Storch, H., Santer, B. D., Cubasch, U., and Jonas, P. (1996) Detecting greenhouse-gas-induced climate change with an optimal fingerprint method. *J. Climate* **9**, 2281–306.

Hegerl, G. C. and North, G. R. (1997) Comparison of statistically optimal approaches to detecting anthropogenic climate change. *J. Climate* **10**(5), 1125–33.

Hegerl, G. C., Hasselmann, K., Cubasch, U., Mitchell, J. F. B., Röckner, E., Voss, R., and Waszkewitz, S. M. (1997) Multi-fingerprint detection and attribution analysis of greenhouse gas-plus-aerosol and solar forced climate change. *Climate Dynamics* **13**(9), 613–34.

Helden, U. (1994) Desertification – time to assess. *Methods for Local, Regional and Global Biospheric Monitoring*. Nauka, Moscow, pp. 44–7 (in Russian).

Helfert, M. and Holz, R. (1985) Multi-source verification of the desiccasion of Lake Chad, Africa. *Adv. Space Res.* **5**(6), 379–84.

Henderson-Sellers, A., Zhang, H., Berz, G., Emanuel, K., Gray, W., Landsea, C., Holland, G., Lighthill, J., Shieh, S.-L., Webster, P., and McGuffie, K. (1998) Tropical cyclones and global climate change: A post-IPCC assessment. *Bull. Amer. Meteorol. Soc.* **79**(1), 19–38.

Herman, J. R., Krotkov, N., Celarier, E., Larko, D., and Labov, G. (1999) Distribution of UV radiation at the Earth's surface from TOMS-measured UV-backscattered radiance. *J. Geophys. Res.* **104**, 12059–76.

Hewitt, C. N. (ed.) (1999) *Reactive Hydrocarbons in the Atmosphere*. Academic Press, San Diego, CA, 319 pp.

Høv, O. (ed.) (1997) Tropospheric ozone research. *Tropospheric Ozone in the Regional and Sub-Regional Context*. Springer for Science, The Netherlands, 455 pp.

Høv, O., Kley, D., Volz-Thomas, A., Beck, J., Grennfelt, P., and Penkett, S. A. (1999) An overview of tropospheric ozone research. In: O. Høv (ed.) *Tropospheric Ozone Research*. Springer-Verlag, Berlin, pp. 1–34.

Hileman, B. (1997) Global climate change. *Chem. Eng. News* **75**(46), 8–16.

Hill, D. C, Allen, M. R., and Stott, P. A. (2001) Allowing for solar forcing in the detection of human influence on tropospheric temperatures. *Geophys. Res. Lett.* **28**, 1555–8.

Hobbs, P. V. (1995) *Basic Physical Chemistry for the Atmospheric Sciences*. Cambridge University Press, 225 pp.

Hollandsworth, S. M., McPeters, R. D., Flynn, L. E., Planet, W., Miller, A. S., and Chandra, S. (1995) Ozone trends deduced from combined Nimbus 7 SBUV and NOAA 11 SBUV/2 data. *Geophys. Res. Lett.* **22**(8), 905–8.

Hólm, E. V., Untch, A., Simmons, A., Saunders, R., Bouttier, F., and Andersson, E. (1999) *The SODA EU Project 1996–1999*. Final Report from ECMWF, Abstract, 6 July 1999, 12 pp., Reading, UK.

Hood, L. L., Sonkharev, B. E., Fromm, M., and McCormick, V. P. (2001) Origin of extreme ozone minima at middle to high northern latitude. *J. Geophys. Res.* **106**(D18), 20925–40.

Houghton, J. (2000).The IPCC Report 2001. *Proc. of the 1st Solar and Space Weather Euroconference: The Solar Cycle and Terrestrial Climate, Santa Cruz de Tenerife, 25–29 September 2000*, Noordwijk.

Houghton, J. (2001) Global climate and human activities: In: *Our Fragile World: Challenges and Opportunities for Sustainable Development*. (Vol. 1, pp. 103–115). Eolls Publ. Co. Ltd, Oxford, UK.

Houghton, J. T., Meira Filho, L. G., Callander, B. A., Harris, N., Kattenberg, A., and Maskell, K. (eds) (1995) *The Science of Climate Change*. Cambridge University Press, 572 pp.

Houghton, R. A., Hackler, J. L., and Lawrence, H. T. (1999) The US carbon budget: Contributions from land-use change. *Science* **285**, 574–8.

Hu, Q., Woodruff, C. M., and Mudrick, S. E. (1998) Interdecadal variations of annual precipitation in the Central United States. *Bull. Amer. Meteorol. Soc.* **79**(2), 221–9.

Hu, Z.-Z., Latif, M., Roeckner, E., and Bengtsson, L. (2000) Intensified Asian summer monsoon and its variability in a coupled model forced by increasing greenhouse gas concentrations. *Geophys. Res. Lett.* **27**, 2681–4.

Hulme, M., Barrow, E. M., Arnell, N. W., Johus, T. C., and Downing, T. E. (1996) Relative impacts of human-induced climate change and natural variability. *Nature* **397**, 688–91.

Hunt, B. G. (1998) Natural climatic variability as an explanation historical climatic fluctuations. *Clim. Change* **32**(2), 133–57.

Idso, S. B. (1989) *Carbon Dioxide and Global Change*. IBR Press, Tempe, AZ, 292 pp.

Indermühle, A., Stocker, T. F., Joos, F., Fischer, I., Smith, H. J., Wahlen, M., Deck, B., Mastroianni, D., Tschums, J., Blunier, T., Meyer, R., Staaffer, B. (1999) Holocene carbon-cycle dynamics based on CO_2 trapped in ice at Taylor Dome, Antarctica. *Nature* **398**(6723), 121–6.

International Conference on Global Problems as a Source of Catastrophe Situations. Progress, Moscow, 320 pp.

Investigations of Forestry with the Use of Remote Sensing Techniques (1997) Nauka, Novosibirsk, 290 pp.

IPCC Special Report Land-Use Change, and Forestry (2000) (ed. by R. T.Watson *et al.*), Cambridge University Press, 377 pp.

IPCC Third Assessment Report Vol. 1 (2001) *Climate Change 2001. The Scientific Basis*. Cambridge University Press, 881 pp.

Isaksen, I. S. A. (1998) *Tropospheric Ozone. Regional and Global Scale Interactions*. D. Reidel, Dordrecht, Holland, 425 pp.

Isayev, A. S., Sukhikh, V. I., Kalashnikov, E. N. *et al.* (1991a) *Aerospace Monitoring of Forests*. Nauka, Moscow, 240 pp. (in Russian).

Isayev, A. S., Kiselev, V. V., and Kondakov, Yu. P. (1991b) Forest pathology monitoring. *Aerospace Monitoring of Forests*. Nauka, Moscow, pp. 135–53 (in Russian).

Isayev, A. S., and Ryapolov, V. Ya. (1988) Remote sensing technique to control and forecast forest enthomological conditions of taiga areas. *Studying the Earth from Space* **1**, 48–55.

Ivanova, N. S., Kruchenitsky, G. M., and Chernikov, A. A. (1999) Creation of the first stage UV-radiation monitoring system in Russia. *Optics Atmos. Ocean* **12**(1), 5–9.

Izrael, Yu. A., Pavlov, A.V., and Anokhin, Yu. A. (1998) Analysis of the current and expected future changes in climate and cryolithozone in the northern regions of Russia. *Meteorol. Hydrol.* **3**, 18–27 (in Russian).

Izrael, Yu. A., Kalabin, G. V., and Nikonov, V. V. (eds) (1999) *Anthropogenic Impacts on the Nature of the North and its Ecological Consequences*. CSC Russian Academy Science, Apatity, 313 pp. (in Russian)

Jackson, T. (1994) Joint implementation and the climate convention. *Renewable Energy Dev.* **7**(3), 5–6.

Jaeglé, L., Webster, C. R., May, R. D., Scott, D. E., Stimpfle, R. M., Kohn, D. W., Wennberg, P. O., Hamisco, T. F., Cohen, R. C., Profitt, M. H., Kelley, K. K., Elkins, J. Baumgardner, D., Dye, J. E., Wilson, J. C., Pueschel, R. F., Chan, K. R., Salawitch, R. J., Tuck, A. F., Houde, S. J., and Yung, Y. L. (1997) Evolution and stoichiometry of heterogenous processing in the Antarctic stratosphere. *J. Geophys. Res.* **102**(D11), 13213–43.

Jaeglé, L., Jacob, D. J., Brune, W. H., Tan, D., Faloona, I. C., Weinheimer, A. S., Ridely, B. A., Campos, T. L., and Saches, G. W. (1998) Sources of NO_x and production of ozone in the upper troposphere over the United States. *Geophys. Res. Lett.* **25**(10), 1705–8.

Jaeschke, W., Salkowski, T., Dierssen, J. P., Trumbach, J. V., Krischke, U., and Glinther, A. (1999) Measurements of trace substances in the Arctic troposphere as potential precursors and constituents of Arctic Haze. *J. Atmos. Chem.* **34**(3), 291–319.

James, P. M. (1998). An interhemispheric comparison of ozone minihole climatologies. *Geophys. Res. Lett.* **25**(3), 301–4.

Jasinski, J., and Karnovitz, A. (1985) Weather satellites: Tracking the draught. *Weatherwise* **38**(2), 79, 98–99.

Jastrow, R., Nierenberg, W., and Seitz, F. (1990) *Scientific Perspectives on the Greenhouse Problem.* Marshall Press, Jameson Books, Ottawa, Vol. III, 254 pp.

Jaworowski, Z (1996) Reliability of ice core records for climatic projections. In: J. Emsley (ed.) *The Global Warming Debate.* Bourne Press, Bournemouth, UK, pp. 95–105.

Jaworowski, Z. (2000) The global warming. *21st Century Sci. Technol.* Winter 1999–2000, **12**(4), 64–75.

Jepma, C. J. and Munasinghe, M. (1998). *Climate Change Policy: Facts, Issues and Analysis.* Cambridge University Press, 349 pp.

Jepma, C. J., Munasinghe, M., Bruce, J. P., and Lee, H. (1997) *Climate Change Policy. Facts, Issues and Analyses.* Cambridge University Press, 304 pp.

Jeuken, A. B. M., Eskes, H. J., van Velthoven, P. J. F., Kelder, H. M., and Holm, E. V. (1999) Assimilation of total ozone satellite measurements in a three-dimentional tracer transport model. *J. Geophys. Res.* **104**, pp. 5551–63.

Jiang, Y., Yung, Y.L., and Zurek, R.W. (1996) Decadal evolution of the Antarctic ozone hole. *J. Geophys. Res.* **101**(D4), 8985–8999.

Jin, F. F., Hu, Z. Z., Latif, M., Bengtsson, L., and Roeckner, E. (2001) Dynamical and cloud radiation feedbacks in El Niño and greenhouse warming. *Geophys. Res. Lett.* **28**, 1539–1542.

Johannessen, O. M., Shalina, E. V., and Miles, M. W. (1999) Satellite evidence for an Arctic Sea ice cover in transformation. *Science* **286**, 1937–9.

Johns, T. C., Gregory, J. M., Stott, P. A., and Mitchell, J. F. B. (2001) Correlations between patterns of 19th and 20th century surface temperature change and the CM2 climate model emsembles. *Geophys. Res. Lett.* **28**, pp. 1007–10.

Johnson, G. R., Rao, A. V., Orsini, J. B. (1987) The use of NOAA AVHRR data as a tool for operational agroclimatic assessment. Proc. 10th Asian Conf. on Remote Sensing. Kuala Lumpur (Malaysia)

Jones, P. D., Osborn, T. J., Wigley, T. M. L., Kelly, P. M., and Santer, B. D. (1997) Comparisons between the microwave sounding unit temperature record and the surface temperature record from 1979 to 1996: Real differences of potential discontinuities? *J. Geophys. Res.* **102**(D25), 30135–45.

Jones, P. D., New, M., Parker, D. E., Martin, S., and Riger, I. G. (1999) Surface air temperature and its changes over the past 150 years. *Rev. Geophys.* **37**, 173–99.

Joos, F., Plattner, G. K., Stocker, T. F., Marchal, O., and Schmittner, A. (1999) Global warming and marine carbon cycle feedbacks on future atmospheric CO_2. *Science* **284**(5413), 464–8.

Jouzel, J., Petit, J. R., Souchez, R., Barkov, N. I., Lipenkov, V. Ya., Raynard, D., Stievenard, M., Vassiliev, N. I., Verbeke, V., and Vimeux, F. (1999) More than 200 meters of lake ice above subglacial Lake Vostok, Antarctica. *Science* **12**(1), 46–53.

Kadigrov, V. E. and Jadin, E. A. (1999) Anomalies and trends in ozone content in 1979–1992. *Optics Atmos. Ocean* **12**(1), 46–53.

Kahl, J. D., Charlevoix, D. J., Zaitseva, N. A., Schnell, R.,C., and Serreze, M. C. (1993) Absence of evidence for greenhouse warming over the Arctic Ocean in the past years. *Letters to Nature* **361**, 335–7.

Kahl, J. D., Zaitseva, N. A., Khattatov, V., Schnell, R. C., Bacon, D. M., Bacon, J., Radionov, V., and Serreze, M. C. (1999) Radiosonde observation from the former Soviet North Pole series of drifting ice stations, 1954–90. *Bull. Amer. Meteorol. Soc.* **80**(10), 2019–26.

Kane, R. P. Sahai, Y., and Casiccia, C. (1998) Latitude dependence of the quasi-biannual oscillation and quasi-triannual oscillation characteristics of the total ozone measured by TOMS. *J. Geophys. Res.* **103**(D7), 8477–90.

Kapitsa, A. P. (1996) Contradictions in the theory of ozone holes formation. *Optics of the Atmosphere and Ocean* **9**(9), 1164–7.

Kaplin, P. A. and Ignatov, E. I. (eds) (1997) *Geology of the Near-Caspian Area*, Vol. 1. Izd. MSU, Moscow, 190 pp. (in Russian).

Karl, T. R. and Knight, R. W. (1998) Secular trends of precipitation amount, frequency, and intensity in the United States. *Bull. Amer. Meteorol. Soc.* **79**(2), 231–41.

Karl, T. R., Knight, R. W., and Christy, J. R. (1994) Global and hemispheric temperature trends : uncertainties related to inadequate spatial sampling. *J. Climate* **7**(7), 1144–63.

Karl, T. R., Knight, R. W., Easterling, D. R., and Quayle, R. G. (1996) Indices of climate change for the United States. *Bull. Amer. Meteorol. Soc.* **77**, 279–92.

Karl, T. R., Nicholls, N., and Ghazi, A. (eds) (1999) *Weather and Climate Extremes – Changes, Variations and a Perspective from the Insurance Industry*. Kluwer Academic, Dordrecht, The Netherlands, 349 pp.

Karl, T. R. and Gleckler, P. J. (2001) Tracking changes in AMIP model performance. *Abstracts 8th Sci. Assembly of IAMAS, Innsbruck, 10–18 July 2001*, p. 8.

Kato, H., Nishizawa, K., Kadokura, S., Oshima, N., and Giorgi, F. (2001) Performance of Reg CM2.5/NCAR–CSM nested system for the simulation of climate change in East Asia caused by global warming. *J. Meteorol. Soc. Jap.* **79**, 99–121.

Katsambas, A., Varotsos, C. A., Vezirgianni, G., and Antoniou, Ch. (1997) Surface solar ultraviolet radiation: A theoretical approach of the SUVR reaching the ground in Athens, Greece. *Environ. Sci. Pollut. Res.* **4**(2), 69–73.

Kawa, S. R. and Anderson, D. E. (1999) Assessing the impact of aircraft emissions on the stratosphere. *SPARC Newsletter* **13**, 21–6.

Kaye, J. A (1998) Summary of Total Ozone Mapping Spectrometer (TOMS) science team meeting. *Earth Observer* **10**(3), 15–22.

Kaye, J. A. and Rind, D. (1998) Summary of chemistry/climate modeling meeting. *Earth Observer* **10**(3), 27–36.

Keeling, R. F., Piper, S. C., and Heinmann, M. (1996) Global and hemispheric CO_2 sinks deduced from changes in atmospheric CO_2 concentrations. *Nature* **381**, 218–21.

Keen, A. B. and Murphy, J. M. (1997) Influence of natural variability and the cold start problem on the simulated transient response to increasing CO_2. *Climate Dynamic* **13**(12), 847–64.

Kennedy, D. (2001) Going it alone. *Science* **293**, 1221.

Kerr, R. A. (2001a) World starts taming the greenhouse. *Science* **293**, 583.

Kerr, R. A. (2001b) Rising global temperature, rising uncertainty. *Science* **292**, 192–4.

Khalil, M. A. K. (1999) *Atmospheric Methane*. Springer, Berlin, 335 pp.

Khalil, M. A. K. and Rasmussen, R. A. (1997) Chlorine-containing gases in Antarctica. *Antarct. J. (US)* **32**(5), 187–9.

Kharin, N. G. and Kiriltseva, A. A. (1988) New data on the area of desertified lands in the arid zone of the USSR. *Problems of Deserts Development* **4**, 3–8 (in Russian).

Kheshgi, H. S., Schlesinger, M. E., and Lapenis, A. G. (1997) Comparison of paleotemperature reconstructions as evidence for the paleo-analog hypothesis. *Clim. Change* **35**(1), 123–9.

Khosravi, R., Brasseur, G. P., Smith, A. K., Rusch, D. W., Waters, J. W., and Russel, J. M. III (1998) Significant reduction in the stratospheric ozone deficit using a three-dimensional model constrained with UARS data. *J. Geophys. Res.* **103**(D13), 16203–20.

Kim, J. H. and Newchurch, M. J. (1998) Biomass-burning influence on tropospheric ozone over New Guinea and South America. *J. Geophys. Res.* **103**(D1), 1455–62.

Kim, W., Arai, T., Kanae, S., Oki, T., and Musiake, K. (2001) Application of the Simple Biosphere Model (SiB2) to a paddy field for a period of growing season in GAME – Tropics. *J. Meteorol. Soc. Jap.* **79**, 387–400.

King, M. D., Kaufman, Y. S., Tanré, D., and Nakajima, T. (1999) Remote sensing of tropospheric aerosols from space: past, present and future. *Bull. Amer. Meteorol. Soc.* **80**(11), 2229–59.

Kirk-Davidoff, D. B., Hintsa, E. J., Anderson, J. G., and Keith, D. W. (1999) The effect of climate change on ozone depletion through changes in stratospheric water vapour. *Nature* **402**, 399–401.

Knudsen, D. M., Larsen, N., Mikkelson, I. S., Morcette, J. J., Braathen, G. O., Kyrö, E., Fast, H., Gernandt, H., Kanazawa, H., Nakane, H., Dorenkov, V., Yushkov V., Hansen, G., Gil, M., and Shearman, R. J. (1998) Ozone depletion in and below the Arctic vortex for 1997. *Geophys. Res. Lett.* **25**(5), 627–30.

Koehler, J. and Hajost, S. A. (1990) The Montreal Protocol. A dynamic agreement for protecting the ozone layer. *Ambio* **19**(2), 82–6.

Komhyr, W. D., Grass, K. D., Evans, R. D., Leonard, R. K., Quincy, D. M., Hofmann, D. J., and Koenig, G. L. (1994) Unprecedented 1993 ozone increase over the United States from Dobson spectrophotometre observations. *Geophys. Res. Lett.* **21**(3), 201–4.

Kondratyev, K. Ya. (1988) *Climate Shocks: Natural and Anthropogenic*. John Wiley & Sons, New York, 296 pp.

Kondratyev, K. Ya. (1989) Global ozone dynamics. *Geomagnetism and Upper Atmospheric Layers*. VINITI, Moscow, 212 pp. (in Russian).

Kondratyev, K. Ya. (1990) *Key Issues of Global Ecology*, Vol. 9. VINITI, Moscow, 454 pp. (in Russian).

Kondratyev, K. Ya. (1992a) *Global Climate*. Nauka, St Petersburg, 359 pp. (in Russian).

Kondratyev, K. Ya. (1992b) Framework convention on climate change: Problems and perspectives. *Izv. Russian Acad. Sci. Ser. Geogr.* **2**, 52–63 (in Russian).

Kondratyev, K. Ya. (1993a) Letter to the editor. *Climatic Change* **24**, 383–4.

Kondratyev, K. Ya. (1993b) The second UN conference on the environment and development : some results and perspectives. *Izv. Russian Geogr. Soc.* **125**(3), 1–8 (in Russian).

Kondratyev, K. Ya. (1996) (ed.) *Ecodynamics and Ecological Monitoring of St. Petersburg Region in the Context of Global Changes*. Nauka, St Petersburg, 443 pp.

Kondratyev, K. Ya. (1997a) Global ecodynamics and sustainable development: Natural and scientific aspects and human dimension. *Izv. Russian Acad. Sci.* **129**(6), 1–12 (in Russian).

Kondratyev, K. Ya. (1997b) Key issues of global change at the end of the second millenium. *Energy Environ.* **8**, 279–89.

Kondratyev, K. Ya. (1997c) *Volcanoes and Climate*, WCP-54 (WMO/TD), No. 166. WMO, Geneva, 103 pp.

Kondratyev, K. Ya. (1997d) New stage of satellite remote sensing of the middle atmosphere. *Studing the Earth from Space* **2**, 127–34 (in Russian)

Kondratyev, K. Ya. (1998a) The outcome of the UN General Assembly Special Session, *Vestnik Russian Acad. Sci.* **68**, 30–40 (in Russian).

Kondratyev, K. Ya. (1998b) *Multidimensional Global Change*. Wiley/Praxis, Chichester, UK, 761 pp.

Kondratyev, K. Ya. (1998c) Ecodynamics and geopolicy: From global to local scale. Izv. *Russian Geogr. Soc.* **130**, 41–57 (in Russian).

Kondratyev, K. Ya. (1998d) Aerosol and climate: Some results and perspectives of remote sensing. 1. Multidimenstional climate change and diverse aerosol properties. *Ecol. Chemistry* **7**, 11–24.

Kondratyev, K. Ya. (1999a) *Climate Effects of Aerosols and Clouds*. Springer/Praxis, Chichester, UK, 264 pp.

Kondratyev, K. Ya. (1999b) *Ecodynamics and Geopolitics. 1. Global problems*. SRCES RAS, St Petersburg, 1040 pp. (in Russian).

Kondratyev, K. Ya. (2000a) Studying the Earth from space: Scientific plan of EOS system. *Studying the Earth from Space* **2**.

Kondratyev, K. Ya. (2000b) Global changes at the millenium threshold. *Vestnik Russian Acad. Sci.* **70**, 788–96 (in Russian).

Kondratyev, K. Ya. (2001a) Key issues of global change at the end of the second millenium. *Our Fragile World: Challenges and Opportunities for Sustainable Development*. EOLSS, Vorrunner, Vols 1–2.

Kondratyev, K. Ya. (2001b) Possible impacts of climate changes on ecosystems and economy of the USA. *Izv. Rus. Geogr. Soc.* **133**(6) (in Russian).

Kondratyev, K. Ya. (2002a) Aerosols as a climate-forming component of the atmosphere. 1. Chemical composition and optical properties. *Optics of the Atmosphere and Ocean* **15**(1) (in press).

Kondratyev, K. Ya. (2002b) Aerosols as a climate-forming component of the atmosphere. 2. Direct and indirect climatic impact. *Optics of the Atmosphere and Ocean* **15**(2) (in press).

Kondratyev, K. Ya. (2002c) Global Climate Change: A reality, suggestions, and illusions. *Studying the Earth from Space* **1**, 3–23 (in Russian).

Kondratyev, K. Ya. and Cracknell, A. P. (1998) *Observing Global Climate Change*. Taylor & Francis, London, 562 pp.

Kondratyev, K. Ya. and Cracknell, A. P. (2001) Global climate change. Socioeconomic aspects of the problem. In: A. P. Cracknell (ed.) *Remote Sensing and Climate Change. The Role of Earth Observations*. Springer/Praxis, Chichester, UK, pp. 37–79.

Kondratyev, K. Ya. and Demirchian, K. S. (2000) Global climate changes and carbon cycle. *Izv. Rus. Geogr. Soc.* **132**(4), 1–20 (in Russian).

Kondratyev, K. Ya. and Demirchian, K. S. (2001) Global climate change and Kyoto Protocol. *Vestnik Russian Acad. Sci.* **71**(11).

Kondratyev, K. Ya. and Galindo, I. (1997) *Volcanic Activity and Climate*. A. Deepak, Hampton, VA, 382 pp.

Kondratyev, K.Ya. and Grigoryev, Al. A. (2000) Natural and anthropogenic ecological disasters. 3. Meteorological disasters and catastrophes. *Studying the Earth from Space* **4**.

Kondratyev, K. Ya. and Moskalenko, N. I. (1983) Greenhouse effect of atmospheres of planets. *Outer Space Research*, Vol. 10. VINITI, Moscow, 156 pp.

Kondratyev, K. Ya. and Varotsos, C. A. (1995) Atmospheric ozone variability in the context of global change. *Int. J. Remote Sensing* **16**(10), 1851–81.

Kondratyev, K. Ya. and Varotsos, C. A. (1997) A review of greenhouse effect and ozone dynamics in Greece. In: C. Varotsos (ed.) *Atmospheric Ozone Dynamics*. Springer-Verlag, Berlin, pp. 175–228.

Kondratyev, K. Ya. and Varotsos, C. A. (1998) Energy production development and the environment. *Conf. on Climate Change in the Mediterranean, Metsovo, Greece*, pp. 340–82.

Kondratyev, K. Ya. and Varotsos, C. A. (1999) Total and tropospheric ozone changes: observations and numerical modeling. *Il Nuovo Cimento* **22C**(2), 219 46.

Kondratyev, K. Ya. and Varotsos, C. A. (2000a) *Atmospheric Ozone Variability: Implications for Climate Change, Human Health, and Ecosystems*. Springer-Praxis, Chichester, UK, 657 pp.

Kondratyev, K. Ya. and Varotsos, C. A. (2000b) Investigations of tropospheric chemistry in Europe. *Meteorol. Hydrol.* **10**, 105–112.

Kondratyev, K. Ya. and Varotsos, C. A. (2001a) Global tropospheric ozone dynamics. Part I: Tropospheric ozone precursors. *Environ. Sci. Pollut. Research* **8**(1), 57–62.

Kondratyev, K. Ya. and Varotsos, C. A. (2001b) Global tropospheric ozone dynamics. Part II: Numerical modeling of tropospheric ozone variability. *Environ. Sci. Pollut. Research* **8**(2), 113–19.

Kondratyev, K. Ya., Moskalenko, N. I., and Pozdnyakov, D. V. (1983a) *Atmospheric Aerosol*. Hydrometeoizdat, Leningrad, 224 pp.

Kondratyev, K. Ya., Grigoryev, Al. A., Pokrovsky, O. M., and Shalina, E. V. (1983b) *Satellite Remote Sensing of Atmospheric Aerosol*. Hydrometeoizdat, Leningrad, 218 pp.

Kondratyev, K. Ya., Buznikov, A. A., and Pokrovsky, O. M. (1996) *Global Change and Remote Sensing*. Wiley-Praxis, Chichester, UK, 396 pp.

Krajick, K. (2001) Tracking icebergs for clues to climate change. *Science* **292**, 2244–5.

Krapivin V. R., and Kondratyev, K. Ya. (2002). *Global Environmental Changes: Ecoinformatics*. St Petersbury State University, St Petersburg, 721 pp. (in Russian).

Krasilnikov, A. A., Kulikov, K. Yu., Mazur, A. B., Riskin, V. G., Serov, N. V., Fedoseyev, L. I., and Shvetsov, A. A. (1997) Detection of ozone clouds in the upper stratosphere of the Earth by milimeter radiometry method. *Geomagnetism Aeronomy* **37**(3), 174–83 (in Russian).

Kravtsova, V. I. (1993) Aerospace investigations of the littoral zone of Dagestan seashore of the Caspian Sea under elevation of its level. *Studying the Earth from Space* **5**, 96–101.

Kruchenitsky, G. M., Bekoryukov, V. I., Voloshchuk, V. M., Zvyagintsev, A. M., Kadigrov, N. E., Kadigrova, T. V., and Perov, S. P. (1996) On the contribution of dynamic processes to formation of anomally low values of total ozone content in the northern hemisphere. *Optics of the Atmosphere and Oceean* **9**(9), 1233–42.

Kukla, G. (2000) The last inteerglacial. *Science* **287**, 987–8.

Kukla G (2001). Did the last interglacial end with global warming? *First Int. Conf. on Global Warming and the Next Ice Age, Dalhousie University, 19–24 August 2001*. Vancouver, p. 257.

Kukla, G., McManus, J. F., Rousseau, D. D., and Chuine, I. (1997) How long and stable was the last interglacial? *Quaternary Sci. Rev.* **16**(6), 605–12.

Kulik, N. F., Petrov, V. I., and Gusinov, A. F. (1980) Studying deflation processes on sandy soils of near Caspian region from satellite imagies. *Problems of Desert Development* **4**, 95–6 (in Russian).

Kuzhelka, R., Seacrest, S., and Leonard, R. (2001) Climate change: Who is listening? Who is Planning? *First Int. Conf. on Global Warming and the Next Ice Age, Dalhousie University, 19–24 August 2001*. Vancouver, pp. 207–10.

Labitzke, K. G. and van Loon, H. (1999) *The Stratosphere*. Springer, Berlin, 179 pp.

Labitzke, K. G. (1997) Sunspots, the QBO, and the stratospheric temperature in the North Polar Region. *Geophys. Res. Lett.* **14**, 535–7.

La déforestation survey par satellite. *Science et vie* **912**, 29 (1992).

Landgraf, J. and Crutzen, P. S. (1998) An efficient method for online calculations of photolysis and heating rates. *J. Atmos. Sci.* **55**(5), 863–78.

Lanzerotti, L. J. (1999) Position statement adopted on climate and greenhouse gases. *Earth Observations from Space* **80**(5), 49.

Latif, M. (1997) Dynamics of interdecadal variability in coupled ocean–atmosphere models. *Int. WOCE Newsletter* **26**, 17–24.

Lawler, A. (1999) Terra launch spotlights NASA Observing System. *Science* **286**(5447), 2064–5.

Lawrence, M. G., Crutzen, P. J., and Rasch, P. J. (1999) Analysis of the CEPEX ozone data using a 3D chemistry–meteorology model. *Quart. J. Roy. Meteorol. Soc.* **125**(560), Part B, 2987–3010.

Lawrimore, J. H., Halpert, M. S., Bell, G. D., Menne, M. J., Lyon, B., Schnell, R. C., Gleason, K. L., Easterling, D. R., Thiaw, W., Wright, W. J., Heim, R. R. Jr, Robinson, D. A., and Alexander, L. (2001) Climate Assessment for 2000. *Bull. Amer. Meteorol. Soc.* **86**, S1–S55.

Lean, J. (1997) The Sun's variable radiation and its relevance to earth. *Earth. Annu. Rev. Astron. Astrophys.* **36**, 33–67.

Lee, S. H., Akimoto, H., Nakane, H., Kumoshenko, S., and Kinjo, Y. (1998) Lower tropospheric ozone trend observed in 1989–1997 at Okinawa, Japan. *Geophys. Res. Lett.* **25**(10), 1637–40.

Lempert, R. J., Schlesinger, M. E., Bankes, S. C., and Andronova, N. G. (2000) The impacts of climate variability on near-term policy and the value of information. *J. Clim. Change* **45**, 129–61.

Levitus, S., Antonov, J. I., Wang, J., Delworth, T. L., Dixon, K. W., and Broccoli, A. J. (2001) Anthropogenic warming of Earth's climate system. *Science* **292**, 267–70.

Levintanus, A. Yu. (1991) Lessons from Lake Chad. *Nature* **10**, 24–9.

Liang, L., Horowitz, L. W., Jacob, D. S., Wang, Y., Fiore, A. M., Logan, J. A., Gardner, G. M., and Mungred, M. (1998) Seasonal budgets of reactive nitrogen species and ozone over the United States, and export flixes to the global atmosphere. *J. Geophys. Res.* **103**(D11), 13435–50.

Lin, X., Trainer, M., and Hsie, E.Y. (1998) A modeling study of tropospheric species during the North Atlantic Regional Experiment (NARE). *J. Geophys. Res.* **103**(DII), 13593–613.

Lindzen, R. S. (1994) Climate dynamics and global change. *Ann. Rev Fluid Mech.* **26**, 353–78.

Lindzen, R. S. and Giannitsis, C. (1998) On the climatic imlications of volcanic cooling. *J. Geophys. Res.* **103**(D6), 5929–5942.

Liou, K. N. (1992) *Radiation and Cloud Processes in the Atmosphere. Theory, Observations and Modeling*. Oxford University Press, New York, 487 pp.

Logan, J. A. (1999a) An analysis of ozonesonde data for the troposphere: Recommendations for testing 3D models and development of a gridded climatology for tropospheric ozone. *J. Geophys. Res.* **104**(D13), 16115–49.

Logan, J. A. (1999b) An analysis of ozonesonde data for the lower stratosphere : Recommendations for testing models. *J. Geophys. Res.* **104**(D13), 16151–70.

Loginov, V. F. and Mikitskii, V. S. (2000) Assessment of the anthropogenic signal in city climate. *Izv. Rus. Geogr. Soc.* **132**(1), 23–31 (in Russian).

Lovelock, J. (1995) *The Ages of Gaia. A Biography of Our Living Earth.* W. W. Norton, New York, 255 pp.

Low, P. S. (1998) Tropospheric ozone: The missing GHG in the UNFCCC equation. *Linkages Journal* **3**(4), 32–4.

Lowe, J. J. (1998) What can we learn from the last episode of rapid global warming (at the Pleistocene–Holocene transition) about the mode of operation of global climate change? *The Globe* **41**, 4–6.

Lubin, D., Jensen, E. H., and Gies, H. P. (1998) Global surface ultraviolet radiation climatology from TOMS and ERBE data. *J. Geophys. Res.* **103**(D20), 26061–91.

Lucic, D., Harris, N. P. R., Pylc, J. A., and Jones, R. L. (1999) A technique for estimating polar ozone loss: Results for the northern 1991/92 winter using EASOE data. *J. Atmos. Chem.* **34**(3), 365–83.

Mabbut, A. (1981) Climatic periodicity and landscape variability as environmental factors in desertification. *Combating Desertification through Integrated Development.* Fan, Tashkent, pp. 21–23.

Macalady, D. L. (ed.) (1998) *Perspectives in Environmental Chemistry.* Oxford University Press, 512 pp.

McCauley, J. F., Breed, C. S., Groiler, M. J., and McKinnon, D. J. (1981) *The US Dust Storm of February 1977*, Special Paper No. 186. Geological Society of America, Boulder, CO, pp. 123–47.

McCormick, M. P. (1998) Summary of the SAGE III (Stratospheric Aerosol and Gas Experiment III) science team meeting. *Earth Observer* **10**(3), 7–14.

McCormick, M. P., Veiga, R. E., and Chu, W. P. (1992) Stratospheric ozone profile and total ozone trends deriver from the SAGE I and SAGE II data. *Geophys. Res. Lett.* **19**(3), 269–72.

McDermid, I. S., Bergwerff, J. B., Bodeker, G., Boyd, I. S., Brinksma, E. J., Connor, B. J., Farmer, R., Gross, M. R., Kimvilakani, P., Matthews, W. A., McGee, T. J., Ornel, F. T., Parrish, A., Sing, U., Swart, D. P. J., Tsou, J. J., Wang, P. H., and Zawodny J. (1998a) OPAL: Network for the Detection of Stratospheric Change. Ozone Profiler Assessment at Lauder, New Zealand, 1, Blind intercomparison. *J. Geophys. Res.* **103**(D22), 28683–92.

McDermid, I. S., Bergwerff, J. B., Bodeker, G., Boyd, I. S., Brinksma, E. J., Connor, B. J., Farmer, R., Gross, M. R., Kimvilakani, P., Matthews, W. A., McGee, T. J., Ornel, F. T., Parrish, A., Sing, U., Swart, D. P. J., Tsou, J. J., Wang, P. H., and Zawodny, J. (1998b) OPAL: Network for the Detection of Stratospheric Change. Ozone Profiler Assessment at Lauder. New Zealand, 2, Intercomparison of Revised Results. *J. Geophys. Res.* **103, No. D22, pp. 28693-28700.**

McElroy, C. N., McLinden, C. A., and McConnel, J. C. (1999) Evidence for bromine monoxide in the free troposphere during the Arctic polar sunrise. *Nature* **397**(6717), 338–41.

MacKenzie, I. A., Harwood, R. S., Scott, P. A., and Watson, G. C. (1999) Radiative-dynamic effects of the Antarctic ozone hole and chemical feedback. *Quart. J. Roy. Meteorol.* **125**(558), Part B, 2171–204.

McKenzie, C., Schiff, S., Aravena, R., Kelly, C.St., Lonis, V. (1998a) Effect of temperature on production of CH4 and CO_2 from peat in a natural and flooded boreal forest wetlands. *Clim. Change* **40**(2), 247–66.

McKenzie, R. L., Paulin, K. J., Madronich, S. (1998b) Effects of snow cover on UV irradiance and surface albedo: A case study. *J. Geophys. Res.* **103**(D22), 28785–92.

McKirtrick, R. (2002) Heteroscedasticity of air temperature time series. *Proc. Russian Geogr. Soc.* **134**(2) (in press).

McLeod, N. H. (1974a) *Use of ERTS imagery and other space data for rehabilitation and development programs in West Africa*, Preprint presented at the COSPAR Seminar on Space Applications of direct interest to developing countries. São José dos Campos, Brazil, 21 pp.

McLeod, N. H. (1974b) Remote sensing experiment in West Africa. *Proc. Third Symp. Earth Resources Technology Satellite, Vol. 1*, Washington, DC, pp. 247–66.

McNamara, W. (2001) *Background on Kyoto Protocol. The Week that Was – 14 July 2001*. Washington, DC, pp. 5–8.

Madronich, S. (1998) WMO ad hoc Scientific Steering Committee on UV monitoring: Activities and scientific issues. *SPARC Newsletter* **10**, 19–26.

Maduro, R.A. and Schauerhammer, R. (1992) *The Holes in the Ozone Scare. The Scientific Evidence that the Sky Isn't Falling*. 21st Century Science Associates, Washington, DC, 356 pp.

Maier, J. (2001) Climate: Last chance in Bonn? *Network 2002* **July 2001**, 7, 8.

Mailhot, J., Strapp, J. W., MacPherson, J. I., Benoit, R., Belair, S., Donaldson, N. R., Froude, F., Benjamin, M., Zawadzki, I., and Rogers, R. R. (1998) The Montreal '96 Experiment on Regional Mixing and Ozone (MERMOZ): An overview and some preliminary results. *Bull. Amer. Meteorol. Soc.* **79**(3), 439–42.

Mainguet, M., Canon-Cossus, L., and Chemin, M. C. (1979) Dégradation dans les régions centrales de la Republique de Niger. *Trav. Inst. Geogr. de Reims* **34–40**, 61–73.

Malingreu, J. P. (1988) Large-scale deforestation in the south-eastern Amazon basin of Brazil. *Ambio* **17**(1), 49–55.

Malysheva, N. V. and Sukhikh, V. I. (1991) *Forecast Mapping of Potential Entomopests in Forest Patology Monitoring System. Aerospace Forest Monitoring*. Nauka, Moscow, pp. 154–62 (in Russian).

Manabe, S. and Stouffer, R. J. (1997) Climate variability of a coupled ocean–atmosphere–land surface model: Implication for the detection of global warming, the Walter Orr Roberts Lecture. *Bull. Amer. Meteorol. Soc.* **78**(6), 1177–85.

Mann, M. E., Bradley, R. S., and Hughes, M. K. (1998) Global-scale temperature patterns and climate forcing over the past six centuries. *Nature* **392**(392), 779–87.

Mann, M. E., Bradley, R. S., and Hughes, M. K. (1999) Northern Hemisphere temperature during the past millenium: inferences, uncertainties, and limitations. *Geophys. Res. Lett.* **26**, 759–62.

Manual of Remote Sensing (1983) NASA, Washington, DC, 820 pp.

Marchuk, G. I., Kondratyev, K. Ya., Kozoderov, and V. V., Khvorostyanov, V. I. (1986) *Clouds and Climate*. Hydrometeoizdat, Leningrad, 512 pp.

Marchuk, G. I., Kondratyev, K. Ya., Aloyan, A. E., and Varotsos, C. A. (1999) Changes in the total stratospheric and tropospheric ozone amount: Observations and numerical modeling. *Studying the Earth from Space* **5**, 12–30.

Marland, G., Molnar, S., Sankowski, A., and Wisniewski, J. (eds) (1995) Greenhouse gas emissions and response in Central and Eastern Europe. *Időiárás* **95**, 139–452.

Martens, P. (1998) *Health and Climate. Modelling the Impacts of Global Warming and Ozone Depletion*. Earthscan, London, 224 pp.

Mauzerall, D. L., Logan, J. A., Jacob, D. S., Anderson, B. E., Blake, D. R., Bradshaw, J. D., Heikes, B., Sachse, G. W., and Talbot, B. (1998) Photochemistry in biomass burning plumes and implications for tropospheric ozone over the tropical Southern Atlantic. *J. Geophys. Res.* **103**(D7), 8401–24.

Maximov, E. V. (1992) *Historical Geography of Mountaineous Lakes of Middle Asia*. Izv. SPB University, St Petersburg, 304 pp. (in Russian).

Mayer, B., Fischer, C. A., and Madronich, S. (1998) Estimation of surface actinic flux from satellite (TOMS) ozone and cloud reflectivity measurements. *Geophys. Res. Lett.* **22**(5), 4321–4.

Meehl G.A. and Washington, W. M. (1996) El Niño-like climate change in a model with increased atmospheric CO_2 concentrations. *Nature* **382**, 56–60.

Meehl, G. A., Washington, W. M., Wigley, T. H. L., Arblaster, J. M., and Dai, A. (2001) Solar and anthropogenic forcing and climate response in the 20th century. *Abstracts 8th Sci. Assembly of IAMAS, Innsbruck, 10–18 July 2001*, 12 pp.

Meleshko, V. P., Katsov, V. M., Sporyshev, P. V., Vavulin, S. V., and Govorkova, V. R. (1999) Sensitivity of the climate model of the Main Geophysical Observatory to atmospheric CO_2 concentration changes. In: M. E. Berlyand and V. P. Meleshko (eds) *Present-day studies of the Main Geophysical Observatory*. Gidrometeoizdat, St Petersburg, pp. 3–32 (in Russian).

Michaels, P. J., Singer, S. F., Knappenberger, P. C., Kent, J. B., and McElroy, C. N. (1994) Analysing ultra-violet-B radiation: Is there a trend? *Science* **264**, 1341–3.

Michaels, P. J., Knappenberger, P. C., and Weber, G. R. (1996) Human effect on global climate? Comments on the paper by B. D. Santer *et al.* [Author's reply in Santer, B. D., Boyle, J. S., and Parker, D. E. (1996) *Nature (UK)* **384**(6609), 522–24.)

Michaels, P. J., Knappenberger, P. C., and Davis, R. E. (2001) Integrated projections of future warming base upon observed climate during the greenhouse enhancement. *First Int. Conf. on Global Warming and the Next Ice Age, Dalhousie University, 19–24 August 2001*. Vancouver, pp. 162–7.

Mitchell, J., Johns, T. C., Gregory, J. M., Eagles, M., Jenkins, G., Viner, D., Hulme, M., Jones, P., Nwe, M., and Barrow, E. (1997) *How Much More Will Climate Change in the Future?* The Met Office, Bracknell, UK, pp. 6–7.

Mo, T., Goldberg, M. D., Crosby, D. S., and Cheng, Z. (2001) Recalibration of the NOAA-microwave sounding unit. *J. Geogr. Res.* **106**, 10145–50.

Molemaker, M. J. and de Arellano, J. V. G. (1998) Control of chemical reactions by convective turbulence in the boundary layer. *J. Atmos. Sci.* **55**(4), 568–79.

Molina, M. J. and Rowland F. S. (1974) Stratospheric sink for chlorofluoromethanes – chlorine atom catalyzed destruction of ozone. *Nature*, **249**, 810.

Monod, A. and Carlier, P. (1999) Impact of clouds on the tropospheric ozone budget: Direct effect of multiphase photochemistry of soluble organic compound. *Atmos. Environ.* **33**, 4431–46.

Montzka, S. A., Butler, J. H., Elkins, J. W., Thompson, T. M., Clarke, A. D., and Lock, L. T. (1999) Present and future trends in the atmospheric burden of ozone-depleting halogens. *Nature* **398**(6729), 690–4.

Moody, J. L., Muller, J. W., Goldstein, A. H., Jacob, D. S., and Wofsy, S. C. (1998) Harvard forest regional-scale air mass composition by Patterns in Atmospheric Transport History (PATH) *J. Geophys. Res.* **103**(DII), 13181–94.

Moore, G. A., Roulet, N. T., and Waddington, J. M. (1998) Uncertainty in predicting the effect of climate change of the carbon cycling of Canadian peatlands. *Clim. Change* **40**(2), 229–45.

Mordvintsev, I. N. and Petrosyan, V. G. (1994) Application of satellite telemetry and geoinformation systems for the study of big mammals ecology. *Studying the Earth from Space* **2**, 119–24.

Morris, G. A., Kawa, S. R., Douglas, A. R., Schoeberg, M. R., Froidevaux, L., and Waters, J. (1998) Low-ozone pockets explained. *J. Geophys. Res.* **D3**, 3595–610.

Moulik, M. D. and Milford, J. B. (1999) Factors influencing ozone chemistry in subsonic aircraft plumes. *Atmos. Environ.* **33**(6), 869–80.

Muehlberger, W. R. and Wilmarth, U. R. (1977) The Shuttle era: A challenge to the Earth scientist. *American Sci.* **65**(2), 153–8.

Müller, M. J. W (1997) *Klimalüge? Wissenschaft–Politik–Zeitgeist.* Irene Müller Verlag, ENER, Munich, 256 pp.

Munro, R., Siddans, R., Reburn, W. S., and Kerridge, B. S. (1998) Direct measurement of tropospheric ozone distributions from space. *Nature* **65**(2), 168–71.

Nadelhoffer, K. L., Emmet, B. A., Gundersen, P., Kjonaas, O. J., Koopmans, C. J., Schleppi, P., Tietem, A., and Wright, R. F. (1999) Nitrogen deposition makes a minor contribution to carbon sequestration in temperate forests. *Nature* **398**(6723), 145–8.

Naden, P. S. and Watts, C. D. (2001) Estimating climate–induced change in soil moisture at the landscape scale: An application to five areas of ecological interest in the UK. *Clim. Change* **49**, 411–40.

Nair, H., Alien, M., Froidevaux, L., and Zurek, R. W. (1998) Localized rapid ozone loss in the northern winter stratosphere: An analysis of UARS observations. *J. Geophys. Res.* **103**(D1), 1555–71.

Nagurny, A. P. and Shirochkov, A. V. (1989) Variability of general ozone content in the northern near-polar space from the data of expedition on Siberia ice-breaker (May–June 1987). *Papers of Russian Acad. Sci.* **308**(5), 1099–1104 (in Russian).

Natural Calamities: Studies and Fighting Techniques (1978) (translated from English). Progress, Moscow, 440 pp.

Navarra, A. (eds) (1998) *Beyond El Niño. Decadal Variability in the Climate System.* Springer for Science, The Netherlands, 374 pp.

New, M., Hulme, M., and Jones, P. (1999) Representing twentieth-century space-time climate variability. Part 1: Development of a 1961–90 mean monthly terrestrial climatology. *J. Climate* **12**, 829–56.

Newell, R. E. and Weare, B. C. (1976a) Ocean temperature and large-scale atmospheric variations. *Nature* **262**, 40–1.

Newell, R. E. and Weare, B. C. (1976b) Factors governing tropospheric mean temperature.- *Science* **194**, 1413–14.

Newell, R. E., Browell, E. V., Davis, D. D., and Liu, Sh. E. (1997) Western Pacific tropospheric ozone and potential vorticity: Implications for Asian pollution. *Geophys. Res. Lett.* **24**(22), 2733–6.

Nicholls, N. (2001) The insignificance of significance testing. *Bull. Amer. Meteorol. Soc.* **82**, 981–6.

Nikonov, A. A. (1999) Environmental impacts of earthquakes. *Russian Acad. Sci. Vestnik* **69**(12), 1107–11 (in Russian).

North, G. R. and Stevens, M. (1998) Detecting climate signals in the surface temperature record. *J. Climate* **11**(4), 563–77.

Obasi, G. O. P. (2000) World Meteorological Day 2000. WMO – 50 Years of Service. *WMO Bull.* **49**(1), 3–7.

Oldfield, F. (1998) Past global changes and their significance for the future. *The Globe* **41**, 1, 3, 4.

Oldfield, F. (2000) Out of Africa. *Nature* **403**(6768), 370–1.

Olszyna, K. J., Parkhurst, W. J., and Meagher, J. F. (1998) Air chemistry during the 1995 SOS/Nashville intensive determined from level 2 network. *J. Geophys. Res.* **103**(D23), pp. 31143–31154.

Oltmans, S. J., Lebone, A. S., Scheel, H. E., Harris, J. M., Levy, H., Galbally, I. E., Brunke, E. G., Meyer, C. P., Lathorp, J. A., Johnson, B. S., Shadwick, D.,S., Cuevas, E., Schmidlin, F. J., Tarasick, D. W., Claude, H., Kerr, S. B., Uchino, O., and Mohnen, V. (1998) Trends of ozone in the troposphere. *Geophys. Res. Lett.* **25**(2), 143–6.

O'Neill, B. C., MacKellar, F. L., and Lutz, W. (2001) *Population and Climate Change.* Cambridge University Press.

Oppo, D. W., McManus, J. F., amd Cullen, J. L. (1998) Abrupt climate events 500,000 to 340,000 years ago: Evidence from subpolar North Atlantic sediments. *Science* **279**, 1335–8.

O'Riordan, T. and Jager, J. (eds) (1996) *Politics of Climate Change. A European Perspective.* Routlege, London, 396 pp.

Otterman, J. (1971) Monitoring surface albedo change with Landsat. *Geophys. Res. Lett.* **9**(September), 13–19.

Our Changing Planet/The FY 1998 US Global Change Research Program An investment in Science for the Nation's Future. A Supplement to the President's Fiscal Year 1998 Budget. Washington, DC, 118 pp.

Ourgensen, A. F. and Nuhr, H. (1996) The use of satellite images for mapping of landscape and biological diversity in the Sahel. *Int. J. Remote Sensing* **17**(1), 91–109.

Overpeck, J. T. (1996) Warm climate surprisees. *Science* **271**, 1820–1.

Overpeck, J. T. (1989) *Ozone Depletion, Greenhouse Gases, and Climate Chnage.* National Academy Press, Washington, DC, 122 pp.

Ozone Depletion, Greenhouse Gases, and Climate Change (1989) National Academy Press, Washington, DC, 122 pp.

Pachauri, R. K. (1998) *The Geopolitics of Climate Change.* Presented at Nehru Cultural Center, Embassy of India. Moscow, 23 April 1998, 22 pp.

Parikh, J. K. (1998) The Emperor needs new clothes: Long-range energy use scenarios by ILASA-WEC and OPCC. *Energy* **23**, 69–70.

Parry, M. and Carter, T. (1997) *Climate Impact and Adaptation Assessment. The IPCC Method.* Earthscan, London, 192 pp.

Parsons, M. (1995) *Global Warming: The Truth Behind the Myth.* Plenum Press, New York, 271 pp.

Paul, Ch. K. (1980) Satellites and world food resources. *Space World* **Q-10-202**(October), 111–19.

Pengracz, R. and Bartholy, J. (2000) Statistical linkage between ENSO, NAO and regional climate. *Idöiárás* **104**, 1–20.

Penkett, S. A., Plane, J. M. C., Comes, F. J., and Clemitshaw, K. C. (1999) The Weyborne Atmospheric Observatory. *J. Atmos. Chem.* **33**(2), 107–10.

Penner, J. E. and Rotstayan, L. (2000) Indirect aerosol forcing. Response by T. J. Crowley. *Science* **290**, 407 pp.

Penner, J. E., Lister, D. H., Griggs, D. J., Dokken, D. J., and McFarland, M. (eds) (1999) *Aviation and the Global Atmosphere*. IPCC Special Report Nos 61 and II. Cambridge University Press, 373 pp.

Peters, D. and Entzian, G. (1999) Longitude-dependent decadal changes of total ozone in boreal winter months during 1979–92. *J. Climate* **12**(44), 1038–48.

Peters, A. J. and Eve, M. D. (1995) Satellite monitoring of desert plant community response to moisture availability. *Int. Symp. on Desertificatification in Developed Countries, Tucson, AZ, 24–29 October 1993. Environ. Monit. Assess.* **37**(1–3), 273–87.

Pielke, R. A. Sr, Eastman, J., Chase, T. N., Knaff, J., and Kittel, T. G. (1998) Trends 1973–1998 in the depth-averaged tropospheric temperature. *J. Geophys. Res.* **103**(D14), 16927–33.

Pielke, R. A. Sr, Liston, G. E., and Robock, A. (2000) Insolation-weighted assessment of Northern Hemisphere snow-cover and sea-ice variability. *Geophys. Res. Lett.* **27**, 3061–4.

Pierce, R. B., Fairlie, T. D., Remsberg, E. E., Russel, J. M., and Grose, W. L. (1997) HALOE observations of the Arctic vortex during the 1997 spring: Horizontal structure in the lower stratosphere. *Geophys. Res. Lett.* **24**(22), 2705–8.

Pinty, B., Szejwach, G., and Desbois, M. (1984) Investigation surface albedo variations over West Africa from Meteostat. *Conf. Satell. Remote Sensing and Applications, Clearwater Beach, FL, 25–28 June 1984. Boston, MA*, pp. 76–9.

Pittock, A. B. (1999) The question of significance. *Nature* **397**, 657–8.

Plumb, I. C. and Ryan, K. R. (1998) Effect of aircraft on ultraviolet radiation reaching the ground. *J. Geophys. Res.* **103**(D23), 31231–40.

Ponater, M., Sausen, R., Feneberg, B., and Roeckner, E. (1999) Climate effect of ozone changes caused by present and future air traffic. *Climate Dynamics* **15**(9), 631–42.

Pouyaud, P. and Colombani, J. (1989) Les variations extrêmes du lac Tchad? L'assèchement est-il possible? *Annales de Geographie* **545**, 1–23.

Prabhakara, C., Yoo, J. M. Y., Maloney, S. P., Nuccarone, J. J., and Arking, A. (1996) Examination of Global atmospheric temperature monitoring with satellite microwave measurement: 2. Analysis of satellite data. *Clim. Change* **33**(4), 1927–30.

Prather, M. S. (1998) Time scales in atmospheric chemistry: Coupled perturbations to N_2O, NO and O_x. *Science* **279**(5355), 1339–42.

Prather, M., Midgley, P., Rowland, S. R., and Stolarski, R. (1996) The ozone layer: The road not taken. *Nature*, **381**(6583), 551–4.

Prinn, R., Jacoby, H., Sokolov, A., Wang, C., Xiao, X., Yang, Z., Eckhaus, R., Stone, P., Ellerman, D., Melillo, J., Fitzmaurice, J., Kicklinhter, D., Holian, G., and Liu, Y. (1999) Integrated Global System Model for climate policy assessment: Feedbacks and sensitivity studies. *Clim. Change* **41**, 469–546.

Prinn, R. G. (ed.) (1994) *Global Atmospheric–Biospheric Chemistry*. Plenum Press, New York, 261 pp.

Project Performance Report. Global Environmental Facility (1998) Washington, DC, 89 pp.

Pszenny A. and Brasseur G. (1997) Tropospheric ozone: An emphasis on IGAC Research. *Global Change Newsletter* **30**, 2–10.

Pyle, S. A. and Harris, N. R. D. (1991) *Polar Stratospheric Ozone*. Air Pollution Research Report 34. Commission of the European Communities, Brussels, 306 pp.

Qian, W. and Zhu, Y. (2001) Climate chnage in China from 1880 to 1998 and its impact on the environment condition. *Climate Change* **50**, 419–44.

Quayle, R. G., Peterson, T. C., Basist, A. N., and Godfrey, C. S. (1998) The NCDS Global Temperature Index. *Earth System Monitor* **8**(3), 6–8.

Rafikov, A. A. and Tetyukhin, G. F. (1981) *Aral Sea Level Decline and Changes in Natural Conditions of Amu-Darya Lowlands*. Fan,Tashkent, 200 pp. (in Russian).

Ramade, F. (1987) *Les catastrophes écologiques*. McGraw-Hill, Paris, 318 pp.

Ramanathan, V. and Vogelman, A. M. (1997) Atmospheric greenhouse effect, excess solar absorption and the radiation budget: From the Arrhenius/Langley era to the 1990s. *Ambio* **26**, 38–46.

Randall, D., Curry, J., Battisti, D., Flato, G., Grumbine, R., Hakkinen, S., Martinson, D., Preller, R., Walsh, J., and Weatherly, J. (1998) Status of and outlook for large-scale modeling of atmosphere–ice-ocean interactions in the Arctic. *Bull. Amer. Meteorol. Soc.* **79**(2), 197–219.

Randel, W. J. and Wu, F. (1999) Cooling in the Arctic and Antarctic polar stratospheres due to ozone depletion. *J. Climate* **12**(5), Part 2, 1467–79.

Rao, U. R. (1996) *Space Technology for Sustainable Development*. Tata McGraw-Hill, New Delhi, 564 pp.

Rapp, A. (1974) A review of desertification in Africa. *Water, Vegetation and Man*. University of Lund, Stockholm, 77 pp.

Rapp, A. and Helden, U. (1979) *Research on Environmental Monitoring Methods for Land Use Planning in African Drylands*. University of Lund, Sweden, 116 pp.

Ratcliffe, R. A. S. (1995) Back to basics: Is our climate changing? *Weather* **50**(2), 54.

Raven, P. H. (1987) *The Global Ecosystem in Crisis*. A MacArthur Foundation Occasional Paper, Chicago, 24 pp.

Ravetta, F., Ancelett, G., Kowol-Anten, J., Wilson, R., and Nedeljkovic, D. (1999) Ozone, temperature, and wind field measurements in a tropopause fold: Comparison with a mesoscale model simulation. *Mon. Weather. Res.* **127**(II), 2641–53.

Ravishankara, A. R., Hancock, G., Kawasaki, M., and Matusumi, Y. (1998) Photochemistry of ozone surprises and recent lessons. *Science* **280**, 60–1.

Reconciling Observations of Global Temperature Change (2000) National Academy Press, Washington, DC, 85 pp.

Rees, M., Condit, R., Crawley, M., Pacala, S., and Tilman, D. (2001) Long-term studies of vegetation dynamics. *Science* **293**, 650 pp.

Reilly, J., Stone, P. H., Forest, C. E., Webster, M. D., Jacoby, H. D., and Prinn, R. G. (2001) Uncertainty and climate change assessments. *Science* **2983**, 430–3.

Renu, J. (2001) Intercomparison of stationary waves in AMIP–2 GCM's and their maintenance mechanisms. *Abstracts 8th Sci. Assembly by IAMAS. Innsbruck, 10–18 July 2001*, p. 8.

Rex, M., von der Gathen, P., Harris, N. R. P., Lucic, D., Knudsen, B. M., Braathen, G. O., Reid, S. J., De Backer, H., Claude, H., Fabian, R., Fast, H., Gil, M., Kyr, M., Mikkelsen, I. S., Rummikainen, M., Smit, H., Staehelin, J., Varotsos, C., and Zaitsev, I. (1998) In situ measurements of stratospheric ozone depletion rate in the Arctic winter 1991/92: A Lagrangrian approach. *J. Geophys. Res.* **103**(5), 5843–53.

Richter, A., Wittrock, F., Eisinger, M., and Burrows, J. P. (1998) GOME observations of tropospheric BrO in Northern Hemispheric spring and summer 1997. *Geophys. Res. Lett.* **25**(14), 2683–6.

Ridley, B. A., Walega, S. G., Lamarque, J. F., Grahek, F. E., Trainer, M., Hubler, G., Lin, X., and Fehsenfeld, F. C. (1998) Measurements of reactive nitrogen and ozone to 5-km altitude in June 1990 over the southeastern United States. *J. Geophys. Res.* **103**(3), 8365–88.

Riebsame, W. E. (1991) Drought: Opportunities for impact mitigation. *Episodes* **14**(1), 62–4.

Rind, D. (1998) Latitudinal temperature gradients and climate change. *J. Geophys. Res.* **103**(D6), 5943–71.

Rind, D. and Overpeck, J. (1994) Hypothesized causes of modeled decade to century scale variability. *Quart. Sci. Rev.* **12**, 357–74.

Rinsland, C. P., Salawich, R. J., Gunson, M. R., Solomon, S., Zander, R., Mahien, E., Goldman, A., Newchurch, M. J., Irion, E. W., and Cheng, A. Y. (1999) Polar stratospheric descent of NOY and CO and Arctic denitrification during winter 1992–1993. *J. Geophys. Res.* **104**(D1), 1847–61.

Ritson, D. M. (2000) Gearing up for IPCC-2001. *Clim. Change* **45**, 471–88.

Roberts, J. M., Bertman, S. B., Parrish, D. D., Fehsenfeld, F. C., Jobson, B. T., and Niki, H. (1998) Measurement of alkyl nitrates at Cheboque Point, Nova Scotia during the 1993 North Atantic Regional Experiment (NARE) intensive. *J. Geophys. Res.* **103**(D11), 13569–80.

Robock, A., Vinnikov, K. Y., Srinivasan, G., Entin, J. K., Hollinger, S. E., Speranskaya, N. A., Liu, S., and Namkhai, A. (2000). The global soil moisture data bank. *Bull. Amer. Meteorol. Soc.* **81**, 1281–99.

Röckmann, T., Brenninkmeijer, C. A. M., Neeb, P., and Crutzen, P. J. (1998) Ozonolysis of nonmethane hydrocarbons as a source of mass independent oxygen isotope enrichment in tropospheric, IO. *J. Geophys. Res.* **103**(D1), 1463–70.

Roemer, M. (1999) There are no trends like ozone trends. In: P. M. Borell and P. Borell (eds) *Proc. of EUROTRAC Symposium 98*. WIT Press, Southampton, UK, pp. 208–13.

Rose, A. (1998) Global warming policy: Who decides what is fair? *Energy Policy* **28**, 1–3.

Rosinski, J. and Kerrigan, T. C. (2001) The role of extraterrestrial particles in the formation of the ozone hole. *Il Nuovo Cimento* **24C**(6), Ser. 2, 815–42.

Rozanov, E., Zubov, V., Schlesinger, M., Yang, F., and Andronova, N. (1999) Three-dimensional simulations of ozone in the stratosphere and comparison with UARS data. *Phys. Chem. Earth (C)* **24**(5), 459–63.

Russel, J. M. III and Lastvicka, J. (1998) *The IAGA/ICMA Symposium Session 2.18 on Middle Atmosphere Trends and Low Frequency Variability, Uppsala, Sweden, 14 August 1997. SPARC Newsletter* **10**, 25–6.

Rychagov, G. I. (1996) Ecological aspects of unstable Caspian Sea level. *Arid Ecosystems* **2**(2), 74–81.

Sabziparvar, A. A., Shine, K. P., and Foster, P. M. de F. (1999) A model-derived global climatology of UV irradiation of the earth's surface. *Photochem. Photobiol.* **69**(2), 193–202.

Sadov, A. V. (1981) Dynamics of ecosystems of Aral basin from materials of space photography. *Studying the Earth from Space* **4**, 18–26.

Sahm, P. and Moussiopoulos, N. (1999) The OFIS model: A new approach in urban scale photochemical modelling. *EUROTRAC Newsletter* **21**, 22–8.

Salardino, D. H. and Caroll, J. J. (1998) Correlation between ozone exposure and visible foliar injury in Ponderosa and Jeffney pines. *Atmos. Environ.* **32**(17), 3001–10.

Salawitch, R. J. (1998) Ozone depletion: A greenhouse warming connection. *Nature* **392**(6676), 551–2.

Saloranta, T. M. (2001) Post-normal science and the global climate change issue. *Clim. Change* **50**, 395–404.

Santer, B. D. (1996) On the detection of climate change. *Exchanges* **1**(1), 4–6.

Santer, B. D., Wigley, T. M. L., Boyle, J. S., Gaffen, D. J., Hnilo, J. J., Nychka, D., Parker, D. E., and Taylor, K. E. (2000) Statistical significance of trends and trend differences in layer-average atmospheric temperature time series. *J. Geophys. Res.* **105**, 7337–56.

Sarma, K. M. (1998) Protection of the ozone layer: A success story of UNEP. *Linkages J.* **3**(3), 6–10.

Sasano, Y., Bodeker, G., and Kreher, K. (1998) Preliminary ILAS observations of O_3, HNO, and N_2O in the Arctic stratosphere during the winter of 1996/97. *SPARC Newsletter* **10**, pp. 18–19.

Satellites will be urban planning tools (1972) *Eng. News Rec.* **188**(20), 43–51.

Saunders, M. A. and Harris, A. R. (1997) Statistical evidence links exceptional 1995 Atlantic hurricane season to record sea warming. *Geophys. Res. Lett.* **24**, 1255–8.

Scheider, S. R., McGinis, D. F., and Stephens, G. (1985) Monitoring Africa's Lake Chad basin with Landsat and NOAA satellite data. *Int. J. Remote Sensing* **6**(1), 59–73.

Scherman, R. (2001) Let the sun rise on the Kyoto Protocol. *New York 2002* **September 2001 issue**, 1–2.

Schimel, D. S. (1999) The carbon equation. *Nature* **393**(6682), 208–9.

Schindler, D. W. (1999) The mysterious missing sink. *Nature* **398**(6723), 105, 107.

Schlesinger, M. E. and Andronova, N. (2000) Temperature changes during the 19th and 20th centuries. *Geophys. Res. Lett.* **27**, 2137–40.

Schlesinger, M. E. and Andronova N. (2001) Climate sensitivity. *J. Climate* (in press).

Schlesinger, M. E. and Ramankutty, N. (1994) An oscillation in the global climate system of period 65–70 years. *Nature* **367**, 723–6.

Schlesinger, M. E., Ramankutty, N., and Andronova, N. (2000) Temperature oscillations in the North Atlantic. *Science* **289**, 547.

Schneider, S. H. (1998) Kyoto Protocol: The unfinished agenda. An editorial essay. *Clim. Change* **39**(1), 1–21.

Schneider, S. H. (2001) What is dangerous climate change? *Nature* **411**, 17–19.

Schneider, S. H. and Mesirow, L. E. (1997) *The Genesis Strategy*. Climate and Global Survival, New York, 419 pp.

Scholes, B. (1999) Will the terrestrial carbon sink saturate soon? *Global Change Newsletter* **37**, 2–3.

Scholes, D. and Tucker, C. G. (1994) *Decrease in theArea of Tropical Forests and Increase in the Area of Settlements in the Amazon Basin*, Satellite data from 1978 up to 1988, Methods of local, regional and global monitoring. Nauka, Moscow, pp. 41–4.

Sedjo, R. A. (1995) Forests. Conflicting signals. In: R. Bailey (ed.) *The State of the Planet*. Free Press, New York, pp. 177–209.

Schröder, W. (ed.) (2000) *Long and Short Term Variability in Sun's History and Global Change*, Science Edition, Bremen-Roennebeck, Germany, 63 pp.

Schrope, M. (2001) Consensus science, or consensus politics? *Nature* **412**, 112–14.

Schultz, M. G., Jacob, D. J., Wang, Y., Logan, J. A., Atlas, E. L., Blake, D. R., Blake, N. J., Bradshaw, J. D., Browell, E. V., Fenn, M. A., Flocke, F., Gregory, G. L., Heikes, B. G., Sachse, G. W., Sandholm, S. T., Shetter, R. E., Singh, H. B., and Talbot, R. W. (1999) On the origin of tropospheric ozone and NOX over the tropical South Pacific. *J. Geophys. Res.* **104**(D5), 5829–43.

Seinfeld, J. H. and Pandis, S. N. (1998) *Atmospheric Chemistry and Physics*. John Wiley & Sons, New York, 1326 pp.

Sen, O. L., Bastidas, L. A., Shuttleworth, W. J., Yang, Z. L., Gupta, H. V., and Sorooshian, S. (2001) Impact of field-calibrated vegetation parameters on GCM climate simulations. *Quart. J. Roy. Meteorol. Soc.* **127**, Part B, 1199–223.

Senior, C. A. and Mitchell, J. B. (2000) The time dependence of climate sensitivity. *Geophys. Res. Lett.* **27**, 2685–8.

Sevastyanov, D. V. (ed.) (1991) *History of lakes Sevan, Issyk-kul, Balkhash, Zaisan, and Aral.* Nauka, Leningrad, 302 pp.

Severinghaus, J. P., Sowers, T., Brook, E. J., Ally, R. B., and Bender, M. L. (1998) Timing of abrupt climate change at the end of the Younger Dryas interval from thermally fractionated gases in polar ice. *Nature* **391**(6663), 141–6.

Shackley, S., Young, P., Parkinson, S., and Wynne, B. (1998) Uncertainty, complexity and concepts of good science in climate change modelling: Are DCMs the best tools? *Clim. Change* **38**, 159–205.

Shanklin, J. (1998) *The Antarctic Ozone Hole.* NERC-British Antarctic Survey, London, 10 pp.

Sharonenko, S. J. (1998) Crisis of Aral sea: The 20th century. *Ecology* **4**, 11–16 (in Russian).

Shindell, D. T., Rind, D., and Lonergan, P. (1998) Increased polar stratospheric ozone losses and delyed eventual recovery owing to increasing greenhouse concentration. *Nature* **392**, 589–92.

Singer, S. F. (1997) *Hot Talk, Cold Science: Global Warming's Unfinished Debate.* Independent Institute, Oakland, CA, 120 pp.

Singer, S. F. (1998) Unfinished business: The scientific case against the Global Climate Treaty. *Energy Environment* **9**, 617–32.

Singer, S. F. (1999) Human contribution to climate remains questionable. *Earth Observed from Space* **80**, 183, 186, 187.

Sinha, A. and Harris, J. (1995) Water vapour and greenhouse trapping: The role of far infrared absorption. *Geophys. Res. Lett.* **22**(16), 2147–50.

Sinnhuber, B.-M., Müller, R., Langer, J., Bovensmann, H., Eyring, V., Klem, U., Trentmann, J., Burrows, J. P., and Kunzi, K. F. (1999) Interpretation of mid-stratospheric Arctic ozone measurements using photochemical box-model. *J. Atmos. Chem.* **34**(3), 281–90.

Sklyarov, Yu. A., Dvinskikh, V. A., Brichkov, Yu. I., and Kotuma, A. I. (1998) Solar variability and its terrestrial manifestations. *Studying the Earth from Space* **6**.

Slingo, A., Webb, M. J. (1997).The spectral signature of global warming. *Quart. J. Roy. Meteorol. Soc.* **123**(538), Part B, 293–307.

Smith, K. (1997) *Environmental Hazards. Assessing Risk and Reducing Disasters.* Longman London, 389 pp.

Smith, S. J., Wigley, T. M. L., and Edmonds, J. (2000) A new route toward limiting climate change? *Science* **290**, 1109–10.

Smith, W. L. and Stamnes, K., (eds) (1997) *IRS '96: Current Problems in Atmospheric Radiation. Proc. of the Int. Radiation Symp. Fairbanks, Alaska, 19–24 August 1996.* A. Deepak Publ., Hampton, VA., 1067 pp.

Sokolov, A. P. and Stone, P. H. (1998) A flexible climate model for ise in integrated assessment. *Clim. Dyn.* **14**, 291–303.

Somerville, R. C. J. (2001). Forecasting climate change: Prospects for improving models. *Abstracts. 8th Sci. Assembly of IAMAS, Innsbruck, 10–18 July 2001*, p. 9.

Sonechkin, D. M. (1998) Climate dynamics as a nonlinear Brownian motion. *Int. J. Bifurcation and Chaos* **8**, 799–803.

Sonechkin, D. M. and Datsenko, N. M. (2000) Wavelet analysis of nonstationary and chaotic time series with an application to the climate change problem. *Pure Appl. Geophys.* **157**, 653–77.

Sonechkin, D. M., Datsenko, N. M., and Ivashchenko, N. N (1997) Estimates of the global warming trend by means of wavelet analysis. *Izv. Rus. Acad. Sci. Physics Atmos. Ocean* **33**(2), 184–94.

Sonechkin, D. M., Astafyeva, M. M., Datsenko, N. M., Ivashchenko, N. N., and Jakubiak, B. (1999) Multiscale oscillations of the global climate system as revealed by wavelet transform of observational data time series. *Theoret. Appl. Climatol.* **64**, 131–42.

Soon, W., Baliunas, S., Kondratyev, K. Ya., Idso, S. B., and Postmentier, E. (2000) Calculating the climatic impacts of increased CO_2: The issue of model validation. *Proc. of the 1st Solar and Space Weather Euroconference – The Solar Cycle and Terrestrial Climate, Santa Cruz de Tenerife, 25–29 September*. European Space Agency, Noordwijk, The Netherlands, ESA SP No. 463), pp. 243–54.

Soon, W., Baliunas, S., Demirchian, K. S., Kondratyev, K. Ya., Idso, S. B., and Postmentier, E. S. (2001). Impact of anthropogenic CO_2 emissions on climate: unsolved problems. *Izv. Rus. Geogr. Soc.* **133**(2), 1–19 (in Russian)

Sorokhtin, D. M. and Ushakov, S. A. (1998) Accumulation of CO_2 in the atmosphere: Harmful or useful? *Gas Industry* **6**, 44–8 (in Russian)

Sorokhtin, D. M. and Ushakov, S. A. (1999a) Impact of the ocean on atmospheric composition and climate of the Earth. *Oceanology* **6**, 928–37 (in Russian).

Sorokhtin, D. M. and Ushakov, S. A. (1999b) Greenhouse effect and global evolution of the Earth's climates. *Izv. Sect. Earth Sciences Rus. Acad. Natur. Sci.* **3**, 82–101 (in Russian).

Soros, M. S. (2000) *Preserving the Atmosphere as a Global Common*, Environmental Change and Security Project Report No. 6. The Woodrow Wilson Center, Washington, DC, pp. 149–55.

Spänkuch, D., Döhler, W., Gülldner, J., and Schulz, E. (1998) Estimation of the amount of tropospheric ozone in a cloudy sky by ground-based Fourier-transform infrared emission spectroscopy. *Appl. Optics.* **37**(15), 3133–42.

Spencer, R. W. (1998) The state of climate change science. *Global Warming, The Continuing Debate.* ESEF, Cambridge, pp. 23–8.

Stafford, J. M., Wendler, G., and Curtis, J. (2000) Temperature and precipitation of Alaska: 50 year trend analysis. *Theor. Appl. Clim.* **67**, 33–44.

Sterin, A. M. (1999) Analysis of linear trends in temperature series of the free atmosphere for 1958–1997. *Meteorol. Hydrol.* **5**, 52–68.

Stone, T. A. and Woodwell, G. M. (1985) Analysis of deforestation in Amazonia using Shuttle–imaging radar. *Int. Geoscience Remote Sensing Symp. (IGARSS-85), Amherst, MA*, Vol. 2, 574 pp.

Stone, T. A., Schlesinger, P., Houghton, A. *et al.* (1994) A map of the vegetation of South America based on satellite imagery. *Photogramm. Eng. Remote Sensing* **60**(5), 541–51.

Stott, P. (2001) External control of 20th century temperature by natural and anthropogenic forcing. *Abstracts 8th Sci. Assembly of IAMAS, Innsbruck, 10–18 July 2001*, p. 12.

Stott, P. A., Tett, S. F., Jones, G. S., Allen, M. R., Ingram, W. J., and Mitchell, J. F. (2001) Attribution of twentieth century temperature change to natural and anthropogenic causes. *Clim. Dyn.* **17**, 1–21.

Stratospheric Processes and Their Role in Climate (1998) Implementation Plan WCRP-105, WMO/TD No. 914. World Meteorological Organization, Geneva, 131 pp.

Stuhler, H. and Frohn, A. (1998) The production of NO by hypersonic flight. *Atmos. Environ.* **32**(18), 3153–5.

Stockwell, D. Z. and Chipperfield, M. P. (1999) A tropospheric chemical transport model: Development and validation of the model transport schemes. *Quart. J. Roy. Meteorol. Soc.* **125**(557), Part A., 1747–83.

Sugihara, S., Ishiyama, T., and Naya, M. (1994) Desertification monitoring based on the satellite data. *PIKEN Rev.* **5**, 15–16.

Takigawa, M., Takahashi, M., and Akiyoshi, H. (1999) Simulation of ozone and other chemical species using a Center for Climate System Research/National Institute for Environmental Studies atmospheric GCM with coupled stratospheric chemistry. *J. Geophys. Res.* **104**(D11), 14003–18.

Taupin, F. G., Bessafi, M., Baldy, S., and Bremaud, P. J. (1999) Tropospheric ozone above the southwestern Indian Ocean is strongly linked to dynamical conditions prevalling in the tropics. *J. Geophys. Res.* **104**(D7), 8057–66.

Taylor, P. K. (ed.) (2000) *Intercomparison and Validation of Ocean-Atmosphere Energy Flux Fields*, World Climate Research Programme No. 1036. World Meteorological Organization, Geneva, 303 pp.

Tett, S. F. B., Johns, T. C., and Mitchell, J. F. B. (1997a) Global and regional variability in a coupled AOGCM. *Climate Dynamics* **13**(5), 303–25.

Tett, S., Stott, P., Mitchell, J., Johns, T., Ingram, W., Sexton, D., Rayner, N., Folland, C., Parker, D., Gordon, M., Cullum, D., Horton, B., Lavery, J., O'Donnell, M., and Jenkins G. (1997b) Is humankind already changing global climate? *Climate Change and Its Impacts: A Global Research Programme.* Department of the Environment, Transport and the Regions/The Met. Office, Bracknell, UK, pp. 4–5.

The Atmospheric Sciences Entering the Twenty-First Century (1998) National Academy of Sciences, Washington, DC, 364 pp.

The Earth Charter (1998) Zeleny Mir (The Green World), Moscow, 24 pp. (in Russian)

The Greenhouse Effect and Climate Change. A Briefing from the Hadley Center (1999) The Met. Office, Bracknell, UK, 28 pp.

The Kyoto Protocol: A milestone on the approach to sustainability (1998) *WMO Bull.* **47**, 161–3.

The Kyoto Protocol to the Convention on Climate Change (1998) Climate Change Secretariat. Bonn, 34 pp.

The Study of Forests by Aerospace Methods (1978) Nauka, Novosibirsk (in Russian).

The World Bank and the Environment (1991) A Progress Report, Fiscal Year 1991. The World Bank, Washington, DC, 131 pp.

The World Bank Energy and Environment Strategy Paper (1998) The World Bank, Washington, DC, 59 pp.

The World Meteorological Organization in the Service of Humankind: A vision for the 21st Century (2000) *WMO Bull.* **49**(1), 13–16.

Tol, R. S. J. (2000) International climate change policy: An assessment. *IHDP Update* **3**, 11–12.

Tolba, M. (1978) Desertification–problem common to all mankind. *Problems of Desert Development* **3**, 7–11 (in Russian).

Tourpali, K., Tie, X. X., Zerefos, C. S., and Brasseur, G. (1997) Decadal evolution of total ozone decline: Observations and model results. *J. Geophys. Res.* **102**(D20), 23955–62.

Transboundary Photooxidation Air Pollution in Europe: Calculations of Tropospheric Ozone and Comparison with Observations (1998) The Norwegian Meteorological Institute Research Report No. 67/EMEP/MSC-W Report 2/98, Oslo, Norway, 74 pp. (Appendices).

Trenberth, K. E. (2001) Climate variability and global warming. *Science* **293**, 48–9.

Trenberth, K. E., Haar, T. J. (1997) El Niño and climate change. *Geophys. Res. Lett.* **24**, 3057–60.

Tropospheric Chemistry and Space Observations (1998) N. Chaumerliac, G. Megie, N. Papinea, and G. Angeletti (eds), Air Pollution Research Report No. 65, V.C. Directorate-General Science, Research and Developoment. EOR 17790, Luxembourg, 277 pp.

Tuck, A., Watson, R., and Toon, B. (eds) (1989) The Airborne Antarctic Ozone Experiment (AAOE). *J. Geophys. Res.* **94**(D9), 11179–737; (D14), 16434–858.

Tucker, C. J. and Choudhury, B. J. (1987) Satellite remote sensing of draught conditions. *Remote Sensing Environ.***23**(243).

Udelhofen, P. M., Gies, P., Randel, W. J. (1999) Surface UV radiation over Australia, 1979–1992: Effects of ozone and cloud cover changes on variations of UV Radiation. *J. Geophys. Res.* **104**(D16), 19135–59.

UNEP Scientific Assessment of Ozone Depletion (1994) WMO Global Ozone Research and Monitoring Project Report No. 37. World Meteorological Organization, Geneva.

UNEP (1991) *Environmental Data Report*, 3rd edn. Basil Blackwell, London, 408 pp.

Unni, N. V. M., Roy, R. S., Jadhav, R. W. *et al.* (1991) IRS-1A applications in forestry. *Current Sci.* **61**(3–4), 189–92.

US National Academy of Sciences Report (2001) *Climate Change Science: An Analysis of Some Key Issues*. US National Academy of Sciences, Washington, DC, 28 pp.

van der Sluijs, J.P. (1998) Anchoring amid uncertainty: On the managing of uncertainties in risk assessment of anthropogenic climate change. *Linkages J.* **3**(2), 2–6.

Varotsos, C. A. (ed.) (1997) *Atmospheric Ozone Dynamics. Observations in the Mediterranean Region*, NATO ASI Series. Springer-Verlag, Berlin, 336 pp.

Varotsos, C. A. and Cracknell, A. P. (1998) Total ozone depletion over Greece as deduced from satellite observations. *Int. J. Remote Sensing* **19**(17), 3317–26.

Varotsos, C. A. and Kondratyev, K. Ya. (1995) Ozone dynamics over Greece as derived from satellite and in situ experiments. *Int. J. Remote Sens.* **16**(10), 1777–98.

Varotsos, C. A. and Kondratyev, K. Ya. (2000a) On the seasonal variation of the surface ozone in Athens, Greece. *Atmos. Environ.* (in press).

Varotsos, C. A. and Kondratyev, K. Ya. (2000b) A new evidence for ozone depletion over Athens, Greece. *Int. J. Remote Sensing* (in press).

Varotsos, C. A., Helmis, C., and Cartalis C. (1992) Annual and semiannual waves in ozone as derived from SBUV vertical global ozone profiles. *Geophys. Res. Lett.* **19**(9), 1777–98.

Varotsos, C. A., Chronopoulos, G. J., Cracknell, A. P., Jonson, B. E., Katsambas, A., and Philippou, A. (1998a) Total ozone and solar ultraviolet radiarion, as derived from satellite and ground-based instrumentation at Dundee, Scotland. *Int. Remote Sens.* **19**(17), 3301–6.

Varotsos, C. A., Ghosh, S. S., Chronopoulos, G. J., Katsikis, S. C., Cracknell, and A. P. (1998b) Total ozone measurements over Athens: Intercomparison between Dobson, TOMS (version 6) and SBUV measurements. *Int. J. Remote Sensing* **19**(17), 3327–34.

Varotsos, C. A., Kondratyev, K. Ya., Alexandris, D., and Chronopoulos, G. J. (2000a) Aircraft observations of UVR irvadiance vertical gradient. *Proc. Int. Workshop on UVR Exposure, Measurement, and Protection, 18–20 October 1999, St Catherine College, Oxford.*

Varotsos, C. A., Kondratyev, K. Ya., Katsambas, A., Feretis, H., and Efstathio, M. (2000b) On the risk of eye and skin deseases resulting from SUVR exposure of humans. *Proc. Int. Workshop on UVR Exposure, Measurement, and Protection, 18–20 October 1999, St Catherine College, Oxford.*

Vay, S. A., Anderson, B. E., Sachse, G. W., Collins, J. E., Podolske, J. R., Twohy, C. H., Gandrud, B., Chan, K. R., Baughcum, S. L., and Wallio, H. A. (1998) DC-8-based observations of aircraft CO, CH_4, N_2O and $H_2O(g)$ emission indices during SUCCESS. *Geophys. Res. Lett.* **25**(10), 1717–20.

Verschuren, D., Laird, F. R., and Cumming, B. F. (2000) Rainfall and draught in equatorial east Africa during the past 1,100 years. *Nature* **403**(6768), 410–14.

Victor, D. G. (2001) *The Collapse of the Kyoto Protocol and the Struggle to Slow Global Warming*. Princeton University Press, Princeton, 192 pp.

Victorov, A. S. and Plotnikov, Y. A. (1995) Application of satellite information for investigation of the landscape changes in the Caspian Sea region. *Abstracts Global Changes and Geography: IGU Conf., Moscow, 14–18 August 1995*, Moscow, 371 pp.

Viner, D. and Hulme, M. (1997) *The Climate Impact LINK Project: Applying Results from the Hadley Center's Climate Change Experiments for Climate Change Impact Assessment*. Climatic Research Unit, UEA, Norwich, UK, 17 pp.

Vinnikov, K. Y., Robock, A., Stouffer, R. J., Walsh, J. E., Parkinson, C. L., Cavalieri, D. S., Mitchell, J. F. B., Garrett, D., and Zacharov, V. F. (1999) Global warming and Northern Hemisphere sea ice extent. *Science* **286**, 1934–7.

Vinogradov, B. V. (1980) Indicators for desertification and their aerospace monitoring. *Problems of Deserts Development* **1**, 17–24 (in Russian).

Vinogradov, B. V., Cherkashin, A. K., Gornov, A. Yu., and Kulik, K. N. (1990) Dunamic monitoring of degradation and rehabilitation of pastures in black lands of Kalmykia. *Problems of Deserts Development* **1**, 10–14 (in Russian).

Visvader, H. and Burton, Ya. (1978) *Natural Calamities and Measures to Control Them in Canada and the USA. Natural Calamities: The study and Control Methods*, translated from English. Progress, Moscow, pp. 301–22 (in Russian).

Volkovitsky, O. A., Gavrilov, A. A., and Kaidalov, O. V. (1966) Is man really destroying the ozone layer of the planet? *Vestnik Rus. Acad. Sci.* **66**(9), 783–6.

von Storch, H. and Floser, G. (eds) (1999) *Anthropogenic Climate Change*. Springer, Berlin, 351 pp.

von Storch, H. and Zwiers, F. W. (1999) *Statistical Analysis in Climate Research*. Cambridge University Press, 484 pp.

Waddington, J. M., Griffis, T. J., and Rouse, W. R. (1998) Northern Canadian wetlands: Net ecosystem CO_2 exchange and climate change. *Clim. Change* **40**(2), 267–75.

Waibel, A. E., Peter, T., Carslaw, K. S., Oelhaf, H., Etzel, G., Crutzen, P. J., Pöschl U., Tsias, A., Reimer, E., and Fisher, H. (1999) Arctic ozone loss due to denitrification. *Science* **283**, 2064–8.

Wallace, J. M., Zhang, Y., and Renwick, J. A. (1995) Dynamic contribution to hemispheric mean temperature trends. *Science* **270**, 780–3.

Wang, Ch., Prinn, R. G., and Sokolov, A. A. (1998c) A global interactive chemistry and climate model: Formulation and testing. *J. Geophys. Res.* **103**(D3), 3399–417.

Wang, Y. and Jacob, D. J. (1998) Anthropogenic forcing on tropospheric ozone and OH since preindustrial times. *J. Geophys. Res.* **103**(D23), 31123–35.

Wang, H. S., Cunnold, D. M., and Bao, X. A. (1996) A critical analysis of Stratospheric Aerosol and Gas Experiment ozone trends. *J. Geophys. Res.* **101**(D7), 12495–514.

Wang, Y., Jacob, D. J., and Logan, J. A. (1998a) Global simulation of tropospheric O_3–NO_2–hydrocarbon chemistry. 1. Model formulation. *J. Geophys. Res.* **103**(D9), 10713–25.

Wang, Y., Logan, J. A., and Jacob, D. J. (1998b) Global simulation of tropospheric O_3–NO_2–hydrocarbon chemistry. 2. Model formulation. *J. Geophys. Res.* **103**(D9), 10727–55.

Wang, W. C. and Isaksen, I. S. A. (eds) (1995) *Atmospheric Ozone as a Climate Gas: General Circulation Model Simulations*. Springer-Verlag, Berlin, 459 pp.

Wardle, D. I., Kerr, J. B., McElroy, C. T., and Francis, D. R. (eds) (1997) *Ozone Science: A Canadian Perspective on the Changing Ozone Layer*. University of Toronto Press, 119 pp.

Washington, W. M. and Meehl, G. A. (1996) High-altitude climate change in a global coupled ocean-atmosphere-sea-ice model with increased atmospheric CO_2. *J. Geophys. Res.* **101**(D8), 12795–801.

Watanabe, M. (2000) *Mechanisms of the Decadal Climate Variability in the Midlatitude Atmosphere–Ocean System*. Center for Climate System Research, University of Tokyo, No. 12, 157 pp.

Weatherhead, E. C. (2001) Detecting climate change. *First Int. Conf. on Global Warming and the Next Ice Age, Dalhousie University, 19–24 August 2001*. Vancouver, pp. 168–71.

Weber, G. D. (1997) Spatial and temporal variations of 300 hPa temperatures in the Northern Hemisphere between 1966 and 1993. *Int. J. Climatology* **17**, 1–15.

Weisenstein, D. K., Ko, M. K. W., Dyominov, I. G., Pitari, G., Riccardulli, L., Visconti, G., and Bekki, S. (1998) The effect of sulfur emissions from HSCT aircraft: A 2-D model intercomparison. *J. Geophys. Res.* **103**(D1), 1527–47.

Wessel, S., Aoki, S., Winkler, P., Weller, R., Herber, R., Gernandt, H., and Schrems, O. (1998) Tropospheric ozone depletion in polar regions: A comparison of observations in the Arctic and Antarctic. *Tellus* **50B**, 34–50.

Wetheralds, R. T., Stouffer, R. J., and Dixon, K. W. (2001) Commited warming and its implications for climate change. *Geophys. Res. Lett.* **28**, 1535–8.

Wellington, G. M., Glynn, P. W., Strong, A. E., Navarette, A., Wieters, E., and Hubbard, D. (2001) Crisis on coral reefs linked to climate change. Earth Observed from Space **82**(1), 1, 5.

Weyant, J. (ed.) (1999) The costs of the Kyoto Protocol: A multi-model evaluation. *Energy Journal* **Special Issue**, 448.

Wigley, T. M. L. (1998) The Kyoto Protocol: CO_2, CH_4 and climate implications. *Geophys. Res. Lett.* **25**, 2285–8.

Wigley, T. M. L. (1999) *The Science of Climate Change. Global and US Perspectives*. Pew Center on Global Climate Change. Arlington, VA, 48 pp.

Wigley, T. M. L. and Raper, S. C. B. (2001) Interpretation of high projections for global-mean warming. *Science* **293**, 451–5.

Wigley, T. M. L, Jones, P. D., and Raper, S. C. B. (1997) The observed global warming record: What does it tell us? *Proc. Natl Acad. Sci. USA*. 94.

Wigley, T. M. L, Smith, R. L., and Santer, B. D. (1998) Anthropogenic influence on the autocorrelation structure of hemispheric-mean temperature. *Science* **282**(5394), 1676–9.

Wiin-Nielsen, A. A. (1997) A note on hemispheric and global temperature changes. *Atmosfera* **10**(3), 125–36.

Wiin-Nielsen, A. A. (1998) Limited predictability and the greenhouse effect. A scientific review. *Energy Environment* **9**(6), 633–46.

Wilby R. L. (2001) Cold comfort. *Weather* **56**, 213–15.

Wilkinson, G. G., Ward, N. R., Milford J.,R., and Dugdol, G. (1983) Remote Sensing Rangeland Monitor. and Manag. *Proc. 9th Ann. Conf. Remote Sensing Society, Silsoe, 21–23 September 1983*, Reading, pp. 94–107.

Williams, V. and Toumi, R. (1999) The radiative-dynamical effects of high cloud on global ozone distributio. *J. Atmos. Chem.* **9**(1), 1–22.

Wojick, D. E. (2001) *The UN IPCC Artful Bias*, Glaring omissions, false confidences and misleading statistics in the Summary for Policymakers, July 2001, 11 pp.

WMO (2000) Programmes in 2000 and beyond. *WMO Bull.* **49**(1), 22–50.

WMO (1999) *Statement on the Status of the Global Climate in 1998*, WMO Publ. No. 896. World Meteorological Organization, Geneva, 11 pp.

WMO/UNEP (1999) *Scientific Assessment of Ozone Depletion*. World Meteorological Organization, Geneva.

Woodcock, A. (1999) Global warming: A natural event? *Weather* **54**, 162–3.

Woodcock, A. (2000) Global warming: The debate heats up. *Weather* **55**, 143–4.

UNEP (1992) *World Atlas of Desertification*. UNEP, Nairobi, 92 pp.

World Resources 1996–1997 (1996) *A Guide to the Global Environment*. Oxford University Press, 365 pp.

Wu, Z. X. and Newell, R. E. (1998) Influence of sea surface temperature on air temperatures in the tropics. *Climate Dynamics* **14**(4), 275–90.

Wuebbles, D. J. and Penner, J. E. (1988) *Sensitivity of Urban/Regional Chemistry to Climate Change*, Report of the Workshop, Lawrence Livermore Nat. Lab. Preprint. UCRL, Livermore, CA, 17 pp.

Xinhua, X., Dahui, W., Hong, J., and Haixiang, S. (1999) Study on greenhouse gas emissions in Jangsu province. *Water Air Soil Pollut.* **109**(1–4), 293–301.

Yates, S. R., Wang, D., Gan, J., Ernst, F. F., and Yury, W. A. (1998) Minimizing methyl bromoide emissions from soil fumigation. *Geophys. Res. Lett.* **25**(10), 1633–6.

Zaletayev, V. S. (1996) Ecological nature of desertification as a phenomenon of natural environment destabilization. *Arid Ecosystem* **3**(6–7), 29–34 (in Russian).

Zaletayev, V. S. (1997) World network of water and land ecotons, its function in the biosphere and role in global changes. *Ecotons in the Biosphere*. Russian Academy Agriculture Science, Moscow, pp. 77–89.

Zanis, P., Schuepbach, E., Scheel, H. E., Baudenbacker, M., and Buchmann, B. (1999) Inhomogenities and trends in the surface ozone record (1988–1996) at Jungfraujoch in Swiss Alps. *Atmos. Environ.* **33**(23), 3777–86.

Zerefos, C. and Ghazi, A. (eds) (1985) Atmospheric ozone. *Proc. Quadriannual Ozone Symp., Halkidiki, Greece, 3–7 September 1984*. D. Reidel, Dordrecht, The Netherlands, 842 pp.

Zerefos, C., Meleti, C., Balis, D., Tourpali, K., and Bais, A. F. (1998). Quasibiennial and longer-term changes in clear-sky UV-B solar irradiance. *Geophys. Res. Lett.* **25**(15), 4345–8.

Zhang, H., Wallace, J. M., and Battisti, D. C. (1997) ENSO like interdecadal variability: 1990–93. *J. Climate* **10**, 1020–40.

Zhang, H., Henderson-Sellers, A., and McGuffe, K. (2001) The compunding effects of troppical deforestation and greenhouse warming of climate. *Clim. Change* **49**, 309–38.

Ziemke, J. R., Chandra, S., and Bhartia, P. K. (1998a) The new methods for deriving tropospheric column ozone from TOMS measurements: Assimilated UARS MLS/HALOE and convective-cloud differential technique. *J. Geophys. Res.* **103**(D17), 22115–27.

Ziemke, J. R., Herman, J. R., Stanford, J. L., and Bhartia, P. K. (1998b) Total ozone/UVB monitoring and forecasting: Impact of clouds and horizontal resolution of satellite retrievals. *J. Geophys. Res.* **103**(D4), 3865–71.

Zillman, J. W. (1997a) A critical review of the Intergovernmental Panel of Climate Change (IPCC) Second Assessmenrt. *National Academies Forum, Summary of Proceedings, Australians and Our Changing Climate*. IPCC, Canberra, pp. 10–21.

Zillman, J. W. (1997b) Atmospheric science and public policy. *Science* **276**, 1084–6.

Zillman, J. W. (2000) The challenges ahead. *WMO Bull.* **49**(1), 8–13.

Ziomas, I. C., Gruning, S.-E., and Bornstein, R. D. (1998a) The Mediterranean Campaign of Photochemical Tracers–Transport and Chemical Evolution (MEDCAPHOT–TRACE). *Atmos. Environ.* **32**(12), 2043–326.

Ziomas, I. C., Tzoumaka, P., Balis, D., Melas, D., Zerefos, C. S., and Klemm, O. (1998b) Ozone episodes in Athens, Greece. A modelling approach using data from the MEDCAPHOT–TRACE. *Atmos. Environ.* **32**(12), 2313–21.

Zonn, I. S. (1986) *Black Soils of Kalmyk Republic*, Development of arid territories and desertification fighting: Comprehensive approach. Nauka, Moscow, pp. 130–3 (in Russian).

Zonn, I. S. (1977) UN Conference in Nairobi: Desertification problems after 20 years. *Arid ecosystems* **3**(6–7), 10–19 (in Russian).

Zuev, V. V. (1998) Grond-level ozone layer behavior: Possible explanations. *Optics Atmos. Ocean* **11**(12), 1356–7.

Zuev, V. V., Marichev, V. N., and Khryapov, P. A. (1999) Pecularities of stratospheric ozone vertical distribution over Tomsk. *Optics Atmos. Ocean* **12**(7), 632–4.

6

Perspectives of environmental monitoring

There is no doubt that special satellite monitoring of disasters is essential. However, plans for developing satellite remote sensing at an internationally agreed level have not yet been concluded. Such problems should be tackled under a general programme of environmental monitoring. Since they have been discussed in detail in monographs in Kondratyev (1998c, 1999a, 1999b) and in Kondratyev and Varotsos (2000), here we shall briefly examine only the results of developing and implementing the international EOS programme of observations of the Earth from space. The goal of EOS is to complete a complex study of the planet's environmental and natural resources (Kondratyev, 2000; Greenstone and King, 1999).

By 1991, under the aegis of NASA, a full-scale EOS programme had been prepared. It was planned to come on stream within 5 to 7 years (Greenstone and King, 1999) but the first EOS satellite, *Terra*, was actually launched on 19 December 1999. Its mission, its relationship to the problems of global climate change and the development of its instrumentation have been discussed in Marchuk *et al.* (1986); King *et al.* (1990); Houghton *et al.* (1996); Kondratyev (1992, 1993, 1998a, 1998b, 1998c, 1999b, 2000a, 2000b); Nikonov (1999); and Kondratyev and Varotsos (2000).

6.1 GENERAL GOALS

The EOS programme was described in *The State of Science in the EOS Program* (Greenstone and King, 1999). In the introduction there is a historical outline of the development of environmental and natural-resources remote sensing observations, especially those of importance identified in the World Climate Research Programme (WCRP) and the International Geosphere–Biosphere Programme (IGBP) (see Houghton *et al.*, 1996; Kondratyev, 1982, 1990, 1999b). From 1985 onwards, the National Research Council in the USA published a series of reports (*Our Changing*

Planet, 1998), which included substantiation of the strategy for observational systems. These reports were prepared in the context of prioritizing environmental change studies and have been widely discussed in many publications. For example: Kondratyev and Galindo (1987); Kondratyev (1988); Kondratyev (1990, 1992, 1993, 1998a, 1998b, 1998c, 1999b); Kondratyev and Varotsos (2000). In 1988, the Earth System Science Committee was set up on the initiative of NASA. It called for a programme of long-term observation of the planet's environmental characteristics by means of a global system of satellite meteorological observations (Kondratyev, 1982, 1983, 1990; Choudhury, 1990) and was the starting point for such observations from space. It should be emphasized that complex observations are necessary for solving key environmental problems and substantiating models of the Earth's systems, as well as developing the information systems (such as data archives) that make the information accessible.

EOS is now a part of the US Global Climate Research Program (USGCRP), the four key components of which are the study of:

1 Climate variability on a seasonal-to-interannual timescale.
2 Climate variability on a decadal-to-centenary scale.
3 Changes in total ozone, ultraviolet radiation and chemical processes in the atmosphere.
4 Changes in land surface properties and the dynamics of continental and marine ecosystems.

In addition to these four components, due regard for the 'human dimensions' of global change, the interaction between nature and society – that is, interactive anthropogenic impacts on the environment – is an important goal of the USGCRP.

Unfortunately this latter aspect has not been included in *The State of Science in the EOS Program*. Monitoring of the values and processes mentioned below is of key importance in implementing the EOS programme; the description of relevant instruments for remote sensing and explanation of the abbreviations can be found in Kondratyev (1990, 1992, 1999b).

Atmosphere

Cloud cover characteristics (amount, top and bottom height, optical properties); radiation budgets of the Earth and the surface; precipitation; chemical composition of the troposphere (ozone and its gaseous precursors); chemical composition of the stratosphere (O_3, ClO, BrO, OH and other minor gas constituents); the properties of tropospheric and stratospheric aerosols; meteorological parameters (temperature and humidity; lightning (spatial-temporal variability, the structure of flashes).

Extra-atmospheric solar radiation

The total and spectral (especially in the UV spectrum region) solar 'constant' (SC).

Land

Changes in land surface and land use; vegetation cover dynamics; surface temperature, forest and other fires, volcanic eruptions; surface soil moisture.

Ocean

Sea surface temperature; content of phytoplankton and dissolved organic matter; wind field at the sea surface, oceanic surface topography (waves, ocean level).

Cryosphere

Land ice cover (glacier dynamics); sea ice (extension, water equivalent).

As for the atmospheric part, the EOS data should contain information on processes from cloud systems to regional and global spatial scales, and the range of temporal resolution should cover the intervals from those shorter than 24 hours to 100 years. Looking at the climate change problem, particular emphasis should be placed on studying cloud–climate feedbacks, as well as aerosol–cloud–climate interactions.

Data analysis on GHGs and chemical processes in the atmosphere would open the possibilities for more adequate understanding of processes of tropospheric ozone formation with consideration paid to, in particular, biomass burning and processes on the land surface. The stratospheric data would enable better understanding of the processes of ozone layer formation and ozone impact on climate. Monitoring of volcanic eruption consequences (mainly, stratospheric aerosol changes) is of importance for climate studies.

Observations of the oceans are targeted to obtain data on physical and biological processes in the World Ocean, including general circulation, bioproductivity and their role in gas exchange between the ocean and the atmosphere (this particularly refers to carbon dioxide).

Observations of the dynamics of land ecosystems and hydrological processes would open perspectives for obtaining unique global information on vegetation cover and its characteristics and, hence, the dynamics of land biosphere. New glaciological information would play an important role in solving 'cryosphere and climate' problems, as well as in the understanding of ocean level variability. The implementation of EOS on the whole would make a significant contribution to the developments under WCRP and IGBP.

The strategy of observations involves a necessity for:

- Simultaneous complex observations (preferably with the aid of one satellite or through spatial–temporal coordination of different observations).
- Data overlapping (in time) for successively used instruments with the aim of substantiating intercalibration and the homogeneity of long-term observation series.

- Frequently repeated observations, to observe rapidly changing natural systems (such as cloud cover).
- High-quality calibration of instruments.

Of course, the above list is far from being complete but, more importantly, it is not oriented at the problems on the basis of adequately based priorities – indeed, it could be provided, if previous experience was taken into account (see Kondratyev, 1990, 1998a, 1999b).

It was mentioned earlier (Kondratyev, 1998c, 1999b) that programme developments in the field of global change carried out in the USA are somehow centred on the problems of global climate changes. It is very strange that the conceptual part of *The State of Science in the EOS Program* does not even mention such important problems – especially as they are intensively discussed elsewhere in scientific literature identifying natural and anthropogenically-induced climate changes; and, in this connection, identification of 'anthropogenic signals' in climate change. It would be an important exercise to conduct, particularly against the background of the Kyoto Protocol (1997) recommendations on reducing the emissions of greenhouse gases into the atmosphere, a substantiation of the conclusion (see Houghton *et al.*, 1996) that the climatic 'anthropogenic signal' already shows up in observations. A very important paper on these problems was published by Barnett *et al.* (1999); they underlined the high degree of uncertainty of the results obtained. And, in *The State of Science in the EOS Program*, there is no mention of the important problem of solar activity effect on climate, which is dealt with in a voluminous, though contradictory, literature. Nor is there, unfortunately, any mention of the results of Russian developments devoted to substantiating the priorities of global change; the authors of *The State of Science in the EOS Program* did not even mention them. The Russian developments show clearly that the main priority, as compared with global-scale processes, is the problem of closed global biochemical cycles and in this connection a programme for studying global changes should be based on the concept of environmental biotic regulation (Gorshkov, 1995; Kondratyev, 1990, 1999b). Such an approach makes it clear that the EOS approach to the problem of global climate change (though very important) cannot be considered as the first priority.

One of the important aspects of global change ignored by *The State of Science in the EOS Program* is a range of real problems on pollution of the environment (atmosphere, land, water basins) and, in this context, the transboundary transport of pollutants. It is of particular importance to study the consequences of economic decline in the former USSR countries and in some other countries in terms of their impact on the environment. For example, the repercussions of the Chernobyl accident still deserve consideration. A further serious omission is that the conceptual part of *The State of Science in the EOS Program* does not consider the problems of natural and anthropogenic disasters – earthquakes, floods, tropical storms, tsunami, and so on.

Here we look at separate sections of the *The State of Science in the EOS Program.*

For our own convenience, we abbreviate this publication to the 'Plan' in what follows.

6.2 RADIATION, CLOUDS, WATER VAPOUR, PRECIPITATION AND ATMOSPHERIC CIRCULATION

The first section of the 'Plan' is focused on climatic problems and mainly deals with the problems of the Earth's radiation balance (ERB) (Kondratyev, 1982, 1983, 1990), beginning with measurements of extra-atmospheric solar radiation – the 'solar constant' (SC). Long-term satellite SC monitoring was started in late 1978 under a programme of ERB observations, using a cavity pyrheliometer (ACRIM) mounted on board the *UARS* satellite. Measurements of SC from the Shuttle spacecraft have also made a significant contribution. According to ACRIM data, the SC values varied from 1,367 W/m^2 in 1986 (during the minimum of the solar activity 11-year cycle) to 1,369 W/m^2 in 1992 (the solar cycle maximum). Such SC variations can be an essential factor of global climate changes on timescales from decades to centuries. However, it is not doubted that the problem of solar activity effect on climate is not limited to the impact of SC variations (Kondratyev, 1999a).

Three problems are most urgent in the context of the climate-forming role of radiation processes in the atmosphere: cloudiness effect (cloud–radiation feedbacks), atmospheric greenhouse impact, and atmospheric aerosol influence. The authors of the 'Plan' state that 'clouds are the second after greenhouse gases in terms of their effect on climate'. There is no doubt now that the dynamics of cloud cover and cloud–radiation feedbacks play the principal role in this respect (Kondratyev, 1982, 1992, 1999; Marchuk *et al.*, 1986). This is illustrated, for example, by the well-known fact that the level of uncertainties in existing assessments of cloud radiative forcing (CRF) considerably exceeds the absolute values of CRF due to greenhouse gases (greenhouse effect) and other factors.

That is why the EOS programme involves various long-term observations of CRF characteristics, cloud amount and aerosol using multi-channel scanning radiometers CERES, MISR and EOSP. (Descriptions of a set of the EOS satellites instruments and explanations of abbreviations can be found in Kondratyev, 1982, 1992, 1999, among others.) Further progress in the development of methods is important for retrieving the components of surface radiation budget (SRB) from satellite observations (Kondratyev, 1982, 1992, 1999a). There are two independent approaches to retrieve SRB components:

1. Using the data on outgoing shortwave and longwave radiation (OSR, OLR) in combination with the results of remote sensing of the atmosphere and estimation of cloud-cover characteristics for calculating the SRB components.
2. Using the empirical correlation between simultaneously measured values of OSR, OLR and SRB.

In the context of the 'Plan', the absence of a special observation programme which would enable the 'filtering out' of the anthropogenic (greenhouse, aerosol) component of climate change (Kondratyev, 1998a, 1999a, 1999b) is perplexing in view of the high priority of relevant problems.

The data on cloud-cover dynamics are of importance not only in terms of its effect on the radiation regime, but also with respect to its influence on the convective transport of energy and water vapour. The interaction between clouds and cloud-free atmosphere is a key factor determining the water content of a cloud-free atmosphere and precipitation intensity. The convection process depending on cloud conditions contributes substantially to the formation of heat exchange between the surface and the atmosphere. The instrumentation on EOS satellites (radiometers AIRS, AMSU, HIRDLS, HSB, MLS, SAGE III) would allow acquisition of sufficiently complete information on cloud-cover characteristics and atmospheric water vapour content, which (in combination with wind field data) would give a more adequate understanding of the processes of atmospheric water balance formation.

The data on precipitation and its dynamics occupy a special place since latent heat plays a key role in forming atmospheric general circulation. The solution of the problem of obtaining sufficiently complete information on precipitation becomes seriously complicated due to the high space–time changeability of precipitation. This can be on a scale from hundreds of metres to thousands of kilometres and from minutes to several years. The complicating factor is the inadequacy of numerical models that can predict the intensity of global precipitation within a 10–20% error rate.

Observations of cloud cover are important for validation of techniques for numerical modelling of atmospheric circulation on space–time scales from local to global and from 1 day to 100 years, the requirements for information on clouds being widely different. This concerns, for example, the data on anvils, tropical convective clouds and the mechanisms of their formation, the evolution of cirrus clouds, mid-latitude cloud systems, cumulus-cloud clusters, as well as many other specific features of cloud cover.

Bearing in mind the key significance of the interaction between the atmosphere and the ocean as a climate-forming factor, it is important to ensure adequate observations of the atmosphere and ocean. In the latter case this concerns sea surface temperature, the components of sea surface energy budget, wind fields and wind shear at sea surface using both active (scatterometry) and passive microwave remote sensing. A special place belongs to obtaining the data on meridional heat transport (it is expected that the absolute error of retrievals would be not more than 10 W/m^2).

The interaction between the atmosphere and the land surface, with account taken of climatically significant processes on land surface, is important. As to the achievable reliability of observations, one can assume that the absolute (relative) errors of measuring the solar constant would amount to 0.1% (0.001%), which is enough for climatic monitoring. The expected cloud radiative forcing (CRF) changes due to SC variations would come to about 0.25 W/m^2, while doubling CO_2 concentration should cause the 'greenhouse' change of CRF of about 4 W/m^2. However, much higher are the changes of radiation fluxes depending on geographical conditions, interannual variations and other natural factors, which sometimes reach 100 W/m^2. Since the errors in the numerical modelling of radiation fluxes may be up to several tens of W/m^2, one can consider as acceptable the practically achievable absolute errors of satellite measurements of the order of 5 W/m^2 (OSR) and 2.5 W/m^2 (OLR). So far, it has proved difficult to assess reliably errors of SRB component retrievals.

The adequate interpretation of satellite observation data is possible only when performing a system of simultaneous observations at the level of the Earth's surface and in a free atmosphere using both remote sensing and direct (*in situ*) measurements. Such observations are required:

1 To estimate satellite measurement errors.
2 To check the reliability of satellite instrumentation calibration.
3 As an important additional source of necessary information. In this context it is very important to implement such WCRP key projects as GEWEX and CLIVAR, as well as field observational programmes similar to ARM (Kondratyev, 1998c, 1999b). The functioning of the global network of baseline actinometric stations (BSRN) and the network of lidar sounding of the atmosphere should play a substantial role.

6.3 GENERAL CIRCULATION, BIOPRODUCTIVITY AND THE EXCHANGE BETWEEN THE OCEAN AND THE ATMOSPHERE

The general purpose of the developments discussed above consist of: studying ocean–atmosphere interaction, including assessments of the ocean's role in forming the heat balance of the Earth and the global water cycle; the effect of ocean circulation on its level; the ocean functioning as a biological system (mainly in terms of forming the global carbon cycle).

Three aspects of the dynamics of the 'ocean–atmosphere–land surface' interactive system are most urgent:

1 The exchange of momentum, heat, and radiation.
2 The income of freshwater due to evaporation and precipitation fall.
3 The growth and melting of ice cover.

Momentum exchange is important as a climate-forming factor since it influences general ocean circulation, determining the formation of principal sea currents that are responsible for heat transfer from low- to high-latitudes and the SST field, as well as making an impact on vertical mixing in the upper ocean layer. This latter governs the mixing of warm surface waters with the lower cold waters. The variability of thermohaline ocean circulation in a time range from tens to hundreds of years is also explained by variations of heat and water exchange within the atmosphere.

The ocean loses approximately 10% more water through evaporation than it obtains via precipitation. This 10% is in part river run-off waters. Freshwater dynamics on the oceanic surface is likely to be a factor in the formation and variability of the El Niño/Southern Oscillation phenomenon. The above-listed and other circumstances motivated the launch of the *TRMM* satellite in 1998 for measurements of rainfall in the tropics and subtropics (see Kondratyev, 1999a).

The following are the three most important factors in determining the formation of oceanic general circulation:

1 Horizontal heat transfer.
2 SST field dynamics controlling heat exchange between the atmosphere and ocean.
3 The transport of nutrients, different chemical components and biota, which governs biogeochemical processes.

The horizontal transport of heat in the ocean causes about half of heat transport to high latitudes, which participates in the global energy balance determining climate. Meridional heat transfer reaches the maximum in the 'raging' latitudes (30°–75° S and N) where currents form at the western boundaries of the continents.

The transport of nutrients, chemical components and biota controls the bio-productivity of the ocean and its chemical interaction with the atmosphere.

The problem of World Ocean level variability depends on the dynamics of the melting of the Antarctic and Greenland glaciers. The solution to this most pressing of problems is still a complicated one and clouded by many uncertainties, including incomplete observation data.

Due to the vast extent of the World Ocean, it can assimilate about 40% of anthropogenic emissions of carbon dioxide into the atmosphere. In this connection three basic questions arise:

1 What is the effect on the carbon cycle in the upper oceanic layer of those changes in heat flux, momentum and nutrients on the oceanic surface?
2 What are the fluxes of carbon dioxide, carbon monoxide and dimethyl *sulphide* across the 'atmosphere–ocean' interface?
3 Which factors determine ocean bioproductivity?

So far, there are no reliable answers to these questions. The above-mentioned, and many other problems, determine the principal objectives of observations on general circulation, bioproductivity and the exchange between the ocean and the atmosphere.

6.4 GREENHOUSE GASES AND CHEMISTRY OF THE ATMOSPHERE

The goals of the relevant developments include studying the processes (receiving necessary information) and determining the GHGs content in the atmosphere. Such processes can be referred to as biogeochemical since they are involved in the interaction of biological, geochemical, and photochemical processes on a global scale in which such GHGs as carbon dioxide, carbon monoxide, methane, nitrous oxide, and tropospheric ozone are present.

Six aspects of these problems are of high priority:

1 The effect of land surface changes on GHG release into the atmosphere.
2 The effect of climate interannual variability on biogeochemical processes.
3 The influence of changing hydrological conditions and soil moisture on the fluxes of methane and carbon dioxide in wetland regions.
4 The spatial distribution and budget of tropospheric ozone.
5 The vertical distribution of tropospheric ozone.
6 The interaction of those processes that control carbon assimilation by the ocean.

Changes in land surface contribute substantially to the formation of the GHG budget (Gorshkov, 1995; Kondratyev, 1999a). If, as a result of fossil fuel burning and emissions from the cement industry, 5.5 gigatons (Gt) of carbon are released into the atmosphere every year, then the contribution caused by changes in land surface amounts to 1–2 Gt C/year. Changes of such kind also influence nitrogen dioxide and nitrous oxide emissions into the atmosphere. The interannual variability of climate determines the presence of interannual variations in land ecology and atmospheric chemical processes. In estimating climate variability factors, an important contributor is the availability of data on the vertical profile of ozone content in the troposphere. The field of tropospheric ozone content depends on its precursors forming during fossil fuel and biomass burning, as well as the action of nutrient sources in soils and vegetation cover (Kondratyev and Varotsos, 2000). As for carbon assimilation by the ocean, it is explained by the combined effect of physical, chemical and biological processes.

The EOS data would enable important information concerning the formation of the global carbon cycle (with consideration for the processes on land and in ocean), as well as the fields of GHGs and other minor gas constituents to be obtained.

6.5 LAND ECOSYSTEMS AND HYDROLOGICAL PROCESSES

These investigations are aimed at studying three categories of the impact processes have on climate. They are connected with the interactions of such processes as climate dynamics, hydrological processes and vegetation cover changes. While GHGs influence climate directly, the changes of land surface characteristics affect biophysical aspects of energy and mass exchange with atmosphere and hydrological balance. The combined effect of climate and hydrological processes controls bioproductivity which, of course, is the source of resources essential to people such as food and fuel, etc.

The EOS data are intended to supply the required information to solve the above-mentioned and other problems, together with numerical modelling results. In connection with climate, the key part is played by the use of interactive models for the processes on land surface (LSM) and for atmospheric general circulation (AGCM) that are, however, far from being adequate. In the field of hydrology the priorities belong to:

1 Identification and quantitative estimation of the most important hydrological characteristics over a wide range of spatial–temporal scales.
2 Development or improvement of hydrological processes models. To solve the appropriate problems, a voluminous set of observation data is required (see Kondratyev, 1998d, 1999b), including information on such extreme hydrological events as heavy storms, floods, and droughts.

Biospheric problems include a wide range from the carbon balance of boreal forests (their contribution in the formation of the global carbon cycle) to the primary productivity of the biosphere. In view of a comparatively rapid response of vegetation cover to climate changes, the EOS data cover a range of phenomena – from the interannual variability of phenology in spring and the annual CO_2 balance variation at the level of surface, to the annual variation and interannual variability of net bioproductivity (NPP). A set of proper instruments will include, in particular, the videospectroradiometer MODIS developed specially for determining NPP.

6.6 CRYOSPHERIC PROCESSES

The section of the 'Plan' under discussion concerns such components of the Earth's system as sea, lake, and river ice, snow cover, glaciers, ice sheets and caps, and frozen soil (including permafrost). The cryosphere consisting of these components is an interactive component of the global climate system due to the presence of numerous feedbacks because of energy and water exchange, cloud and precipitation dynamics, hydrological processes, and the general circulation of the atmosphere and the ocean. What does attract particular attention in the problem of man-made climate change (what we call global warming), is the processes of glaciers melting and the World Ocean level rise. Attracting attention to climate problems as well, the authors of the 'Plan' emphasize the special urgency of studying the processes in the high latitudes of the Northern Hemisphere, where, according to the well-known stereotype, global warming growth should take place. However, there is observation evidence that such a stereotype is not correct (see Kondratyev, 1999b).

The urgency of the data on cryosphere dynamics has determined the development of ICESat to monitor ice and snow cover dynamics. Its function is to monitor ice cover, clouds and land topography – the central preoccupations in this case being the solution of the problem of retrieving glacier topography data from the radio altimeter and lidar aboard ICESat, working out and using techniques for numerical modelling of the dynamics of the cryosphere and its effect on climate.

Analysis of available observation data has shown that, although the extent of ice cover is at its greatest in the Northern Hemisphere, the main ice mass is concentrated in the Antarctic glaciers. The most important properties of ice cover regulating energy exchange between surface and atmosphere are surface albedo, heat conductivity and latent heat. Critical to satellite observations interpretation is information on surface roughness, emissivity and ice dielectric characteristics.

Snow cover is the most extensive (with a maximum in the annual variation of about 47,000,000 km^2) of all the above-mentioned components of the cryopshere. Satellite observation results indicate the presence of a correlation between the mean hemispheric air surface temperature and the snow cover extent (SCE), which determines the importance of analysing SCE dynamics in the Northern Hemisphere for observing climate change. An interesting result was the discovery of the effective penetration of shortwave radiation through snow cover – which ensures a rather high level of photosynthetically active radiation (PAR) present, enough for algae to develop under the ice.

Monitoring of the sea ice cover extent (SICE), using satellite microwave sensing data from late October 1978 to late 1996, has revealed a decrease of SICE in the Arctic. The decrease amounts to 2.9% for 10 years whereas in the Antarctic growth has been observed equal to 1.3% per year. An important characteristic of SICE dynamics is polynyas playing the dominating part in heat and moisture exchange with the atmosphere in winter. Ensuring the high level of PAR under the ocean surface, polynyas contribute substantially to the primary productivity of sea waters.

Lake and river ice is characterized by particularly high variability (up to and including a complete disappearance). The distinctive feature of the Northern Hemisphere is permafrost occupying about 20% of the complete land area of the whole hemisphere with a layer depth of 600 m along the Arctic coast in the north-east of Siberia and in Alaska. Permafrost can and does form even at the present time of glacier retreat.

Ice sheets are the main reservoir of freshwater on the planet. They accumulate some 77% of water, which is equivalent to an 80 m thick global layer, 90% of this being water in the Antarctic and only 10% in Greenland. Climatic changes produce only weak variations in ice sheets, but they cause a great change in mountain glaciers with a typical response time in the order of 10–50 years. Although the variability of the retreat of glaciers does not have a practical influence on global climate, glacier melting in the twentieth century has resulted in 33–50% of the observed rise in the World Ocean level.

In the understanding of cryosphere dynamics, three points are critical:

1 Parameterization of cryospheric processes in the models of climate and hydrological variability.
2 Analysis of cryospheric–climatic feedbacks.
3 Study of those factors that determine cryosphere variability.

As to the first of the above, the most important factor is the inadequacy of consideration for the properties of sea ice cover (snow albedo, heat storage, heat conductivity) in climate models. One should include in the numerical modelling of climate, in particular, the advection and deformation of sea ice cover occurring during freshwater formation. These changes are due to the melting of ice because of the advection effect from the Arctic to the Greenland and Norwegian Seas, as well as the processes in polynyas. Recent developments have demonstrated that even relatively small changes in the parameterization of thermodynamic processes produce high variability in ice depth and, thus, affect the results of climate numerical modelling.

Reliable parameterization is required for the following processes relevant to snow cover dynamics: the metamorphism of snow cover (in the course of ageing), taking account of vertical gradients and distinguishing between the roles of solid and liquid precipitation. Undoubtedly, an adequate parameterization of permafrost dynamics in the models of hydrological processes is important. There are serious uncertainties in simulating the variability of ice sheets and reliable estimation of water resources is impossible in the absence of reliable river run-off simulations.

Cryospheric feedbacks can be separated into two categories of relationship: that functioning through the freshwater cycle and that determined by the heat balance of the surface. The interrelationship of the former with the cryosphere depends on the storage and transport of water in solid phase over a wide time range (this refers, respectively, to effects on climate as well). As for heat balance mechanisms, they show up effectively in two ways: on the one hand, through properties of ice and snow cover isolating the atmosphere and the ocean and, on the other hand, in the well-known form of albedo feedback (Kondratyev, 1992) and OLR variability.

6.7 OZONE AND STRATOSPHERIC CHEMISTRY

Since the problems of stratospheric ozone have been thoroughly discussed in our earlier monograph (see Kondratyev and Varotsos, 2000), here we examine the appropriate problems only very briefly.

Although the most acute problem concerns the spring minimum of TOZ, referred to as the 'ozone hole' in the Antarctic, considerable TOZ depletions (revealed in 1996) in the Arctic are comparable with those in the Antarctic. Also observed in recent decades are interesting episodes of small, but statistically significant, TOZ depletions in the mid-latitudes. The dynamics of the stratospheric ozone layer are determined by many factors, including chemical processes, transport, interaction with aerosol and polar stratospheric clouds (PSC), as well as interactions with UV solar radiation and high-energy particles.

At present, the chemical processes of ozone layer formation, including heterogeneous chemical reactions on the surface of aerosol particles – particularly volcanic aerosols and PSCs – have been established rather reliably. And a comparatively reliable estimate has been made of the role of CFCs and bromocarbons, mainly of anthropogenic origin, in the destruction of the ozone layer. However, the understanding of the effect of atmospheric circulation on the ozone layer remains inadequate, though it is not in doubt that the annual variation in TOZ is caused by a transport effect and that long waves in the stratosphere in the winter high latitudes do have a substantial influence on daytime TOZ. And extra-atmospheric UV radiation variations, arising in the course of the 11-year solar activity cycle on the stratospheric ozone, have also been studied for their causal effect.

Scientific development should be directed at finding out the relationships between natural and anthropogenic causes of TOZ variability and based upon observations and numerical modelling results (the latter are still far from adequate). Natural variations and trends in TOZ are explained by such causes as: the interannual and long-term variability of stratospheric circulation; external effects (extra-atmospheric UV solar radiation and high-energy particles, volcanic eruptions); changes in PSC amount; the interannual variations of stratospheric temperature in high latitudes.

Anthropogenic effects are connected with the emissions of CFCs and other ozone-destroying compounds, as well as aircraft emissions of nitrogen oxides. In this respect the requirements may be characterized as acceptable error in the retrieval of vertical profiles (with a resolution of 1–2 km) of necessary parameters. The latter will include meteorological parameters (the acceptable level of error is given in parentheses), as: temperature (1 K); wind velocity (2–5 m/s); and components of atmospheric composition made up of O_3 (0.2 ppm), H_2O (0.5 ppm), CFC-11 and 12 (0.2 ppb), N_2O (20 ppb), CH_4 (0.1 ppm), HCl (0.1 ppb), $ClONO_2$ (0.1 ppb), HNO_3 (1.0 ppb), NO_2/NO (0.2 ppb), ClO (50 ppt), BrO (5 ppt), OH (0.5 ppt), N_2O_5 (0.2 ppb), aerosol particles (with a total surface of 10%), and extra-atmospheric UV radiation in the wavelength interval 100–400 nm (4%). The description of the corresponding instrumentation can be found in recent monographs (see Kondratyev, 1998d, 1999a).

Unfortunately, the 'Plan' – as in some other cases – does not discuss the urgent problem of substantiating the necessary observation programme in the context of the Montreal Protocol (and the supplements to it) as to reducing the level of emissions of ozone-destroying components and an assessment of the consequences of such reductions. Although such consequences are long term (tens of years), recent observational data indicate a possible presence of those first signs which would indicate the TOZ depletion trend is slowing down (Kondratyev, 1999c; Kondratyev and Varotsos, 2000). Complex observation programmes (using observations gathered by satellite, aircraft, balloon, and ground-based means) are very important in the study of ozone layer dynamics in the polar regions and the mid-latitudes.

6.8 VOLCANIC ERUPTIONS AND AEROSOL IMPACT ON CLIMATE

As with the problems discussed above, these too have been the subject of recent monographs (see Kondratyev, 1988; Kondratyev and Galindo, 1997). Aspects of the relevant developments are those related to the need to obtain more complete information on volcanic emissions of sulphur gaseous compounds (mainly sulphur dioxide), subsequent gas-to-particle reactions of forming sulphate aerosol and its spread in the stratosphere, and the effect on climate and ozone layer. A specific place is occupied by the problems of tropospheric aerosol and their effect on climate (King *et al.* 1999; Kondratyev, 1999a), as well as the use of satellite information for observing the consequences of explosive volcanic eruptions. The purpose here is to minimize such consequences (Kondratyev and Galindo, 1997). On the whole, the EOS data would become a more representative source of information on stratospheric and tropospheric aerosol than the information obtained earlier. The data does not accentuate, however, the necessity to develop and implement a complex programme using both satellite and the usual observational facilities for monitoring the consequences of explosive eruptions.

6.9 CONCLUSION

The State of Science in the EOS Program is a very comprehensive document. It contains a lot of useful information and shows perspectives of satellite remote sensing. However, some parts of the document are not homogenous (due to the large number of its authors) in terms of presentation of material and, more importantly, its scientific level. Its principal defect is the absence of a proper concept – one which would have combined the fragments of a single whole. This opinion is the same as that mentioned earlier with respect to the USGCRP programme (see Kondratyev, 1998c).

The substantiation of priorities for the problems of global change are of vital importance. Such substantiation is absent in *The State of Science in the EOS Program*; perhaps the authors could not, or did not wish to, use results published earlier (Marchuk *et al.*, 1986; Gorshkov, 1995; Kondratyev, 1998b, 1999a). Had they done so the inadequate concentration on climate problems (for all the importance of it) would have become clear (Jaworowski, 2000).

The absence of a necessary concept has determined the fragmentary nature of *The State of Science in the EOS Program*. It does not contain, for instance, a chapter on the basic importance of global biochemical cycle problems and the dynamics of the biosphere (the biodiversity problem requires particular consideration). Nor is there mention of biotic regulation of the environment (Gorshkov, 1995; Kondratyev, 1990; 1999a) which should be the basis for global environmental safety. Although the authors have, for some reason (or maybe by chance), avoided the popular, but contradictory, notion of sustainable development (Kondratyev, 1998a; 1999b), the problem of environmental safety should have been central to their concept. In this latter connection, a special chapter would have been appropriate, a systematic examination of the problems of natural and anthropogenic disasters. Similar consideration should be undertaken with respect to the problem of polluting natural environments (atmosphere, land, water basins).

Although 'human dimensions' cannot be the direct purpose of remote sensing (RS), the general analysis of the contribution of RS to solving the problems of socio-economic development of society would be without doubt helpful (arable land resources, drought and flood consequences, population concentrations in cities and coastal areas; biosphere resources, and many others). Unfortunately, none of the problems of non-renewable natural resources (oil, coal, natural gas) and the consequences of their use under conditions of extreme urgency in the modern situation and in terms of the efficiency of using global natural resources and economic potential (Kondratyev, 1998a, 1999a) are discussed.

In summing up this discussion, the growing importance of coming to terms with the problems of natural and anthropogenic disasters should be emphasized once more. An illustration of this is the implementation of the new project 'Natural Catastrophes and Developing Countries' (CAT). Begun in the International Institute of Applied System Analysis in Luxemburg, Austria and supported by the World Bank (Freeman *et al.*, 1999), the long-term goal of the project is substantiation of scenarios that can realistically demonstrate the effects of natural calamities on different branches of the economy and open up possibilities of minimizing their destructive effects. On the other hand, it has allowed the World Bank and other international creditors to render more effectively the required financial assistance to those countries that need it. Under the CAT Project, it is planned to develop scenarios that will produce a basis for assessment of direct and indirect material losses as a result of natural catastrophes in developing countries. The requirement is to determine those factors that have the strongest influence in alleviating conditions of socio-economic damage when floods, earthquakes and other disasters occur. The methodology of the CAT Project includes two key components:

1. Analysis of critically important factors that determine the progress of a developing country and its liability to the effect of natural calamities.
2. Simulated 'shocks' due to natural disasters affecting the results of macro-economic modelling. The first of these simulations have already produced evidence that even rough estimations of losses can be a basis for useful macro-economic analysis of the consequences of natural disasters.

6.10 BIBLIOGRAPHY

Adequacy of Climate Observing Systems (1999) National Academy Press, Washington, DC., 51 pp.

Barnett, T. P., Hasselmann, K., Chelliah, M., Delworth, T., Hegerl, G., Jones, P., Rasmasson, E., Roekuer, E., Ropelewski, C., Santer, B., and Tett, S (1999) Detection and attribution of recent climate change: A Status Report. *Bull. Amer. Meteorol. Soc.* **80**(12), 2631–59.

Bottorf, H. and Mayney, T. (1994) GPS, GTS and the Great Flood of 1993. *Satell. Commun.* **18**(1).

Choudhury, A. M. (1980) Study of floods in Bangladesh and India with the help of meteorological satellites. Contribution to *Space Observations of Water Resource Management Proc. Symp. 22nd Plenary Meeting COSPAR, Bangalore, 1979*. Oxford, pp. 227–30.

Dousset, B., Flament, P., and Bernstein, R. (1993) Los Angeles fires seen from space. *Earth Observations from Space* **74**(3), 37–8.

ESA (1995) *New Views of the Earth. Scientific Achievements of ERS-1*. European Space Agency, Brussels, 162 pp.

Freeman, P., Martin, L., Mechler, R., and Warner, K. (1999) Natural catastrophes, infrastructure, and poverty in developing countries. *Options* **Fall/Winter**, 6–16.

Gorshkov, V. G. (1995) *Physical and Biological Basis for Sustainable Life*. VINITI, Moscow, 470 pp.

Greenstone, R. and King, M. D. (eds) (1999) *EOS Science Plan. The State of Science in the EOS Program*. NASA Goddard Space Flight Center, Freenbelt, MD, 397 pp.

Grigoryev, Al. A. (1975) *Satellite Indication of the Earth Landscapes*. LGU, Leningrad, 160 pp.

Grigoryev, Al. A. (1982) *Cities and the Environment. Satellite Investigations*. Mysl, Moscow, 160 pp.

Grigoryev, Al. A. (1985) *Anthropogenic Impacts on the Environment from Satellite Observations*. Nauka, Leningrad, 239 pp.

Grigoryev, Al. A. (1991) *Ecological Lessons of the Historical Past and Present Time*. Nauka, Leningrad, 250 pp.

Grigoryev, Al. A. and Kondratyev, K. Ya. (1994) Modern state of the global environment and natural resources. *Izv. Rus. Geogr. Soc.* **3**, 12–26.

Grigoryev, Al. A. and Zhogova, M. L. (1992) Heavy dust outbreaks in Priaralye in 1985–1990. *Papers Rus. Acad. Sci.* **324**(3), 672–5.

Houghton, J. T., Meira-Vieho, L. G., Callander, B. A., Harris, N., Kattenberg, A., and Maskell, K. (eds) (1996) *Climate Change 1995. The Science of Climate Change.* Cambridge University Press, 572 pp.

Jaworowski, Z. (2000) The global warming folly. *21st Century Sci. Technol.* **12**(4), 64–75.

Jiang, J., Cao, S. (1990) *All Weather Flood Monitoring and Loss Estimation by Using Space Technique*, Preprint. Space Science and Application Research Center, Academia Cinica, Beijing, 10 pp.

King, M. D., Kaufman, Y. J., Tanré, D., and Nakajama, T. (1999) Remote sensing of tropospheric aerosols from space: Past, present and future. *Bull. Amer. Meteorol. Soc.* **80**(11), 2229–59.

Kondratyev, K. Ya. (1982) World program for climate investigations: Modern state and perspectives. Outcomes of science and technology. *Meteorology and Climatology*, Vol. 8. VINITI, Moscow, 315 pp.

Kondratyev, K. Ya. (1983) *Satellite Climatology*. Hydrometeoizdat, Leningrad, 264 pp.

Kondratyev, K. Ya. (1985) Volcanoes and climate. Outcomes of science and technology. *Meteorology and Climatology*, Vol. 13. VINITI, Moscow, 205 pp.

Kondratyev, K. Ya. (1988) *Climate Shocks: Natural and Anthropogenic.* New York, Wiley, 296 pp.

Kondratyev, K. Ya. (1990) Key problems of global ecology. Outcomes of science and technology. *Theoretical and General Problems of Geography*, Vol. 9. VINITI, Moscow, 454 pp.

Kondratyev, K. Ya. (1992) *Global Climate*. Nauka, Leningrad, 359 pp.

Kondratyev, K. Ya. (1993) Complex monitoring of Pinatubo volcano eruption. *Studying the Earth from Space* **1**, 111–22.

Kondratyev, K. Ya (1998a) *Multidimentional Global Change*. Wiley/Praxis, Chichester, UK, 761 pp.

Kondratyev, K. Ya. (1998b) Ecological risk: Actual and hypothetical. *Izv. Rus. Geogr. Soc.* **130**(3), 13–24.

Kondratyev, K. Ya. (1998c) Ecodynamics and geopolicy: From global to local scales. *Izv. Rus. Geogr. Soc.* **130**(5).

Kondratyev, K. Ya. (1998d) Global natural resources and economical potential: Efficiency of its use. *Izv. Rus. Geogr. Soc.* **130**(6), 1–10.

Kondratyev, K. Ya. (1999a) *Climatic Impacts of Aerosols and Clouds.* Springer/Praxis, Chichester, UK, 264 pp.

Kondratyev, K. Ya. (1999b) *Geodynamics and Geopolicy*, Vol. 1, *Global Problems*. SPB, St Petersburg, 1032 pp.

Kondratyev, K. Ya. (2000a) Global changes at the threshold of two millennia. *Izv. Rus. Geogr. Soc.* **133**(4).

Kondratyev, K. Ya. (2000b) Studying the Earth from space: Scientific plan of EOS system. *Izv. Rus. Geogr. Soc.* **1**.

Kondratyev, K. Ya. and Galindo, I. (1997) *Volcanic Activity and Climate*. A. Deepak, Williamsberg, VA, 282 pp.

Kondratyev, K. Ya. and Varotsos, C. A. (2000) *Atmospheric Ozone Variability: Implications for Climate Change, Human Health, and Ecosystems*. Springer/Praxis, Chichester, UK, 761 pp.

Kuroda, T. and Koizumi, S. (1992) A plan for the world environment and disaster observation system. *World Space Congress, 43rd Cong. Int. Astronautical Federation and 29th Plenary Meeting Communications in Space Research, Washington, DC.*

Nikonov, A. A. (1999) Consequences of earthquakes for the environment. *Vestnik Rus. Acad. Sci.* **69**(12), 1107–11.

Marchuk, G. I., Kondratyev, K. Ya., Kozoderov, V. V., and Khvorostyanov, V. I. (1986) *Clouds and Climate*. Hydrometeoizdat, Leningrad, 512 pp.

Morozova, L. I. (1993). Atmospheric indication of earthquakes in Near East. *Studying the Earth from Space* **6**, 81–3.

Our Changing Planet/Fiscal Year 1998, US Global Change Research Program. An Investment in Science for the Nation's Future (1998) A Supplement to the President's Fiscal Year 1998 Budget. Washington, DC, 118 pp.

Pichugin, A. P. (1985) Radar observations of river floods from 'Cosmos-1500' satellite. *Studying the Earth from Space* **5**, 61–6.

Rango, A. and Anderson, A. T. (1974) Flood hazard studies in the Mississippi river basin using remote sensing. *Water Resourc. Bull.* **10**(5), 1061–81.

Shults, S. S. (1973) *The Earth from Space*. Nedra, Leningrad, 114 pp.

Winokur, R. S. (2000) SAR Symposium Keynote address. *Johns Hopkins ARL Technical Digest* **21**(1), 5–11.

Wentz, F. J. and Schabel, M. (2000) Precise climate monitoring using complementary satellite data sets. *Nature* **403**(6768), 414–16.

Zhang, J. and Tuan, Q. (1990) Research and establishment of the nationwide natural environment information system. *Proc. 2nd Int. Workshop on Geographical Information Systems, Beijing*, pp. 198–208.

Zonjin, M. and Xianglin, C. (1991) *Application of Space Techniques to Earthquake Hazards Reduction in China*, Preprint. Institute of Geology, Beijing, pp. 1–15.

Index